Advances in Comparative and Environmental Physiology 6

Volumes already published

Vol. 1: NaCl Transport in Epithelia
Edited by *R. Greger* (1988)

Vol. 2: From the Contents: Hibernation – Vapor Absorption
Nutrient Transport – Stable Pollutants (1988)

Vol. 3: Molecular and Cellular Basis of Social Behavior in Vertebrates
Edited by *J. Balthazart* (1989)

Vol. 4: Animal Adaptation to Cold
Edited by *L. C. H. Wang* (1989)

Vol. 5: From the Contents: Osmoregulation – Chemoreception
Temperature – Spectroscopy – Metalloproteins
Maternal-Fetal Relationships (1989)

Vol. 6: Vertebrate Gas Exchange: From Environment to Cell
Edited by *R. G. Boutilier* (1990)

Volumes in preparation

Vol. 7: Volume and Osmolality Control in Animal Cells
Edited by *R. Gilles, E. K. Hoffman,* and *L. Bolis*

Advances in

Comparative and Environmental Physiology 6

Vertebrate Gas Exchange

From Environment to Cell

Guest Editor: R. G. Boutilier

With Contributions by

R. G. Boutilier · S. Egginton · N. Heisler · W. K. Milsom · S. F. Perry
J. Piiper · E. R. Swenson · D. Randall · R. M. G. Wells

With 126 Figures

Springer-Verlag
Berlin Heidelberg New York
London Paris Tokyo Hong Kong
Barcelona

Guest Editor:
Dr. Robert G. Boutilier

Department of Biology
Dalhousie University
Halifax, Nova Scotia
Canada B3H 4J1

ISBN-13:978-3-642-75382-4 e-ISBN-13:978-3-642-75380-0
DOI: 10.1007/978-3-642-75380-0

Library of Congress Cataloging-in-Publication Data. Vertebrate gas exchange: from environment to cell / guest editor, R. G. Boutilier; with contributions by R. G. Boutilier ... [et al.]. p. cm. – (Advances in comparative and environmental physiology; 6) Includes bibliographical references. 1. Respiration. 2. Vertebrates – Respiration. 3. Physiology, Comparative. 4. Pulmonary gas exchange. I. Boutilier, R. G. (Robert G.), 1953– . II. Series. QP121.V47 1990 596'.012 – dc20 90-9416

Softcover reprint of the hardcover 1st editiong 1990

Typesetting: K+V Fotosatz GmbH, Beerfelden

2131/3145-543210 – Printed on acid-free paper

Foreword to the Series

The aim of the series is to provide comprehensive, integrated reviews giving sound, critical, and provocative summaries of our present knowledge in environmental and comparative physiology, from the molecular to the organismic level.

Living organisms have evolved a widespread range of basic solutions to cope with the different problems, both organismal and environmental, with which they are faced. A clear understanding of these solutions is of course of fundamental interest for all biologists, zoologically or medically oriented. They can be best comprehended in the framework of the environmental and/or comparative approaches. These approaches demand either wide surveys of animal forms or a knowledge of the specific adaptive features of the species considered. This diversity of requirements, both at the conceptual and technological level, together with the fact that physiology and biochemistry have long been mainly devoted to the service of medicine, can account for the fact that these approaches emerged only slowly amongst the other new, more rapidly growing disciplines of the biological sciences.

The field has now gained the international status it deserves and the organization of a series devoted to it appeared timely to me in view of its actual rapid development and of the interest it arouses for a growing number of biologists, physiologists, and biochemists, independently of their basic, major orientation.

Liège, Belgium *Raymond Gilles*

List of Editors

Those interested in contributing to this series should contact Professor R. Gilles or one of the Series Editors. Review proposals should include a short outline – and a Table of Contents – briefly defining the aims and scopes of the review.

Contents

Chapter 4

Morphometric Analysis of Tissue Capillary Supply
S. Egginton

Chapter 5

Hemoglobin Physiology in Vertebrate Animals:
a Cautionary Approach to Adaptationist Thinking
R.M.G. Wells

Chapter 6

Kinetics of Oxygen and Carbon Dioxide Exchange
E.R. Swenson

Chapter 7

Acid-Base Regulation:
Interrelationships Between Gaseous and Ionic Exchange
N. Heisler

Chapter 8

Control and Co-Ordination of Gas Exchange in Water Breathers
D. Randall

Chapter 9

Control and Co-Ordination of Gas Exchange in Bimodal Breathers
R. G. Boutilier

Chapter 10

Control and Co-Ordination of Gas Exchange in Air Breathers
W.K. Milsom

Chapter 1

Respiratory Gas Tensions in the Environment

R. G. Boutilier[1]

Contents

1 Introduction

The aim of this chapter is to provide a brief introduction to the wide range of environmental oxygen and carbon dioxide partial pressures (PO_2, PCO_2) that certain vertebrates experience as a part of their natural life history. Various biological and physical features of the environment, particularly of aquatic systems, can lead to marked oscillations in the ambient gas tensions both daily and seasonally, and these in turn can have profound effects on the regulation of gas exchange in a number of these animals. In some instances, unfavourable gas concentrations in the environment (e.g. low O_2 levels) can place restrictions on the activity range of an animal or limit the overall success of a species. On the other hand, animals that can tolerate the less than optimal gas composition of certain secluded habitats are often freed from risks of predation. Whereas some animals choose to live in confined, sometimes hypoxic areas (e.g. fossorial mammals), others are subject to periodic extremes of the medium in which they live (e.g. overwintering fish).

2 Natural Variations in Environmental Gas Tensions

Figure 1 illustrates a discrete subdivision of either the aerial or aquatic environment as well as the physical and biological processes that interact to influence its

[1] Department of Biology, Dalhousie University, Halifax, N.S., Canada B3H 4J1

Advances in Comparative and Environmental Physiology
Vol. 6. Edited by R. G. Boutilier

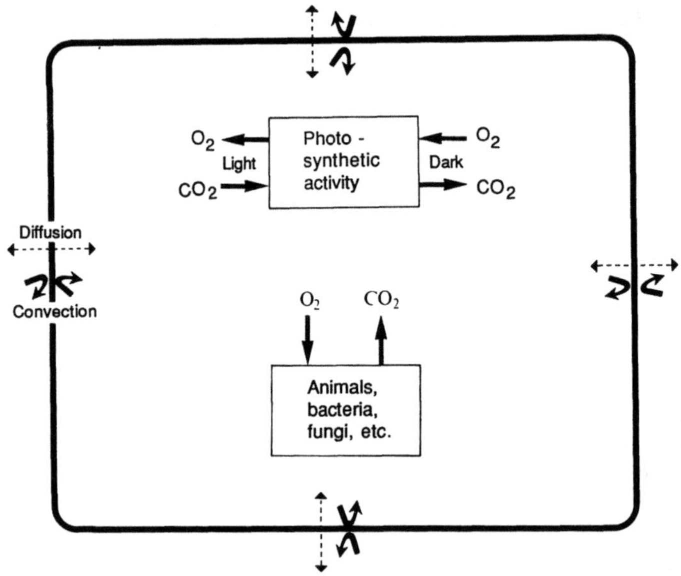

Fig. 1. The physical and biological processes that lead to changes in the partial pressures of oxygen and carbon dioxide in the environment. The *block* represents one region of a terrestrial or aquatic environment. Respiratory gases are distributed from region to region by diffusion and convection (Redrawn from Hill and Wyse 1989)

respiratory gas composition. Interchanges between different segments of the environment occur through convection and diffusion, both of which tend to equalize the gas tensions throughout the system. The balance between O_2 production via photosynthesis and O_2 consumption by respiration determines the net changes in O_2 (or CO_2) in the environment over the course of a day.

In open terrestrial environments, the CO_2 production and O_2 consumption of organisms are comparatively minute in comparison to the immense O_2 and CO_2 capacitances and volumes of the atmosphere (Dejours 1981). Indeed, in the free atmosphere, hypoxia is usually only encountered as a result of changes in ambient pressure due to altitude (Dejours 1981, 1982). Some birds, for example, are known to fly or soar, at least for brief periods of time, at heights of 9000–10000 m, where O_2 partial pressures are approximately one-fourth those at sea level (Black and Tenny 1980). Compared to air, the O_2 and CO_2 levels in aquatic environments are much more subject to change as a result of biological processes. This occurs because of the much slower diffusion of respiratory gases in the aquatic medium and the comparatively weak convection currents in water as opposed to air.

In confined aerial environments, O_2 and CO_2 tensions often change in proportion to the overall respiratory quotient of the animal, owing to the equal capacitances of O_2 and CO_2 in the gaseous phase. Thus, assuming a respiratory quotient of 1.0, a fall in PO_2 of 30 Torr would be accompanied by an equivalent increase in PCO_2. However, if originally normoxic water were to become completely devoid of oxygen due to aerobic metabolism alone, the PCO_2 would only rise to approximately 7 Torr, because of the high CO_2/O_2 solubility ratio (Rahn 1966). The fact that the PCO_2 of many aquatic systems can get much higher (e.g. up to 70 Torr, Ultsch 1976; Heisler et al. 1982), is attributable to the action of anaerobic processes and vegetation. For instance, anaerobiosis can produce CO_2 directly or it can liberate CO_2 when anaerobically produced protons dehydrate water bicarbonate. As shown in several field studies on freshwater environments (Ultsch 1976), vegetation cover can induce marked hypoxia and hypercarbia. In some instances, these unfavourable conditions may be localized within a larger body of water, with adjacent open-water regions being normoxic and normocarbic. Depending on their tolerance to variations in respiratory gas tensions, these factors can dictate the distribution of certain species within the aquatic community (Ultsch 1976).

Whenever plant and animal populations co-exist in poorly mixed waters (e.g. swamps, tidepools), diurnal fluctuations in respiratory gases can be predicted. During daylight hours, for example, O_2 production by photosynthesis can outpace the respiratory demands of the inhabitants of the freshwater ecosystem, and can lead to increases in PO_2 to several hundred Torr. At night, when O_2 production ceases, the O_2 content of these waters can be depleted to extremely low levels by the continued respiration of the resident organisms (Garey and Rahn 1970; Ultsch 1973; Kramer et al. 1978). Similar oscillations in O_2 and CO_2 occur in rockpools of the marine intertidal; e.g. changes in PO_2 from less than 2 Torr at night to more than 500 Torr during the day (Stephenson et al. 1934; Truchot and Duhamel-Jouve 1980). In addition to the changes in oxygen, diurnal oscillations also occur in the HCO_3^-/CO_2 system, titratable alkalinity, salinity and temperature of intertidal rockpools (Fig. 2). Thus, the resident organisms (e.g. fish) within the rockpool can be exposed to changes in PCO_2 and pH, both of which may influence their acid-base balance (see chapters by Heisler and Randall, this volume). As noted earlier, the oscillations in PCO_2 are comparatively small (e.g. Fig. 2) in comparison to PO_2, owing to the high CO_2/O_2 capacitance ratio of water.

In fast flowing streams and rivers, as well as the surface waters of the ocean, the O_2 and CO_2 tensions are often at near equilibrium conditions with the atmosphere, owing to turbulent mixing with air. Respiratory gas tensions in the ocean are also kept at favourable levels even at great depths, owing to the distribution of O_2-rich surface waters by large scale convection by oceanic currents. Even so, at depths of 100 to 1000 m, O_2-poor layers can occur (Harvey 1957). Anaerobic conditions can also exist in landlocked basins and fjords, where circulation patterns are often insufficient to supply O_2 at the rate at which it is consumed (Grasshoff 1975). Also, large O_2 and CO_2 gradients can exist between surface waters and at depth in some coastal estuarine environments (Reid 1961; Harper et al. 1981; Gnaiger 1983; Officer et al. 1984). For example, light penetration for photosynthetic activity and opportunities for equilibration with the atmosphere are greatest at surface waters, but convective mixing with deeper waters is often inhibited by thermal- or salinity-induced layering of water. In coastal areas, the less dense freshwater outflow from rivers may essentially float on top of higher density saline water, effectively inhibiting free mixing of water throughout the water column. This, as well as thermal effects (below), are evidently responsible for

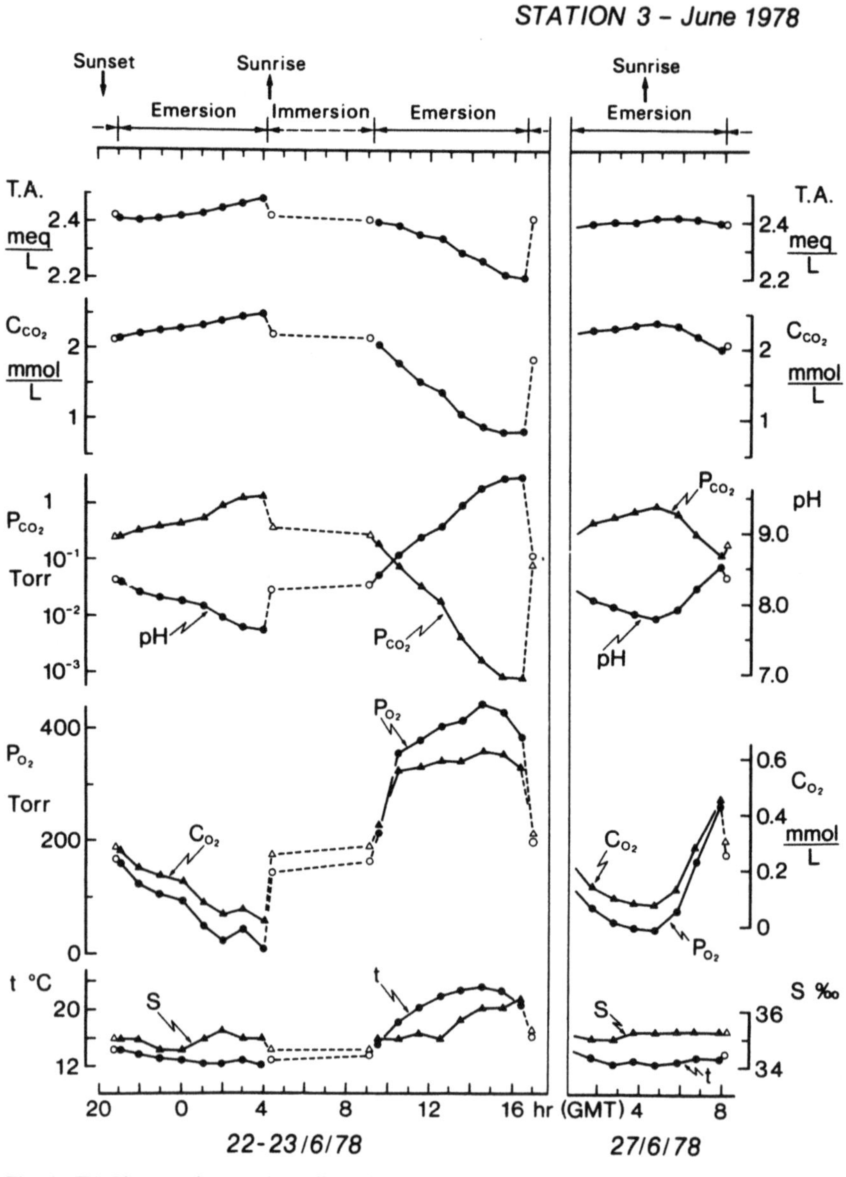

Fig. 2. TA Changes in titration alkalinity; C_{CO_2} total carbon dioxide concentration; P_{CO_2} carbon dioxide partial pressure; *pH*; P_{O_2} oxygen partial pressure; C_{O_2} oxygen concentration; *t*, temperature; and *S* salinity in a rockpool of the marine intertidal. *Left panel:* During spring tides; consecutive emersions, the first one at night (*farthest left*), the second one during the day. *Right panel*: At the onset of neap tides; the emersion starts at night and ends in the morning. *Closed symbols* and *continuous lines* are for pool water; *open symbols* and *dashed lines* are for water of the open sea (Truchot and Duhamel-Jouve 1980)

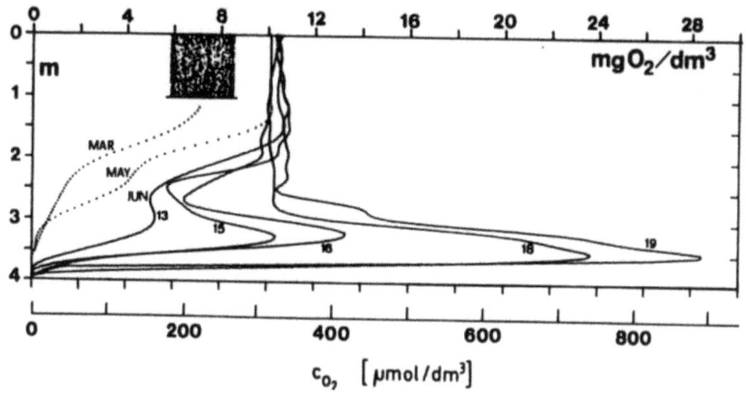

Fig. 3. Oxygen profiles before and after icebreak during spring stratification in a shallow mountain lake. Data for March and May are shown as *dashed lines*, when winter cover extended to approximately 1 m depth (as indicated by the *dark bar*). Profiles are also shown for several days in June, following icebreak on June 11th (Gnaiger 1983)

the marked hypoxia that has been reported in many coastal areas throughout the world (Seki et al. 1974; Harper et al. 1981; Dethlefson and von Westernhagen 1983; Officer et al. 1984; Seliger et al. 1985).

Thermal stratification occurs when surface water is heated and becomes less dense than the deeper, cooler waters. As a consequence, the surface water effectively floats on top of the deeper water, and this limits convective mixing between the two thermal strata. Because light penetration into the colder, deeper water is often insufficient to enable photosynthesis to meet the O_2 demands of the inhabitants, O_2 levels can become severely depleted at depth (e.g. Fig. 3). In most instances, mobile animals are able to avoid these regional differences in respiratory gas composition by simply moving away. However, the possibility exists that some species may be unable to tolerate either of the two thermal layers; i.e. the low O_2 tensions at depth or the high temperatures of the surface waters.

When the temperatures of surface waters decrease, as they do during the fall season, convective mixing is restored, and the respiratory gas tensions within the body of water become more uniform. If, during winter months, the pond or lake becomes covered with ice and snow, this inhibits both equilibration of surface waters with the atmosphere, as well as light penetration for photosynthesis. Under such 'closed' conditions, the waters can become severely hypoxic or anoxic both spatially and temporally (Babin and Prepas 1985; Baird et al. 1987), and may lead to 'winter kills' of various organisms (e.g. Barica and Mathias 1979; Gnaiger 1983). During the spring thaw, the respiratory gas composition of freshwater systems can change rather abruptly, within a matter of a few days. In the example shown in Fig. 3 (Gnaiger 1983), an initial chemical stratification following icebreak prevented convective mixing of water within the lake. The marked increases in O_2 levels at the 3.5-m depth (June 13–19, Fig. 3) were caused by a large concentration of algae, which also set a limit to light penetration. Thus, a steep gradient was set up between the photosynthetic source of O_2 and the anoxic mud (Fig. 3). These data illustrate the large differences in respiratory gas tensions that can exist between horizontal strata within a body of fresh water.

3 Animals in Secluded Environments: Impact on Ambient Gas Tensions

3.1 Gas Exchange in Burrows

Many vertebrates spend considerable periods of time in subterranean or subaquatic burrows, either in walled chambers or within the substrata itself. Such burrowing behaviour often results in limited access to the free atmosphere or water system. The barrier to gaseous diffusion is much greater in water than in air, owing to the marked differences in gas diffusion properties between the two media. As well, convective exchange of gases within the aquatic substratum is much reduced in comparison to air, due to the greater viscosity and density of the medium. Thus, in terrestrial environments, the interstitial spaces in highly porous soil may remain quite close to atmospheric levels to considerable depths, provided that the spaces remain filled with air. However, the water-filled spaces of rain-soaked soil or of subaquatic burrow systems can constitute such a considerable barrier to gaseous diffusion that mass flow of the respiratory medium may be required in order to support the metabolic rates of the inhabitants.

3.1.1 Terrestrial Tunnels and Cavities

Many birds and mammals construct and/or occupy tunnels or cavities which they utilize either seasonally (e.g. for incubation), or throughout the year, as a refuge from predation or environmental pressures. Burrows such as these are often constructed as nest chambers with tunnels leading to the surface. Air renewal in some mammalian burrows, such as those of the prairie-dog, can be promoted by prevailing winds at the surface (Vogel et al. 1973). The connection between the surface and the tunnel itself is of minor importance as a passageway for gas exchange, since most of the diffusion of gases occurs through the walls of the burrow (Withers 1978). Thus, plugging the passageways of the burrow with soil, as fossorial mammals often do, will have little consequence on gas exchange.

Effective mixing of gas between the burrow and the open atmosphere is not always sufficient to replace the O_2 that is being consumed by the inhabitants of the burrow and, as a consequence, the respiratory gases in these confined spaces often deviate significantly from the open air. In many instances, the decreases in environmental oxygen occur in proportion to the increases in carbon dioxide, approximating the respiratory exchange quotient of the animal (Arieli 1979; Wickler and Marsh 1981). Oxygen concentrations from 20% to as low as 12.1% have been recorded in avian burrows, with carbon dioxide concentrations ranging from

1.2% to as high as 7.9% (White et al. 1978; Birchard and Kilgore 1980; Furilla 1980; Wickler and Marsh 1981). In certain mammalian burrows, CO_2 levels have been reported as high as 10%–15% (Kennerley 1964; Studier and Baca 1968; Williams and Rausch 1973). Fossorial endotherms may experience even more severe O_2 and CO_2 levels during conditions such as crowding (e.g. Hayward 1966) or when changes in soil porosity due to rain (e.g. Withers 1978) restrict the paths for diffusional and microconvective gas exchange with the open atmosphere. Thus, several parameters may interact to contribute to the variations in respiratory gas tensions; the effective soil porosity, passive air movements, burrow depth and convective currents. Evidently, the latter can be set up either by thermal gradients or by movements of the animals themselves (Withers 1978; Birchard and Kilgore 1980; Wickler and Marsh 1981).

Recent studies on birds that occupy nest cavities excavated in trees have shown that gaseous conditions within such cavities are less severe than in subterranean burrows (White et al. 1984; Howe and Kilgore 1987). The moderate levels found in such cavities (e.g. O_2 20.4%, CO_2 0.4%; Howe and Kilgore 1987) indicate that there is considerable turnover of air between the nest cavity and the open atmosphere. This turnover is enhanced even more so by the fact that the temperatures of the cavity nests are often several degrees higher than the free atmosphere (Kendeigh 1961; White et al. 1984), thereby promoting thermally-induced convective exchange (Howe and Kilgore 1987).

3.1.2 Subterranean and Aquatic Incubation

A particularly striking example of subterranean incubation in an ectotherm is that of the sea turtle nests of *Chelonia* and *Caretta*, excavated in sandy beaches in the tropics. These nests (25 cm diameter, 50 cm deep) are stereotypically constructed by the female, and the eggs then deposited in the chamber, which is free of sand (Carr 1967). Even after being buried, the nest retains its shape and does not fill with sand, so that the 100 or so eggs remain in a subterranean chamber with air pockets between them. Under these conditions, and throughout the 60-day period of time until hatching, the PO_2 of the nest chamber decreases and PCO_2 increases as shown in Fig. 4 (Ackerman 1977). The prevailing porosity of the sand will undoubtedly affect the gas exchange characteristics of the nest, as will the way in which the nest is constructed; i.e. with the eggs free of sand. Indeed, Ackerman (1977) has shown, through model calculations, that if the spaces between the eggs were filled with sand rather than air, a greater hypoxia and hypercapnia would develop in the centre of the egg mass, thereby creating different respiratory environments for individuals. By constructing the nest in the way she does, the sea turtle ensures that there is a uniform respiratory environment for the entire clutch of eggs throughout development. It is noteworthy that a similar uniformity of gas tensions is facilitated in submerged egg masses of some frogs by intense convection of fluids within each egg. Such movements are generated by the ciliated epithelium of the embryo and act as a convective 'ventilation' that effectively reduces the diffusion distance between the centre and periphery of the gelatinous egg mass (Burggren 1985).

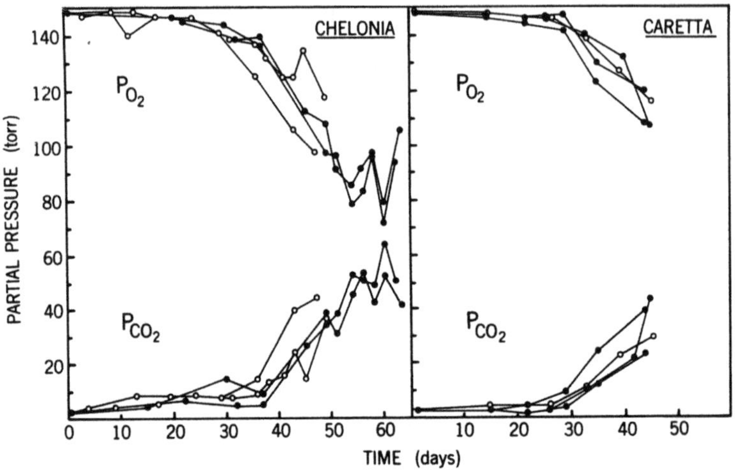

Fig. 4. Changes in the partial pressures of oxygen (P_{O_2}) and carbon dioxide (P_{CO_2}) with time in the centre of 100-egg subterranean nests of the sea turtles, *Chelonia mydas* and *Caretta caretta*. *Open circles* natural nests in beach sand; *closed circles* nests constructed artificially in the sand and containing freshly laid eggs (Ackerman 1977)

3.1.3 Subaquatic Burrows

Certain species of Amazonian air-breathing fish (e.g. *Symbranchus*) and apodan amphibians (e.g. *Typhlonectes*) excavate burrows along the banks of rivers that are subject to large changes in water level between the rainy and dry seasons (Lüling 1975; Moodie 1978). Such burrows, which often contain water in their lower reaches, are purported to be hypoxic and hypercapnic (Kramer et al. 1978; Toews and Macintyre 1978), as is known to be the case in the burrows of some land crabs (Wood and Boutilier 1985).

In addition to burrows at the air-water interface along riverbanks, certain marine fish construct and occupy subaquatic burrows. The burrowing behaviour of one such animal, the red band fish (*Cepola rubescens*), has been described in both field and laboratory studies (Atkinson 1976; Atkinson et al. 1977; Pullin et al. 1980). The burrow itself is constructed in the ocean floor as a vertical shaft with a terminal chamber, the overall depth of which can range from 0.1 to 0.75 m (Pullin et al. 1980). Field observations have shown that the animals spend prolonged periods of time curled up in the terminal chamber, sometimes partially buried in mud and sand. In situ PO_2 measurements of occupied burrows have shown there to be far less O_2 in the terminal chamber than at the aperture. In the laboratory, animals will construct burrows and remain in seclusion even when burrow PO_2 levels get as low as 40–60 Torr. In fact, as shown in Fig. 5, the PO_2 of the burrow is maintained within this PO_2 range by body movements and/or visits to the aperture. The displacement of water caused by such movements always raised the PO_2 of the burrow water, sometimes by as much as 44% (Pullin et al. 1980).

Because the transfer of gases by diffusion is extremely slow in water as compared to air, the turnover of gases in the interstitial spaces of aquatic sediments

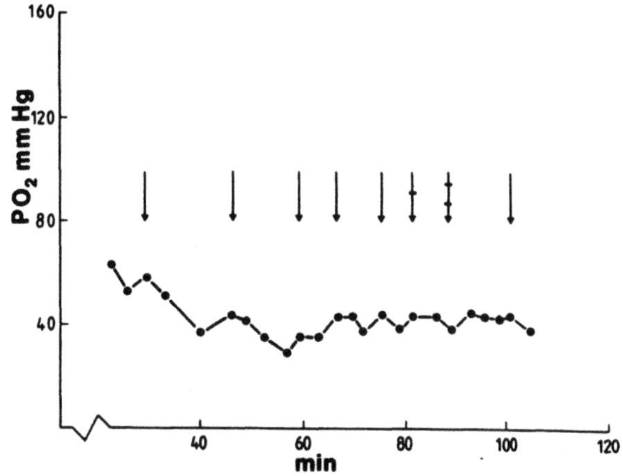

Fig. 5. Oscillations in the partial pressure of oxygen (*PO_2*) measured in water samples taken from the subaquatic burrow (ca. 0.1 m depth) of a captive burrowing fish (*Cepola rubescens*). *Arrows* indicate animal movements, which displaced some of the burrow water and thereby led to a rise in PO_2. Animals will periodically visit the aperture of the burrow and withdraw either head first (*single-dashed arrow*) or tail first (*double-dashed arrow*) (Pullin et al. 1980)

are not sufficient to support the O_2 consumption of the inhabitants of aquatic burrows. Thus, unlike the situation in air, exchange of gases in subaquatic systems must be achieved primarily through changes in mass flow (Withers 1978; Fig. 5).

3.2 Burrowing into the Substrata

Many ectothermic vertebrates burrow into the terrestrial or aquatic substrata, either as part of their routine behaviour or in escape from environmental pressures such as seasonal drought or extreme temperatures. Some, like the Amphisbaenians (squamate reptiles) are active predators that remain underground their entire life, burrowing through compacted soils (e.g. Gans 1969). While providing a unique habitat, the characteristics of gas exchange through soil, particularly wet soil, is such that O_2 tensions can reach quite low levels (Rawitcher 1944; Withers 1978). Certain air breathing fish and amphibians that burrow in response to drought, excavate soil burrows and form an estivation cocoon (Smith 1961; Lee and Mercer 1967) where they remain dormant over entire seasons or even longer. Under such conditions, the effective porosity of the soil (e.g. whether dry or wet), can have a profound influence on the diffusion of gases and therefore the environmental (i.e. inspired) levels of O_2 and CO_2. For example, when the toad *Bufo marinus* was induced to burrow into wet sand, the animals experienced a hypercapnic acidosis (Boutilier et al. 1979), presumably as a result of a decreased capacity for extrapulmonary CO_2 removal (Boutilier, this volume) attributed to diffusive and convective limitations for gas exchange through the soil (Withers 1978).

Various freshwater turtles (e.g. *Chrysemys*) and frogs (e.g. *Rana*) undergo seasonal periods of dormancy where they remain at least partially buried in the mud at the bottom of lakes or ponds during the winter (Penny 1987). The green sea turtle (*Chelonia mydas*) is also known to overwinter in a dormant state, partially buried, at the sea bottom (Felger et al. 1976). The sediment into which such animals might lie dormant are often severely hypoxic or even anoxic below only a few mm depth (Revsbech et al. 1980a, b; Revsbech 1983; Reimers et al. 1986).

4 Concluding Remarks

As a part of their natural behaviour, many terrestrial vertebrates occupy secluded environments including subterranean tunnels, above-ground cavities, or the soil itself. In some instances, gas exchange between these confined spaces and the free atmosphere can be restricted so much so that the metabolic activities of the inhabitants can outpace air renewal, leading to hypoxic and hypercarbic conditions. Vertebrates also inhabit various types of waters, ranging in salinity from freshwater to hypersaline, each of which is unique with respect to its gas-carrying properties (Dejours 1981). In addition, marked differences in gas tensions can exist between regions of a water body, due to poor vertical mixing of the water column (e.g. thermal stratification) or to temporary isolation (e.g. intertidal rockpools) from larger, more stable regions. Vertebrates that live in such environments often display a greater tolerance to both hypoxia and/or hypercapnia than e.g. comparably sized non-burrowing species (Bentley et al. 1967; Darden 1972; Parer and Hodson 1974; Bennett and Dawson 1976; Maclean 1978; Arieli and Ar 1979; Boutilier et al. 1979; Gatten 1981; Boggs and Kilgore 1983; Jackson 1986). Adaptations to these hypoxic environments apparently include increases in capillary density (Arieli and Ar 1981), haemoglobin oxygen affinity (Johansen et al. 1976; Bridges et al. 1982) and myoglobin concentrations (Weber et al. 1981). The capacity of some ectothermic vertebrates to endure entire seasons of hypoxia (e.g. overwintering frogs and turtles) can be attributed to marked depressions in aerobic metabolic rate (Hochachka and Somero 1984), as the animals ultimately must adapt to the amount of O_2 that is available in the environment.

Acknowledgements. The author's work is supported by grants from the Natural Sciences and Engineering Research Council of Canada.

References

Ackerman RA (1977) The respiratory gas exchange of sea turtle nests (*Chelonia*, *Caretta*). Respir Physiol 31:19–38

Arieli R (1979) The atmospheric environment of the fossorial mole rat (*Spalax ehrenbergi*): effects of season, soil texture, rain, temperature and activity. Comp Biochem Physiol 63A:569–575

Arieli R, Ar A (1979) Ventilation of a fossorial mammal (*Spalax ehrenbergi*) in hypoxic and hypercapnic conditions. J Appl Physiol 47:1011–1017

Arieli R, Ar A (1981) Blood capillary density in heart and skeletal muscles of the fossorial mole rat. Physiol Zool 54:22–27

Atkinson RJA (1976) Some preliminary field observations of the burrows of the Red Band fish, *Cepola rubescens* L. J Fish Biol 9:181–183

Atkinson RJA, Pullin RSV, Dipper FA (1977) Studies on the Red Band fish, *Cepola rubescens*. J Zool (Lond) 182:369–384

Babin J, Prepas EE (1985) Modelling winter oxygen depletion rates in icecovered temperate zone lakes in Canada. Can J Fish Aquat Sci 42:239–249

Baird DJ, Gates TE, Davies RW (1987) Oxygen conditions in two prairie pothole lakes during winter ice cover. Can J Fish Aquat Sci 44:1092–1095

Barica J, Mathias JA (1979) Oxygen depletion and winterkill risk in small prairie lakes under extended ice cover. J Fish Res Board Can 36:980–986

Bennett AF, Dawson WR (1976) Metabolism. In: Gans C, Dawson WR (eds) Biology of the Reptilia, vol 5. Physiology. Academic Press, New York, pp 127–233

Bentley PJ, Herreid CF, Schmidt-Nielsen K (1967) Respiration of a monotreme, the echidna, *Tachyglossus aculeatus*. Am J Physiol 212:957–961

Birchard GF, Kilgore DL (1980) Conductance of water vapor in eggs of burrowing and non-burrowing birds: Implications for embryonic gas exchange. Physiol Zool 53:284–292

Black CP, Tenny SM (1980) Oxygen transport during progressive hypoxia in high-altitude and sea-level waterfowl. Respir Physiol 39:217–239

Boggs DF, Kilgore DL Jr (1983) Ventilatory responses of the burrowing owl and bobwhite to hypercarbia and hypoxia. J Comp Physiol B 149:527–533

Boggs DF, Kilgore DL Jr, Birchard GF (1984) Respiratory physiology of burrowing mammals and birds. Comp Biochem Physiol 77A:1–7

Boutilier RG, Randall DJ, Shelton G, Toews DP (1979) Acid-base relationships in the blood of the toad *Bufo marinus*. III. The effects of burrowing. J Exp Biol 82:357–365

Bridges CR, Taylor AC, Atkinson RJA (1982) Respiratory properties of the blood of the burrowing red band fish *Cepola rubescens* L. J Exp Mar Biol Ecol 59:51–60

Burggren W (1985) Gas exchange, metabolism, and "ventilation" in gelatinous frog egg masses. Physiol Zool 58:503–514

Carr AF (1967) So excellent a fishe. Natural History Press, New York, 248 pp

Darden TR (1972) Respiratory adaptations of a fossorial mammal, the pocket gopher (*Thomomys bottae*). J Comp Physiol 78:121–137

Dejours P (1981) Principles of comparative respiratory physiology, 2nd edn. North Holland/Elsevier, Amsterdam New York, 265 pp

Dejours P (1982) Mount Everest and beyond: breathing air. In: Taylor CR, Johansen K, Bolis L (eds) A companion to animal physiology. Cambridge University Press, Cambridge, pp 17–30

Dethlefsen V, von Westernhagen H (1983) Oxygen deficiency and effects on bottom fauna in the eastern German Bight 1982. Meeresforschung 30:42–53

Felger RS, Cliffton K, Regal PJ (1976) Winter dormancy in sea turtles: independent discovery and exploitation in the Gulf of California by two local cultures. Science 191:283–285

Furilla RA (1980) Ecological and physiological correlates of tunnel nesting in the bank swallow, *Riparia riparia*, with reference to the cliff swallow, *Petrochelidon pyrrhonata*. M Sc Thesis, University of Alaska, Fairbanks, 54 pp

Gans C (1969) Amphisbaenians – reptiles specialized for a burrowing existence. Endeavour 28 (105):146–151

Garey WF, Rahn H (1970) Gas tensions in the tissues of trout and carp exposed to diurnal changes in oxygen tension of the water. J Exp Biol 52:575–582

Gatten RE Jr (1981) Anaerobic metabolism in freely diving painted turtles (*Chrysemys picta*). J Exp Zool 216:377–385

Gnaiger E (1983) In situ measurement of oxygen profiles in lakes: microstratifications, oscillations, and the limits of comparison with chemical methods. In: Gnaiger E, Forstner H (eds) Polarographic oxygen sensors. Springer, Berlin Heidelberg New York Tokyo, pp 245–264

Grasshoff K (1975) The hydrochemistry of landlocked basins and fjords. In: Riley JP, Skirrow G (eds) Chemical Oceanography, Vol 2, 2nd edn. Academic Press, New York, pp 456–598

Harper DE Jr, McKinney LD, Salzer RR, Case RJ (1981) The occurrence of hypoxic bottom water off the upper Texas coast and its effects on the benthic biota. Contrib Mar Sci 24:53–79

Harvey HW (1957) The chemistry and fertility of sea waters. Cambridge University Press, London, 240 pp
Hayward J (1966) Abnormal concentrations of respiratory gases in rabbit burrows. J Mammal 47:723–724
Heisler N, Forcht G, Ultsch GR, Anderson JF (1982) Acid-base regulation in response to environmental hypercapnia in two aquatic salamanders, *Siren lacertina* and *Amphiuma means*. Respir Physiol 49:141–158
Hill RW, Wyse GA (1989) Animal physiology, 2nd edn. Harper and Row, New York, 656 pp
Hochachka PW, Somero GN (1984) Biochemical adaptation. Princeton University Press, Princeton, 537 pp
Howe S, Kilgore DL Jr (1987) Convective and diffuse gas exchange in nest cavities of the northern flicker (*Colaptes auratus*). Physiol Zool 60:707–712
Jackson DC (1986) Acid-base regulation of reptiles. In: Heisler N (ed) Acid-base regulation in animals. Elsevier Scientific, Amsterdam, pp 235–263
Johansen K, Lykkeboe G, Weber RE, Maloiy GMO (1976) Respiratory properties of blood in awake and estivating lungfish, *Protopterus amphibius*. Respir Physiol 27:335–345
Kendeigh SC (1961) Energy of birds conserved by roosting in cavities. Wilson Bull 73:140–147
Kennerley TE (1964) Microenvironmental conditions of the pocket gopher burrow. Texas J Sci 16:395–441
Kramer DL, Lindsey CC, Moodie GEE, Stevens ED (1978) The fishes and the aquatic environment of the central Amazon basin, with particular reference to respiratory patterns. Can J Zool 56:717–729
Lee AK, Mercer EH (1967) Cocoon surrounding desert-dwelling frogs. Science 157:87–88
Lüling KH (1975) Ichthyologische und gewässerkundliche Beobachtungen und Untersuchungen an der Yarina Cocha, in der Umgebung von Pucallpa und am Rio Pacaya (mittlerer und unterer Ucayali, Ostperu). Zool Beitr 21:29–96
Maclean GS (1978) Respiratory physiology of a semi-fossorial mammal, *Tamias striatus*. Physiologist 21:75
Moodie GEE (1978) Observations on the life history of the caecilian *Typhlonectes compressicaudus* (Dumeril and Bibron) in the Amazon basin. Can J Zool 56:1005–1008
Officer CB, Biggs RB, Taft JL, Cronin LE, Tyler MA, Boynton WR (1984) Chesapeake Bay anoxia: origin, development, and significance. Science 223:22–27
Parer JT, Hodson WA (1974) Respiratory studies of monotremes. IV. Normal respiratory function of echidnas and ventilatory response to inspired oxygen and carbon dioxide. Respir Physiol 21:307–316
Penny DG (1987) Frogs and turtles: Different ectotherm overwintering strategies. Comp Biochem Physiol 86A:609–615
Pullin RSV, Morris DJ, Bridges CR, Atkinson RJA (1980) Aspects of the respiratory physiology of the burrowing fish *Cepola rubescens* L. Comp Biochem Physiol 66A:35–42
Rahn H (1966) Aquatic gas exchange: theory. Respir Physiol 1:1–12
Rawitcher FK (1944) Problemas de fitoecologia com consideracoes especiais sobre o Brazil meridional. Bol Fac Filos Cienc Univ São Paulo 4:1–153
Reid GK (1961) Ecology of inland waters and estuaries, D. Van Nostrand, New York, 375 pp
Reimers CE, Fischer KM, Merewether R, Smith KL, Jahnke RA (1986) Oxygen microprofiles measured in situ in deep ocean sediments. Nature 320:741–744
Revsbech NP (1983) In situ measurement of oxygen profiles of sediments by use of oxygen microelectrodes. In: Gnaiger E, Forstner H (eds) Polarographic oxygen sensors. Springer, Berlin Heidelberg New York Tokyo, pp 265–273
Revsbech NP, Sorensen J, Blackburn TH, Lomholt JP (1980a) Distribution of oxygen in marine sediments measured with microelectrodes. Limnol Oceanogr 25:403–411
Revsbech NP, Jorgensen BB, Blackburn TH (1980b) Oxygen in the seabottom measured with a microelectrode. Science 207:1355–1356
Seki H, Tsuji T, Hattori A (1974) Effects of zooplankton grazing on the formation of the anoxic layer of Tokyo Bay. Estuarine Coastal Mar Sci 2:145–151
Seliger HH, Boggs JA, Biggley WH (1985) Catastrophic anoxia in the Chesapeake Bay in 1984. Science 228:70–73

Smith HW (1961) From Fish to Philosopher. The Natural History Library, Anchor Books, Doubleday, Garden City NY, 293 pp
Stephenson TA, Zoond A, Eyre J (1934) The liberation and utilization of oxygen by the population of rock-pools. J Exp Biol 11:162–172
Studier EH, Baca TP (1968) Atmospheric conditions in artificial rodent burrows. Southwest Nat 13:401–410
Toews DP, Macintyre DH (1978) Respiration and circulation in an apodan amphibian. Can J Zool 56:998–1004
Truchot JP, Duhamel-Jouve A (1980) Oxygen and carbon dioxide in the marine intertidal environment: diurnal and tidal changes in rockpools. Respir Physiol 39:241–254
Ultsch GR (1973) The effects of water hyacinths (*Eichhornia crassipes*) on the microenvironment of aquatic communities. Arch Hydrobiol 72:460–473
Ultsch GR (1976) Eco-physiological studies of some metabolic and respiratory adaptations of sirenid salamanders. In: Hughes GM (ed) Respiration of Amphibious Vertebrates. Academic Press, New York, pp 287–312
Vogel S, Ellington CP, Kilgore DL (1973) Wind-induced ventilation of the burrow of the prairie-dog, *Cyanomys ludovicianus*. J Comp Physiol 85:1–14
Weber RE, Johansen K, Abe AS (1981) Myoglobin from the burrowing reptile *Amphisbaena alba*. Concentrations and functional characteristics. Comp Biochem Physiol 68A:159–165
White FN, Bartholomew GA, Kinney JL (1978) Physiological and ecological correlates of tunnel-nesting in the European bee-eater, *Merops apiaster*. Physiol Zool 51:140–148
White FN, Kinney J, Siegfried WR, Kemp AC (1984) Thermal and gaseous conditions of hornbill nests. Natl Geogr Res Rep 17:931–936
Wickler SJ, Marsh RL (1981) Effects of nestling age and burrow depth on CO_2 and O_2 concentrations in the burrows of bank swallows (*Riparia riparia*). Physiol Zool 54:132–136
Williams DD, Rausch RL (1973) Seasonal carbon dioxide concentrations in the dens of hibernating mammals (Sciuridae). Comp Biochem Physiol 44:1227–1235
Withers PC (1978) Models of diffusion-mediated gas exchange in animal burrows. Am Nat 112:1101–1112
Wood CM, Boutilier RG (1985) Osmoregulation, ionic exchange, blood chemistry, and nitrogenous waste excretion in the land crab *Cardisoma carnifex*: A field and laboratory study. Biol Bull 169:267–290

Chapter 2

Modeling of Gas Exchange in Lungs, Gills and Skin

J. Piiper[1]

Contents

1 Introduction

The aim of this review is to discuss the models developed for analysis of gas transfer in vertebrate gas exchange organs. In Section 2, which deals with steady state gas exchange, the models are those previously reviewed by Piiper (1982a) and Piiper and Scheid (1977, 1982, 1989). The Section 3, which addresses gas exchange in unsteady state, is inspired by the seminal studies on pulmonary gas exchange and CO_2 stores by Farhi and Rahn (1955a), and includes discussion of experimental data recently obtained in intermittently breathing amphibians and reptiles.

[1] Department of Physiology, Max Planck Institute for Experimental Medicine, D-3400 Göttingen, FRG

Advances in Comparative and Environmental Physiology
Vol. 6. Edited by R. G. Boutilier

2 Gas Exchange in Steady State

2.1 General Model

Steady state gas transfer in all vertebrate gas exchange organs can be described using a general model (Fig. 1). Gas is transferred between medium and blood by three processes arranged in series: ventilation, medium/blood transfer by diffusion, and perfusion.

The overall performance of a gas exchange organ, as well as that of three component processes, can be described in terms of conductances (or their reciprocals, resistances). Conductance (G) is defined as transfer rate M (e.g. O_2 uptake, CO_2 output), divided by the effective partial pressure difference $P_1 - P_2$ (Piiper et al. 1971; Piiper and Scheid 1972, 1975):

$$G = \frac{M}{P_1 - P_2} \ . \tag{1}$$

For convective transport by medium flow or ventilation, the transfer rate $\dot{M}$ is equal to the product of flow (V), effective solubility in the medium (β_m) and inspired-expired partial pressure difference ($P_i - P_e$). Since $P_i - P_e$ corresponds to $P_1 - P_2$ in Eq. (1), one obtains for the ventilation conductance, G_{vent}:

$$G_{vent} = \dot{V} \cdot \beta_m \ . \tag{2}$$

In an analogous manner, for convective transport by blood flow or perfusion, $\dot{M}$ is given by the product of blood flow ($\dot{Q}$), effective solubility in blood (β_b) and arterial-venous partial pressure difference ($P_a - P_v$). The perfusion conductance, G_{perf}, is:

$$G_{perf} = \dot{Q} \cdot \beta_b \ . \tag{3}$$

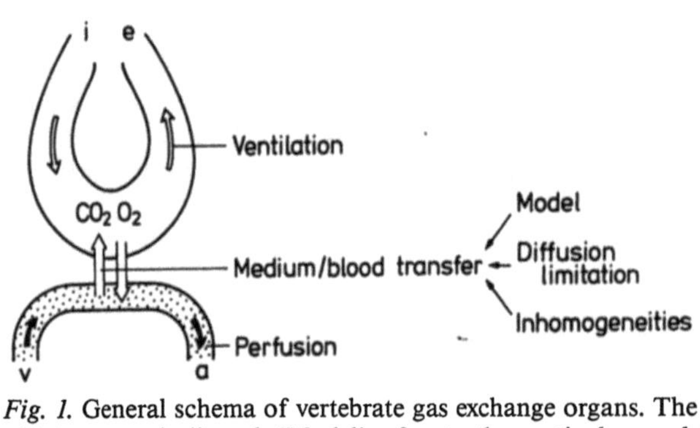

Fig. 1. General schema of vertebrate gas exchange organs. The factors determining gas transfer efficiency are indicated. 'Model' refers to the particular models shown in Fig. 2. For explanations, see text

The effective solubility, also termed capacitance coefficient, β, is defined as increment in concentration (ΔC) per increment in partial pressure (ΔP) (Piiper et al. 1971):

$$\beta = \frac{\Delta C}{\Delta P} . \tag{4}$$

For a gaseous medium, β is the same for all gases and equal to 1/(RT) (R, gas constant; T, absolute temperature). For water, β is generally equal to the physical solubility; but for CO_2, β may exceed the physical solubility due to buffering. For blood, β for gases that are chemically bound (CO_2, O_2, CO) is equal to the slope of the dissociation curve (plotted as concentration against partial pressure), whereas β for inert gases is identical with physical solubility.

In the case of non-linear dissociation curves, β_b for perfusion conductance is defined as the slope of the straight line intersecting the dissociation curve at the arterial and venous partial pressures. In the analysis of diffusion processes, β may be taken as the slope of the tangent to the curve or as the slope averaged between two partial pressures according to the particular conditions (see below).

An important inference from Eqs. (2) and (3) is that flow and effective solubility are of equal importance in convective transport.

The transfer between medium and blood occurs by diffusion. It is customary to apply the simplest model for diffusion, i.e. a uniform flat sheet. It follows from the Fick diffusion equation that the diffusion conductance or diffusing capacity, D, is given by the relationship:

$$D = d \cdot \alpha \frac{F}{x} , \tag{5}$$

where d = diffusion coefficient, α = physical solubility, F = surface area, x = thickness; all these parameters are for the diffusion barrier.

2.2 Particular Models

Besides G_{vent}, G_{perf} and D, the arrangement of blood and medium flows in the gas exchange organ is of importance for the overall gas transfer, and defines the particular model. The four models used for the comparative analysis of gas transfer in vertebrate gas exchange organs are shown in Fig. 2 (Piiper and Scheid 1972, 1975).

1. Open Model. This, the simplest of the four models, is used for the analysis of gas exchange between ambient medium and cutaneous blood capillaries in amphibian skin. There is usually a strong diffusion limitation because the mechanical protective function of skin imposes a thick diffusion barrier. Consequently, arterial blood P_{O_2} is much below that of ambient air or water P_{O_2} (Piiper et al. 1976).

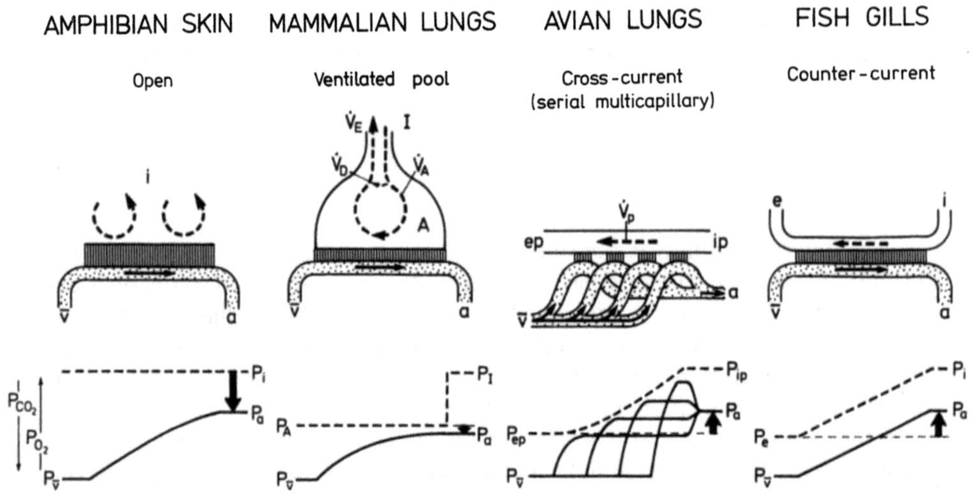

Fig. 2. Types of vertebrate gas exchange organs and their models. *Upper row*: Models. The slightly differing symbols used with each model are indicated. *Lower row*: Schematic O_2 and CO_2 partial pressure profiles. The partial pressure differences between the medium exiting from the organ and arterialized blood are marked by *arrows* (for homogeneous models with diffusion limitation)

2. Ventilated Pool. The well-known model for mammalian alveolar lungs is analogous to the open model as far as gas exchange between alveolar gas and pulmonary capillary blood is concerned. But the alveolar gas differs from the medium due to finite G_{vent}. In contrast to amphibian skin, the diffusion barrier is very thin, allowing highly efficient diffusive gas/blood equilibration.

3. Cross-Current (or Serial Multicapillary) Model. This model has been proposed for bird lungs in which relatively long parabronchi are surrounded by an intertwining air capillary-blood capillary network where gas exchange takes place (Scheid 1978, 1979). There is a considerable drop of P_{O_2} (and increase of P_{CO_2}) along the parabronchus. The arterial blood is a mixture of end-capillary blood of differing P_{O_2}, and its P_{O_2} may be higher and PCO_2 lower than their respective gas tensions in exiting (end-parabronchial) gas. This confers upon bird lungs a higher intrinsic gas exchange efficiency compared to mammalian (ventilated pool) lungs.

4. Counter-Current Model. This model is used for simulating gas exchange in fish gills where water flows between the secondary lamellae in a direction opposite to blood perfusing the secondary lamellae (Piiper and Scheid 1983b, 1984; Piiper 1982b). It provides the potential for the highest gas exchange efficiency of all of the models, since it allows equilibration of arterialized blood with inspired medium. In all observed cases, however, the actual gas exchange efficiency has been found to be much lower, apparently due to strong diffusion limitation both in the water/blood barrier and in the interlamellar water, and to ventilation/perfusion inhomogeneity and shunting (see below).

2.3 Diffusion Limitation

The role and extent of diffusion limitation in medium/blood gas transfer is an important issue. In three of the four gas exchange models (not in the counter-current model) a medium of constant composition is in gas exchange contact with capillary blood in which the partial pressures of CO_2 and O_2 change from the mixed venous to the arterialized values (Fig. 2). Therefore, it is useful to study the gas exchange characteristics of a model in which blood flow ($\dot{Q}$) is separated from a medium of constant composition by a barrier with diffusing capacity D. It is easily shown that for constant β_b the relative equilibration deficit, $(P_A - P_a)/(P_A - P_{\bar{v}})$, is a simple function of the ratio $D/(\dot{Q} \cdot \beta_b)$, which is the diffusion/perfusion conductance ratio (Piiper 1982c; Piiper and Scheid 1981, 1983a):

$$\frac{P_A - P_a}{P_A - P_v} = e^{-\frac{D}{\dot{Q} \cdot \beta_b}} . \tag{6}$$

In Fig. 3 the model is illustrated as applied to O_2 transfer. The blood capacitance coefficient, β_{O_2}, varies with P_{O_2}. At the P_{O_2} corresponding to nearly full saturation of hemoglobin with O_2, β_{O_2} is very small, practically equaling the physical solubility in blood. With falling P_{O_2}, β_b rises, reaching a maximum below the P_{O_2} of half-saturation (P_{50}), thereafter declining slightly (when the O_2 dissociation curve is sigmoid). Accordingly, in human (mammalian) lungs diffusion limi-

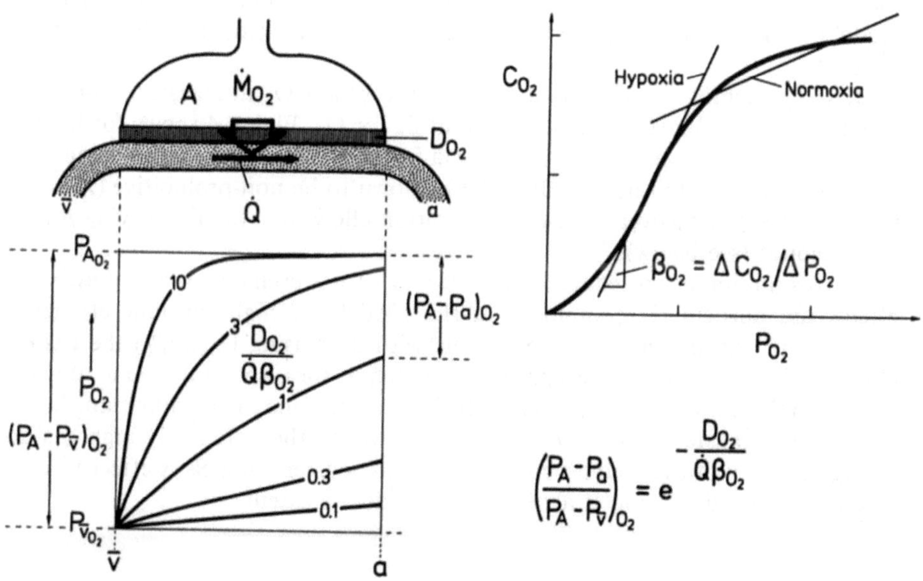

Fig. 3. Model for O_2 diffusion in alveolar lungs with P_{O_2} profiles in pulmonary capillaries and the resulting equilibration deficit $(P_A - P_a)_{O_2}/(P_A - \bar{P}_P)_{O_2}$ as determined by the ratio $D_{O_2}/(\dot{Q} \cdot \beta_{O_2})$. β_{O_2}, the slope of blood O_2 dissociation curve, is shown to be higher in hypoxia as compared to normoxia

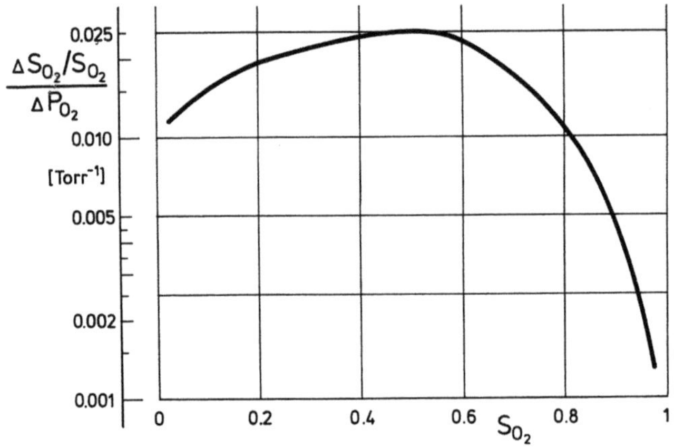

Fig. 4. Variation of the O_2 capacitance coefficient (effective solubility, slope of dissociation curve) in human blood with hemoglobin O_2 saturation (S_{O_2}). Ordinate: $\Delta S_{O_2}/(S_{O_2} \cdot \Delta P_{O_2})$ as relative measure for the capacitance coefficient $\beta_b(O_2)$, on logarithmic scale. $\beta_b(O_2) = \alpha_{O_2} + Cap_{O_2} \cdot \Delta S_{O_2}/(S_{O_2} \cdot \Delta P_{O_2})$ (α_{O_2}, physical solubility of O_2 in blood; Cap_{O_2}, O_2 capacity). Based on normal human blood at pH = 7.4

tation for O_2 is usually considered negligible in hyperoxia and also in normoxia (at least in resting conditions), but it becomes evident in hypoxia. Thus it is understandable that diffusion limitation may be neglected for all inert gases. For CO, diffusion limitation is prominent, and it is expected to play some role in CO_2 transfer (see below). The relationships should be similar in avian lungs where the model of Fig. 3 can be applied to every cross-sectional element of the parabronchus.

It is frequently stated that in hypoxia the blood O_2 dissociation curve becomes practically linear, meaning constant β_b for O_2. Figure 4 shows for human blood that this is a very rough approximation. But the errors due to using constant average $\beta_b(O_2)$ in hypoxia have been shown to be non-prohibitive (Kobayashi et al. 1987). Moreover, the Bohr integration allows one to take varying $\beta_b(O_2)$ into account (see below).

Diffusion limitation reduces the efficiency of gas exchange in all models. Its effects are particularly prominent in the models of high inherent efficiency (counter-current and cross-current). Thus with decreasing $D/(\dot{Q} \cdot \beta_b)$ the gas exchange efficiency of all models progressively diminishes, and at the same time becomes similar for all models, when diffusion has become the predominant factor limiting gas transfer (Piiper 1982a). In other words, the potentially high gas exchange efficacy of the counter-current and cross-current models is effective only when diffusion limitation is absent or of moderate extent.

2.4 Diffusion Barrier

In simplified models of gas exchange organs it is usually assumed that the resistance to diffusion is located in a distinct and homogeneous tissue barrier between

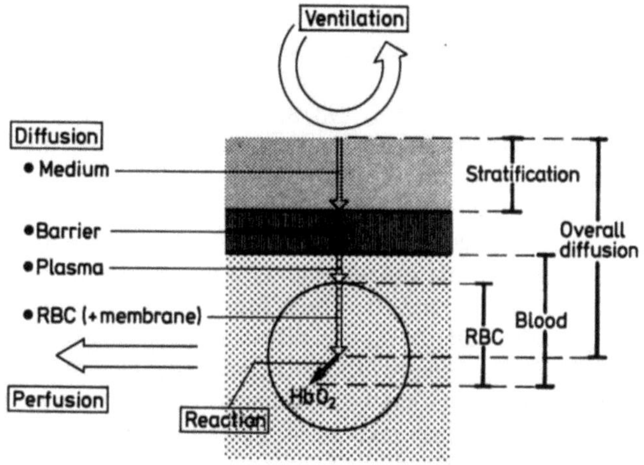

Fig. 5. The various steps involved in medium/blood transfer of O_2. The designations used for component processes of O_2 uptake are indicated. *RBC* red blood cell

blood and medium. But in real gas exchange organs other and multiple sites must be considered (Fig. 5):

1. Convective transport by ventilatory flow does not directly reach the surface of the barrier separating medium and blood: it has to be supplemented by diffusion across a still medium layer. In alveolar lungs, diffusion limitation to some degree has been shown to occur in the distal airways of mammalian lungs (Georg et al. 1965; Cumming et al. 1967; Piiper 1979; Scheid and Piiper 1980a) and in the air capillaries of avian lungs (Scheid 1978). This aspect is particularly important in water breathing. In the interlamellar spaces of fish gills, where water flow is laminar, a functionally stagnant layer must be present due to the parabolic flow profile (Scheid and Piiper 1971; Piiper et al. 1986; Scheid et al. 1986). The same must be true for skin breathing in still water (Feder and Burggren 1985). Although not specifically included in this article, the case of avian eggs is of interest. There, the main resistance to gas transfer by diffusion resides in the air-filled pores of the calcareous egg shell (Wangensteen et al. 1970/71; Paganelli 1980). The term 'stratification' is sometimes used to designate the CO_2 and O_2 gradients in the medium due to diffusion resistance.
2. The 'tissue barrier' is composed of a number of anatomical structures and may vary in thickness locally. Certain averaging methods in morphometric studies have attempted to take this into account (Weibel 1963, 1973).
3. On the blood side of the barrier, one has to consider functionally stagnant layers of plasma. This is particularly important in mammalian (and avian) lungs where the tissue barrier separating gas and blood is very thin.
4. Diffusion within red blood cells may be facilitated by some convective mixing and by hemoglobin diffusion, but it may still constitute a major part of the overall resistance to O_2 uptake in mammalian lungs (Moll 1962; Thews 1963).
5. The finite reaction velocity of the chemical bonding of O_2 to hemoglobin may exert some limiting effect which would be included in an overall 'diffusing capaci-

ty'. However, recent studies with the stopped-flow technique have yielded no clear evidence that a reaction limitation plays a significant role in the overall intraerythrocyte diffusion limitation of O_2 uptake by mammalian or avian red blood cells (Yamaguchi et al. 1985, 1987a, b; Nguyen Phu et al. 1986).

6. In the elimination of CO_2, particular problems arise due to the slowness of the dehydration reaction $H_2CO_3 \rightarrow CO_2 + H_2O$ (since the acceleration by carbonic anhydrase in the red cells may be insufficient) and of the concomitant HCO_3^-/Cl^- exchange between red cells and plasma. It is probably due to these factors that the effective diffusing capacity for CO_2 as obtained from equilibration of C-labeled CO_2 in human lungs has been found to be definitely lower than predicted from D_{O_2} on the basis of Krogh diffusion coefficients for O_2 and CO_2 (Piiper et al. 1980). In accordance with this hypothesis, the effective D for CO_2 from O-labeled CO_2 equilibration has been found to be much higher (Schuster 1985).

In fish gills, an analogous limitation arises additionally in the medium because, if uncatalyzed, the hydration rate of CO_2 in the interlamellar water of gills should manifestly curtail the efficiency of blood/water CO_2 equilibration. Wright et al. (1986) concluded from their experimental results on rainbow trout that the hydration of CO_2 at the gill surface was catalyzed. Mucus of the body surface was shown to possess carbonic anhydrase activity.

7. The equilibria of CO_2 and O_2 are interdependent (Bohr and Haldane effects), and the gas species which is slower in equilibrating may limit the equilibration of the intrinsically more rapidly equilibrating gas (cf. Wagner and West 1972).

Considering all these complicating factors, the 'diffusing capacity' D must be regarded as an 'overall equilibration conductance' that includes chemical reactions besides diffusion processes. Since the chemical reactions are not of first order and since diffusion occurs in complex geometry and in moving media, assumption of a constant overall D can represent an approximation only. In spite of these problems, D and the ratio $D/(\dot{Q} \cdot \beta_b)$ have proved to be useful parameters in estimating the overall diffusion (and reaction) processes in limiting external gas exchange.

2.5 Determination of Diffusing Capacity

Instead of O_2, usually CO is used as test gas for measurement of diffusing capacity, D. The very important advantage of CO over O_2 is its very high β_b value (the CO/O_2 ratio for β_b amounting to about 150 for human blood). Thereby $\dot{Q} \cdot \beta_b$ becomes so high that perfusion limitation disappears. Then Pa_{CO} is practically equal $P\bar{v}_{CO}$, and in many conditions both Pa_{CO} and $P\bar{v}_{CO}$ are very nearly zero. Thus D_{CO} can be obtained from CO uptake and P_{CO} in the medium of the gas exchange organ, e.g. lung alveolar gas:

$$D_{CO} = \frac{\dot{M}_{CO}}{P_{A_{CO}}} \quad . \qquad (7)$$

The disadvantage of CO is its relatively slow reaction rate with hemoglobin leading to difficulties in predicting D_{O_2} from the measured D_{CO}.

For determination of D_{O_2} from analysis of gas exchange in steady state, hypoxia is necessary (at least in mammals with relatively high D_{O_2}) to achieve a measurable P_{O_2} deficit due to diffusion limitation (in mammals: alveolar-to-arterial P_{O_2} difference). According to Eq. (5) one obtains:

$$D_{O_2} = \dot{Q} \cdot \beta_{bO_2} \cdot \ln \frac{(P_A - P_{\bar{v}})_{O_2}}{(P_A - P_a)_{O_2}} . \tag{8}$$

For avian lungs the cross-current model yields the following relationship (P_{ip} and P_{ep} refer to initial-parabronchial and end-parabronchial gas, respectively) (Scheid 1979):

$$D_{O_2} = - \frac{\dot{M}_{O_2}}{(P_a - P_{\bar{v}})_{O_2}} \cdot \ln \left[1 + \frac{(P_a - P_{\bar{v}})_{O_2}}{(P_{ip} - P_{ep})_{O_2}} \cdot \ln \frac{(P_{ep} - P_{\bar{v}})_{O_2}}{(P_{ip} - P_{\bar{v}})_{O_2}} \right] . \tag{9}$$

For fish gills one obtains on the basis of the counter-current model (Piiper and Scheid 1984):

$$D_{O_2} = \frac{\dot{M}_{O_2}}{(P_i - P_a)_{O_2} - (P_e - P_{\bar{v}})_{O_2}} \cdot \ln \frac{(P_i - P_a)_{O_2}}{(P_e - P_{\bar{v}})_{O_2}} . \tag{10}$$

Theoretically more accurate is the well-known Bohr integration method (Bohr 1909), which takes the actual shape of the blood O_2 dissociation curve into account (Fig. 6A). By this method, the D_{O_2} for a cross-sectional element of the capillary, ΔD_{O_2}, is calculated from the local O_2 uptake, $\Delta \dot{M}_{O_2}$, (which is proportional to the O_2 content difference across the element, ΔC_{O_2}) and the local effective gas/blood P_{O_2} difference, $(P_A - P_b)_{O_2}$:

$$\Delta D_{O_2} = \frac{\Delta \dot{M}_{O_2}}{(P_A - P_b)_{O_2}} . \tag{11}$$

The calculation is consecutively performed for all cross-sectional elements, from the venous to the arterial (end-capillary) value. D_{O_2} is obtained by summation:

$$D_{O_2} = \sum_{\bar{v}}^{a} \frac{\Delta \dot{M}_{O_2}}{(P_A - P_b)_{O_2}} . \tag{12}$$

If the ΔM_{O_2} elements are chosen equal, the relationship simplifies to:

$$D = \dot{M}_{O_2} \cdot \sum_{\bar{v}}^{a} \frac{1}{(P_A - P_c)_{O_2}} . \tag{13}$$

For application to the counter-current model of fish gills (Fig. 6B), a sloping straight line for the medium ('medium O_2 line') is drawn in such a manner that the environmental or inspired P_{O_2} (Pi_{O_2}) is at the level of Ca_{O_2} and the expired P_{O_2} (Pe_{O_2}) at the level of Cv_{O_2}. The P_{O_2} difference effective for diffusion is that between the 'medium O_2 line' and the blood O_2 dissociation curve (the slope of the medium O_2 line is equal to $\beta m_{O_2} \cdot \dot{V}/\dot{Q}$). The method has been applied to both O_2 and CO_2 for data obtained from a dogfish (Piiper and Baumgarten-Schumann 1968).

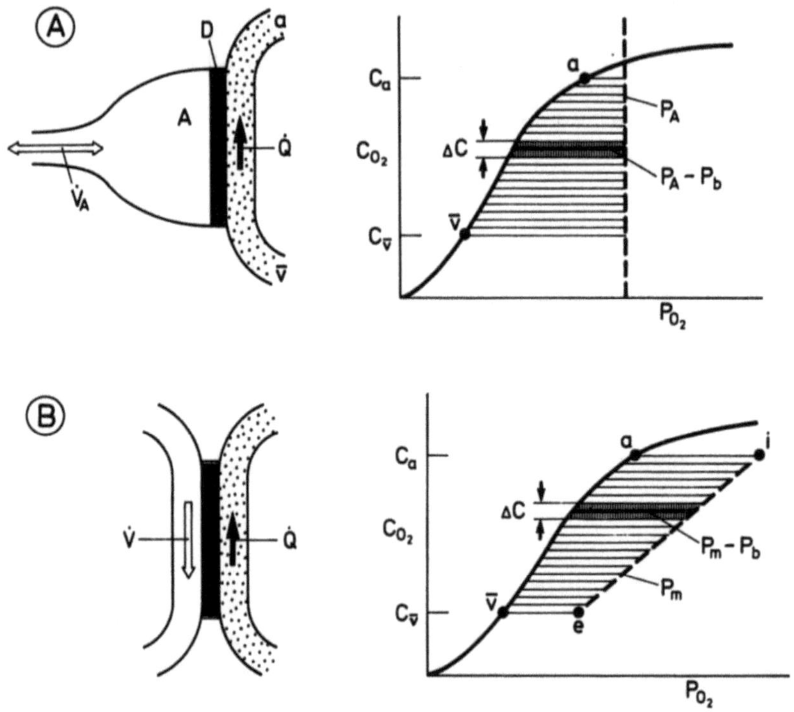

Fig. 6. Bohr integration for calculation of O_2 diffusing capacity, D_{O_2}. *A* Alveolar lung (ventilated pool model). *B* Fish gills (counter-current model). *Left-hand side*: Models. *Right-hand side*: P_{O_2} profiles. For explanation, see text

Application of the Bohr integration to the cross-current model of avian lungs is possible but laborious. Here the classical-type Bohr integration should be performed consecutively for each cross-sectional element of the parabronchus (at constant local parabronchial P_{O_2}). The change in parabronchial P_{O_2} across the element, obtained from air flow and O_2 uptake by the element, determines the P_{O_2} of the next parabronchial element in downstream direction.

The methods described above are based on measurements in steady state (for CO, strictly speaking, in a semi-steady state). Other methods, the single-breath and rebreathing methods, are measurements in unsteady state of the test gas (see below). Whereas the single-breath D_{CO} method has the advantage of simplicity and ease, the great advantage of the rebreathing methods is that during rebreathing, lung gas is efficiently homogenized so that the disturbing inhomogeneity effects are reduced (see below). However, the inherent disadvantage of the breath-holding and rebreathing methods is that they disturb the respiratory steady state regime by maximal inspiration and breath holding, or by increased ventilation and changing alveolar gas during breathing.

To date, only few determinations of the diffusing capacity have been made in non-mammalian vertebrates. The results are summarized in Table 1. To make the values more comparable, the mean diffusional P_{O_2} difference, ΔP_{O_2}(diff), was calculated as

Table 1. Diffusing capacity for $P_2 \cdot \Delta P_{O_2}$ (diff), mean medium-to-blood O_{O_2} difference ($= \dot{M}_{O_2}/D_{O_2}$)

Class	Species	Body mass W (kg)	Body temperature (°C)	Gas exchange organ	$\dot{M}_{O_2}$/W ml/(min·kg)	D_{O_2}/W ml/(min·Torr·kg)	ΔP_{O_2} (diff) (Torr)	Reference	Method for D_{O_2}
Fishes	Larger spotted dogfish (*Scyliorhinus stellaris*)	2.2	17	Gills	0.64	0.0085	75	Piiper et al. (1986)	P_{O_2}, Bohr integration
	Rainbow trout (*Salmo gairdneri*)	0.5 – 1.0	15	Gills	1.43	0.027	53	Randall et al. (1967)	P_{O_2}, algebraic mean
Amphibians	Dusky salamander (*Desmognathus fuscus*)	0.0061	13 – 22	Skin	0.72	0.0067	107	Piiper et al. (1976)	Combined (indirect)
	Bullfrog (*Rana catesbeiana*)	0.26	20	Lungs	0.49	0.027	18	Glass et al. (1981 a)	CO rebreathing[a]
Reptiles	Mediterranean land tortoise (*Testudo graeca*)	0.98	20 – 23	Lungs	1.54	0.084 – 0.140[b]	18 – 11[b]	Crawford et al. (1976)	CO rebreathing[a]
	Painted turtle (*Pseudemys scripta elegans*)	1.55	20 – 23	Lungs	0.90	0.041 – 0.081[b]	22 – 11[b]	Crawford et al. (1976)	CO rebreathing[a]
	Green sea turtle (*Chelonia mydas*)	70	25	Lungs	0.99	0.053	19	Gatz et al. (1987)	CO rebreathing[a]
	Monitor lizard (*Varanus exanthematicus*)	2.2	25 – 27	Lungs	1.35	0.073	19	Glass et al. (1981 b)	CO rebreathing[a]
	Tegu lizard (*Tupinambis teguixin*)	2.2	25 – 27	Lungs	1.20	0.049	24	Glass et al. (1981 b)	CO rebreathing[a]
Birds	Muscovy duck (*Cairina moschata*)	1.8	41	Lungs	10.3	1.24	8	Burger et al. (1979)	P_{O_2}, cross-current model
Mammals	Man	77	37	Lungs	3.90	0.70	6	Meyer et al. (1981)	O_2 equilibration by rebreathing

[a] D_{O_2} calculated as 1.23 D_{CO}.
[b] Range of values due to methodological problems.

$$\Delta P_{O_2}\,(\text{diff}) = \frac{\dot{M}_{O_2}}{D_{O_2}} \,. \tag{14}$$

ΔP_{O_2}(diff) is an indicator of diffusion limitation effect, high values meaning strong diffusion limitation. The highest ΔP_{O_2}(diff) were obtained for cutaneous gas exchange in lungless salamanders and for branchial gas exchange in fishes. The values in Table 1 were obtained in resting animals in normoxic conditions. In hypoxia, at elevated body temperature and during exercise, D_{O_2} is usually found to increase (cf. Piiper and Scheid 1980). An increase in D_{O_2} may be attributed to dilatation and/or recruitment of gas exchange organ capillaries, but reduction of inhomogeneity effects (see below) may also be involved.

2.6 Functional Inhomogeneities

The functional inhomogeneities due to unequal distribution of alveolar ventilation ($\dot{V}_A$) to pulmonary blood flow ($\dot{Q}$) are the main source of gas transfer inefficacy measured as alveolar-to-arterial P_{O_2} difference, $(P_A-P_a)_{O_2}$, in mammals (including man) in normoxia.

The mechanism is schematically illustrated in Fig. 7A for the simplest case of a two-compartment lung. The compartment I is relatively hyperventilated (high $\dot{V}_A/\dot{Q}$), its P_{O_2} is high and P_{CO_2} is low as compared to compartment II, which is relatively hypoventilated (low $\dot{V}_A/\dot{Q}$). For simplicity, complete blood/gas equilibration is assumed in both compartments. The essential point is that mixing of both expired alveolar gas and arterialized blood from the compartments, to mixed alveolar gas (P_A) and to (mixed) arterial blood (P_a), respectively, occurs in a flow-weighted manner. Thus, P_A is biased towards compartment I (high O_2, low CO_2) and P_a towards compartment II (low O_2, high CO_2). In quantitative terms, the resulting $(P_A-P_a)_{O_2}$ and $(P_a-P_A)_{CO_2}$ differences are influenced by the steepness and the shape of the blood dissociation curves, the resulting $(P_A-P_a)_{O_2}$ differences being in most cases larger than the $(P_a-P_A)_{CO_2}$ differences.

In a real lung, there exists a more or less continuous distribution of $\dot{V}_A$ and $\dot{Q}$ as shown by the multiple inert gas infusion method (Wagner et al. 1974a, b; West 1977). For a simple quantification of CO_2 and O_2 exchange in mammalian lungs, the three-compartment model of Riley and Cournand (1949, 1951) has been widely used (Fig. 7B). It consists of (1) a 'functional' of 'effective' compartment (Eff), with finite $\dot{V}_A$ and $\dot{Q}$, (2) a 'shunt' compartment (sh) ($\dot{V}_A = 0$, $\dot{V}_A/\dot{Q} = 0$) and (3) an 'alveolar dead space' compartment (AD) ($\dot{Q} = 0$, $\dot{V}_A/\dot{Q}$ infinity). Since shunt produces $(P_A-P_a)_{O_2}$ but contributes little to $(P_a-P_A)_{CO_2}$, the portions of $(P_A-P_a)_{O_2}$ due to shunt and to alveolar-dead-space ventilation can be estimated.

In addition to diffusion deficit and $\dot{V}_A/\dot{Q}$ inequality it is customary, in mammalian gas exchange physiology, to consider shunt as a third separate factor giving rise to $(P_A-P_a)_{O_2}$ differences. This is justified by the fact that aside from blood flow from low $\dot{V}_A/\dot{Q}$ areas, there may be blood flow from atelectatic lung areas and from branchial and Thebesian veins leading to 'venous admixture' to

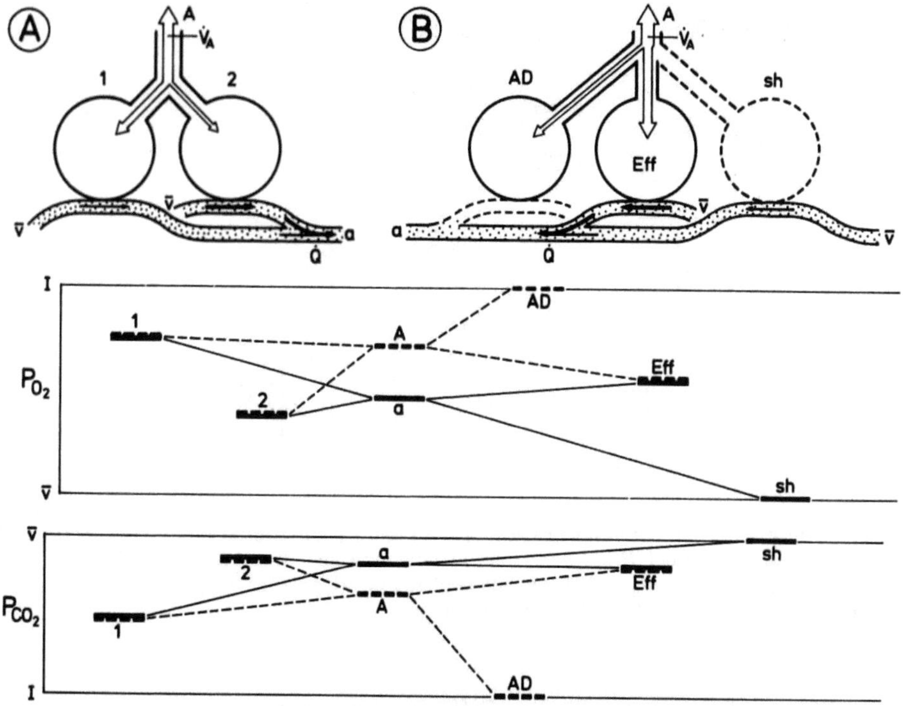

Fig. 7. $\dot{V}_A/\dot{Q}$ inequality generating $(P_A-P_a)_{O_2}$ and $(P_a-P_A)_{CO_2}$ differences in alveolar lung models. *Upper row*: Models. *Lower panels*: P_{O_2} and P_{CO_2} in the individual compartments (*1*, *2*; *AD*, *Eff*, *sh*), in mixed alveolar gas (*A*) and mixed arterial blood (*a*). *Left-hand side*: Two-compartment model. *Right-hand side*: Three-compartment model with effective compartment (*Eff*), alveolar dead space compartment (*AD*) and shunt compartment (*sh*). The variables of the models are chosen in such a manner that both models yield the same P_{O_2} and P_{CO_2} in mixed alveolar gas and arterial blood. For further explanations, see text

arterialized blood. Moreover, in amphibians and reptiles (and in analogous cardiac anomalies in mammals) extensive intracardiac blood mixing may lead to considerable shunting. The functional shunt due to admixture of venous blood that has had no gas exchange contact with pulmonary gas is called a 'true shunt'. Since the shunt due to $\dot{V}/\dot{Q}$ mismatch ('functional shunt'), in contrast to the true shunt, gives rise to an arterial-alveolar P_{N_2} difference (see below), contributes to excretion of the least soluble gas with the multiple gas infusion technique and produces no $(P_A-P_a)_{O_2}$ at pure O_2 breathing, these shunt varieties can be distinguished from each other.

The effects of $\dot{V}/\dot{Q}$ inhomogeneity on gas exchange efficiency in counter-current and cross-current models is basically similar to that in the ventilated pool model. This is shown schematically for a two-parabronchus model of an avian cross-current lung in Fig. 8. With a sufficient degree of inhomogeneity the negative $(P_{E'}-P_a)_{O_2}$ is reversed to positive values. The presence and extent of $\dot{V}/\dot{Q}$ inhomogeneity in avian lungs has been demonstrated by the multiple inert gas infusion technique (Powell and Wagner 1982a, b).

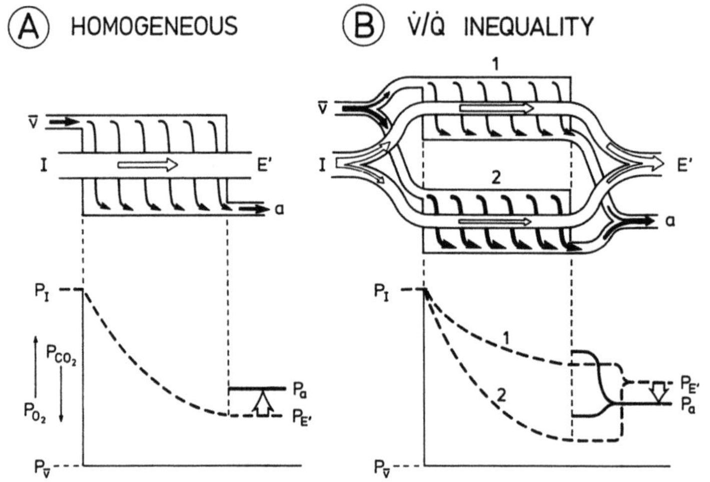

Fig. 8. $\dot{V}_A/\dot{Q}$ inequality in cross-current models for avian lungs. *Upper row*: Models. *Lower row*: P_{O_2} profiles in the parabronchi (*broken line*) and P_{O_2} in arterial blood (P_a). *Left-hand side*: Homogeneous model. *Right-hand side*: Two-compartment model (*1*, *2*) with differing $\dot{V}/\dot{Q}$ ratio. The negative $(P_{E'}-P_a)_{O_2}$ difference is rendered positive by the $\dot{V}/\dot{Q}$ inequality as shown by the *arrows*. *I* inspired; *E'* end-expired; *a* arterial

2.7 Effects of $\dot{V}/\dot{Q}$ Inequality vs. Diffusion Limitation

It follows from the preceding sections that both diffusion limitation and $\dot{V}/\dot{Q}$ inequality may be responsible for a decreased gas exchange efficiency. Indeed, the distinction between the two basic mechanisms is usually difficult. The following approaches have been used in mammals.

1. In hypoxia the $(P_A-P_a)_{O_2}$ due to diffusion limitation is enhanced, whereas that due to $\dot{V}_A/\dot{Q}$ inequality is reduced; in hyperoxia (but not when 100% O_2 is breathed, see below) the effect of strongly underventilated regions (shunt) is much enhanced (Riley and Cournand 1951; Farhi and Rahn 1955b). In attempts to apply this predicted dependence on the oxygenation level in experiments, problems arise from possible changes of $\dot{V}_A/\dot{Q}$ pattern, and D, with experimental changes of the oxygenation level.

2. If N_2 is eliminated from inspired gas at constant arterial P_{O_2} by the corresponding reduction of total pressure, the $(P_A-P_a)_{O_2}$ due to $\dot{V}_A/\dot{Q}$ inequality is expected to disappear, whereas the effect of a true shunt should remain (Haab et al. 1960). The approach requires a hypobaric chamber.

3. The local variations of the exchange ratio R, caused by $\dot{V}_A/\dot{Q}$ inequality, produce local variations of alveolar P_{N_2} which give rise to a negative $(P_A-P_a)_{N_2}$ difference. This P_{N_2} difference is in theory specific for $\dot{V}_A/\dot{Q}$ inequality, but its accurate measurement is rather difficult.

4. The multiple inert gas elimination method (Wagner et al. 1974a, b) is based on the effect of solubility on the $P_a/P_{\bar{v}}$ ratio for inert gases intravenously infused at

constant rate. For inert gases not present in the inspired gas, the following relationship is valid for the $P_a/P_{\bar{v}}$ ratio (called 'retention') in steady-state gas exchange (Farhi 1967) (λ, solubility expressed as blood/gas partition coefficient):

$$\frac{P_a}{P_{\bar{v}}} = \frac{\lambda}{\lambda + \dot{V}_A/\dot{Q}} \; . \tag{15}$$

In a homogeneous lung with a unique $\dot{V}_A/\dot{Q}$ value, $P_a/P_{\bar{v}}$ follows Eq. (15). In a lung model with unequal distribution of $\dot{V}_A$ to $\dot{Q}$, Eq. (15) must be applied to each compartment and the mixed arterial blood partial pressure (P_a) results from flow-weighted mixing (see above). From the pattern of the $P_a/P_{\bar{v}}$ plot against λ the $\dot{V}_A/\dot{Q}$ distribution can be derived. This method, which requires a high degree of accuracy in simultaneous determination of (usually) six inert gases in arterial blood (by gas chromatography or by mass spectrometry), has been successfully applied to mammals, normal humans and patients with various lung diseases (reviewed by West 1977).

Recently the method has also been applied to birds (Powell and Wagner 1982a, b). It was found that the extent of $\dot{V}/\dot{Q}$ inhomogeneity in geese was similar to that in normal man except for the presence of a very high $\dot{V}/\dot{Q}$ compartment. Burger et al. (1979) estimated both D_{O_2} and $\dot{V}/\dot{Q}$ inequality in unidirectionally ventilated ducks by a combination of P_{O_2} and P_{CO_2} measurements, with determination of ventilation shunt ($\dot{V}/\dot{Q} = \infty$) derived from the elimination of $CHCl_3$, and perfusion shunt calculated from the endtidal-arterial P_{O_2} difference in hyperoxia.

5. With increased ventilation, the mean P_{O_2} and P_{CO_2} differences between inspired and end-expired gas are reduced and, therefore, at the same relative $\dot{V}_A/\dot{Q}$ inequality the variance of alveolar P_{O_2} and P_{CO_2} due to $\dot{V}_A/\dot{Q}$ inequality must be proportionately diminished. Hence a $(P_A - P_a)_{O_2}$ difference due to diffusion, which is primarily independent of ventilation, would be less contaminated by effects of $\dot{V}_A/\dot{Q}$ inequalities (Scotto et al. 1987).

2.8 Other Problems in Modeling

Apart from ventilation/perfusion inhomogeneity, there are a number of other factors that may create problems when using the basic ventilation/diffusion/perfusion model in the analysis of gas exchange:

1. In the analysis of steady-state gas exchange it is taken for granted, on the basis of generally accepted physico-chemical principles, that in equilibrium the partial pressures in blood and medium (gas or water) are identical. The validity of this apparently obvious assumption has been cast in doubt for CO_2. Starting with the paper by Gurtner et al. (1969), a number of authors have reported experimental results showing that in rebreathing equilibrium the P_{CO_2} in arterial blood is lower than that in alveolar gas. Several explanations for this remarkable phenomenon have been proposed. The most appealing explanation was based on the effects of a negatively charged surface of pulmonary capillary epithelial cells (Gurtner et al. 1969). Other laboratories, including our own, have been unable to confirm this

finding (see reviews by Scheid and Piiper 1980b; Hlastala and Robertson 1980; Piiper 1986). Evidence for anomalous gas/blood CO_2 equilibrium in steady-state gas exchange in bird lungs was reported by Davies and Dutton (1975). However, further studies by Meyer et al. (1976) showed that, in a cross-current system, the observations could likely be explained on the basis of the Haldane effect (dependence of blood CO_2 dissociation curve upon O_2 saturation of hemoglobin) and that the assumption of an anomalous gas/blood CO_2 equilibrium was unnecessary.

2. There may exist an unequal distribution of diffusing capacity leading to variations of $D/\dot{Q}$ and $D/\dot{V}_A$ ratios. Also, this inhomogeneity leads to decreased gas exchange efficiency and to an underestimation of D when not taken into account (Piiper 1961a, b).

3. The periodic nature of ventilation and blood flow leads to temporal variations of $\dot{V}$ and $\dot{Q}$ (indeed, effective $\dot{V}$ can vary by more than twice mean $\dot{V}$ and zero during a ventilatory cycle). The ensuing temporal variations of P_{O_2} and P_{CO_2} lead to reductions in gas exchange efficiency that are similar to those due to regional $\dot{V}/\dot{Q}$ and $D/\dot{Q}$ variations. Evidently the problem is that of an unsteady state (see below).

4. The ventilation, $\dot{V}$ in the models, means effective ventilation, $\dot{V}_{eff}$, which in lung breathing is lower than total ventilation, $\dot{V}_{tot}$, the difference being dead space ventilation, $\dot{V}_D$:

$$\dot{V}_{eff} = \dot{V}_{tot} - \dot{V}_D \ . \tag{16}$$

The delimitation of $\dot{V}_D$ from $\dot{V}_{eff}$ is often difficult in mammals (in which $\dot{V}_{eff}$ is usually termed alveolar ventilation, $\dot{V}_A$) and is even more difficult in birds, where alternative pathways for gas exist in the lung-air sac system (Hastings and Powell 1986a, b). In particular, the estimation of effective parabronchial ventilation and initial-parabronchial partial pressures is difficult and uncertain (Scheid and Piiper 1970; Powell et al. 1981; Geiser et al. 1984).

5. On the blood side, there are the problems of intracardiac mixing of arterialized and venous blood, which are important in amphibians and reptiles. In order to analyze lung function proper in amphibians and reptiles, arterialized blood should be sampled from pulmonary veins and compared to lung gas. This has been done in two turtle species by Burggren and Shelton (1979) and in the frog *Xenopus laevis* by Boutilier and Shelton (1986).

3 Gas Exchange in Unsteady State

It is customary to study, both in theory and in experiment, the gas transfer systems of the body in steady state, which is defined by constancy in time of the gas transfer parameters (ventilation, perfusion, partial pressures, transfer rates, etc.). Steady state, however, is a special and extreme case which usually is only imperfectly achieved. Moreover, unsteady state is unavoidable in many physiological conditions, and may be purposely produced in experiments for determination of parameters involved in gas transfer. This is illustrated by the following examples:

1. The partial pressures in alveolar gas, considered as constant in simple lung models (see above), must vary within the breathing cycle (DuBois 1952; DuBois et al. 1952). Under conditions like exercise, such variations become important and their neglect may lead to serious errors.

2. Breathing in mammals, birds and fish is in many cases rather regular, allowing application of steady-state models. But in amphibians and reptiles, periodic breathing (i.e. alternation of breathing periods with prolonged apneic periods) is commonplace, particularly in habitually diving species. In such cases steady, state models are of little use (Shelton and Boutilier 1982).

3. The time course of changes in gas transport parameters during transitions from one (relatively steady) state to another, like from rest to exercise, from normoxia to hypoxia, etc., are of particular physiological interest. Model simulations of such changes should be useful in the analysis of the underlying transport mechanisms, their adjustments and their physiological control.

4. A number of methods for determination of basic gas transfer variables (ventilation, lung volume, cardiac output, diffusing capacity) are based on a rapid disturbance of the steady state and subsequent measurement of the kinetics of return to steady state. Particularly useful have been methods based on rebreathing equilibration of various test gases (reviewed by Piiper and Scheid 1980).

3.1 Basic Principles

In the analysis of steady-state gas transfer, the required parameters are transfer rates, partial pressures (or concentrations) and conductances (or resistances). In unsteady states, capacitances are additionally needed (Piiper 1982a). Capacitance, B, is defined as increase in amount of gas (ΔM) per increase in partial pressure (ΔP). Capacitance per volume is the capacitance coefficient or effective solubility (β) which is equal to increment in concentration (ΔC) per increment in partial pressure (ΔP):

$$B = \frac{\Delta M}{\Delta P} \; ; \tag{17}$$

$$\beta = \frac{B}{V} = \frac{\Delta C}{\Delta P} \; . \tag{18}$$

The basic relationship for analysis of transitional states is that which exists between capacitances (B), conductances (G) and the (exponential) rate constants (k). The rate constant is defined as the relative change of partial pressure (or concentration) difference ($P-P_O$) per unit time (t):

$$k = -\frac{d(P-P_O)/dt}{P-P_O} \; . \tag{19}$$

Instead of the rate constant (k), the time constant (τ) and the half time ($t\frac{1}{2}$) may be used:

$$\tau = 1/k \; ; \tag{20}$$

$$t\tfrac{1}{2} = \tau \cdot \ln 2 = 0.693 \; \tau \; . \tag{21}$$

In a system comprising a single capacitance and a single conductance, the following simple relationship is valid:

$$k = \frac{G}{B} \; . \tag{22}$$

In more complex multicompartment models, the equilibration is a composite exponential process, the number of k values being equal to, or less than, the number of compartments. In all cases, however, an increase in any component of G will increase, whereas an increase in any component of B will decrease the speed of equilibration (in qualitative accordance with Eq. 22).

3.2 Examples of Applications

The models required for simulation of the function of real gas exchange organs may be complex. They may contain multiple functional compartments that are interconnected, and connected to venous blood and to the environment, by diffusive and convective conductances. Furthermore, the conductances and/or capacitances may be variable (O_2 and CO_2 dissociation curves of blood and tissues) or may change as a result of physiological adjustment (change in ventilation or perfusion). In all such cases the kinetics will deviate from the monoexponential characteristic. From the complex kinetics, it may be possible to determine the various contributing parameters and their variations. On the other hand, in many cases a complex system may be predominantly determined by a single G and B, leading to a relatively constant k in a given time interval. For instance:

1. When an insoluble test gas is washed out from a continuously and homogeneously ventilated lung model, the rate constant, k, is equivalent to the ratio of alveolar ventilation ($\dot{V}_A$) to alveolar volume (V_A), $k = \dot{V}_A/V_A$, because $B = V_A \cdot \beta_g$ and $G = \dot{V}_A \cdot \beta_g$ (Fig. 9A). Thus, from k either $\dot{V}_A$ or V_A can be obtained when the other parameter is known. In real lungs, complications arise from cyclic ventilation, dead space and unequal distribution of tidal volume to lung volume.

2. The pulmonary diffusing capacity for CO (D_{CO}) can be determined from CO kinetics by rebreathing or single breath methods (Fig. 9B). The following relationships apply $B = V \cdot \beta_g$ (V is alveolar volume for breath-holding, total lung + rebreathing bag volume for rebreathing); $G = D_{CO}$; $k = D_{CO}/(V \cdot \beta_g)$. The rate constant k is determined from P_{CO} recorded continuously (rebreathing) or from initial and final value (breath-holding). In practice, the rebreathing bag/lung homogenization may be impossible because sufficiently high ventilation may not be achieved. In this case the equilibration is bi-exponential (Meyer et al. 1981).

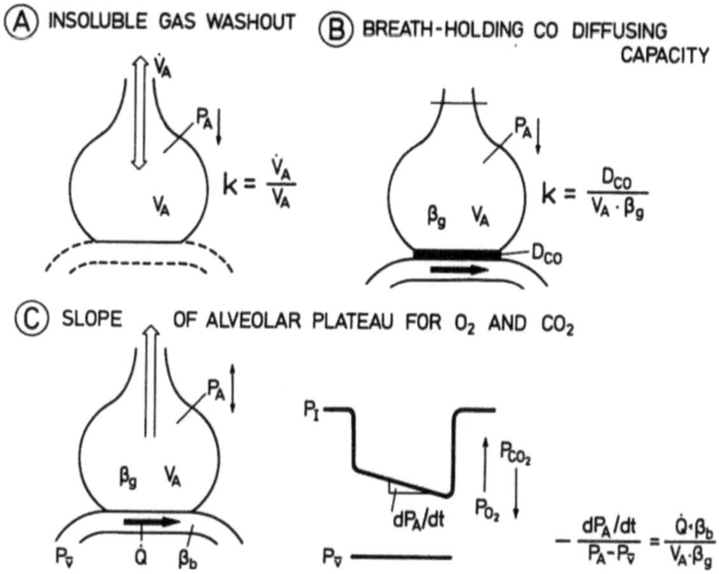

Fig. 9. Examples for gas exchange in unsteady state. Homogeneous alveolar lung models. *A* Kinetics of washout of an insoluble gas, determined by the ratio $\dot{V}_A/V_A$. *B* Absorption rate of CO during breathholding, determined by the ratio D_{CO}/V_A ($\beta_g = (R \cdot T)^{-1}$, capacity coefficient for gas phase). *C* The slope of alveolar plateaus for soluble gases, determined by the perfusive conductance/capacitance ratio, $\dot{Q} \cdot \beta_b/(V_A \cdot \beta_g)$. For explanations, see text

3. The alveolar slope of an expirogram may be viewed as indicating the temporal changes of alveolar gas composition (Fig. 9C). For a homogeneous lung model without diffusion limitation, the following relationship holds for the rate of change of the alveolar partial pressure (CO_2 or O_2) per unit time (dP_A/dt) (Fig. 9C):

$$\frac{dP_A}{dt} = \frac{\dot{Q} \cdot \beta_b}{V_A \cdot \beta_g} \cdot (P_A - P_{\bar{v}}) \ . \tag{23}$$

$\dot{Q} \cdot \beta_b$ is the perfusion conductance; $V_A \cdot \beta_g$ is the capacitance of alveolar gas. When comparing O_2 and CO_2, it is evident that β_b (including its dependence on P) and $P_A - P_{\bar{v}}$ must be considered, as must the capacitance ($V_A \cdot \beta_g$), which, for a highly soluble gas like CO_2, comprises a tissue equivalent (DuBois 1952; DuBois et al. 1952).

3.3 Effects of Changes in Ventilation and Cardiac Output

The effects that stepwise changes in ventilation and cardiac output have on gas transfer have been studied in experiments on dogs and in models by Farhi and Rahn (1955 a). After a sudden increase in ventilation, delivered by a respirator, the inspired-alveolar differences ($P_I - P_A$ for both CO_2 and O_2) approached the new, smaller steady-state value with the following rate constants: $k(CO_2) = 0.165$/min;

$k(O_2) = 1.39$/min. As the new equilibrium is approached, the alveolar ventilation and the cardiac output are the transfer mechanisms concerned, and their conductances are additive to the total conductance (G_{tot}):

$$G_{tot} = Q \cdot \beta_b + \dot{V}_A \cdot \beta_g \ , \tag{24}$$

$$k = \frac{G_{tot}}{B} \ . \tag{25}$$

At the beginning of the equilibration process, the k for CO_2 is expected to be higher than that for O_2, because β_b is much higher for CO_2 and B is close to equal for both gases, being determined by the alveolar gas volume. But after the onset of recirculation (i.e. after about 15 s) P_{CO_2} and P_{O_2} in the venous blood and tissue start to change. Because tissue β_{CO_2} is much higher than β_{O_2}, the effective B_{CO_2} is much higher than the effective B_{O_2}. Thereby the CO_2 kinetics is slowed down, leading to $k_{CO_2} < k_{O_2}$ as shown by the experimental data.

For the whole body, the effective capacitance is the sum of capacitances of compartments resulting from capacitance coefficients and volumes:

$$B = \sum V_x \cdot \beta_x \ . \tag{26}$$

At the very least, the following compartments must be discerned: lung gas, blood and tissue. Since β for blood depends on the partial pressures according to blood CO_2 and O_2 dissociation curves, it may be useful to consider arterial and venous blood compartments separately. A pioneer model published by Farhi and Rahn (1955a) is shown in Fig. 10. For O_2, B is relatively high in lung gas, is variable

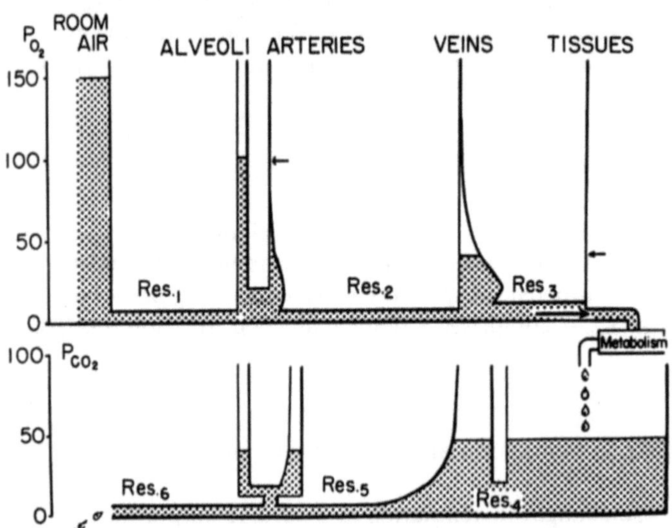

Fig. 10. Model representing the compartments of the body in transport and storage of O_2 and CO_2. Height of the columns: P_{O_2} and P_{CO_2}, respectively. Breadth of the columns: capacitance (B). Area of the column represents amount (M) of gas (i.e. $M = B \cdot P$) in the compartments. *Res* Resistances to transfer (= *1*/conductance) (After Farhi and Rahn 1955a)

in blood being determined by the dissociation curve, and is low in tissue according to low physical solubility (but increased at low P_{O_2} by myoglobin in muscle).

For CO_2, β in blood and tissue is determined by both the physical solubility, α (which is about 25 times higher than for O_2), and by chemical binding of CO_2 as bicarbonate:

$$\beta_{CO_2} = \alpha_{CO_2} + \frac{d[HCO_3^-]}{dP_{CO_2}} \ . \tag{27}$$

The change in bicarbonate depends on (non-bicarbonate) buffering. The buffering power is conventionally characterized by the buffer value, b, defined as the change in pH per addition of free OH^- or H^+ ions per unit volume: $b = dpH/d[OH^-] = -dpH/d[H^+]$. For a homogeneous buffer system equilibrated with CO_2, the following relationship can be derived from the Henderson-Hasselbalch equation:

$$\frac{d[HCO_3^-]}{dP_{CO_2}} = \frac{1/P_{CO_2}}{1/[HCO_3^-]+2.3/b} \ . \tag{28}$$

Thus, $d[HCO_3^-]/dP_{CO_2}$ is higher with high b, high $[HCO_3^-]$, and low P_{CO_2}. Due to the combined effects of high solubility and buffering, β_{CO_2} is higher in blood and tissue than in gas.

Furthermore, it is important to consider that the effective B_{CO_2}, measured by determining the amount of CO_2 bound or set free upon variation of P_{CO_2}, is time-dependent (Farhi and Rahn 1955a). At the beginning, B_{CO_2} is determined by the alveolar gas (and lung tissue), then tissues start contributing, at first well-perfused tissues, followed by less-perfused tissues. Finally, bone with very high B_{CO_2} may start contributing days after a step change in P_{CO_2}.

3.4 Gas Exchange Ratio (Respiratory Quotient)

In air breathers, the respiratory gas exchange ratio (R_R), i.e. the ratio of CO_2 output and O_2 uptake, reacts very sensitively to deviations from steady state. Indeed, R_R values lower or higher than the metabolic respiratory quotient (R_M) are generally considered as sensitive indicators of unsteady state in the gas transfer system.

At the beginning of hyperventilation, CO_2 output is considerably increased above metabolic CO_2 output, whereas the O_2 uptake is only slightly increased above the O_2 consumption. Hence R_R is increased (Fig. 11A). This is due to $\beta_{CO_2} > \beta_{O_2}$ in blood and particularly in tissue. When the new steady state is reached (i.e. CO_2 output has again become equal to CO_2 production and O_2 uptake equal to O_2 consumption), R_R is again equal to R_M. In the CO_2-O_2 diagram, R_R can be visualized as the sum of two components, R_M and R_S, the R for changes in stored amounts of CO_2 and O_2 (Fig. 11A). During hyperventilation, R_S gradually increases and comes close to infinity when the new equilibrium for O_2 has been reached but the amount of CO_2 is still changing. This behavior is also responsible for the different rate constants (see above).

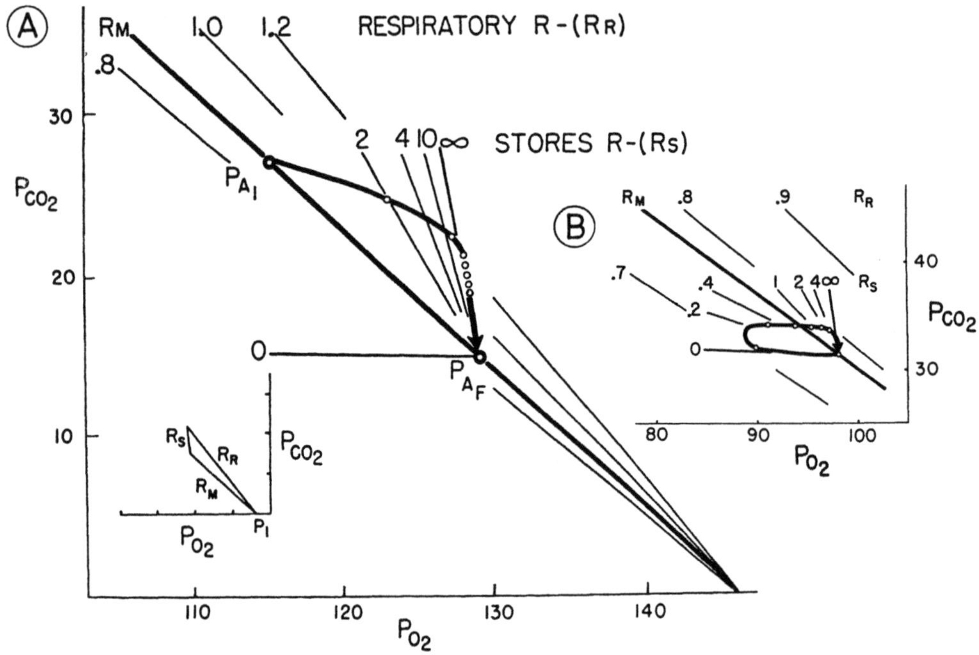

Fig. 11. Gas exchange ratios (R) during unsteady state. After experimental data in anesthetized dogs. *A* Stepwise increase in ventilation. End-expired P_{CO_2} and P_{O_2} change from P_{A_I} (control) to P_{A_F} (final, new steady state). R lines are drawn for pulmonary exchange (R_R) and for store change (R_S). Calculation of R_S is shown in the *insert*. *B* Increase of cardiac output. The R values move along the loop and return to the initial value. For further explanations, see text (After Farhi and Rahn 1955a)

The converse takes place after onset of hypoventilation: the storage of CO_2 causes CO_2 output to fall below CO_2 production, whereas the O_2 uptake is only slightly below O_2 consumption.

Immediately after an increase of cardiac output, and before the onset of recirculation, alveolar P_{O_2} should decrease and alveolar P_{CO_2} should increase, but only slightly, leading to a decrease of R_R (Fig. 11B). The changes should be similar to those occurring after the onset of hypoventilation, both states being characterized by a decrease of the $\dot{V}_A/\dot{Q}$ ratio. But later, as the P_{O_2} of mixed venous blood increases and P_{CO_2} drops, and the arterial values return towards control values, R_R increases due to considerable liberation of CO_2 from stores, far exceeding the storage of O_2. Finally, R_R approaches R_M as the new equilibrium is attained.

The rise in R_R due to hyperventilation in air breathing is basically due to the CO_2/O_2 capacitance ratio of blood and tissue being higher than that in the medium. In water the β_{CO_2}/β_{O_2} ratio is about 25, and higher still in carbonated water (hard water, sea water), leading to a β_{CO_2}/β_{O_2} ratio for the body that is equal to, or lower than, that of water. Thus, hyperventilation in fish is expected to have little effect on R_R, or may decrease R_R rather than increase it.

3.5 Periodic Breathing

In periodically breathing reptiles and amphibians, CO_2 and O_2 in lung gases and in blood show large variations. In recent years, a number of experimental studies have been performed in periodically breathing turtles (Ackerman and White 1979; Burggren and Shelton 1979), crocodiles (Glass and Johansen 1976) and frogs (Shelton and Boutilier 1982; Boutilier and Shelton 1986). The following features have emerged.

1. The P_{O_2} variations in lung gas, end-expired gas and arterial blood are much larger than those of P_{CO_2} (Fig. 12). This is mainly due to the high capacitance of the body for CO_2 as compared to O_2.

2. During a ventilatory period following apnea, end-tidal P_{CO_2} falls much more slowly than P_{O_2} rises, and a considerable increase of the R_R value occurs with progressing ventilation (Fig. 13). This is again due to the high capacitance of tissue for CO_2 as compared to O_2. The increasing R_R during the ventilatory period means that the mean R_R of a breath or a ventilatory period can vary depending on the length of the preceding apneic period and the vôlume of fresh gas inspired during the ventilation period. This is a plausible explanation for the large variation of ventilation period R_R that was observed in *Pseudemys scripta* by Ackerman and White (1979).

3. During prolonged apnea, lung gas P_{CO_2} tends to become stable whereas P_{O_2} keeps falling. In part this is due to very high CO_2 capacitance of tissue, in part to cutaneous CO_2 elimination. Cutaneous CO_2 transfer is much more extensive than that of O_2, due to high Krogh diffusion constant of tissue for CO_2.

In presenting their lung gas P_{CO_2} and P_{O_2} data during or at the end of breath-holding, Ackerman and White (1979; Fig. 14) and Boutilier and Shelton (1986; Fig. 15) applied the O_2-CO_2 diagram with the conventional R lines converging to the inspired gas in estimating the exchange ratio. These R lines designate the ex-

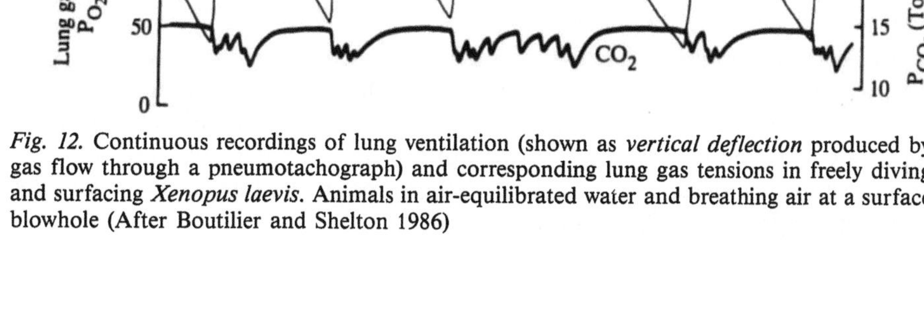

Fig. 12. Continuous recordings of lung ventilation (shown as *vertical deflection* produced by gas flow through a pneumotachograph) and corresponding lung gas tensions in freely diving and surfacing *Xenopus laevis*. Animals in air-equilibrated water and breathing air at a surface blowhole (After Boutilier and Shelton 1986)

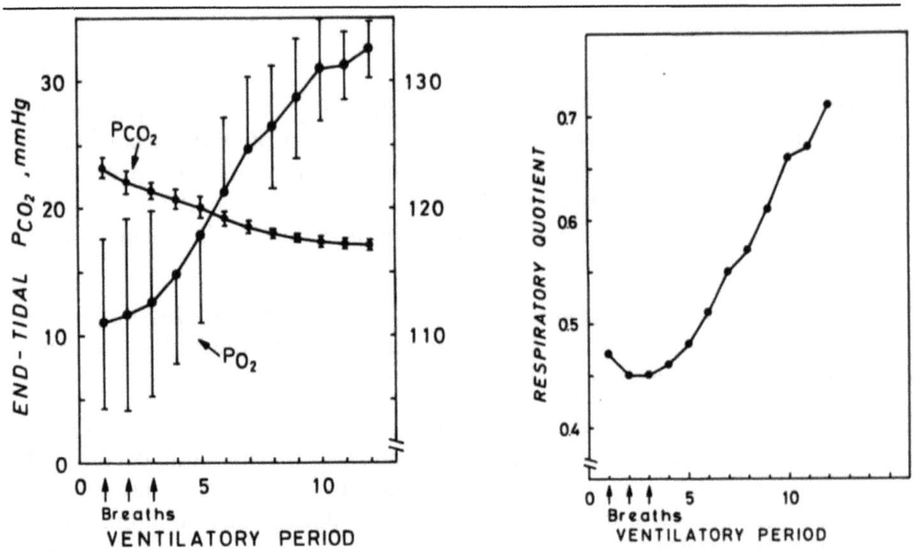

Fig. 13. Changes in end-tidal P_{CO_2} and P_{O_2} (*left*) and in the gas exchange ratio (*right*) in *Crocodylus niloticus.* A total of 165 expirations measured in two animals are grouped according to the position in the ventilatory periods. Mean values (± SD) (After Glass and Johansen 1979)

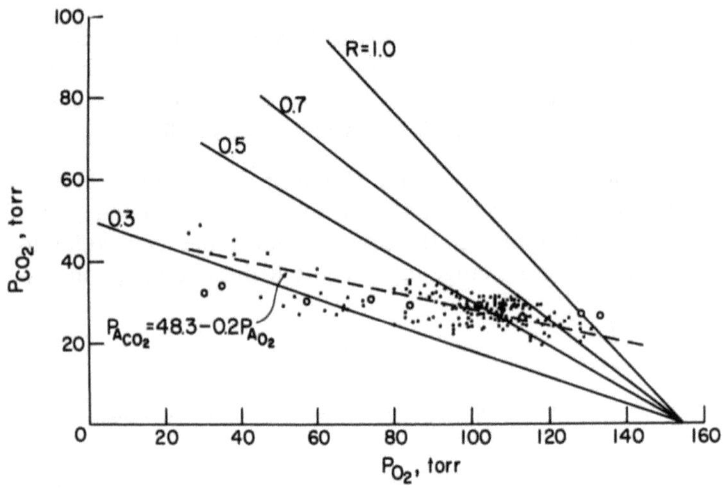

Fig. 14. Partial pressure relationships for O_2 and CO_2 in lung gas of *Pseudemys scripta* determined at various apneic times from 10 s to 30 min post-ventilation. Data include apneas during voluntary air breathing and face submersion simulating diving. *Broken line* Regression. *Open circles* Data of Wilson (1939) (After Ackerman and White 1979)

change ratio of the first breath following apnea. But during apnea an R value defined in this manner is meaningless because no air is inspired. To describe comparatively CO_2 and O_2 exchange during apnea, the 'instantaneous' exchange ratio, R_i, should be used, which is close to the slope of the line, s, describing the simultaneous P_{CO_2} and P_{O_2} changes:

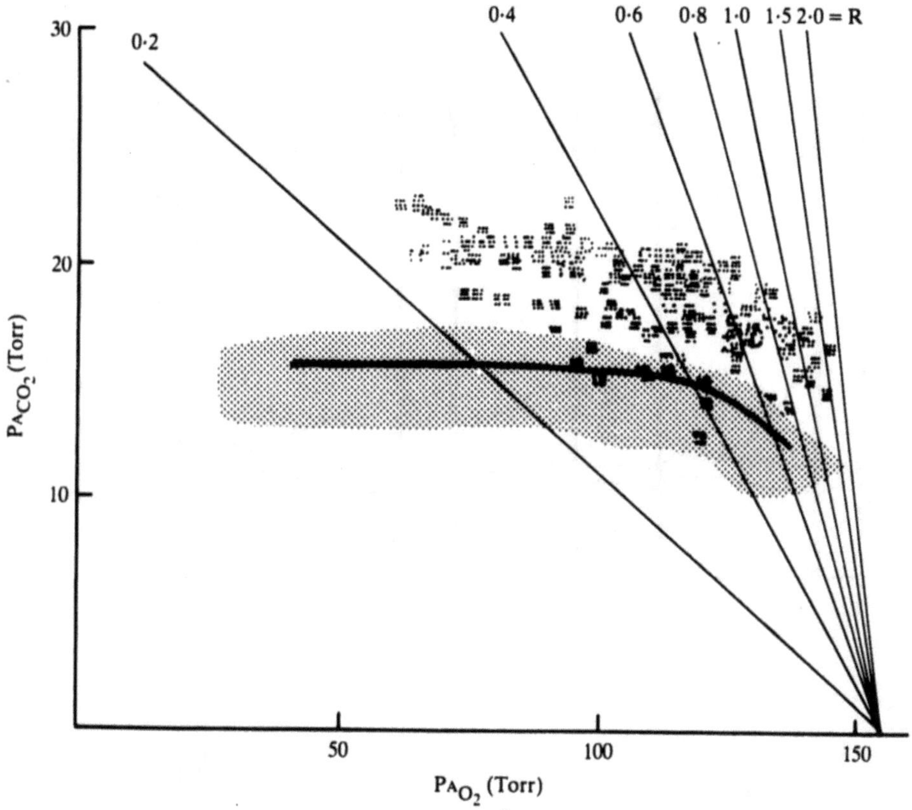

Fig. 15. Pulmonary P_{O_2}-P_{CO_2} relationships in a specimen of *Xenopus laevis* obtained by a closed, extracorporeal air loop containing P_{O_2} and P_{CO_2} electrodes during breathing (values at the *righ-hand margin*) and during breath holding (values to the *left*). *Continuous line* (mean) and *shading* (range): during rest. *Points above the line*: during activity. (After Boutilier and Shelton 1986)

$$s = \frac{-\Delta P_{ACO_2}}{\Delta P_{AO_2}} . \tag{29}$$

Assuming that there is no net N_2 exchange, R_i can be obtained from the relationship (Kim et al. 1966):

$$R_i = \frac{s - F_{AO_2} \cdot s - F_{ACO_2}}{1 - F_{AO_2} \cdot s - F_{ACO_2}} , \tag{30}$$

where F_A designate the instantaneous fractional CO_2 and O_2 concentrations in alveolar gas. It is obvious from Eq. (30) that R_i deviates from 1.0 more than s does.

The gas transport system in amphibians and reptiles is more complex than that of mammals, birds and fishes, due to intracardiac arterio-venous mixing and presence of cutaneous gas exchange. Intracardiac mixing of arterialized and venous

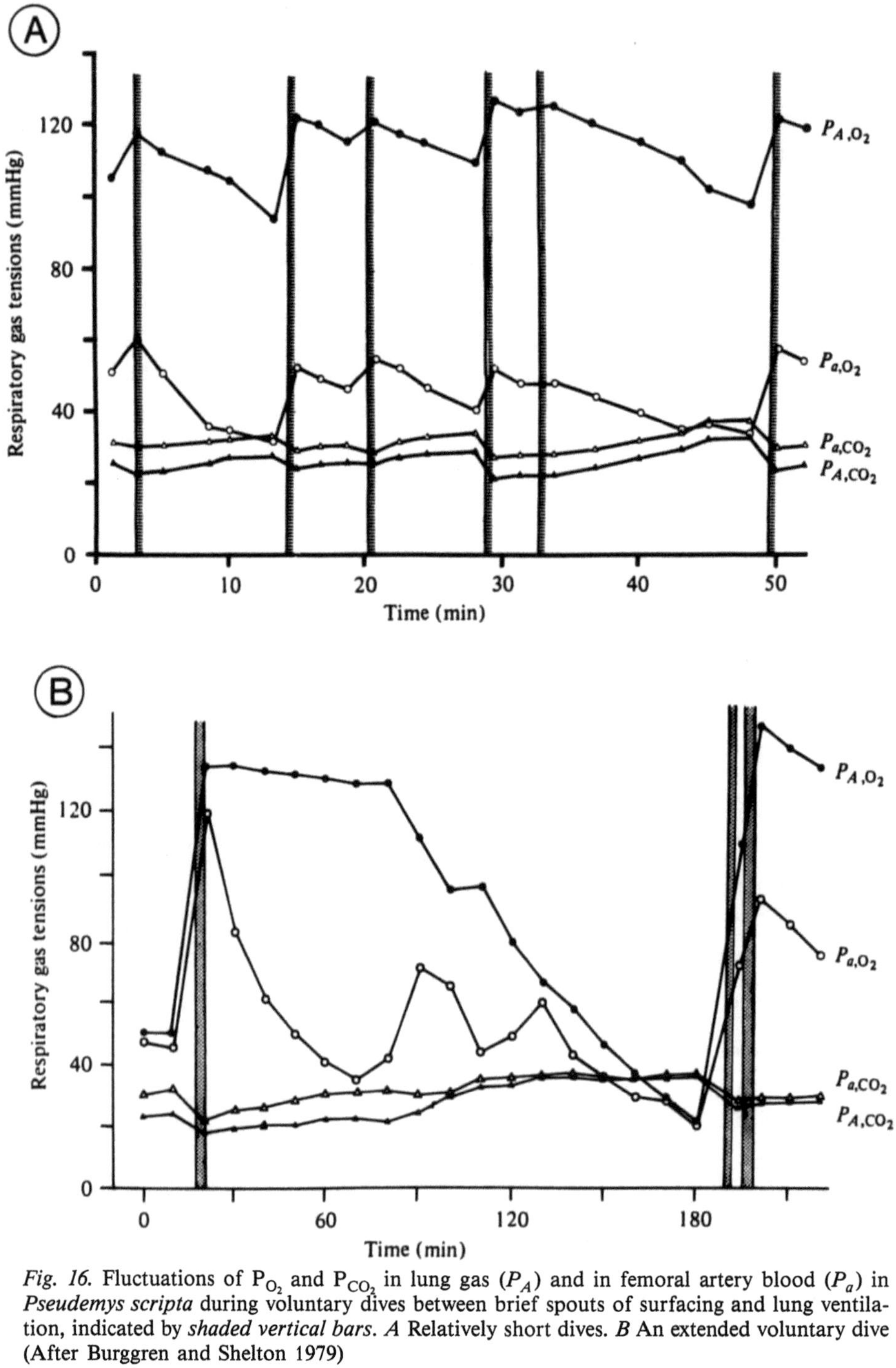

Fig. 16. Fluctuations of P_{O_2} and P_{CO_2} in lung gas (P_A) and in femoral artery blood (P_a) in *Pseudemys scripta* during voluntary dives between brief spouts of surfacing and lung ventilation, indicated by *shaded vertical bars*. *A* Relatively short dives. *B* An extended voluntary dive (After Burggren and Shelton 1979)

blood does not lead to a complete mixing. In frogs possessing a single ventricle, the extent of mixing may be unexpectedly small (Tazawa et al. 1979). When more pronounced, intracardiac mixing leads to a depression of P_{O_2} in systemic arterial blood as compared to lung gas or left auricular P_{O_2} and to corresponding P_{CO_2} differences of reversed sign. These relationships have been studied in detail by Boutilier and Shelton (1986) in the frog *Xenopus*. Very pronounced P_{CO_2} and P_{O_2} differences between lung gas and arterial blood have been found in the water turtle *Pseudemys scripta* by Burggren and Shelton (1979; Fig. 16). The changing relationships between P_A and P_a during breath-holding may be attributed to changing patterns of intracardiac mixing (Fig. 16B).

Cutaneous gas exchange can lead to considerable cutaneous loss of CO_2 as the P_{CO_2} of body fluids rises during apnea. In frogs (Boutilier and Shelton 1986) and sea snakes (Seymour and Webster 1975; Seymour 1978) a steady state is apparently approached as the total metabolic CO_2 is eliminated through the skin. Since the cutaneous diffusing capacity for O_2 is about 25 times smaller than that for CO_2, no such steady state is reached for O_2 during apnea, although some cutaneous O_2 uptake does occur, enhanced by the low P_{O_2} in the blood perfusing the skin.

Boutilier and Shelton (1986) have adapted the model of Farhi and Rahn (Fig. 10), incorporating cardiac mixing and cutaneous gas exchange, to the analysis of O_2 transfer in intermittently breathing amphibians. In the model, the capacitances of lung gas, pulmonary venous blood, systemic arterial and venous blood and cutaneous venous blood are included, whereas the tissue capacitance is neglected. It would be of considerable interest to extend this model to CO_2 equilibria (for which tissue capacitance has to be incorporated) and acid-base balance. This may be a promising approach to the analysis of gas exchange and transport in unsteady state, to be followed up in future.

References

Ackerman RA, White FN (1979) Cyclic carbon dioxide exchange in the turtle *Pseudemys scripta*. Physiol Zool 52:378–389

Bohr C (1909) Über die spezifische Tätigkeit der Lungen bei der respiratorischen Gasaufnahme und ihr Verhalten zu der durch die Alveolarwand stattfindenden Gasdiffusion. Skand Arch Physiol 22:221–280

Boutilier RG, Shelton G (1986) Gas exchange, storage and transport in voluntarily diving *Xenopus laevis*. J Exp Biol 126:133–155

Burger RE, Meyer M, Graf W, Scheid P (1979) Gas exchange in the parabronchial lung of birds: experiments in unidirectionally ventilated ducks. Respir Physiol 36:19–37

Burggren WW, Shelton G (1979) Gas exchange and transport during intermittent breathing in chelonian reptiles. J Exp Biol 82:75–92

Crawford EC Jr, Gatz RN, Magnussen H, Perry SF, Piiper J (1976) Lung volumes, pulmonary blood flow and carbon monoxide diffusing capacity of turtles. J Comp Physiol 107:169–178

Cumming G, Horsfield K, Jones JG, Muir DCF (1967) The influence of gaseous diffusion on the alveolar plateau at different lung volumes. Respir Physiol 2:386–398

Davies DG, Dutton RE (1975) Gas-blood P_{CO_2} gradients during avian gas exchange. J Appl Physiol 39:405–410

DuBois AB (1952) Alveolar CO_2 and O_2 during breath holding, expiration, and inspiration. J Appl Physiol 5:1–12
DuBois AB, Brett AG, Fenn WO (1952) Alveolar CO_2 during the respiratory cycle. J Appl Physiol 4:535–548
Farhi LE (1967) Elimination of inert gas by the lung. Respir Physiol 3:1–11
Farhi LE, Rahn H (1955a) Gas stores of the body and the unsteady state. J Appl Physiol 7:472–484
Farhi LE, Rahn H (1955b) A theoretical analysis of the alveolar-arterial O_2 difference with special reference to the distribution effect. J Appl Physiol 7:691–703
Feder ME, Burggren WW (1985) Cutaneous gas exchange in vertebrates: design, patterns, control, and implications. Biol Rev 60:1–45
Gatz RN, Glass ML, Wood SC (1987) Pulmonary function of the green sea turtle (*Chelonia mydas*). J Appl Physiol 62:459–463
Geiser J, Gratz RK, Hiramoto T, Scheid P (1984) Effects of increasing metabolism by 2,4-dinitrophenol on respiration and pulmonary gas exchange in the duck. Respir Physiol 57:1–14
Georg J, Lassen NA, Mellemgaard K, Vinther A (1965) Diffusion in the gas phase of the lungs in normal and emphysematous subjects. Clin Sci 29:525–532
Glass ML, Johansen K (1976) Control of breathing in *Acrochordus javanicus*, an aquatic snake. Physiol Zool 49:328–340
Glass ML, Johansen K (1979) Periodic breathing in the crocodile, *Crocodylus niloticus*: Consequences for the gas exchange ratio and control of breathing. J Exp Zool 208:319–326
Glass ML, Burggren WW, Johansen K (1981a) Pulmonary diffusing capacity of the bullfrog (*Rana catesbeiana*). Acta Physiol Scand 113:485–490
Glass ML, Johansen K, Abe AS (1981b) Pulmonary diffusing capacity in reptiles (relations to temperature and O_2-uptake). J Comp Physiol 142:509–514
Gurtner GH, Song SH, Farhi LE (1969) Alveolar to mixed venous P_{CO_2} difference under conditions of no gas exchange. J Appl Physiol 30:619–624
Haab P, Piiper J, Rahn H (1960) Attempt to demonstrate the distribution component of the alveolar-arterial oxygen pressure difference. J Appl Physiol 15:235–240
Hastings RH, Powell FL (1986a) Single breath CO_2 measurements of dead space in ducks. Respir Physiol 63:139–149
Hastings RH, Powell FL (1986b) Physiological dead space and effective parabronchial ventilation in ducks. J Appl Physiol 60:85–91
Hlastala MP, Robertson HT (1980) Evidence for active elimination of carbon dioxide from the lung. In: West JB (ed) Pulmonary gas exchange, vol II. Organism and Environment. Academic Press, New York, pp 241–273
Kim TS, Rahn H, Farhi LE (1966) Estimation of true venous and arterial P_{CO_2} by gas analysis of a single breath. J Appl Physiol 21:1338–1344
Kobayashi H, Piiper J, Scheid P (1987) Effect of the alinearity of the O_2 dissociation curve on determination of pulmonary diffusing capacity. Pflügers Arch 408, Suppl 1:R25
Meyer M, Worth H, Scheid P (1976) Gas-blood CO_2 equilibration in parabronchial lungs of birds. J Appl Physiol 41:302–309
Meyer M, Scheid P, Riepl G, Wagner HJ, Piiper J (1981) Pulmonary diffusing capacities for O_2 and CO measured by a rebreathing technique. J Appl Physiol 51:1643–1650
Moll W (1962) Die Oxygenation der Erythrocyten in der Lunge durch Diffusion, Reaktion und spezifischen Transport. Pflügers Arch 275:420–438
Nguyen Phu D, Yamaguchi K, Scheid P, Piiper J (1986) Kinetics of oxygen uptake and release by red blood cells of chicken and duck. J Exp Biol 125:15–27
Paganelli CV (1980) The physics of gas exchange across the avian eggshell. Am Zool 20:329–338
Piiper J (1961a) Unequal distribution of pulmonary diffusing capacity and the alveolar-arterial P_{O_2} differences: theory. J Appl Physiol 16:493–498
Piiper J (1961b) Variations of ventilation and diffusing capacity to perfusion determining the alveolar-arterial O_2 difference: theory. J Appl Physiol 16:507–510
Piiper J (1979) Series ventilation, diffusion in airways and stratified inhomogeneity. Fed Proc 38:17–21

Piiper J (1982a) Respiratory gas exchange at lungs, gills and tissues: mechanisms and adjustments. J Exp Biol 100:5–22

Piiper J (1982b) Diffusion in the interlamellar water of fish gills. Fed Proc 41:2140–2142

Piiper J (1982c) A model for evaluating diffusion limitation in gas-exchange organs of vertebrates. In: Taylor CR, Johansen K, Bolis L (eds) A comparison to animal physiology. Cambridge University Press, Cambridge, pp 49–64

Piiper J (1986) Blood-gas equilibrium of carbon dioxide in lungs: a continuing controversy. J Appl Physiol 60:1–8

Piiper J, Baumgarten-Schumann D (1968) Effectiveness of O_2 and CO_2 exchange in the gills of the dogfish (*Scyliorhinus stellaris*). Respir Physiol 5:338–349

Piiper J, Scheid P (1972) Maximum gas transfer efficacy of models for fish gills, avian lungs and mammalian lungs. Respir Physiol 14:115–124

Piiper J, Scheid P (1975) Gas transport efficacy of gills, lungs and skin: theory and experimental data. Respir Physiol 23:209–221

Piiper J, Scheid P (1977) Comparative physiology of respiration: functional analysis of gas exchange organs in vertebrates. In: Widdicombe JG (ed) Respiratory Physiology II. Int Rev Physiol Ser Vol XIV. Univ Park Press, Baltimore, MD, pp 219–253

Piiper J, Scheid P (1980) Blood-gas equilibration in lungs. In: West JB (ed) Pulmonary Gas Exchange, vol I. Ventilation, Blood Flow, and Diffusion. Academic Press, New York, pp 131–171

Piiper J, Scheid P (1981) Model for capillary-alveolar equilibration with special reference to O_2 uptake in hypoxia. Respir Physiol 46:193–208

Piiper J, Scheid P (1982) Models for a comparative functional analysis of gas exchange organs in vertebrates. J Appl Physiol 53:1321–1329

Piiper J, Scheid P (1983a) Comparison of diffusion and perfusion limitations in alveolar gas exchange. Respir Physiol 51:287–290

Piiper J, Scheid P (1983b) Physical principles of gas exchange in fish gills. In: Houlihan DF, Rankin JC, Shuttleworth TJ (eds) Gills. Cambridge University Press, London, pp 45–62

Piiper J, Scheid P (1984) Model analysis of gas transfer in fish gills. In: Hoar WS, Randall DJ (eds) Fish physiology, vol XA. Academic Press, Orlando, pp 229–262

Piiper J, Scheid P (1989) Gas exchange – theory, models and experimental data. In: Wood SC (ed) Comparative pulmonary physiology: current concepts. Lung biology in health and disease, vol 39. Marcel Dekker, New York, pp 369–416

Piiper J, Dejours P, Haab P, Rahn H (1971) Concepts and basic quantities in gas exchange physiology. Respir Physiol 13:292–304

Piiper J, Gatz RN, Crawford EC Jr (1976) Gas transport characteristics in an exclusively skin-breathing salamander, *Desmognathus fuscus* (Plethodontidae). In: Hughes GM (ed) Respiration of amphibious vertebrates. Academic Press, New York, pp 339–356

Piiper J, Meyer M, Marconi C, Scheid P (1980) Alveolar-capillary equilibration kinetics of $^{13}CO_2$ in human lungs studied by rebreathing. Respir Physiol 42:29–41

Piiper J, Scheid P, Perry SF, Hughes GM (1986) Effective and morphometric oxygen-diffusing capacity of the gills of the elasmobranch *Scyliorhinus stellaris*. J Exp Biol 123:27–41

Powell FL, Wagner PD (1982a) Measurement of continuous distributions of ventilation-perfusion in non-alveolar lungs. Respir Physiol 48:219–232

Powell FL, Wagner PD (1982b) Ventilation-perfusion inequality in avian lungs. Respir Physiol 48:233–241

Powell FL, Geiser J, Gratz RK, Scheid P (1981) Airflow in the avian respiratory tract: variations of O_2 and CO_2 concentrations in the bronchi of the duck. Respir Physiol 44:195–213

Randall DJ, Holeton GF, Stevens ED (1967) The exchange of oxygen and carbon dioxide across the gills of rainbow trout. J Exp Biol 46:339–348

Riley RL, Cournand A (1949) 'Ideal' alveolar air and the analysis of ventilation-perfusion relationships in the lungs. J Appl Physiol 1:825–847

Riley RL, Cournand A (1951) Analysis of factors affecting partial pressures of oxygen and carbon dioxide in gas and blood of the lungs: theory. J Appl Physiol 4:77–101

Scheid P (1978) Analysis of gas exchange between air capillaries and blood capillaries in avian lungs. Respir Physiol 32:27–49

Scheid P (1979) Mechanisms of gas exchange in bird lungs. Rev Physiol Biochem Pharmacol 86:137–186
Scheid P, Piiper J (1970) Analysis of gas exchange in the avian lung: theory and experiments in the domestic fowl. Respir Physiol 9:246–262
Scheid P, Piiper J (1971) Theoretical analysis of respiratory gas equilibration in water passing through fish gills. Respir Physiol 13:305–318
Scheid P, Piiper J (1980a) Intrapulmonary gas mixing and stratification. In: West JB (ed) Pulmonary Gas Exchange, vol I. Ventilation, Blood Flow, and Diffusion. Academic Press, New York, pp87–130
Scheid P, Piiper J (1980b) Blood-gas equilibrium of carbon dioxide in lungs. A review. Respir Physiol 39:1–31
Scheid P, Hook C, Piiper J (1986) Model for analysis of counter-current gas transfer in fish gills. Respir Physiol 64:365–374
Schuster KD (1985) Kinetics of pulmonary CO_2 transfer studied by using labeled carbon dioxide $C^{16}O^{18}O$. Respir Physiol 60:21–37
Scotto P, Ichinose Y, Patané L, Meyer M, Piiper J (1987) Alveolar-capillary diffusion of oxygen in dogs exercising in hypoxia. Respir Physiol 68:1–10
Seymour RS (1978) Gas tensions and blood distribution in sea snakes at surface pressure and at simulated depth. Physiol Zool 51:388–407
Seymour RS, Webster MED (1975) Gas transport and blood acid-base balance in diving sea snakes. J Exp Zool 191:169–182
Shelton G, Boutilier RG (1982) Apnoea in amphibians and reptiles. J Exp Biol 100:245–273
Tazawa H, Mochizuki M, Piiper J (1979) Respiratory gas transport by the incompletely separated double circulation in the bullfrog, *Rana catesbeiana*. Respir Physiol 36:77–95
Thews G (1963) Die theoretischen Grundlagen der Sauerstoffaufnahme in der Lunge. Ergeb Physiol Biol Chem Exp Pharmakol 53:42–107
Wagner PD, West JB (1972) Effects of diffusion impairment on O_2 and CO_2 time courses in pulmonary capillaries. J Appl Physiol 33:62–71
Wagner PD, Laravuso RB, Uhl RR, West JB (1974a) Continuous distributions of ventilation-perfusion ratios in normal subjects breathing air and 100% O. J Clin Invest 54:54–68
Wagner PD, Saltzman HA, West JB (1974b) Measurement of continuous distributions of ventilation-perfusion ratios: theory. J Appl Physiol 36:588–599
Wangensteen OD, Wilson D, Rahn H (1970/71) Diffusion of gases across the shell of the hen's egg. Respir Physiol 11:16–30
Weibel ER (1963) Morphometry of the human lung. Springer, Berlin Göttingen Heidelberg
Weibel ER (1973) Morphological basis of alveolar-capillary gas exchange. Physiol Rev 53:419–495
West JB (1977) Ventilation-perfusion relationships. Am Rev Respir Dis 116:919–943
Wilson JW (1939) Some physiological properties of reptilian blood. J Cell Comp Physiol 13:315–326
Wright P, Heming T, Randall DJ (1986) Downstream pH changes in water flowing over the gills of rainbow trout. J Exp Biol 126:499–512
Yamaguchi K, Nguyen Phu D, Scheid P, Piiper J (1985) Kinetics of O_2 uptake and release by human red blood cells studied by a stopped-flow technique. J Appl Physiol 58:1215–1224
Yamaguchi K, Jürgens KD, Bartels H, Piiper J (1987a) Oxygen transfer properties and dimensions of red blood cells in high-altitude camelids, dromedary camel and goat. J Comp Physiol B 157:1–9
Yamaguchi K, Glahn J, Scheid P, Piiper J (1987b) Oxygen transfer conductance of human red blood cells at varied pH and temperature. Respir Physiol 67:209–223

Recent Advances and Trends in the Comparative Morphometry of Vertebrate Gas Exchange Organs

S. F. Perry[1,2]

Contents

1 Introduction

The functional morphologist asks, "Just how structurally well suited are gills and lungs for extracting oxygen from the environment and delivering it to the blood?" The physiologist wants to know, "What is the oxygen conductance of the gas exchanger?" Both the morphologist and the physiologist are addressing the oxygen diffusing capacity of the gas exchange organ, and each can answer the question in his own way.

For the physiologist, oxygen diffusing capacity (D_{O_2}) – the physical capacity for oxygen transfer per unit time and driving pressure (Dejours 1981) – is most

[1] Universität Oldenburg, D-2900 Oldenburg, FRG
[2] Present address: Respiratory Physiology Research Group, Faculty of Medicine, University of Calgary, Calgary, Alberta, Canada T2N 4N1

Advances in Comparative and Environmental Physiology
Vol. 6. Edited by R. G. Boutilier

readily approached in terms of Fick's first law of diffusion: the respiratory physiologist's equivalent of Ohm's law. In the comparison below, we see the similarity of the two equations, and that D_{O_2} is really the reciprocal of the resistance to oxygen transfer:

$$\text{Fick's first law } \dot{V}_{O_2} = D_{O_2} \cdot \overline{\Delta P_{O_2}} ,$$

$$\text{Ohm's law } \quad I = 1/R \cdot U ,$$

where $\dot{V}_{O_2}$ is the oxygen consumption rate; $\overline{\Delta P_{O_2}}$ is the mean partial pressure gradient for O_2, usually measured as the alveolar-capillary P_{O_2} difference; I is current; R is resistance and U is potential difference (voltage).

In order to determine the total resistance of an electrical circuit in series, the resistances are added and the total conductance (diffusing capacity) is given by the sum of the reciprocals of the individual components:

$$1/D_{O_2} = 1/D1_{O_2} + 1/D2_{O_2} + \ldots + 1/Dn_{O_2} ,$$

where $D1_{O_2}$, $D2_{O_2}$ and Dn_{O_2} are the oxygen diffusing capacities of the component parts of the diffusion barrier.

Physiologically, two components are discernable: the "membrane" diffusing capacity and the rate of oxygen combination with capillary blood, θVc (Cotes 1968). Morphometrically, however, it is possible to break down the "membrane" diffusing capacity into its epithelial, interstitial, endothelial and plasma components according to the following relationship:

$$Di_{O_2} = Ki_{O_2} \, Si/\tau i ,$$

where Di_{O_2}, Ki_{O_2}, Si and τi are the O_2 diffusing capacity, Krogh's constant of diffusion for oxygen in the appropriate tissue, the surface area and the diffusion distance of the appropriate component, respectively. "τ" is given by the harmonic mean distance across the barrier in question (Weibel and Knight 1964). The harmonic mean (the reciprocal of the mean of the reciprocals) weights short distances in a skewed distribution and is thus a more attractive measure of central tendency than the arithmetic mean as an indicator of diffusion distance. It expresses mean tissue "thinness" rather than "thickness."

Thus, if the morphologist knows the appropriate Ki_{O_2} he can theoretically determine the diffusing capacity not only of the whole system, but also of any of its component parts. There are various approaches to reporting these morphometric data. Some authors choose to calculate what they believe to be the morphometric diffusing capacity of the whole gas exchange organ: $D_{G_{O_2}}$ for the gill or $D_{L_{O_2}}$ for the lung (Hughes 1972; Weibel 1970/71). This takes into account the entire resistance to oxygen transfer, from the alveolar surface to the erythrocyte and also includes the erythrocytes themselves. $D_{G_{O_2}}$ and $D_{L_{O_2}}$ are therefore the morphological equivalent of the physiologically determined D_{O_2}, but, for reasons discussed in Section 7, are not identical to the latter. In other cases they calculate the morphological "membrane diffusing capacity," $D_{M_{O_2}}$, which does not include the erythrocyte component.

Still other authors (e.g., Hughes 1980; Perry 1978a, 1983; Stinner 1982) prefer to report only the diffusing capacity of the air-blood tissue barrier, Dt_{O_2}. If Kt_{O_2}

is not known or is variable because the animals are ectothermic, the purely morphological component of Dt_{O_2} is used as a basis for comparison with other species. This ratio of the respiratory surface area to the harmonic mean thickness of the air-blood tissue barrier has been termed the "anatomical diffusion factor," ADF (Perry 1978), and will serve as the primary basis of comparison among different species in the present review.

Until the mid-1960s, the ADF of a lung could not be accurately measured: the alveolar surface area was usually determined by multiplying the surface area of a model alveolus by the estimated number of alveoli in the lung, and these values varied widely. For the human lung, for example, estimates ranged from 30 m^2 (Schulze 1871; von Hayek 1953) to 200 m^2 (Arthus 1927). In addition, the existence of an epithelial lining layer in the lungs of all tetrapod classes had only been recently demonstrated (Low and Sampaio 1957; Okada et al. 1962, 1964, 1965) by electron microscopy, which was still in its early stages of biological application, and accurate and reproducible methods for measuring the thickness of the diffusion barrier (Weibel and Knight 1964) were not yet available.

With E. R. Weibel's publication of *Morphometry of the Human Lung* in 1963, a revolution in the quantitative morphology of gas exchange organs began. This does not mean, of course, that there were no quantitative studies of gills and lungs before 1963. Monro (1787) is cited by Riess (1881) as providing an early estimate of gill surface area in fish, and in 1731 Hales had already estimated the surface area of the calf lung. Rather, the significance of Weibel's (1963) contribution lies in the application of mathematically sound and scrupulously objective stereological methods to the traditionally descriptive disciplines of histology and ultrastructural morphology. Measurement of the surface area of gas exchange organs became, for the first time, reproducible and largely independent of the investigator, and new methods for determining the diffusion distance between the external, oxygen-carrying medium and the capillary blood of the respiratory organ were introduced (Weibel and Knight 1964). Other authors quickly followed, modifying stereological methods to suit fish gills (Hughes 1972), reptilian lungs (Perry 1981b) and avian lungs (Duncker 1971; Abdalla 1977). For further references on the ensuing flood of morphometric studies on gills and lungs the reader is referred to Gehr et al. (1981), Hughes (1984a, 1989), Maina (1988), Perry (1983) and Weibel (1979a, 1984). The focus here will remain on recent advances and trends in the morphometric study of vertebrate gas exchange organs.

2 Models

Underlying most morphometric studies of diffusing capacity is the assumption that the structure of a gas exchange organ in some way reflects the oxidative, metabolic capabilities of the organism. The most explicit statement of this belief is the principle of *symmorphosis*, coined by Taylor and Weibel (1981). In the words of these authors,

"The principle states that no more structure is formed and maintained than is required to satisfy functional needs; this is achieved by regulating morphogenesis during growth and during main-

tenance of structures. ... From this concept of close structure-function correlation we derive the following hypotheses:
1. *The structural design is a rate-limiting fator for O_2 flow at each level.* For example, the maximal O_2 flow rate into the sink depends on the number of respiratory chain units available, and hence on the surface of the inner mitochondrial membrane into which they are packed. Likewise, the O_2 transfer rate from air to blood in the lung depends on the air-blood contact area established at the lung's tissue barrier.
2. *The structural design is optimized,* i.e., there is just enough structure at each level to support the maximal O_2 flow rate, perhaps with some minimal safety factor. This hypothesis derives from the notion that maintaining biological structures with their often high turnover rates is costly. In fact, there are numerous well-known examples where structures become quantitatively reduced when they are not required to work.
3. *The structural design is adaptable,* at least within certain limits. This hypothesis is clearly a consequence of the two others, because if structure is (a) rate-limiting and (b) optimized to functional demand, then an adaptive change of function must entail an adaptive response of structure. Specifically, if O_2 requirements on the part of muscles are altered, the structural design properties of all parts of the respiratory system, should become adapted" (Taylor and Weibel 1981).

Taylor and Weibel (1981) thus extend their model of morphometric determination of diffusing capacity beyond the lung to include body tissues. The validity of the morphometric calculation of $D_{L_{O_2}}$ is further discussed in Section 7.

Other workers (e.g., Hughes 1980a; Perry 1978a, 1983; Stinner 1982), tend to regard the morphometric diffusing capacity or the ADF of the gill or lung tissue primarily as an indicator of that organ's degree of specialization for gas exchange without necessarily assuming that the locomotor apparatus is equally well adapted for *sustaining* the maximum oxygen flux predicted by analysis of the gas exchange organ. This "liberal" model is most suitable for comparison of low-energy strategists or multimodal breathers.

Still others (e.g., Czopek 1965; de Jager et al. 1977; Jakubowski 1982, 1989; Pohla et al. 1985; Smith and Rapson 1977) prefer to employ only the extent of vascularity as an indicator of the respiratory surface area and thus of the gas exchange potential of the skin, gills or lungs.

3 Methods

3.1 Stereological Methods

A number of excellent texts and articles on the theory and practice of stereological morphometry are available (DeHoff and Rhines 1968; Underwood 1970; Weibel 1969, 1979b, 1980; Weibel and Elias 1967). Since some knowledge of the premises upon which these important methods are based is necessary in order to appreciate the difficulties which some respiratory organs present the morphometrist, the most important concepts are summarized below.

Stereological methods were originally designed for and applied to geological and metallurgical systems, and not biological ones. The basic problem was to deduce from a polished surface of stone or metal alloy the number, volume and surface area of included particles. Of secondary importance were the shape and dimensions of the particles. The practical techniques designed to solve these problems in biological systems employ tissue sections, which are analyzed at the light or electron microscopic level using a "test lattice" of known dimensions. On the basis of the number of points of the test lattice falling over tissue profiles of interest or the number of intersections of the test lines with profile boundaries, the volume or surface area of the structure of interest per unit tissue volume can be calculated (see above references for details).

Stereological analysis does not work equally well on all systems. Optimally, the biological system to be measured should meet the following criteria:

1. The profiles being analyzed as well as the embedding matrix must be opaque. Alternatively, if dealing with sections of finite thickness (e.g., in light or electron microscopy) the section thickness must negligible in relation to the size of these structures being measured.
2. The distribution of profiles of interest within the embedding matrix must be nonbiased: i.e., such that a set of samples can adequately represent the whole.
3. The orientation of profiles and their relevant surfaces must be random relative to the plane of sectioning and the direction of measuring their dimensions.
4. The system must be free of regional and directional distortion. If distortion (e.g., shrinkage) occurs, this must be the same in all parts of the sample, thus allowing the reconstruction of the distortion-free state by the use of correction factors.

The proper use of osmotically adjusted fixative solutions (Bullock 1984) and methacrylate or epoxy embedding media permits compliance with conditions (1) and (4). Synthetic resins can be cut sufficiently thin as to allow accurate measurements of the thinnest air-blood barrier electron microscopically without danger of systematic overestimation (Dubach 1981). Light microscopy of synthetic resin sections 0.5 – 1 μm in thickness is sufficient for some teleost gills (Hughes and Perry 1976; Soivio and Tuurala 1981) and is much less labor-intensive than electron microscopy. Where paraffin embedding provides advantages (opaqueness, large, compression-free lung profiles; Perry 1978b, 1981a) and is still in use, the shrinkage must be carefully measured.

In mammals and birds, the alveolar lungs and the blood-air-capillary net, respectively, conform well at the parenchymal level to conditions (2) and (3). The inhomogeneous distribution of large airways and blood vessels, however, requires a large sample area for accurate measurement (Weibel et al. 1981). Reptilian lungs and fish gills present significant problems at all levels with respect to conditions (2) and (3).

3.2 Nonstereological Methods

Although seldom used for measuring the morphometric D_{O_2} of lungs, nonstereological methods still find application in systems such as fish gills, which conform poorly to the constraints of stereology. As outlined in Section 5, these methods usually consist of determining the surface area of a model or "average" gas exchange unit and multiplying it by the total number of such units. The surface area of this average unit, if not determined by point counting, is measured by the cut and weigh method (in which a camera lucida tracing is actually cut out and weighed), by mechanical planimetry or by computer-assisted planimetry.

Barrier thicknesses in lower vertebrate lungs and in fish gills are seldom reported as harmonic mean values (Meban 1980), but more often occur as incidental measurements, usually taken at what the observer deems to be typically thin locations (Dube and Munshi 1974; Hughes and Gray 1972; Meban 1980; Welsch 1979), or as "preferred channel" means (Hughes 1970; Hughes and Wright 1970). In the teleost gill, the latter technique provides a close approximation of the harmonic mean measured according to stereological guidelines (Hughes 1972).

Another common nonstereological method for estimating parameters relevant to diffusing capacity bases upon simple, geometrically regular models. Such systems can serve as a second-method test for possible order-of-magnitude computational errors in stereological studies and are also useful as a basis for mathematical models (Hughes 1966; Piiper et al. 1986).

Casts of hollow structures represent a real supplement to stereological investigations. With the aid of these techniques, which have been practiced since the dawn of microscopy, the continuity and branching patterns in airways and vascular systems can be studied (Haefeli-Bleuer and Weibel 1988). By allowing casting material to be drawn in under negative pressure, even noncollapsible spaces such as pneumatic bones of birds can be filled and the volume measured by weighing and/or by water displacement (Duncker 1971). Synthetic resin or elastomer casts can be studied at all levels from gross anatomy to scanning electron microscopy. Such injection techniques have been used not only for studying the anatomy of the vascular bed in lower vertebrate lungs (Farrell 1981; Perry 1972; Smith and Campbell 1976) and fish gills (for references see Laurent 1984), but also for measuring its physical properties (Farrell 1980, 1981; Maloney and Castle 1969). In addition, injection and corrosion techniques recently have been put forth as a reliable method for demonstrating the secondary lamellar surface area in fish gills (Pohla et al. 1985).

Three-dimensional reconstruction of microscopic structures on the basis of serial sections, while not necessarily reliable as a quantitative technique, can be a valuable orientation help, particularly for modelling the continuity of noninjectible tissue components such as connective tissue or nerves, or for demonstrating the spatial distribution of discontinuous objects. Recently the ancient procedure of melting together wax-plate tracings of serial section profiles has been superceded by more complex but infinitely more versatile computer techniques, for which relatively inexpensive software for personal computers is now available.

4 Skin

Although significant cutaneous gas exchange has been demonstrated physiologically in fish, amphibians and reptiles (for a recent review see Feder and Burggren 1985), morphometric diffusing capacity has been measured only in a single species of fish (*Heteropneustes fossilis*) and in the lungless salamander, *Desmoganathus fuscus* (Gatz et al. 1975; Hakim et al. 1978). In *D. fuscus* the morphometric and physiological diffusing capacities are virtually identical (Piiper and Scheid 1975; Piiper et al. 1976).

In fish, injection and clearing techniques have been used for visualizing and quantifying vascularization in the skin and gills (for references see Hughes 1984a, 1989; Jakubowski 1982, 1989; Videler 1968). Similar methods have also been applied to the skin, lungs and buccal cavity of amphibians (for references see Czopek 1965; Smith and Rapson 1977). Unfortunately, however, few comparable measurements of the diffusion barrier are available for the skin (Czopek 1965). For the enterprising morphometrist it would be well worth the time invested to supplement the extensive data of Czopek, Jakubowski and Videler (see above) and to provide the missing harmonic mean barrier thicknesses, thus allowing at least an internally comparable index of the partitioning of ADF among the potential respiratory organs in fish and amphibians.

Active recruitment of existing skin capillaries ("blushing") has been demonstrated in teleosts (Connolly 1926; Sacca and Burggren 1982), amphibians (Burggren and Moalli 1984; Moalli et al. 1981; Poczopko 1957) and reptiles (Bartholomew 1982; Courtice 1981), and the density of the cutaneous capillary bed increases in the bullfrog (*Rana catesbeiana*) tadpole by a factor of nearly three, following exposure to moderate hypoxia. Dramatic seasonal fluctuations in skin vascularity have also been reported in other amphibians, notably in the so-called hairy frog, *Astylosternus robustus* (Noble 1925). Furthermore, when on land, animals may mechanically impair their cutaneous gas exchange by resting on the ventral body surface: a region particularly well vascularized in anurans (Czopek 1965). The physiological and morphological patency of the cutaneous respiratory apparatus clearly points up the inherent difficulty in applying the "symmorphosis" model to bimodal or multimodal breathers.

5 Fish Gills

The gas exchange units of all adult fish – from lampreys to fast-swimming percomorph teleosts – consist of plate-like secondary lamellae that extend like fern leaves from the "primary lamellae," the gill filaments (Hughes 1984a; Hughes and Morgan 1973; Mallatt 1984). The folded gill structure of hagfish, the threadlike gills of some larval elasmobranchs and teleosts and the external gills of larval lungfish (Bütschli 1934; Mallatt and Paulsen 1986; Rauter 1935) notwithstanding, all fish gills are difficult to deal with stereologically. The main reason lies in the isotropic disposition of the gas exchange surfaces in space.

Each gill arch supports two hemibranchs, the filaments of each lying in a single plane. Furthermore, the length of the filaments varies in a regular way with their position on the gill arch, and the size and shape of the secondary lamellae, in turn, differ systematically according to their position along the filament (Hughes and Gray 1972; Hughes et al. 1986; Niimi and Morgan 1980). Thus, obtaining histological sections which are oriented randomly with respect to the secondary lamellar surface, while at the same time satisfying the criteria for representative sampling for the whole gill, is an extremely tedious and time-consuming procedure.

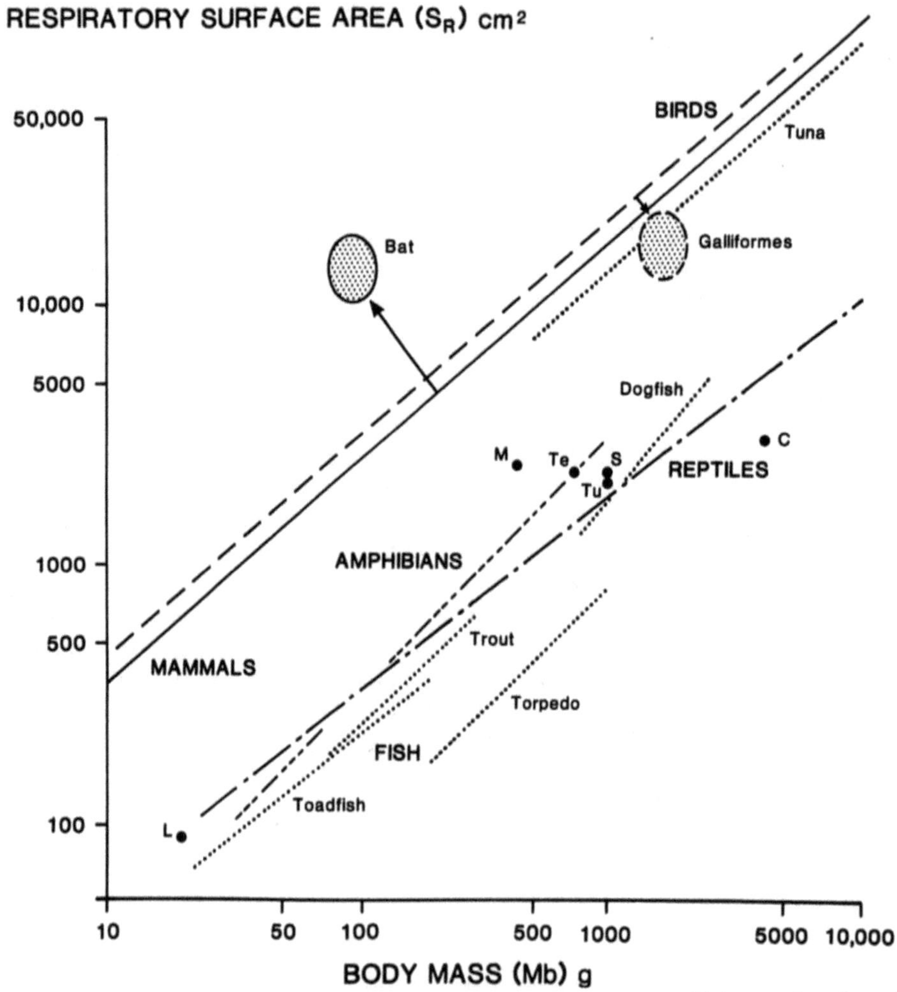

Fig. 1. Double logarithmic representation of respiratory surface area (S_R) as a function of body mass (*Mb*) in the lungs of mammals (*solid line*), birds (*dashed line*) and reptiles (*alternating dashes and dots, filled circles*) as well as in the gills of selected fish (*dotted lines*). Designated fish species are (elasmobranchs) the lesser spotted dogfish, *Scyliorhinus canicula*; the torpedo, *Torpedo marmorata*; (teleosts) the toadfish, *Opsanus tau*; the rainbow trout, *Salmo gairdneri*, and the yellowfin tuna, *Thunnus albacares*. Designated data points for reptiles are *C* the nile crocodile, *Crocodylus niloticus*; *L Lacerta* spp.; *M* the savanna monitor, *Varanus exanthematicus*; *S* a colubrid snake, *Pituophis melanoleucus*; *Te* the teju lizard, *Tupinambis nigropunctatus* and *Tu* the red-eared turtle, *Pseudemys* = *Trachemys scripta elegans*. *Bat* is the fruit bat, *Epomophorus wahlbergi*, and *Galliformes* is the mean of the domestic fowl, *Gallus gallus*; the red jungle fowl, *Gallus gallus*; the domestic Guinea fowl, *Numida meleagris*; and the domestic turkey, *Meleagris gallopavo* (Data from Cragg 1975; Gehr et al. 1981; Hughes 1980a; Maina 1988; Maina et al. 1982; Perry 1978a, 1983, in press; Stinner 1982; and Tenney and Tenney 1970)

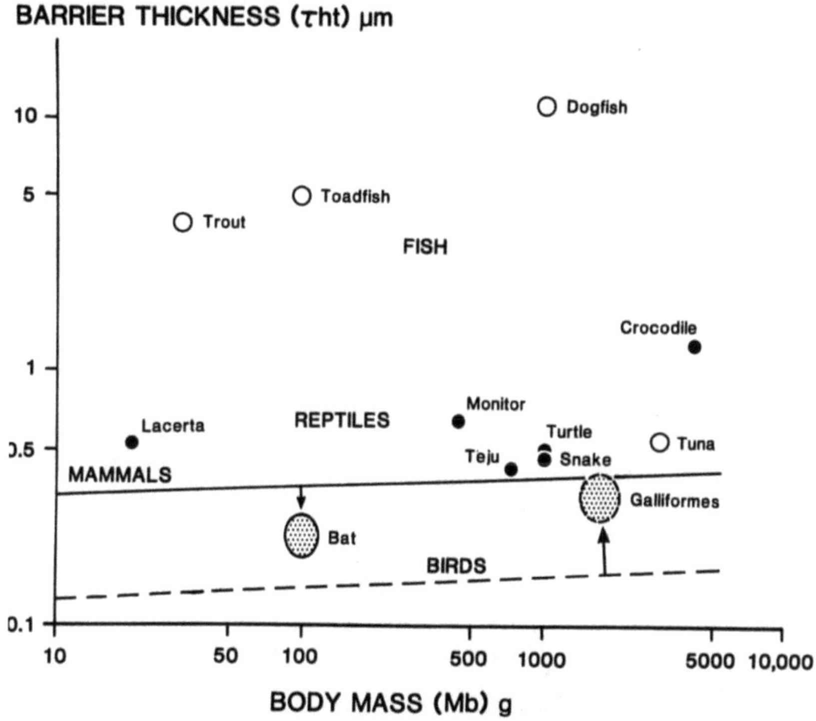

Fig. 2. Double logarithmic representation of the harmonic mean thickness of the air-blood tissue barrier (τht) as a function of body mass (Mb) in the lungs of mammals (*solid line*), birds (*dashed line*), and reptiles (*filled circles*). Except for the harmonic mean in the trout, the thickness of the water-blood tissue barrier in the gills of selected fish (*open circles*) is the arithmetic mean of preferred channel or orthogonal measurements. For an explanation of *abbreviations* see Fig. 1 (Data from Cragg 1975; Gehr et al. 1981; Hughes 1984a; Hughes and Gray 1972; Maina 1988; Perry 1978a, 1983, in press; and Stinner 1982)

For these reasons the determination of diffusing capacity of fish gills based upon the stereological morphometry of randomly oriented sections (Hughes 1972) has not met with universal acceptance. Rather, various authors have repeatedly attempted to utilize (1) the mathematical predictability of gill heterogeneity (Hughes et al. 1986; de Jager et al. 1977; Videler 1968) in order to determine secondary lamellar number, (2) injection and clearing (for references see Byczkowska-Smyk 1959; Jakubowski 1982b) or scanning electron microscopic observation of vascular corrosion casts (Farrell 1980; Pohla et al. 1985; Palzenberger 1989) for determination of secondary lamellar surface area, or (3) orthogonal measurements for obtaining the water-blood diffusion distance (Hughes and Perry 1976; Hughes et al. 1986).

Mathematically sound stereological procedures have recently been developed for dealing with oriented structures, using perpendicular sections for measuring the surface area (Baddely et al. 1986; Gundersen et al. 1978). Until now these methods have been applied only to gills in cephalopod molluscs (Eno 1987): it is not clear why they have not yet been used for fish gills.

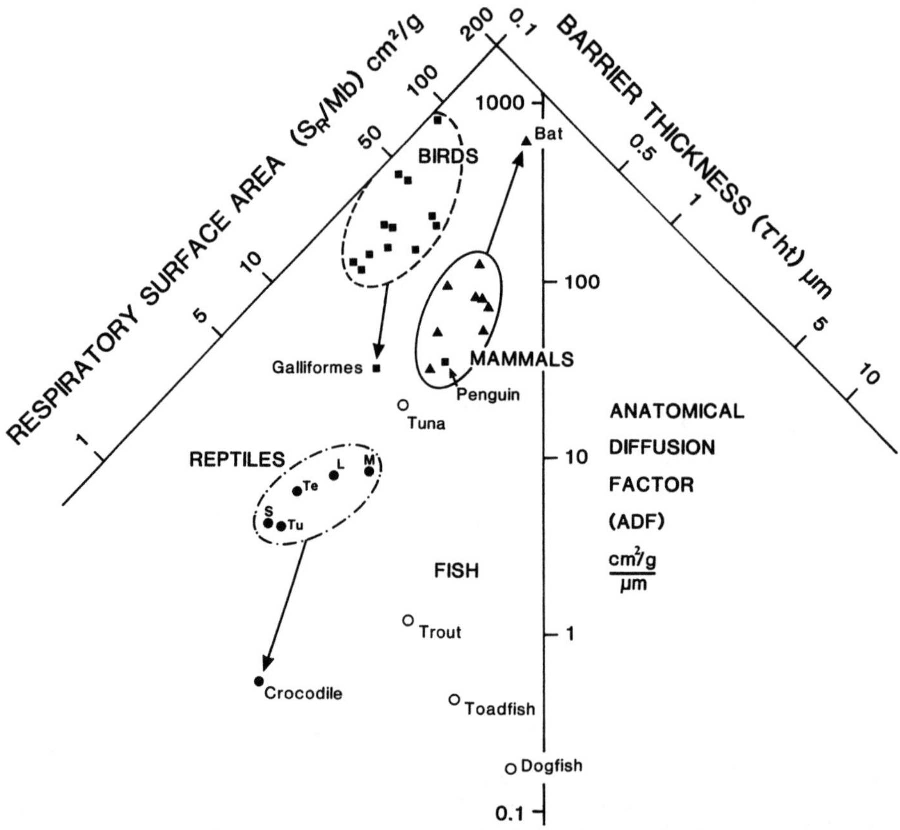

Fig. 3. Triple logarithmic representation of body-mass specific respiratory surface area (S_R/Mb), harmonic mean thickness of the air (water)-blood tissue barrier (τht) and ADF ($S_R/Mb/\tau ht$). Each data point is read off perpendicularly against the appropriate axis. *Filled squares* represent means for avian orders, *filled triangles* are means for mammalian families, *filled circles* are means for reptilian species and *open circles* represent the middle of the Mb range for species of fish. *Penguin* is the Humboldt penguin, *Spheniscus humboldti*. (or other *symbols* and *abbreviations* and for data sources see Figs. 1, 2)

Gill surface area is usually measured by multiplying the total length of gill filaments by the secondary lamellar frequency and the surface area of an average secondary lamella. Although there is reasonably good agreement among various authors on filament length and secondary lamellar frequency, the estimates of average secondary lamellar surface area vary significantly (Hughes 1984a). There may be several reasons for this, including incomplete removal of the secondary lamellae for surface area determination, distortion of the delicate tissue during fixation and the manner in which an "average" secondary lamella is derived from the very heterogeneous population at hand. In addition, there may be real differences among specimens of the same species and body mass (Mb). In the rainbow trout, *Salmo gairdneri*, for example, the interlamellar tissue shrinks following chronic exposure of the fish to hypoxic water, thus exposing a greater portion of each secondary lamella (Soivio and Tuurala 1981). The thickness of the water-blood barrier also decreases under these conditions (Soivio and Tuurala 1981).

Taking into account possible procedural differences in morphometric methods, several general tendencies still remain: (1) relative to Mb, slow-moving, bottom-dwelling species tend to show a smaller lamellar frequency (i.e., wider spacing of secondary lamellae), smaller secondary lamellar surface area, thicker water-blood barrier and smaller ADF than do active, pelagic species (Figs. 1, 2, and 3; de Jager and Dekkers 1975; de Jager et al. 1977; Gray 1954; Hughes 1980a, 1984a;. (2) elasmobranchs (e.g., Dogfish and Torpedo in Fig. 1; Dogfish in Figs. 2, 3) tend to have "coarser" gills (i.e., larger secondary lamellae, thicker water-blood barrier) than do their teleostome ecological counterparts. Otherwise phylogenetic considerations shows little relationship to the measured morphometric parameters (Hughes 1972; Hughes 1980b; Hughes and Morgan 1973; Hughes and Wright 1970); and (3) the respiratory surface area of the most active fish (Fig. 1; Hughes 1984b; Muir 1969) is greater than that of similar sized reptiles (Perry 1983) and approaches the regression line for captive mammals (Gehr et al. 1981). Since the thickness of the diffusion barrier (e.g., 0.533 μm in the yellowfin tuna *Thunnus albacares*) in these fishes is even less than in some mammalian lungs, the ADF of the fish gill shows a similar degree of morphological adaptation for its gas exchange function, as does the lung of terrestrial vertebrates (Figs. 2, 3).

6 Lungs

6.1 Low-Performance Lungs

In light of the generally lower metabolic rate of amphibians and reptiles as compared with birds and mammals, it is not surprising that this is reflected in differences in lung structure. Lung volume is extremely variable among amphibians and, at least in aquatic species, depends primarily on the buoyancy requirements of the animal. The secondarily lungless state of plethodontid salamanders, for example, is commonly attributed to the origin of this family in fast-flowing upland waters, where a low specific gravity would be of negative survival value. The Olympic salamander, *Rhyacotriton olympicus*, which tends to have vestigial lungs, appears to be undergoing a similar process (Porter 1972).

In reptiles, on the other hand, the lung volume tends to be large, ranging from 5 to 50 times that of a similar sized rodent per unit Mb (Perry 1983). Within this volume, however, it contains only about one-tenth of the lung surface area expected for a mammal. There are many solutions to the problem of distributing a small surface area in a large volume (see Duncker 1978a; Milani 1894), but most of these involve the heterogeneous distribution of parenchyma (gas exchange tissue and its enclosed air spaces). Commonly, the caudal and ventral lung regions tend to form a thin-walled sack. The parenchyma consists of open-ended cubicles (ediculae), which in many species of snake and lizard develop to highly specialized honeycomb-like units called faveoli (Duncker 1978a) and tend to be concentrated in the cranial two-thirds of the lung (Perry 1983; Wolf 1933).

The gas exchange capillaries are typically disposed as a double net: one network on each side of an interedicular/interfaveolar septum. Consequently, only one side of a capillary forms a thin air-blood barrier, while the other faces the connective tissue central leaflet of the septum. Occasionally anastomosing capillaries are seen, which unite the two nets and demonstrate a thin air-blood barrier on both sides. The capillaries form a predominantly hexagonal array (Perry 1972). As might be expected in analogy to the heterogeneous parenchymal distribution, the mesh of this array may vary within the lung (Perry 1972, 1978a). In some portions of reptilian lungs, such as the caudal sac-like portion of snake lungs, respiratory capillaries are totally absent (Luchtel and Kardong 1981; Stinner 1982).

Low performance lungs are technically an even more difficult system to morphometrically evaluate than are fish gills. Some of the major problems and their practical solutions are summarized below:

1. *Nearly half the volume of many such lungs at a standard state of inflation (three-fourths the total lung capacity) consists of a tissue-free central lumen* (Perry 1978a, 1983, 1988a). Care must be taken during preparation for morphometry not to distort this space. Technically, this requirement is met by conducting the first stages of volumetry and surface area determination on the microtome-planed tissue block rather than on sections (Perry 1978a, b, 1981a).
2. *The degree of heterogeneity of parenchymal distribution appears to differ from species to species and could be altered by experimental exposure to hypoxia or other stimuli.* It is possible to quantify this heterogeneity by primarily dividing the lung into a series of transversely oriented samples of standard thickness, each including the entire lung profile. This allows a morphometric breakdown along the long axis of the lung. Additionally, the primary point and intersection counts are performed using a transparent mask, which divides the lung profile into quadrants, into concentric rings relative to the intrapulmonary bronchus (if present), or both. In this way, the surface area of various parts of lungs (e.g., cranial third, dorsal half, portion nearest the intrapulmonary bronchus) can be evaluated separately. If tissue samples for electron microscopy are taken from similar portions, the regional ADF within a given lung can be calculated (Perry 1981b).
3. *The sometimes extreme heterogeneity of the parenchyma confounds the determination of harmonic mean barrier thickness. The small diameter (sometimes less than 0.2 mm) and often folded walls of the faveoli as well as strongly bulging capillaries makes perpendicular sections of the faveoli impossible, whereas random sectioning of straight-walled eidulae which may be up to 1 cm in diameter is impractical because of the large number of individual thin sections required for each lung region.* Two models have been proposed for the determination of harmonic mean barrier thickness in low-performance lungs: one similar to that used for mammalian lungs when random sections are unavoidable (Perry 1981b), and one employing orthogonal measurements where perpendicular sections are feasible (Perry 1976). Comparison of both methods in transitional cases yielded similar results (Martina Dubach-Günthert, pers. communi.).
4. *The mesh of the capillary net is in some cases so great that inclusion of the extremely long diffusion distances according to the model of Weibel (1970/1971) results in an apparent decrease in diffusing capacity.* When perpendicular sections

and orthogonal measurements are employed (which is usually the case in those lung regions in which the capillary spacing is greatest), cutoff limits for barrier thickness measurements are used (Perry 1976). Those measurements which are greater than the designated cutoff limit are considered to pertain to the "nonrespiratory" portion of the lung surface. For cases in which randomly oriented sections and measurements are employed, the surfaces are usually sufficiently curved and the capillary profiles are close enough together that extremely long barrier thickness measurements are rare. In such cases, the nonrespiratory portion of the lung surface is defined as that proportion intersected by transects which do not contact a capillary surface (Perry 1983). It should be noted that the two definitions of nonrespiratory surface presented here are different and that the degree of congruity between them has not yet been established.

6.1.1 Fish Lungs

Lungs and swimbladders are both derivatives of the most caudally located gill pouches: the swimbladder developing from dorsal, and the lungs from ventral anlagen (for references see Perry 1988b). In spite of their different origins, the swimbladders and lungs of air-breathing fish and the lungs of anuran amphibians are remarkably similar in their gross and parenchymal structure (Starck 1982).

Also quantitatively, fish swimbladders and lungs can be similar to low-performance lungs in other vertebrate taxa. The respiratory surface area of the air sac per gram Mb in the siluroid teleost *Saccobranchus (Heteropneustes) fossilis* (145.9 mm^2: Hughes et al. 1974) is approximately double that of the Nile crocodile (Perry 1990) and the harmonic mean thickness of the air-blood diffusion barrier in the lungfish *Protopterus aetheopicus* (0.37 μm: Maina and Maloiy 1985) is less than that of reptiles (Meban 1980; Perry 1983) and similar to that of a mammal of similar Mb (Gehr et al. 1981).

6.1.2 Amphibian Lungs

The ADF and Dt_{O_2} of the amphibian lung have not yet been determined. Following the pioneer works of Czopek (see Czopek 1965 for references), which compare the total capillary length in the various respiratory surfaces of amphibians, the only quantitative work of significance is that of Tenney and Tenney (1970), in which the pulmonary surface area of dried lung preparations is determined stereologically. The latter study shows the mean edicular diameter to be greater in amphibians than in reptiles and the lung volumes to be similar in the two groups in the Mb range of 50 to 500 g. The lung surface area in both groups is similar (approximately 10 cm^2/g Mb) within the range of 10 to 100 g and increases in proportion to the 0.75 power of Mb in amphibians: in nearly direct proportion to Mb in reptiles (Fig. 1).

Available harmonic mean barrier thickness measurements in amphibian lungs (Meban 1980; Goniakowska-Witalinska 1980) indicate a diffusion distance in excess of 2 μm in some species, and a mean for ten species of 1.7 μm: significantly greater than that in reptilian lungs (Cragg 1975; Meban 1980; Perry 1983; Stinner 1982; Welsch 1979). It is therefore likely that the ADF of amphibian lungs, with the possible exception of those of the pipid anurans, which show a particularly finely divided parenchyma, large surface area and great total capillary length (Czopek 1955; Tenney and Tenney 1970), is less than that of reptiles in the same Mb range. Morphometric studies of all respiratory surfaces for amphibians, similar to the work of Hughes et al. (1974) for the fish *Saccobranchus* would be extremely valuable.

6.1.3 Reptilian Lungs

To date, the ADF and morphometric D_{O_2} of the pulmonary tissue barrier have been determined in six species of reptile: one turtle, *Pseudemys* = *Trachemys scripta*; three lizards, *Lacerta* spp., *Tupinambis nigropunctatus, Varanus exanthematicus*; one snake, *Pituophis melanoleucus* and one crocodilian, *Crocodylus niloticus* (Cragg 1975; Perry 1978a, 1983, in press; Stinner 1982). With the exception of the crocodile, the ADF of these reptiles lies approximately one order of magnitude below that of mammals in the same Mb range (Fig. 3), largely due to the smaller surface area in the reptilian lung. The particularly low ADF of the Nile crocodile (Fig. 3) is the result of both a small surface area (Fig. 1) in its (for a reptile) unusually small lungs and an exceptionally thick air-blood barrier of 1.2 µm (Fig. 2; Perry 1990). The low ADF in this species is in keeping with its low habitual oxygen consumption rate (Bennett and Dawson 1976; Brown and Loveridge 1981).

It is interesting to note that the ADF value in reptiles is independent of the number or structural type of lungs: the snake *P. melanoleucus* has one single-chambered lung; *Lacerta* and the teju *T. nigropunctatus* have two single-chambered lungs; and the turtle *P. scripta*, the monitor lizard *V. exanthematicus* and the Nile crocodile have two multichambered lungs. Those species with multichambered lungs, however, do tend to demonstrate a greater surface area and a thicker air-blood barrier than do those with a more simple primary lung structure. Thus, the respiratory surface area of the monitor is twice that of the teju, but the teju compensates with a thinner air-blood barrier such that the ADF in both species is similar (Fig. 3).

Furthermore, the ADF and Dt_{O_2}, while possibly explaining the peak $\dot{V}_{O_2}$ *attainable*, are not useful in predicting the maximal $\dot{V}_{O_2}$ which is *sustainable* over long periods of time. In spite of similar ADF values, the monitor is capable of sustaining rates of oxygen consumption which quickly exhaust the teju. The very high compliance of the monitor lung relative to that of the teju suggests that the respiratory pump may be a key factor (Perry and Duncker 1978). Experimental studies bearing upon the efficiency of the respiratory pump, such as the long-term effects of training or hypoxia on lungs and respiratory muscles in reptiles, are urgently needed.

6.2 High-Performance Lungs

High-performance lungs are distinguished by a high ratio of respiratory surface area to parenchymal volume, a large respiratory surface area per unit Mb and a thin air-blood barrier (usually less than 0.5 μm), which covers a large proportion of the lung surface. While both avian and mammalian lungs fulfill all of these general criteria, their basic architecture is fundamentally different.

6.2.1 Avian Lungs

In the avian respiratory system, the lungs proper are spatially removed from the air sacs, which serve only in storing and propelling air. The gas exchange surfaces of the lung consist of an interwoven network of blood capillaries and air capillaries, which receive their fresh air supply *en passant* from the parabronchi (Duncker 1971). In the phylogenetically older part of the lung, the paleopulmo, the air flow through the parabronchi is unidirectional (Scheid et al. 1972), whereas in the neopulmo, which lies between the paleopulmo and the abdominal air sacs, air flow is two directional (Maina 1984). Even the neopulmo, however, is a flow-through system. The fact that the avian lung undergoes only minute volume changes during breathing means that little cytoplasmic and membrane reserve is necessary in the air capillary epithelium, and extreme attenuation can be achieved. In addition, since the air capillaries are protected from collapse by the firm attachment of the lung to the concave thoracic wall, diameters of as little as 3 μm are maintained, resulting in a parenchymal surface area to volume ratio some ten times greater than in a comparable mammal (Duncker 1978b).

Following the pioneer work of Duncker (1971, 1972, 1973), little stereological morphometry was performed on avian lungs until the 1980s, when a flurry of papers followed: Abdalla and Maina (1981), Abdalla et al. (1982), Dubach (1981), Maina (1984, 1987), Maina and King (1982, 1987), and Powell and Mazzone (1983). Data from these sources as well as from the Ph. D. thesis of Maina (1982b) are compended and compared with similar studies on mammals, including bats, in a recent comprehensive review (Maina 1988), to which the reader is referred for details.

On the basis of analysis of 87 specimens representing 25 different avian species, Maina (1988) concludes that the Mb-normalized respiratory surface area of avian lungs is 15% greater than that of mammals (Fig. 1). In addition, as illustrated in Fig. 2, "In the lightest birds the [harmonic mean thickness of the air-blood tissue] barrier may be about 56% thinner than that of a terrestrial mammal of the same body weight; in the heavier animals the difference appears to become greater."

Comparing Dt_{O_2} values, Maina (1988) finds an order-of-magnitude difference between the most metabolically active of terrestrial mammals, the shrew *Sorex minutus* (Gehr et al. 1980) and the hummingbird, *Colibri corscans* (Dubach 1981): 27.3 and 282 ml O_2/min/mmHg/kg, respectively. These clear differences are only discernable when galliform birds (with their small and relatively poorly developed lungs) and bats (with their highly specialized lungs) are deleted from the regression analysis (see Fig. 3).

Passerine birds tend to have a greater Mb-specific $\dot{V}_{O_2}$ than do nonpasserine species (Lasiewski and Dawson 1967). This difference is not clearly reflected morphometrically (Maina 1988). In general, however, small birds tend to have more gas exchange surface and a thinner air-blood barrier per unit Mb than do larger ones. Passerine species are often characterized by a small size and conform to the above rule.

The neopulmo, which is particularly well developed in passerine species but by no means restricted to them, is morphometrically indistinguishable from the paleopulmo (Maina 1982a), suggesting that both parts of the lung are also functionally equivalent. In the duck, the neopulmo is in fact well-ventilated even at rest (Holle et al. 1978), but this remains to be demonstrated for passerine species, where it may be brought into play mainly during exercise (Duncker 1971).

6.2.2 Mammalian Lungs

The general anatomy of the broncho-alveolar mammalian lung (Engel 1962) will be familiar to most readers. Of particular significance is that the bronchial tree ends blindly in the alveolar ducts, which have the function not only of exchanging gas, but also of propelling and conducting it. Unlike avian air capillaries, mammalian alveoli change their surface area and barrier thickness during breathing (Weibel et al. 1973), and are prevented from collapsing completely only by a delicate balance between the surface-tension lowering properties of pulmonary surfactant and the centripetal pull of the elastic bundles located in the free margins of the interalveolar septa (Weibel 1984; Weibel and Bachofen 1987; Wilson and Bachofen 1982). In these respects the mammalian lung retains a functional similarity to the reptilian lung: the alveoli resemble miniature ediculae with their myoelastic trabecular suspension system (Engel 1962; Perry and Duncker 1978; Perry et al. 1989).

Since 1970 (Weibel 1970/71), stereological morphometric studies which include measurements of the respiratory surface area, harmonic mean thickness of the air-blood tissue barrier and estimation of the morphometric diffusion capacity of the lung have been conducted on some 40 different mammalian species. The range of species investigated includes shrews (Gehr et al. 1980), bats (Lechner 1984; Maina and King 1984; Maina et al. 1982), and a variety of wild and domesticated rodents, herbivores and carnivores (for references see Gehr et al. 1981 and Maina 1988): the range of Mb is from 2 to 700000 g.

The respiratory surface area in terrestrial mammals lies between 16×10^3 cm^2/kg Mb and 83×10^3 cm^2/kg Mb and, surprisingly, increases in proportion the 0.95 power of Mb and not with the 0.80 power of Mb as predicted by the allometric increase in $\dot{V}_{O_2}$ (Taylor et al. 1981). The barrier thickness increases from 0.26 µm in the shrew *Sorex minutus* to 0.62 µm in man, displaying an allometric slope of 0.05 (Gehr et al. 1980, 1981). Combining exponents, one obtains an effective allometric slope of 0.90 for ADF as a function of Mb: still not the predicted 0.80.

In spite of this apparent overdesigning of the lungs of large terrestrial mammals there are obvious correlations between the observed habitual activity of the species and the ADF. Bats, for example, are known to sustain a $\dot{V}_{O_2}$ similar to that of birds in flight (Thomas 1975; Tucker 1972) and their ADF is accordingly great (Maina 1988). With a respiratory surface area of $87 \times 10^3 cm^2/kg$ and a harmonic mean tissue barrier thickness of 0.206 μm, the pipistrelle (*Pipistellus pipistrellus)* exhibits an ADF of 422×10^7 cm/kg, which is 1.3 times the greatest value reported for any non-chiropteran mammal (Gehr et al. 1980) and nearly half that of the hummingbird, *Colibri coruscans* (Dubach 1981). Also, among bovids (Gehr et al. 1981) and among canids (Weibel et al. 1983), domesticated species show a consistently lower ADF than do undomesticated ones.

The lung is also capable of morphologically adapting to chronically elevated $\dot{V}_{O_2}$ and to chronic hypoxia (Bartlett 1970; Bartlett and Remmers 1971; Burri and Weibel 1971; Burri et al. 1976; Gehr et al. 1978; Hugonnaud et al. 1977; Thompson 1980), although the limits of adaptability appear to be genetically set. In the guinea pig, for example, the alveolar surface area of animals exposed to cold and hypoxia is greater than that of controls only during the period of growth: the lungs of adult animals in both groups are indistinguishable (Lechner et al. 1982).

Taylor et al. (1987) distinguish between "athletic species" and "less athletic species" on the basis of their maximal $\dot{V}_{O_2}$. These differences can be explained only partially on the basis of different ADF values. Although the maximal $\dot{V}_{O_2}$ of the dog exceeds that of the goat by a factor of 2.5, the dog's ADF is only 16% greater than the goat's, and the athletic pony, whose maximal $\dot{V}_{O_2}$ also exceeds that of a similar-sized but less athletic calf by a factor of 2.5, demonstrates an ADF only 6% greater than that of the calf (Weibel et al. 1987). The remaining difference must be attributed to factors other than those directly effecting the diffusing capacity of the lung tissue.

7 Symmorphosis Revisited

The validity of symmorphosis (Taylor and Weibel 1981) as a principle which guides the congruence of structure and function in the respiratory and locomotor systems has recently come under scrutiny. At the heart of this hypothesis is the assumption that the conductance of the oxygen-supplying apparatus must equal that of the oxygen-consuming side, and that both sides of the equation are optimized at maximum $\dot{V}_{O_2}$. It therefore should follow that for a given species, any deviation in $\dot{V}_{O_2}$ from the log-log regression line against Mb would be matched by a deviation of equal sign and magnitude in $D_{L_{O_2}}$. Performing this analysis, Garland and Huey (1987) find no such correlation and conclude:

"Although symmorphosis has heuristic value as a working hypothesis, it should not presently be considered an established principle; the data appear largely contradictory. Moreover ... 'symmorphosis' may be an unlikely evolutionary outcome. Some components within a given system exhibit 'excessive construction,' whereas others constitute 'limiting factors.' Furthermore, given the vicissitudes of evolutionary change, factors that are limiting (or in excess) may well differ among species (or populations)."

One of the authors of symmorphosis has also called for caution regarding the general application of this hypothesis. Pointing out the differences in regression slopes between $\dot{V}_{O_2}$ and D_{LO_2} as well as species differences in the $\dot{V}_{O_2}/D_{LO_2}$ mismatch, Weibel (1987) concludes: "The model must be refined to account for the fact that nature is more complex than our models make us believe."

The possibility also exists that the concept of morphometric diffusing capacity is erroneous. It is based upon the assumption that the movement of oxygen from lung surface to erythrocyte hemoglobin is accounted for only by diffusion. Although this model may hold for the air-blood tissue barrier it is doubtful that it applies to moving blood plasma in the simple form employed in morphometric calculations. This becomes important when comparing animals such as the dog and the goat (Weibel et al. 1987), which differ greatly in erythrocyte size, hematocrit and the harmonic mean thickness of the plasma layer in pulmonary capillaries. Furthermore, since the fixation procedures for lung morphometry do not guarantee constant and physiologically meaningful blood pressure levels or hematocrits (Bur et al. 1985), there may be significant variations in the vascular component within the same species, as has been shown for the budgerigar (Dubach 1981; Maina 1988).

A further source of incongruence is that the model for the exchange barrier is not identical in the morphometric and the physiological expressions for diffusing capacity. The membrane diffusing capacity as a physiological entity includes diffusion through a layer of alveolar air of unknown thickness (Piiper and Scheid 1975). Also, the degree of saturation of the alveolar capillary blood cannot be directly measured, but must be approximated from the saturation of fully arterialized blood. Neither can inhomogeneous aveolar perfusion or ventilation be taken into account in the morphometric model: it is well known that D_{O_2} is not constant, but increases with increasing $\dot{V}_{O_2}$ (Salorinne 1976). For these reasons, until quite recently physiological measurements of D_{O_2} approaching the morphometric D_{LO_2} values in a mammalian lung have not been attainable (Ichinose et al. 1987; Weibel et al. 1983). Even if achieved, the significance of a perfect agreement between morphometric and physiological values for diffusing capacity is uncertain.

It is perhaps fortunate that symmorphosis has been refuted as a general principle before it has had time to become too firmly entrenched in the secondary literature (viz., Schmidt-Nielsen 1984; Weibel 1984). Investigation of the grounds for nonconformation with this attractive principle will fruitfully occupy investigators for many years to come.

8 Summary and Perspectives

8.1 The Bottom Line

After a quarter of a century of systematic, stereological morphometry of vertebrate gas exchangers, two important goals in comparative, functional morphology have been achieved: (1) practical, accurate and reproducible methods are available for estimating those dimensions of mammalian and avian lungs relevant to gas diffusion, and (2) these methods have been applied to some 65 different, wild and domestic species. Thus, a relatively large library of data is available, to which experimental studies on the biomedical aspects of lung growth, differentiation and pathology must also be counted.

Probably the most far-reaching conclusion to be drawn from the comparison of mammalian and avian lungs is that the impressive aerobic feats of some birds (such as the barheaded goose, *Anser indicus*, which migrates over Himalayan peaks that are the ultimate physiological challenge to the most experienced mountaineers) can be attributed to the superior design of the avian lung as a gas exchange organ. Although the body-mass specific ADF of some mammalian species may be greater than that of some avian species (compare bat with domestic fowl, Fig. 3), in general this anatomical index of gas exchange "hardware" is at least 70% greater in birds than in mammals (Fig. 3). The difference lies both in the greater surface area and in the thinner diffusion barrier of the avian lung.

The direct comparison of low-performance reptilian lungs with those of mammals and birds is complicated by the necessity of using somewhat different morphometric techniques. A tentative comparison, however, indicates that the major anatomical advance of the mammalian lung lies only in its greater respiratory surface area: harmonic mean barrier thickness overlaps strongly with that of reptiles (Fig. 3). Lungfish lungs are morphometrically similar to reptilian lungs.

The dimensional comparison of fish gills with tetrapod lungs is perhaps even less justified than the comparison of different lung types. Nevertheless, there are important questions to be addressed. For example, why has the direct conversion of gills into lungs among vertebrates only been successfully achieved among some air-breathing fish (Munshi et al. 1986)? Is the fish gill innately inferior to the lung as a gas exchanger, this putative inferiority being balanced by the great efficiency of the counter-current exchange system?

Addressing the second question first, inspection of Fig. 3 reveals that both the surface area and the diffusion barrier in the most active fish are similar to the corresponding values in tetrapod lungs. Thus, fish gills are not innately inferior to lungs. The reasons for the general failure of fish gills to convert to lungs must therefore be sought elsewhere: inadequate ventilatory mechanism, sticking together of the secondary lamellae in air, dessication of the secondary lamellar surfaces. Still it is unlikely that these considerations would entice a tuna to take up terrestrial life, but does help to explain the success of the walking catfish (Hughes et al. 1974). The reasons behind the origin and evolutionary success of tetrapod lungs in light of the large number of alternative possibilities for air breathing seen in recent bony fishes (Munshi 1985) should be reconsidered.

8.2 Perspectives

8.2.1 Scaling

Allometry, the study of relative size, will certainly continue to fascinate biologists. In a recent major contribution to this field (Schmidt-Nielsen 1984), however, information on the dimensions of respiratory organs of reptiles and amphibians is conspicuously scanty. Certainly one area of future endeavour will be to fill this gap.

The analysis of linear (or logarithmically linearized) regressions and the biological interpretation of the results will also be a source of future discussion, which will certainly deepen our knowledge of the mechanisms guiding the growth, differentiation, adaptation and evolution of respiratory organs. Various authors (Heusner 1982, 1983; Hughes 1984b) have pointed out the pitfalls in regression lines which incorporate different species over a wide range of body mass and energy strategies: (1) the overall slope of the regression line can be influenced by the sampling procedure, particularly by highly specialized species at the upper end the body mass scale, such as scombrid fish (Fig. 1; Hughes 1980), aquatic mammals (Tenney and Remmers 1963) or aquatic and marine reptiles (Tenney and Tenney 1970) and (2) the overall slope may bear little relationship to the slopes for the individual species included. Hughes (1980a), for example, has demonstrated that the allometric slope of gill surface area as a function of body mass in highly active fish approaches 1 (as does lung surface area in mammals and birds; Gehr et al. 1981; Maina 1988), as opposed to nearly 0.6 in some benthic species.

8.2.2 *New Methods*

The application of newly developed stereological methods for (see Cruz-Orive and Weibel 1990) will open new avenues in the morphometry of fish gills, amphibian gills and skin, and may also find application in the morphometry of low-performance lungs. A second trend is the increasing automation of standard stereological morphometry. The solving of form-recognition and contiguity problems, which are presently the major barrier to fully automated stereology, could be a mixed blessing. While on the one hand it would make stereology fast and accessible, on the other hand it could encourage the practice of morphometry by persons who are neither thoroughly acquainted with the structures they are measuring nor the pitfalls of stereology.

With the increasing accessibility and speed of stereology, the importance of nonstereological methods in quantitative morphometry would be expected to decrease. One exception will probably be computer-generated three-dimensional models, which allow "walks" through complex structures as an aid to their visualization. The nonstereological method of greatest potential, however, is the possible application of fractal geometry (Mandelbrot 1977) to problems of growth, differentiation and kinetics of alveolar lungs. This method could provide a major breakthrough by simulating extrapolation of observed directions in dimensional alteration during normal and pathological lung development and evolution.

Concentration on the lung as a static gas exchanger drastically oversimplifies the complexity of this organ. One direction of future morphometric investigation will be the analysis of the degree of compromise in lung structure among the various functions of the lung as gas exchanger, air pump, metabolic factory, endocrine organ, buoyancy organ and organ of communication (Perry 1983, 1988a; Scheuermann 1987; Weibel and Bachofen 1987).

Acknowledgement. The author gratefully acknowledges the financial support of the Alberta Heritage Foundation for Medical Research during the completion of this manuscript.

References

Abdalla MA (1977) Morphometry of the avian lung. J Anat 123:622

Abdalla MA, Maina JN (1981) Quantitative analysis of the exchange tissue of the avian lung (Galliformes). J Anat 133:677–680

Abdalla MA, Maina JN, King AS, King DZ, Henry J (1982) Morphometrics of the avian lung. I. The domestic fowl *Gallus gallus* var. *domesticus.* Respir Physiol 47:267–278

Arthus (1927) quoted in: Bargmann W (1936) Die Lungenalveole. Handb Mikrosk Anat 5:799–859

Baddeley AJ, Gundersen HJG, Cruz-Orive LM (1986) Estimation of surface area from vertical sections. J Microsc 142:259–276

Bartholomew GA (1982) Physiological control of body temperature. In: Gans C, Pough FH (eds) Biology of the reptilia, vol 12. Academic Press, New York, pp 167–212

Bartlett D Jr (1970) Postnatal growth of the mammalian lung: influence of low and high oxygen tension. Respir Physiol 9:59–64

Bartlett D Jr, Remmers JE (1971) Effects of high altitude exposure on young rats. Respir Physiol 13:116–125

Bennett AF, Dawson WR (1976) Metabolism. In: Gans C, Dawson WR (eds) Biology of the reptilia, vol 5. Academic Press, New York, pp 127–223

Brown CR, Loveridge JP (1981) The effect of temperature on oxygen consumption and evaporative water loss in *Crocodylus niloticus.* Comp Biochem Physiol 69A:51–57

Bullock GR (1984) The current status of fixation for electron microscopy: a review. J Microsc 133:1–15

Bur S, Bachofen H, Gehr P, Weibel ER (1985) Lung fixation by airway installation: effects on capillary hematocrit. Exper Lung Res 9:57–66

Burggren WW, Moalli R (1984) "Active" regulation of cutaneous gas exchange by capillary recruitment in amphibians: experimental evidence and a revised model for skin respiration. Respir Physiol 55:379–392

Burri PH, Weibel ER (1971) Morphometric estimation of pulmonary diffusion capacity. II. Effect of P_{O_2} on the growing lung. Respir Physiol 11:247–264

Burri PH, Gehr P, Muller K, Weibel ER (1976) Adaption of the growing lung to increased $\dot{V}_{O_2}$. I. IDPN as inducer of hyperactivity. Respir Physiol 28:129–140

Bütschli O (1934) Atemorgane. In: Hamburger C (ed) Vorlesungen über vergleichende Anatomie. Springer, Berlin

Byczkowska-Smyk W (1959) The respiratory surface of the gills in teleosts. 4. The respiratory surface of the gills in the pike (*Esox lucius* L.), stone perch (*Acerina ceruna* L.) and burbot (*Lota lota* L.). Acta Biol Cracov Ser Zool 2:113–127

Connolly CJ (1926) Vasodilation in *Fundulus* due to color stimulus. Biol Bull 50:207–209

Cotes JE (1968) Lung function. Assessment and application in medicine, 2nd edn. Blackwell Scientific, Oxford

Courtice GP (1981) The effect of temperature on bimodal gas exchange and the respiratory exchange ratio in the water dragon, *Physognathus lesuerii.* Comp Biochem Physiol 68A:437–441

Cragg PA (1975) Respiration and body weight in the reptilian genus *Lacerta*: a physiological, anatomical and morphometric study. PhD thesis, University of Bristol, UK

Cruz-Orive L, Weibel ER (1990) Recent stereological methods for cell biology: a brief survey. Am J Physiol (in press)

Czopek J (1955) Vascularization of the respiratory surfaces in *Leiopelma hochstetteri* Fitzinger and *Xenopus laevis* (Daudin). Acta Anat 25:346–360

Czopek J (1965) Quantitative studies on the morphology of respiratory surfaces in Amphibia. Acta Anat 62:296–323

de Jager S, Dekkers WJ (1975) Relations between gill structure and activity in fish. Neth J Zool 25:276–308
de Jager S, Smit-Onel ME, Videler JJ, van Gils BJM, Uffink EM (1977) The respiratory area of the gills of some teleost fishes in relation to their mode of life. Bijdr Dierkd 46:199–205
DeHoff RT, Rhines FN (1968) Quantitative microscopy. McGraw Hill, New York
Dejours P (1981) Principles of comparative respiratory physiology, 2nd revised edn. Elsevier/North Holland Biomedical, Amsterdam
Dubach M (1981) Quantitative analysis of the respiratory system of the house sparrow, budgerigar and violet-eared hummingbird. Respir Physiol 46:43–60
Dube SC, Munshi JSD (1974) Studies on the blood-water diffusion barrier of secondary gill lamellae of an air breathing fish, *Anabas testudineus* (Bloch). Zool Anz (Jena) 193:35–41
Duncker HR (1971) The lung-air-sac system of birds. A contribution to the functional anatomy of the respiratory apparatus. Adv Anat Embryol Cell Biol 45:1–71
Duncker HR (1972) Structure of avian lungs. Respir Physiol 14:43–63
Duncker HR (1973) Der quantitative Aufbau des Lungenluftsacksystems der Vögel. Verh anat Ges 67:197–204
Duncker HR (1978a) Funktionsmorphologie des Atemapparates und Coelom-Gliederung bei Reptilien, Vögeln und Säugern. Verh Dtsch Zool Ges 1978:99–132
Duncker HR (1978b) General morphological principles of amniotic lungs. In: Piiper J (ed) Respiratory function in birds, adult and embryonic. Springer, Berlin Heidelberg New York, pp 2–15
Engel S (1962) Lung structure. Thomas, Springfield, IL
Eno NC (1987) Oxygen diffusion capacity of cephalopool gills. Soc Exp Biol York Meeting, p 32
Farrell AP (1980) Gill morphometrics, vessel dimensions and vascular resistance in ling cod, *Ophiodon elongatus*. Can J Zool 58:807–818
Farrell AP (1981) Relations between capillary dimensions and transmural pressure in the turtle lung. Respir Physiol 45:13–24
Feder ME, Burggren WW (1985) Cutaneous gas exchange in vertebrates: design, patterns, control, and implications. Biol Rev 60:1–45
Garland T Jr, Huey RB (1987) Testing symmorphosis: does structure match functional requirements? Evolution 41:1404–1409
Gatz RN, Crawford EC, Piiper J (1975) Kinetics of inert gas equilibration in an exclusively skin-breathing salamander, *Desmognathus fuscus*. Respir Physiol 24:15–29
Gehr P, Hugonnaud C, Burri Ph, Bachofen H, Weibel ER (1978) Adaptation of the growing lung to increased V_{O_2}. III. The effect of exposure to cold environment in rats. Respir Physiol 32:345–353
Gehr P, Sehovic S, Burri P, Claasen H, Weibel ER (1980) The lung of shrews: morphometric estimation of diffusing capacity. Respir Physiol 40:33–47
Gehr P, Mwangi DK, Ammann A, Maloiy GMO, Taylor CR, Weibel ER (1981) Design of the mammalian respiratory system. V. Scaling morphometric pulmonary diffusing capacity to body mass: wild and domestic animals. Respir Physiol 44:61–86
Goniakowska-Witalinska L (1980) Ultrastructural and morphometric changes in the lung of the newt, *Triturus crisatus carnifex* Laur. during ontogeny. J Anat 130:571–583
Gray IA (1954) Comparative study of the gill area of marine fishes. Biol Bull Mar Biol Lab Woods Hole 107:219–225
Gundersen HJG, Jensen EB, Osterby R (1978) Distribution of membrane thickness determined by lineal analysis. J Microsc 113:27–43
Haefeli-Bleuer B, Weibel ER (1988) Morphometry of the human pulmonary acinus. Anat Rec 220:401–414
Hakim A, Munshi JSD, Hughes GM (1978) Morphometrics of the respiratory organs of the Indian green snake-headed fish, *Channa punctata*. J Zool, London 184:519–543
Hales S (1731) Vegetable Staticks: Analysis of the Air. pp 240
Heusner AA (1982) Energy metabolism and body size. I. Is the 0.75 mass exponent of Kleiber's equation a statistical artifact? Respir Physiol 48:1–12
Heusner AA (1983) Body size, energy metabolism and the lungs. J Appl Physiol (Resp Environ Exercise Physiol) 54:867–873

Holle JP, Heisler N, Scheid P (1978) Blood flow distribution in the duck lung and its control by respiratory gases. Am J Physiol 234:146–154
Hughes GM (1966) The dimensions of fish gills in relation to their function. J Exp Biol 45:177–195
Hughes GM (1970) Morphological measurements on the gills of fishes in relation to their respiratory function. Folia Morphol (Prague) 18:78–95
Hughes GM (1972) Morphometrics of fish gills. Respir Physiol 14:1–25
Hughes GM (1980a) Morphometry of fish gas exchange organs in relation to their respiratory function. In: Ali MA (ed) Environmental physiology of fishes. Plenum, New York, pp 33–56
Hughes GM (1980b) Ultrastructure and morphometry of the gills of *Latimeria chalumnae*, and a comparison with the gills of associated fish. Proc R Soc Lond B 208:309–328
Hughes GM (1984a) General anatomy of the gills. In: Hoar WS, Randall DJ (eds) Fish physiology, vol XA. Academic Press, New York, pp 1–72
Hughes GM (1984b) Scaling of respiratory areas in relation to oxygen consumption in vertebrates. Experientia 40:519–652
Hughes GM (1989) Morphometry of respiratory systems. In: Techniques in comparative respiratory physiology. An experimental approach. SEB Seminar Series 37. Cambridge University Press, Cambridge, UK, pp 21–48
Hughes GM, Gray IE (1972) Dimensions and ultrastructure of toadfish gills. Biol Bull 143:150–161
Hughes GM, Morgan M (1973) The structure of fish gills in relation to their respiratory function. Biol Rev 48:419–475
Hughes GM, Perry SF (1976) Morphometric study of trout gills: a light-microscopic method suitable for the evaluation of pollutant action. J Exp Biol 64:447–460
Hughes GM, Wright DE (1970) Comparative study of the ultrastructure of the water/blood pathway in the secondary lamellae of teleost and elasmobranch fishes. Z Zellforsch Mikrosk Anat 104:478–493
Hughes GM, Singh BR, Guha BR, Dube SC, Munshi JSD (1974) Respiratory surfaces areas of an air-breathing siluroid fish, *Saccobranchus (Heteropneustes) fossilis*, in relation to body size. J Zool 172:215–232
Hughes GM, Perry SF, Piiper J (1986) Morphometry of the gills of the elasmobranch *Scyliorhinus stellaris* in relation to body size. J Exp Biol 121:27–42
Hugonnaud C, Gehr P, Weibel ER, Burri PH (1977) Adaptation of the growing lung to increased oxygen. II. Morphometric analysis. Respir Physiol 29:1–10
Ichinose Y, Scotto P, Meyer M, Piiper J (1987) Pulmonary diffusing capacity for carbon monoxide by rebreathing in awake dogs. Respiration 50:279–285
Jakubowski M (1982) Size and vascularization of the gill and skin respiratory surfaces in the white amur, *Ctenopharyngodon idella* (Val.) (Pisces, Cyprinidae). Acta Biol Cracov Ser Zool 24:93–106
Jakubowski M (1989) Skin vascularization in fishes compared to that in amphibians. In: Splechtna H, Hilgers H (eds) Trends in vertebrate morphology. g Fischer-Verlag, Stuttgart
Lasiewski RC, Dawson WR (1967) A re-examination of the relation between standard metabolic rate and body weight in birds. Condor 69:12–23
Laurent P (1984) Gill internal anatomy. In: Hoar WS, Randall DJ (eds) Fish physiology, vol XA. Academic Press, New York, pp 73–183
Lechner AJ (1984) Pulmonary design in a microchiropteran bat (*Pipistrellus subflavus*) during hibernation. Respir Physiol 59:301–312
Lechner AJ, Grimes MS, Aquin L, Banchero N (1982) Adaptive lung growth during chronic cold plus hypoxia is age dependent. J Exp Zool 219:285–291
Low FN, Sampaio MM (1957) The pulmonary alveolar epithelium as an entodermal derivative. Anat Rec 127:51–63
Luchtel DL, Kardong KV (1981) Ultrastructure of the lung of the rattlesnake, *Crotalus viridis oreganus*. J Morphol 169:29–47
Maina JN (1982a) Stereological analysis of the paleopulmo and neopulmo respiratory regions of the avian lung (*Streptopelia decaocto*). ICRS Med Sci 10:328

Maina JN (1982b) Qualitative and quantitative observations on the lungs of Aves with comments on the lung of a species of Chiroptera: a morphological study. PhD thesis, University of Liverpool, UK
Maina JN (1984) Morphometrics of the avian lung. 3. The structural design of the passerine lung. Respir Physiol 55:291–307
Maina JN (1987) Morphometrics of the avian lung. 4. The structural design of the chadriiform lung. Respir Physiol 68:99–119
Maina JN (1988) The morphometry of the avian lung. In: King AS, McLelland J (eds) Form and function in birds, vol 4, chapt 7. Academic Press, London, pp 307–368
Maina JN, King AS (1982) Morphometrics of the avian lung. 2. The wild mallard (*Anas platyrhynchos*) and graylag goose (*Anser anser*). Respir Physiol 50:299–310
Maina JN, King AS (1984) Correlations between structure and function in the design of the bat lung: a morphometric study. J Exp Biol 111:43–61.b
Maina JN, King AS (1987) A morphometric study of the lung of the Humboldt penguin (*Spheniscus humboldti*). Zentralbl Veterinärmed C Anat Histol Embryol 16:293–297
Maina JN, Maloiy GMO (1985) The morphometry of the lung of the african lungfish (*Protopterus aethiopicus*): its structural-functional correlations. Proc R Soc B224:399–420
Maina JN, King AS, King DZ (1982) A morphometric analysis of the lung of a species of bat. Respir Physiol 50:1–11
Mallatt J (1984) Early vertebrate evolution: pharyngeal structure and origin of the gnathostomes. J Zool (Lond) 204:169–183
Mallatt J, Paulsen C (1986) Gill ultrastructure of the Pacific hagfish *Eptatretus stouti*. Am J Anat 177:243–269
Maloney JE, Castle BL (1969) Pressure-diameter relations of capillaries and small blood vessels in the frog lung. Respir Physiol 7:150–162
Mandelbrot BB (1977) Fractals. Form, chance and dimension. Freeman, San Francisco
Meban C (1980) Thickness of the air-blood barrier in vertebrate lungs. J Anat 131:299–307
Milani A (1894) Beiträge zur Kenntnis der Reptilienlunge. I. Lacertilia. Zool Jahrb Abt Anat Ontog Tiere 7:545–592
Moalli R, Myers RS, Ultsch GR, Jackson DC (1981) Acid-base balance and temperature in a predominantly skin-breathing salamander, *Cryptobranchus alleganiensis*. Respir Physiol 43:1–11
Monro A (1787) cited by Riess JA (1881) Der Bau der Kiemenblätter bei den Knochenfischen. Arch Naturgesch 47:518–550
Muir BS (1969) Gill dimensions as a function of fish size. J Fish Res Bd Can 26:165–170
Munshi JSD (1985) The structure, function and evolution of the accessory respiratory organs of air-breathing fishes of India. In: Duncker HR, Fleischer G (eds) Functional morphology of vertebrates. G Fischer-Verlag, Stuttgart, pp 353–366
Munshi JSD, Weibel ER, Gehr P, Hughes GM (1986) Structure of the respiratory air sac of *Heteropneustes fossilis* (Bloch) (Heteropneustidae, Pisces) – an electron microscope study. Proc Indian Natn Sci Acad B52:703–713
Niimi AJ, Morgan SL (1980) Morphometric examination of walleye, *Stizostedion vitreum vitreum* (Mitchill) and rainbow trout, *Salmo gairdneri* (Richardson). J Fish Biol 16:685–692
Noble GK (1925) The integumentary, pulmonary and cardiac modifications correlated with increased cutaneous respiration in the Amphibia: a solution to the "hairy frog" problem. J Morphol 40:341–416
Okada Y, Ishiko S, Daido S, Kim J, Ikeda S (1962) Comparative morphology of the lung with special reference to the alveolar epithelial cells. I. Lung of the amphibia. Acta Tuberc Jpn 11:62–63
Okada Y, Ishiko S, Daido S, Kim J, Ikeda S (1964) Comparative morphology of the lung with special reference to the alveolar epithelial cells. II. Lung of the reptilia. Acta Tuberc Jpn 12:1–10
Okada Y, Ishiko S, Daido S, Genka K, Kitano M (1965) Comparative morphology of the lung with special reference to the alveolar epithelial cells. III. Lung of the bird. Acta Tuberc Jpn 14:35–41
Palzenberger M (1989) Estimates of respiratory surface areas of fish gills: a critical view. In: Splechtna H, Hilgers H (eds) Trends in vertebrate morphology. G Fischer-Verlag, Stuttgart

Perry SF (1972) The lungs of the red-eared turtle, *Chrysemys (Pseudemys) scripta elegans*, as a gas exchange organ: a histological and quantitative morphological study. PhD dissertation, Boston University
Perry SF (1976) Model of gas exchange barrier and respiratory surface area in the lung of the tortoise (*Testudo graeca*) and its practical application. Mikroskopie 32:282–293
Perry SF (1978a) Quantitative anatomy of the lungs of the red-eared turtle, *Pseudemys scripta elegans*. Respir Physiol 35:245–262
Perry SF (1978b) Staining method for cut surfaces of whole reptilian lungs suitable for morphometry. Mikroskopie 34:76–78
Perry SF (1981a) Improved method for demonstration of cut surfaces of tissue in paraffin blocks. Mikroskopie 38:13–15
Perry SF (1981b) Morphometric analysis of pulmonary structure: Methods for evaluation of unicameral and multicameral lungs. Mikroskopie 38:278–293
Perry SF (1983) Reptilian lungs. Functional anatomy and evolution. Adv Anat Embryol Cell Biol 79:1–81
Perry SF (1988a) Functional morphology of the lungs of the Nile crocodile, *Crocodylus niloticus* (Laurenti). Nonrespiratory parameters. J Exp Biol 134:99–117
Perry SF (1988b) Mainstreams in the evolution of vertebrate respiratory structures. In: King AS, McLelland J (eds) Form and function in birds. Academic Press, London, pp 1–67
Perry SF (1989) Morphometry of crocodilian lungs. In: Splechtna H, Hilgers H (eds) Trends in vertebrate morphology. G Fischer-Verlag, Stuttgart
Perry SF (1990) Gas exchange strategy in the Nile crocodile: a morphometric study. J Comp Physiol B159:761–769
Perry SF, Duncker HR (1978) Lung architecture, volume and static mechanics in five species of lizards. Respir Physiol 34:61–81
Perry SF, Aumann U, Maloney JE (1989) Intrinsic lung musculature and associated ganglion cells in a teiid lizard, *Tupinambis nigropunctatus* Spix. Herpetologica 45:217–227
Piiper J, Scheid P (1975) Gas transport efficacy of gills, lungs and skin: theory and experimental data. Respir Physiol 23:209–221
Piiper J, Gatz RN, Crawford EC (1976) Gas transport characteristics of an exclusively skin-breathing salamander, *Desmognathus fuscus* (Plethodontidae). In: Hughes GM (ed) Respiration of amphibious vertebrates. Academic Press, New York, pp 339–356
Piiper J, Scheid P, Perry SF, Hughes GM (1986) Physiological and morphometric diffusing capacity of the gills of the elasmobranch *Scyliorhinus stellaris*. J Exp Biol 123:27–41
Poczopko P (1957) Further investigations on the cutaneous vasomotor reflexes in the edible frog in connection with the problem of regulation of the cutaneous respiration in frogs. Zool 8:161–175
Pohla H, Bernroider G, Lamentschwandter A, Goldschmid A (1985) Computerized measurement of gill respiratory area with corrosion casts of gill vasculature. Proc Fifth Congr Europ Ichthyol (Stockholm)
Porter KR (1972) Herpetology. Saunders, Philadelphia
Powell FL, Mazzone RW (1983) Morphometrics of rapidly frozen goose lungs. Respir Physiol 51:319–332
Rauter M (1935) Zur Kenntnis der Myxoiden-Kiemen. Morphol Jahrb 75:613–635
Sacca R, Burggren WW (1982) Oxygen uptake in air and water in the air-breathing reedfish *Calamoichthyes calabaricus*: role of skin, gills and lungs. J Exp Biol 97:179–186
Salorinne Y (1976) Single-breath pulmonary diffusing capacity. Reference values and applications in connective tissue diseases and in various lung diseases. Scand J Respir Dis 96: 1–84
Scheid P, Slama H, Piiper J (1972) Mechanisms of unidirectional flow in parabronchi of avian lungs: measurements in duck lung preparations. Respir Physiol 14:83–95
Scheuermann DW (1987) Morphology and cytochemistry of the endocrine epithelial system in the lung. Int Rev Cytol 106:35–88
Schmidt-Nielsen K (1984) Scaling: Why is animal size so important? Cambridge University Press, 241 pp
Smith DG, Campbell G (1976) The anatomy of the pulmonary vascular bed in the toad *Bufo marinus*. Cell Tissue Res 165:199–213

Smith DG, Rapson L (1977) Differences in pulmonary microvascular anatomy between *Bufo marinus* and *Xenopus laevis*. Cell Tiss Res 178:1–15
Soivio A, Tuurala H (1981) Structural and circulatory responses to hypoxia in the secondary lamellae of *Salmo gairdneri* gills at two temperatures. J Comp Physiol 145:37–43
Schulze FE (1871) Die Lungen. In: Stricker S (ed) Handbuch der Lehre von den Geweben des Menschen und der Thiere. W Engelmann-Verlag, Leipzig, pp 464–486
Starck D (1982) Vergleichende Anatomie der Wirbeltiere, vol 3. Springer, Berlin Heidelberg New York
Stinner JN (1982) Functional anatomy of the lung of the snake *Pituophis melanoleucus*. Am J Physiol 243:R251–R257
Taylor CR, Weibel ER (1981) Design of the mammalian respiratory system. I. Problem and strategy. Respir Physiol 44:1–10
Taylor CR, Maloiy GMO, Weibel ER, Langman VA, Kamau JMZ, Seeherman HJ, Heglund NC (1980) Design of the mammalian respiratory system. III. Scaling maximum aerobic capacity to body mass: wild and domestic mammals. Respir Physiol 44:25–37
Taylor CR, Karas RH, Weibel ER, Hoppeler H (eds) (1987) Adaptive variation in the mammalian respiratory system in relation to energetic demand. Respir Physiol 69:1–127
Tenney SM, Remmers JE (1963) Comparative quantitative morphology of mammalian lungs; diffusing areas. Nature 197:54–56
Tenney SM, Tenney JB (1970) Quantitative morphology of cold-blooded lungs: Amphibia and Reptilia. Respir Physiol 9:197–215
Thomas SP (1975) Metabolism during flight in two species of bats, *Phyllostomus hastatus* and *Pteropus gouldii*. J Exp Biol 63:273–293
Thomas SP, Suthers RA (1972) The physiology and energetics of bat flight. J Exp Biol 57:317–335
Thompson ME (1980) Lung growth in response to altered metabolic demand in hamsters: influence of thyroid function and cold exposure. Respir Physiol 40:335–347
Tucker VA (1972) Respiration during flight in birds. Respir Physiol 14:75–82
Underwood EE (1970) Quantitative stereology. Addison-Wesley, Reading, Mass
Videler JJ (1968) Onderzoek naar het kieuwoppervlak en de huiddoorbloeding van enkele Pleuronectidae. Thesis. Zoologisch laboratorium, Universiteit van Amsterdam
von Hayek H (1953) Die menschliche Lunge. Springer, Berlin Göttingen Heidelberg
Weibel ER (1963) Morphometry of the human lung. Academic Press, New York
Weibel ER (1969) Stereological principles for morphometry in electron microscopic cytology. Int Rev Cytol 26:235–302
Weibel ER (1970/71) Morphometric estimation of pulmonary diffusing capacity. I. Model and method. Respir Physiol 11:54–75
Weibel ER (1979a) Morphometry of the human lung. The state-of-the-art after 2 decades. Bull Eur Pathophysiol Respir 15:999–1014
Weibel ER (1979b) Stereological methods, vol 1. Practical methods for biological morphometry. Academic Press, London
Weibel ER (1980) Stereological methods, vol 2. Theoretical foundations. Academic Press, London
Weibel ER (1984) The pathway for oxygen. Harvard University Press, Cambridge MA
Weibel ER (1987) Scaling of structural and functional variables in the respiratory system. Annu Rev Physiol 49:147–159
Weibel ER, Bachofen H (1987) How to stabilize the pulmonary alveoli: surfactant or fibers? News Physiol Sci 2:72–75
Weibel ER, Elias H (1967) Quantitative methods in morphology. Springer, Berlin Heidelberg New York
Weibel ER, Knight BW (1964) A morphometric study of the thickness of the pulmonary air-blood barrier. J Cell Biol 21:367–384
Weibel ER, Untersee P, Gil J, Zulauf M (1973) Morphometric estimation of pulmonary diffusing capacity. VI. Effect of varying positive pressure inflation of air spaces. Respir Physiol 18:285–308
Weibel ER, Gehr P, Cruz-Orive LM, Muller AE, Mwangi DK, Haussener V (1981) Design of the mammalian respiratory system. IV. Morphometric estimation of pulmonary diffusing capacity: critical evaluation of a new sampling method. Respir Physiol 44:39–59

Weibel ER, Taylor CR, O'Neil JJ, Leith DE, Gehr P, Hoppeler H, Langman V, Baudinette RV (1983) Maximal oxygen consumption and pulmonnary diffusing capacity: a direct comparison of physiologic and morphometric measurements in canids. Respir Physiol 54:173–188

Weibel ER, Marques LB, Constantinpol M, Doffey F, Gehr P, Taylor CR (1987) Adaptive variation in the mammalian respiratory system in relation to energetic demand: VI. The pulmonary gas exchanger. Respir Physiol 69:81–100

Welsch U (1979) Die Stellung der Reptilienlunge in der Phylogenese nach licht- und elektronenmikroskopischen Untersuchungen am Alveolarepithel, am Bindegewebe und an der Innervation. Dr med Dissertation, Universität Kiel

Wilson TA, Bachofen H (1982) A model for mechanical structure of the alveolar duct. J Appl Physiol 52:1064–1070

Wolf S (1933) Zur Kenntnis von Bau und Funktion der Reptilienlunge. Zool Jahrb Abt Anat Ontog Tiere 57:139–190

Chapter 4

Morphometric Analysis of Tissue Capillary Supply

S. Egginton[1]

Contents

[1] Department of Physiology, University of Birmingham Medical School, Birmingham B15 2TJ, England

Advances in Comparative and Environmental Physiology
Vol. 6. Edited by R. G. Boutilier

1 Introduction

1.1 Brief Overview

The physiological importance of tissue capillary supply is generally accepted in broad terms, such as providing adequate oxygen delivery and waste removal in working muscle, delivery of substrates and removal of products in endocrine tissue, and facilitating pathologically rapid growth in tumours. However, when one looks in greater detail at the regulation and interrelationship of tissue phenotype and the microvasculature, it quickly becomes evident that our understanding is still far from complete. There is, therefore, a need for quantitative (morphometric) analyses which adequately describe the anatomical capillary bed, from which estimates of the functional capillary supply may be derived. Although new insights arise constantly, this is in fact a very old area of study. In the 17th century, Malpighi and van Leuwenhoek studied the microcirculation by direct observation, while a qualitative picture of microvascular anatomy has been available for over a century (Ranvier 1874; Spälteholz 1888; Krogh 1919). Recent reviews cover much of the important literature regarding architecture (Wiedeman 1984), adaptation (Hudlická 1984) and angiogenesis (Hudlická and Tyler 1986) in various tissues, and offer an overview of the varied techniques employed. This chapter considers the utility of various quantitative estimators of capillary supply, from which an integrated analysis at differing levels of resolution may be performed. Analytical methods for particular networks, even of larger vessels, may also be applicable in varying degrees for other tissue. Although much of the relevant data comes from skeletal muscle, it is hoped the principles outlined will allow the reader to modify this approach for use with other organs and vascular systems, according to individual requirements and experimental limitations. However, it is important to recognize at the outset that morphometric analyses can only estimate the maximal capacity of a system, and therefore functional characteristics are difficult to assess without additional physiological data such as blood viscosity, local O_2 tension, etc. As a consequence global morphometric indices may reflect higher orders of control, but are open to errors of simplification; on the other hand greater analytical resolution may reveal local levels of control, but ignore cooperativity between different levels of organisation. While analyses will eventually have to be reconciled into a model describing both macro- and microcirculatory phenomena, a possibility that seems at least moderately remote, it is hoped that highlighting a few of the more promising approaches may help clarify those areas most likely to benefit from analytical refinement.

Although many different indices of capillary supply are in use, it is clear that none provide an adequate description of either physical dimensions or functional capacity of the capillary bed. August Krogh (1919) was the first to consider this relationship in a quantitative manner, suggesting there was a critical density of capillaries below which some portion of the tissue must become hypoxic. Although still widely adhered to, experiments in recent years have shown this concept to be an over-simplification in that (1) capillaries perform functions other than oxygen delivery, which in some tissues may be of prime importance, (2) in tissue where capillaries do act primarily as O_2 conduits, there is a hierarchy of physiological adaptations which may obviate or accentuate any change in the capillary bed to match changes in $\dot{V}O_2$. A reappraisal of morphometric indices of capillary supply is required in order to distinguish between these options. Recent advances are discussed that may also permit development of a quantitative theory of blood flow and oxygen transport to tissue by integrating morphometric data and mathematical models of diffusion.

1.2 Basic Principles

Morphometric analyses of the capillary bed often rely on the convenient representation of individual vessels as points – relative density of points in a section; linear networks, points joined by lines; point process statistics, distribution of points; planar analyses, neighbourhood of points. In most recent studies the importance of spatial heterogeneity is recognized, suggesting that an understanding of basic regulatory mechanisms requires more complex indices than simple counts. These will be most useful if they incorporate at least four principles:

1. Generally applicable; a large body of literature is composed of indices only applicable to the original study.

2. Continuously variable; just as the graphical representation of integer-based indices conveys more information when the bin size is reduced, so the resolution will be greatest with a non-integer based index.

3. Local indices; analysis may then establish the importance of regional variations, while still permitting analysis of higher-order relationships.

4. Applicable in 2 or more dimensions; morphometric techniques, stereologic theory and microscopic instrumentation are all becoming more sophisticated. Analyses in the next decade or so will not be so concerned with extrapolating from n to n+1 dimensions, but measuring directly in 2- or 3-D.

1.3 Chapter Outline

In common with most branches of biology the literature is full of different methodological approaches to quantification of the microvascular supply. As these are mainly variations on a handful of basic ideas, I have attempted to highlight more recent applications at the time of writing (1987), but to develop the framework with reference to important earlier work.

Two main approaches have been used in modelling the vascular supply to skeletal muscle. That of August Krogh is still widely used in theoretical studies on exchange processes, as subsequent research has produced no generally accepted alternative. Not surprisingly we find most 'traditional' morphometric indices of capillarization are implicitly based on this model; these will be discussed along with their stereological derivatives in Section 3. An alternative approach is to analyze the geometry of vascular networks by following the successive division of larger to smaller vessels in terms of tree structure. Lineal analyses using graph theory or network principles will be dealt with in Section 4. Finally, Section 5 examines how we may extrapolate from 0- and 1-D analyses to produce a 2-D model of the capillary supply, allowing for interaction of neighbouring capillaries and heterogeneous demand of different cell types. This use of geometric constructions in planar analyses is an exciting development that has the potential of directly incorporating oxygen transport calculations.

2 Practical Considerations

The essential requirements for a successful morphometric study are outlined below. It is important to consider these points while planning experiments: the 6-fold range in reported estimates of capillary density in cat gastrocnemius (Plyley and Groom 1975) is easily explained by problems under one or more of these headings.

2.1 Adequate Preparation

While there is a large body of specialist literature dealing with all aspects of tissue preparation, including the relative degree of shrinkage caused by different fixatives and embedding compounds, there is one aspect that fundamentally affects our view of the microcirculation. Immersion-fixed material is usually assumed to result in smaller capillary diameters, and hence thicker walls than in perfused-fixed material (Hoppeler 1984), often revealing apparently 'collapsed' vessels. However, the sometimes strong opposition to use of immersion-fixed tissue ignores the prevalence of seemingly normal profiles of 'patent' capillaries (Fig. 1). Indeed, some workers find no difference in morphometric measurements made in perfusion- or immersion-fixed material (Porter and Bankston 1987). In addition, the possibility of artificially flattened walls in perfused-fixed vessels may obscure pathological changes, such as endothelial cell oedema (Brown and Egginton 1988).

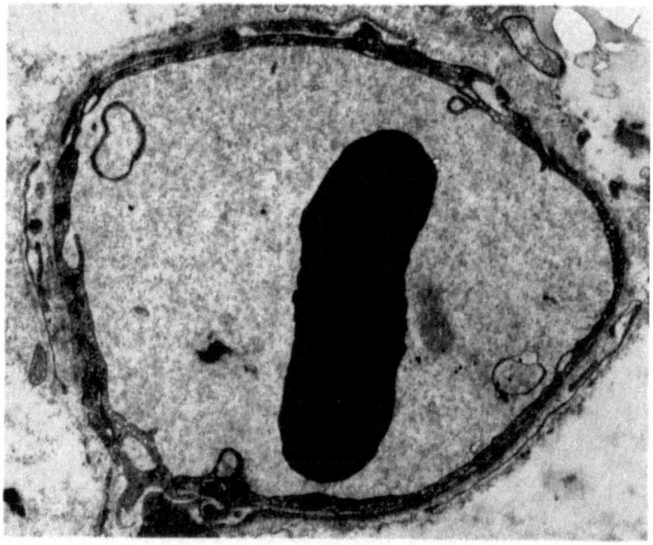

Fig. 1. Adequate preparation. Morphometric analyses of changes in the capillary bed often utilize perfusion-fixed material, on the assumption that immersion fixation produces morphology that is unrepresentative of that in vivo, such as collapsed vessels and unrealistically thick walls. It is possible, however, that the reverse is true and that maximal dilatation of vessels during perfusion with fixative produces artificially narrow walls. Indeed, the majority of capillaries in our studies of mammalian and fish muscle have produced 'open' profiles. The illustration shows the 'normal' appearance of a vessel, apparently patent at the time of immersion into fixative. Large (post-) capillary from slow muscle of striped bass *Morone saxatilis* fixed at resting length by pinning to a strengthened piece of cork in 1% glutaraldehyde, 4% (para) formaldehyde, 0.1% NaN_3, 1% sucrose and 125 mM phosphate buffer pH 7.2 at 20 °C

2.2 Unambiguous Identification

The accuracy of any analysis clearly depends on how reliably and completely the capillary bed may be visualised. A common approach is to fill the vessels in vivo with various compounds, ranging from the earliest attempts with India ink to the latest fluorochromes. Although useful for quantification of the functional (patent/perfused) capillary bed, the inherent uncertainty requires extensive controls before they may be used to assess the total (anatomical) capillary network. In view of both the large sample area available, and the relatively small dimensional changes, in situ localization of capillaries by histochemical staining of frozen sections is a favoured approach. Many reactions are employed, including ATPase, carbonic anhydrase and cholinesterase activities, with the most common being that of alkaline phosphatase (ALP). Although ALP has been shown to provide similar estimates of C:F (numerical capillary to fibre ratio) as morphological identification (Brown and Egginton 1988), recent reports have unfortunately cast doubt on its validity. This appears to be simply a result of a differential sensitivity among staining methods (Fig. 2).

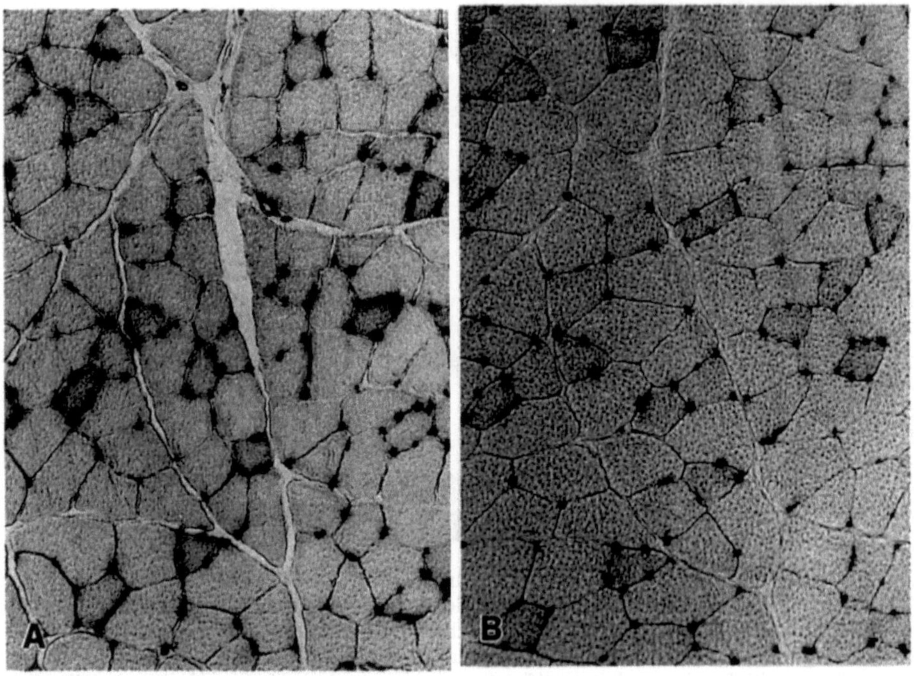

Fig. 2. Unambiguous identification. Capillary counts are often performed using frozen sections, where ICDs approximate those in vivo, by staining capillary endothelium for the enzyme alkaline phosphatase. As with all histological methods, however, the specificity of the reaction will determine the degree to which enzyme activity may be differentiated. *A* was incubated using the indolyl-tetrazolium method (5-bromo, 4-chloro, 3-indolyl phosphate with nitroblue tetrazolium); *B* using the simultaneous azo-coupling method (naphthol AS-MX phosphate with Variamine Blue B), both at pH9.2, using serial sections from rat m. extensor digitorum longus. The number of stained capillaries associated with the central fasicle are 31 and 23 respectively

2.3 Unbiased Sampling

Quantitative indices are only valid if the raw data is not influenced by experimental or analytical procedures. Most statistic textbooks include sections on experimental design, such as determining the level at which greatest variance occurs (usually the level requiring greatest replication). However, while statistical treatment of data is often described in detail, two fundamental principles are often overlooked during initial data collection. A common problem is that of having an inadequate sample size. This oversight may cause alarming errors but is simply avoided by calculation of cumulative means and SEs; a sample is adequate when these parameters fluctuate by $< 10\%$ of final values (Egginton and Johnston 1982).

An equally common problem is ignoring the effects of morphology, such as shape and size of cells. Simply counting all objects within a field of view will bias the data in favour of smaller items. It is therefore usual to collect morphometric data from subsamples of a population within finite boundaries or 'field'. The ef-

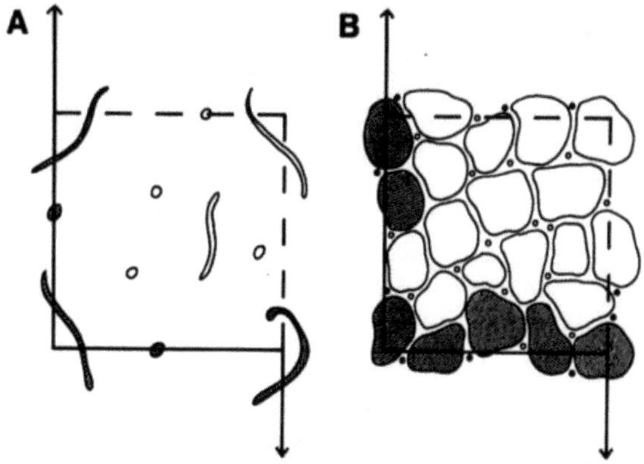

Fig. 3. Unbiased sampling. The influence of capillary size and morphology may be eliminated by adopting, at high magnification (*left*), a counting frame with inclusion (*dotted*) and exclusion edges (*solid lines*). Vessels are only counted if they lie entirely within the frame or if they exclusively intersect one or both of the previously designated adjacent inclusion edges of the frame. Note this forms an extended 'forbidden' zone such that the *lower right* capillary is excluded along with other *shaded* vessels when determining an unbiased count of capillary number. Numerous modifications of this basic counting frame are possible (Gundersen 1977). At lower magnification (*right*), the counting frame can be used to obtain an unbiased sample of both capillary and fibre number; those *stippled* fibres and *filled* capillaries are excluded from the analysis, resulting in an unbiased estimate of C : F = 16/15 = 1.07. In addition to such global estimates of capillary supply, this approach may be used to select those fibres to be included in an analysis of local capillary supply (Sect. 5)

ficacy of sampling using quadrats has been well established in ecological and biostatistical literature; in general, sampling by quadrats placed at random over a tissue will be less efficient (produce greater data variance) than a systematic random procedure, where the first quadrat in a series is placed at random and all subsequent quadrats have a regular, predetermined spatial relationship. However, data bias can only be eliminated when an object can only be counted once with a series of contiguous counting frames (Fig. 3).

2.4 Appropriate Analysis

The type of analysis to be applied to morphometric data clearly depends on the ultimate goal – simple counts are adequate for most studies where gross changes in the global capillary supply are expected (Fig. 4a). Given that most capillary beds appear to have a heterogeneous spacing, various linear analyses may be applied to determine e.g. degree of clustering or variability in intercapillary distances (Fig. 4b). As theoretical studies become more realistic it is increasingly evident that oxygen transport to tissue, particularly the diffusion-limited component, requires a true planar analysis in order to account for interactions between neighbouring capillaries (Fig. 4c).

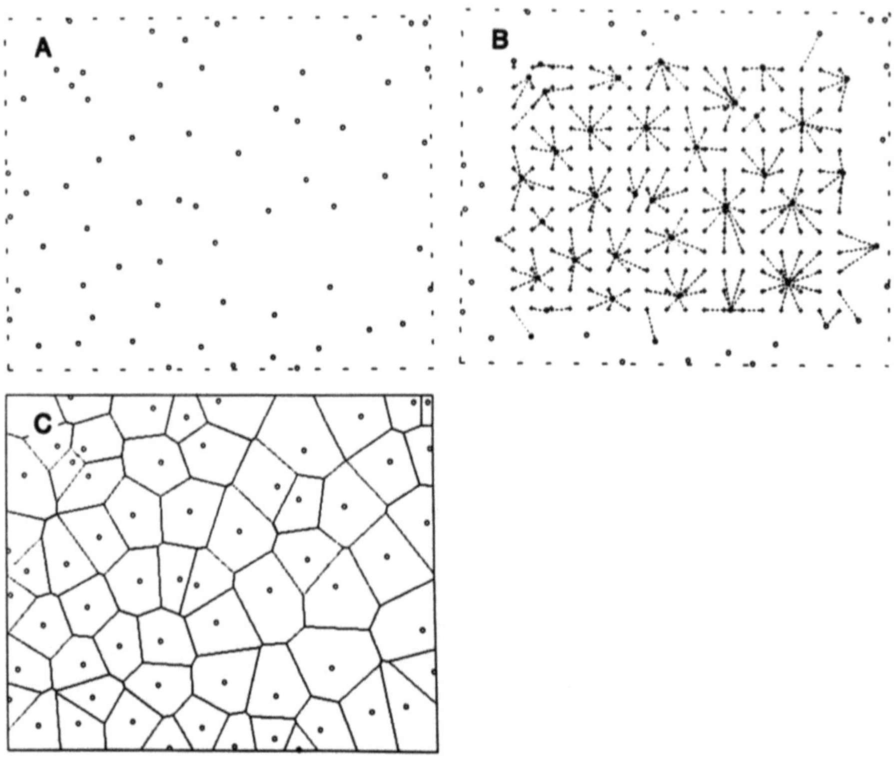

Fig. 4. Appropriate analysis. Given that capillary distribution in tissue cross-sections is heterogeneous, there are three main approaches to quantification: *A* Ignore local heterogeneity and quantify a global index such as CD. For gross changes, e.g. during exercise, this is quite adequate and still permits analysis of regional (field to field) variation (Sect. 3.6). *B* Determine local heterogeneity by analysis of linear separations. In this example, a square lattice sampling frame is superimposed on the field for nearest-neighbour analysis (Sect. 5.2). *C* Integrate all local, linear separations and determine the area of tissue associated with each capillary to provide a true planar analysis of spatial heterogeneity, and estimate the interaction with intervening cell types (Sect. 5.3)

3 0- and 1-D Indices

3.1 Introduction

This section will deal with those indices which probably first come to mind when one reads the chapter title. Much has already been written on this subject (e.g. Plyley and Groom 1975; Gray and Renkin 1978; Maxwell et al. 1980; Mathieu et al. 1983; Egginton and Johnston 1982, 1983; Hudlická 1984, 1985; Hoppeler 1984), so I will restrict coverage to a general overview of relevance to subsequent sections dealing with more recent developments in analytical approach. These indices describe either global or local capillary supply. The former category may reflect total aerobic capacity of a tissue and relative blood flow, but are relatively insensitive to local adjustments in demand. The latter category may have a higher resolution, and are therefore more sensitive to regional variations, but often have limited utility as predictors of in vivo O_2 supply or demand as each capillary or cell is counted more than once, and fail to accommodate the inherent heterogeneity of O_2 demand within tissue or the relationship between adjacent cells or vessels.

3.2 Sharing Factor [SF, $\bar{N}(f,c)$]

A widely used supply index for mammalian capillary networks is the sharing factor, or number of muscle fibres (f) surrounding a capillary:

$$\bar{N}(f,c) = \sum_{i=1}^{i=n} \frac{N(f)_i}{N_c}$$

SF has been confused with the average number of muscle fibres per capillary, the numerical ratio $N_N(f,c)$, which is merely the reciprocal of $N_N(c,f)$. As this is only true for the special case of a homogeneous capillary and fibre distribution pattern, the error caused by this erroneous assumption will increase as capillary density declines and the correct value has to be empirically determined.

In muscles with small fibres most capillaries are shared by three fibres, and relatively few by only two. More capillaries appear around fibres as they grow, such that the proportion of capillaries where SF = 2 increases, while those with SF = 3 declines. SF is therefore a relatively insensitive index, with a reported range of only 1.99 (rat soleus) to 3.22 (dog tongue) in mammalian muscle (Plyley and Groom 1975), although the former value must be an error, either typographical or methodological, as it implies some capillaries supply only one fibre. SF is clearly of limited usefulness as a descriptor of aerobic scope, as it varies little between either subsamples or muscle types. The use of such an index for fish muscle (Kryvi et al. 1980), where red and white fibres are anatomically segregated and many fibres lack direct capillary contact, is particularly surprising, since it takes no account of fibre packing; as with other vertebrates, fish muscle usually has an average close to 3 (Egginton 1982).

3.3 Capillary to Fibre Ratio [C:F, $N_N(c,f)$]

A commonly reported average, or global, estimate of oxygen supply is provided by the numerical ratio of capillary and muscle fibre profiles:

$$N_N(c,f) = N(c)/N(f)$$

C:F has the useful property of being relatively independent of fibre area when comparing different muscles, and normally shows little variation among mammalian muscles ($\bar{x} = 1.4$; Plyley and Groom 1975). It is, however, sensitive to muscle hypertrophy and other adaptive changes; e.g. repetitive stimulation of fast muscle motor nerves at slow muscle frequency results in a 1.5-fold increase (to 2.18) in rabbit EDL (extensor digitorum longus; Brown et al. 1976). In vertebrate locomotory muscles the C:F increases with proportion of oxidative fibres, is higher in homologous muscles from wild than from domestic animals, and appears to decrease in old age; in humans C:F is apparently lower in women (Hudlická 1984). The extreme 100-fold range of reported values for fish muscle (Mosse 1978; Egginton 1982) gives this index a large descriptive potential, although its sensitivity to change in muscle aerobic capacity remains largely untested in lower vertebrates. C:F is perhaps most useful where proliferation of the capillary bed occurs with little or no hyperplasia or hypertrophy of the host tissue, e.g. fast muscle of hyperthyroid rats (Capo and Sillau 1983), in which case increase in C:F truly reflects neovascularization (angiogenesis).

3.4 Mean Number of Capillaries Around a Fibre [CAF; $\bar{N}(c,f)$]

The mean number of capillaries in contact with a fibre is calculated as:

$$\bar{N}(c,f) = \sum_{i=1}^{i=n} \frac{N(c)_i}{N_f}$$

This is thought by many workers to reflect the oxygen demand of muscle fibres. Although not synonymous with $N_N(c,f)$, both indices have been labelled "capillaries per fibre" in the literature, causing some confusion. Again, the range of values found in mammalian muscles is small, 3 to 3.8 across a 23-fold range of fibre size, but shows a good correlation with $N_N(c,f)$ ($r = 0.746$, $n = 16$; Plyley and Groom 1975), suggesting that capillary supply to whole muscles and mean supply to individual fibres are matched. The lack of correlation between $\bar{N}(c,f)$ and fibre area of different muscles may be explained in terms of composition, since the magnitude is likely to depend on both fibre geometry and size distribution. Within any one muscle, however, both C:F and CAF increase linearly with fibre area (Loats et al. 1978). Interestingly, an increased oxygen uptake efficiency associated with the greater surface area: volume ratio of flattened slow muscle fibres in Anchovy (Johnston 1982) is also accompanied by an elevated $\bar{N}(c,f)$, around 12, indicating that capillary supply may indeed be a limiting factor in cellular respiration. Although distribution analysis of supply to individual fibres is possible, this index lacks sufficient sensitivity because it is unable to account for differential supply to adjacent fibres of varying metabolic demand.

Established indices of capillary supply such as capillary density or overall capillary to fibre ratio (C:F) integrate tissue-wide variations, and are most useful with either homogeneous distribution patterns and/or high capillary density. In mammalian muscle this is a serious problem, as the influence of different fibre types on the capillary bed is presumably unequal. It is likely to be less of a problem where tissue is composed of a single fibre type, e.g. cardiac and fish skeletal muscle, but even here use of inappropriate indices may seriously affect estimates of capillary supply (Egginton and Johnston 1982). In mixed muscles a capillary may be surrounded by up to four metabolically distinct fibre types. Such a capillary would be counted, with equal weighting, four times in calculating $\bar{N}(c, f)$ for these fibres without scaling O_2 supply to metabolic demand. Thus, glycolytic fibres may have an apparent excess of capillaries, relative to $\dot{V}O_2$, if they are surrounded by oxidative fibres. An attempt has been made to partition the capillary supply according to fibre type whereby both fibre area and numerical capillary contact are converted to a mean CD for each fibre type. Simply counting the number of capillaries around distinct fibre types fails to reveal more subtle distribution of vessels in either mixed or regionally differentiated muscle; e.g. often capillaries around white fibres preferentially appear close to red or intermediate fibres, therefore each capillary needs to be weighted according to the metabolic character of its neighbours (Gray and Renkin 1978; Flood 1979). Given estimates of relative oxidative capacity, from microhistochemical analyses or biochemical assays from muscles of homogeneous composition, it should then be possible to scale O_2 flux from individual capillaries according to maximal O_2 demand of the surrounding fibres. Selective analysis of capillaries with a homogeneous metabolic neighbourhood (i.e. surrounded by one fibre type) may reveal operation of local bias in capillary distribution. However, erroneous results are likely unless similar numbers of identical neighbours (pure fibre contacts) are available for each fibre type, presumably a rare event, e.g. capillaries are apparently more abundant around white than red fibres in frog muscle (Asmussen and Kiessling 1974). Unfortunately, the increased resolution offered by considering neighbourhood conditions is achieved at the expense of information regarding cell size; both determinants of tissue oxygenation can only be preserved in a true 2-D analysis (Sect. 5).

3.5 Capillary-Fibre Contact Length [$\bar{l}(c,f)$]

Little improvement in descriptive power is afforded by measuring the mean proportion of fibre surface in contact with capillaries:

$$\bar{l}(c,f) = \sum_{i=1}^{i=n} \frac{[b'(c)/b(f_i)]}{n}$$

where b′(c) is the total proportion of capillary boundary length (perimeter) subtending the fibre perimeter, which approximates to b(c)/N(f, c). This index clearly varies in a quantitative manner with N(c, f), as capillary dimensions are not grossly different between tissues in the same animal, but is highly dependent on the degree of tissue shrinkage and experimentor bias. Not surprisingly, there is a correlation between the average contact length and mean mitochondrial volume density in skeletal muscles of fish (Egginton 1982) and mammals (Sullivan and Pitman 1987).

Alternatively, the mean proportion of fibre circumference served by one capillary may be calculated:

$$b(f,c) = \sum_{i=1}^{i=n} \left[\frac{b(f_i)}{N(c,f_i)}\right] / n$$

where $b(f_i)$ = fibre boundary length; $N(c, f_i)$ = number of contiguous capillaries. This index has the advantage of scaling local capillary supply with fibre area, and increasing the apparent differential between oxidative and glycolytic fibres over that given by profile counts (Flood 1979). This approach has an obvious attraction for those seeking to model the capillary supply purely in terms of capacity for O_2 diffusion, although the utility may be deceptive: apart from the limitations shared by many other indices covered in this section, quantitative estimates of maximal O_2 flux require assumptions which are unlikely to be valid, such as equivalent $\dot{V}O_{2max}$ and surface : volume ratios for all fibres.

3.6 Capillary Density [CD; $N_A(c, f)$]

Capillary density is often expressed as capillaries per unit of tissue section (mm^{-2}). This is clearly of limited value, as variations in the volume of interstitial space, or differential shrinkage during processing, may introduce significant errors. When quantifying capillary supply, then, it is best to use the object of that supply as the reference. For example, in muscle the number of capillaries per unit area of fibre is derived from the equation:

$$N_A(c, f) = N(c)/n \cdot \bar{a}(f) \ ,$$

where $\bar{a}(f)$ is mean fibre area.

Using a square lattice counting frame and standard stereological notation this can be simply quantified as:

$$\hat{Q}_A(c, f) = \sum_{j=1}^{j=n} \left[Q(c)_j / \left(d^2 \cdot \sum_{i=1}^{i=n} P_i \right) \right]$$

where Q(c) is the number of capillary profiles per quadrat, sampled using an unbiased counting frame; P_i is the number of points falling on muscle fibres (being $< P_t$, the total number of points on the grid); d is lattice spacing.

Capillary density is considered to be a good index of tissue oxygen consumption in a wide range of animals, reflecting the functional capacity of capillary supply. Unfortunately, there is considerable variation in estimates among workers, indicating a need to standardize the methodological approach. For example, reported values for cat gastrocnemius range from 379 to 2341 capillaries mm^{-2} (Plyley and Groom 1975). Nevertheless, this is the usual index of capillary supply used to support the contention that plasticity in the microvasculature matches the oxidative metabolic demand of muscle. Higher CDs are found in skeletal muscles of wild vs domestic animals (Wachtlová and Parizková 1972), and following exposure to cold (Heroux and St. Pierre 1957), endurance training (Anderson 1975), chronic electrical stimulation (Cotter et al. 1973), high altitude (Valdivia 1958) and hyperthyroidism (Capó and Sillau 1983). During normal hypertrophic growth capillary proliferation is insufficient to match fibre hypertrophy, therefore CD falls while C : F continues to rise (Aquin and Banchero 1981; Hudlická 1985). However, a similar allometric relationship between species is not so clear (Schmidt-Nielsen and Pennycuik 1961). Other data suggest that CD alone is not a sufficient parameter to describe the functional capacity of the capillary bed. For example, in response to endurance exercise some mammalian muscles show an increase in $V_v(mit, f)$ with no adjustment in the capillary supply (Müller 1976), while mammalian muscles having a similar $V_v(mit, f)$ may have a four fold difference in $N_A(c, f)$ (Hoppeler et al. 1981). Such variability causes the correlation between capillary supply and muscle oxidative capacity to be questioned (Schmidt-Nielsen and Pennycuik 1961; Loats et al. 1978; Maxwell et al. 1980; Eginton and Johnston 1983), as "white" muscles may have capillary densities similar to some "red" muscles. This is presumably due to capillaries having other important roles. During anaerobic metabolism, for example, a fast muscle may require an efficient blood supply for the delivery of substrates, or the removal of metabolic products such as lactate and heat.

3.7 Intercapillary Distance (ICD)

This index, particularly favoured by theoreticians quantifying oxygen supply, is intimately linked with CD by mean capillary spacing, as the reciprocal of CD estimates the area of tissue served by one capillary in a cross-section of tissue, although capillary tortuosity is not taken into account (Sect. 3.9). From in vivo observations of mammalian muscles at rest, patent capillaries may be upto 1–2 mm apart, while functional hyperaemia during exercise may open up the entire capillary bed, giving an effective ICD of around 30 µm (Honig et al. 1977). The closer the capillaries are arrayed the more efficient will be O_2 delivery due to reduced intramuscular diffusion distances, which is assumed to improve muscle performance. Estimation of ICD from such longitudinal sections of the capillary bed (Henquell and Honig 1976) may be related to R_K, radius of Krogh's tissue cylinder describing oxygen diffusion, only if the geometrical array of vessels is known and a uniform distribution of capillaries assumed. The regular geometric arrangements most commonly adopted are equilateral triangles, square-array and hexagons with capillaries located at the apices (Fig. 15). Renkin et al. (1981) proposed an alternative array independent of the underlying geometry, using the average circle perimeter supplied by one capillary. Estimations of the shortest distance between adjacent capillaries, R (equivalent to half the mean intercapillary distance) commonly used are:

1. Triangles $R = \dfrac{1000 \cdot \sqrt[4]{3}}{\sqrt{6} \cdot CD}$

2. Squares $R = \dfrac{1000}{2\sqrt{CD}}$

3. Hexagons $R = \dfrac{1000}{3\sqrt{3} \cdot CD}$

4. Circles $R = \dfrac{1000}{\sqrt{\pi} \cdot CD}$

with ICD in µm and CD in mm^{-2}.

Alternatively, the distance to the centre of tissue field surrounded by capillaries may be calculated using the general formula:

$$ICD = \frac{K/CD^{1/2}}{2}$$

(Turek et al. 1986), where K is a constant of proportionality. For the four models outlined this gives corresponding values for K of 538, 500, 439 and 564 for capillary separation; and 620, 707, 877 and 798 for diffusion distances. Note the circle array is simply a realization of Krogh's estimate of mean capillary diffusion radius:

$$R_K = (\pi \cdot CD)^{-1/2}$$

The hexagonal array is particularly popular with modellers of O_2 transport (e.g. Akmal et al. 1978), as this geometric arrangement of capillaries encloses the largest area of tissue, and maximizes ICD, while the triangular and square-ordered array have been assumed to equate with CAF = 3 and 4 respectively (Plyley and Groom 1975).

One might expect ICD in muscles to be proportional to mean fibre area, as CD is reduced and individual capillaries are pushed apart during growth. The longer diffusion distance to cell centres are apparently compensated for by an increased number of adjacent capillaries (CAF), thereby maintaining the area of tissue served by one capillary within characteristic limits (Sillau and Banchero 1977). ICD is known to vary with metabolic activity of the tissue and the ambient O_2 tension in which the animal lives, e.g. ICD is usually less in oxidative muscles than in those with low aerobic capacity. In oxidative muscles from 3 mammals it was shown that ICD_{max} is independent of fibre size, again suggesting that compensation in C:F and CAF is effective in maintaining diffusion distances with reasonable limits for O_2 transport (Loats et al. 1978). However, estimating the radius of an average tissue cylinder from CD provides no information about the heterogeneity of supply or maximum diffusion pathlengths, both of which are important for assessing the capacity for tissue oxygenation. Indeed the maximum diffusion distance (ICD_{max}) may be twice the estimated value of R_K (Loats et al. 1978). These limitations must be considered when quantifying capillary-tissue diffusion distances, and probably explain the discrepancy between ICD and muscle performance (Hudlická et al. 1988).

3.7.1 Triangulation Method

Individual and mean ICD may be measured by drawing lines from each capillary to all surrounding capillaries, but limits need to be imposed in order to include only functionally relevant distances. Renkin et al. (1981) restricted analysis to adjacent capillaries forming a network of non-overlapping triangles. This method produces larger estimates of both mean R_K and heterogeneity of capillary spacing than either concentric circles or closest individual methods (Turek et al. 1987). In fact, this procedure is very similar to a well-defined construction, the Delaunay tessellation (D_t; see Sect. 5.3.4), and estimates the separation of adjacent, rather than nearest capillaries, which probably accounts for those observed discrepancies in heterogeneity. Bias in this 2-D subsample of tissue may be minimized by completing the ICD network within a quadrat by extending lines across two borders to capillaries around the quadrat (Figs. 20, 21). In this case individual ICDs show a log-normal distribution, and the half mean ICD closely mirrors calculated R_K.

3.7.2 Concentric Circles Method

The concentric circles method, first described by Loats et al. (1978), is a simple and rapid method for estimating a comprehensive distribution of diffusion distances in tissue. It is preferable to the use of CD estimates, especially where capillaries are not parallel, as measuring distances to the nearest capillary is insensitive to orientation. A radial stereological counting frame is used for analysis of capillarization; sets of concentric circles (Loats et al. used a relatively coarse frequency distribution of five annuli) are centred on points in a square array, and for each point the largest circle containing no capillaries is recorded. This provides a systematic random estimate of the proportion of tissue beyond or within a given distance from the nearest capillary. Although it does not directly estimate ICD, R_K can be derived by an iterative fitting procedure and has been used to quantify the heterogeneity of capillary supply during heart hypertrophy, and effect on tissue PO_2 (Turek and Rakusan 1981). Kayar et al. (1982a) showed it was also useful in distinguishing between ordered, random or clustered arrays, with an error of $< 10\%$ compared to Monte Carlo simulations.

3.7.3 Closest-Individual (CI) Method

Kayar et al. (1982b) adapted an ecological survey method (Clark and Evans 1954) to measure the complete distribution of distances from tissue to the nearest capillaries. The closest individual method also estimates the distances of points in the tissue to the nearest capillary, but in using a square lattice overlay a continuous distribution of tissue (grid points) at varying distances from the nearest capillary is generated. Again, mean R_K and an estimate of heterogeneity can be calculated. As with other estimates of diffusion distance, measurement ought to be made to the centre of vessels, as capillary radius is a significant factor in overall resistance to diffusion. The closest individual method is related to the concentric circle approach by the common usage of nearest neighbour statistics. Although CI is more laborious, the use of direct measurements rather than estimation of distances can provide a quantitative estimate of capillary distribution patterns with greater resolution. Where capillaries are randomly distributed, distances from a test point placed at random over tissue to the nearest capillary will follow a negative exponential distribution. A cumulative frequency distribution curve of these distances permits calculation of the probability of encountering an individual capillary within a given distance from any arbitrary location. This also provides a benchmark against which other distributions may be compared; e.g. for a random distribution the cumulative frequency distribution is given by:

$$f(r) = 1 - e^{-\pi \cdot r^2 \cdot CD}$$

where r is linear distance.
Here the mean ICD is equal to the mean distance:

$$\overline{ICD} = R = 0.5\sqrt{CD^{-1}}$$

where $R = 2 \cdot R_K$.

However, the maximum diffusion distance appears to be more important in determining tissue oxygenation (Turek and Rakusan 1981), and so distances at the 95th percentile are also calculated:

$$R_{95} = 0.98\ \sqrt{CD^{-1}}$$

Theoretically, if capillaries are in an ordered array R < ICD, while for a clustered array R > ICD, with cumulative frequency distribution curves lying above and below that for random arrays respectively. The degree to which such plots deviate from the benchmark distribution will depend on the underlying pattern. For square and hexagonal arrays the mean and variance terms have been derived in terms of CD, and agree to within 5% of simulations (Kayar et al. 1982b); more exact formulations for moments of the distributions are given elsewhere (Homer 1984). A further advantage of CI is that it can be readily computerized, making data collection more efficient, and may also be used to assess the heterogeneity of capillary spacing (Turek et al. 1987).

This analytical approach may help explain physiological adaptations which are not amenable to study using present experimental techniques. An ordered, approximately hexagonal, array was found in guinea pig gastrocnemius where an increase in R_{95} during maturation was inversely proportional to the allometric decrease in $\dot{V}O_2$, suggesting potential O_2 supply is maintained during cellular hypertrophy (Kayar et al. 1982c). Myocardial capillaries were shown to fit a square-ordered array (Kayar and Banchero 1983) which, in the same tissue at maximal perfusion, would produce a uniform tissue oxygenation and narrow range of PO_2; Turek and Rakusan (1981) on the other hand have shown an asymmetric right-skewed distribution of ICD, which will tend to increase tissue PO_2 heterogeneity. Other discrepancies in the literature may simply be a result of different methodological approaches: R or R_K calculated from CD, assuming either random or ordered capillary arrays, may introduce an error of up to 30%, and using ICD as a basis for predictions would be even worse (Kayar et al. 1982c). However, these calculations were made after measuring ICD between nearest neighbours, and using adjacent (Sect. 3.7.1) or contiguous (Sect. 5.3.3) neighbours may minimize this error. In addition, sampling of tissue distribution between capillaries as a means of predicting O_2 diffusion capacity may be more complete using planar analyses (Sect. 5.3.2).

In both skeletal and cardiac muscle, ICD follows a logarithmic-normal distribution (Renkin et al. 1981; Rakusan et al. 1986). This is one of the most important distribution laws in nature; unlike the normal distribution where the probability density function is symmetric about the mean, in the log-normal distribution it is asymmetric, right-skewed. This is because it is not *differences* of equal magnitude about the mean which are equally likely, but *ratios* of equal magnitude above or below the mean (i.e. it is a multiplicative rather than an additive distribution). In this case the appropriate measure of central tendency is the median, while the standard deviation of log-transformed variables gives rise to a dimensionless index of heterogeneity, $\sigma\log_{10}x$, which may be used to compare capillary distributions in homologous tissue irrespective of the absolute CD. This is an important parameter to quantify, as heterogeneity of capillary spacing can be as important as CD in determining tissue PO_2 distribution, particularly at low values (Rakusan et al. 1984).

3.8 Dimensions of the Capillary Bed

While the density of capillary profiles has an important influence on diffusion distances, the dimension of individual capillaries can also have a marked effect on the efficiency of O_2 delivery: narrow, short capillaries will have a greater contact area with tissue, thereby reducing the resistance to O_2 transport.

3.8.1 Capillary Length (L_c)

This parameter may have some value in analyses of planar networks (Sect. 4) but is generally derived rather than measured, e.g. equated with muscle length, while the effect of orientation and tortuosity are ignored. The wide range of published values from similar tissue emphasizes the subjectivity of this index, which may cause large errors if applied uncritically.

3.8.2 Wall Thickness

The capillary wall is thought to present a significant barrier to transendothelial flux of O_2 in an isolated *rete* (Rasio and Goresky 1979), although whether this applies to capillaries in other tissue is unknown. The diffusing capacity (physiological conductance) for O_2 depends on Krogh's diffusion coefficient and the linear resistance to diffusion. Translational distances across the capillary wall, or barrier thickness, may be quantified using weighted chord lengths to provide a sensitive index of both active and passive transport capacity: Weibel and Knight (1964) used the estimation of harmonic mean thickness to determine the width of lamellar structures, although reciprocal orthogonal intercepts (i.e. width of the capillary wall at 90° to one surface) may represent a more efficient approach (Jensen et al. 1978). Morphometric estimates of diffusive resistance will be covered in more depth in an accompanying chapter (S. F. Perry; Chap. 3).

3.8.3 Mean Capillary Cross-Section [$\bar{a}(c)$]

The dominant factor in determining flow resistance of small vessels is the diameter, possibly explaining observed heterogeneity of velocity among capillaries (Skalak and Chien 1981). Attempts to document differences in capillary resistance between mammalian red and white muscle have led to some confusion: Potter and Groom (1983) showed that capillary diameters in vascular casts of oxidative muscles were larger than those from glycolytic muscle, while Dawson et al. (1987) reached the opposite conclusion from intravital microscope studies. Estimates of the bore size (lumen) often involve electronic image analysis or planimetry to measure capillary 'diameter' – in reality, projection width – from planar IVM (intravital microscopy; direct in vivo observations) images (Safranyos et al. 1983) or vascular casts. Surprisingly little data is published of direct estimates of $\bar{a}(c)$ taken from EM sections (Egginton and Johnston 1983). Most methods seem to rely on convenience rather than accuracy: major and minor axes, automated analysis of enclosed triangle areas (Laursen and Diemer 1980) or even comparison of profiles with circles of known area (Aherne and Dunhill 1982).

Safranyos et al. (1983) found a fourfold range of capillary diameters with a mean of 17 μm in frog sartorius, while a fivefold range and a mean of 5 μm was observed in the cardiac and skeletal muscle of rat (Potter and Groom 1983). Not surprisingly, capillary diameter appears to match both size and deformability of RBCs such that the smallest vessels approximate the minimum RBC diameter, and thus maximize O_2 transport across the capillary wall. Schmidt-Schönbein (1981) has shown how the effect of a discrepancy in capillary size between higher and lower vertebrates on O_2 transport may be deduced from the Fick and Poiseulle equations for diffusion and flow capacity respectively:

$$\frac{\delta \dot{M}}{\delta t} = \frac{\Delta C}{x} \cdot A \cdot D$$

$$BF = \frac{N \cdot r^4 \cdot \Delta P}{8 \cdot \eta \cdot l}$$

where $\dot{M} = O_2$ flux (mass transfer) rate, C = concentration, x = linear distance over which diffusion occurs, A = surface across which O_2 flux is measured, D = diffusivity constant and BF = blood flow, N = number of capillaries, r = capillary radius, P = perfusion pressure, η = blood viscosity, l = capillary length.

If one assumes that mammalian capillaries are one-fourth the size of those from lower vertebrates, but make up the same volume fraction, 16 times more capillaries are required to contain the blood volume. This in turn provides 4 times the surface area (A), and a reduction in half-mean ICD (x) by a similar factor. The rheological behaviour of mammalian erythrocytes enhances the O_2 transport capacity by a factor of 2, increasing the overall efficacy of O_2 diffusion by an impressive 32-fold. In treating capillaries as parallel elements, however, the hydraulic resistance increases and blood viscosity decreases, suggesting that the increase in perfusion pressure required to overcome additional systemic resistance and maintain a similar flow rate is only a factor of 2 to 3, a differential in BP matching that is found during phylogeny. Interestingly, if vessels are considered to be cylinders with constant dimensions, then variations in vessel length and diameters may explain vascular topography, and appear to be of similar importance as blood viscosity in determining pressure and flow distributions (Schmid-Schönbein et al. 1977). However, there are relatively few comparative structural data from IVM observations with comparable physiological recordings, so the validity of such conjecture remains largely untested.

3.9 3-D Derivatives

The spread of values in the literature owes much to systematic error and methodological variation, the most obvious being the assumption that a capillary observed in vivo retains a circular profile, even when non-patent (Casley-Smith et al. 1975). Irregular profiles compound the problem, but perhaps the largest and most common error is to extrapolate from such data to calculate the volume or surface of a single capillary by simple integration:

$$V_c = \pi \cdot r^2 \cdot L_c \text{ or } S_c = \pi \cdot d \cdot L_c$$

Except in the unusual situation of an isolated, truely planar capillary network, these values must be subject to large errors, and owe much to a literal application of the Kroghian cylinder model. These errors may be minimized by sample preparation: where estimates allow for in vivo heterogeneity of perfusion, such as IVM and immersion-fixed EM sections, the form of distribution is similar and may be quite broad (Brown and Egginton 1988). Where perfusion-fixation or casting compounds are introduced, the distribution of values will be narrower and may simply reflect differences in vascular compliance.

Many of the problems referred to with previous indices may be minimized or eliminated if dimensions of the whole capillary bed are considered in 3-D. For example, an increase in tortuosity of the capillaries will not be readily detected in TS section, although it will result in an increase in functional capillary density (Fig. 5).

Stereological principles permit quantitative data to be obtained more efficiently, and possibly more accurately, than with direct planimetry. Using a square lattice counting frame of spacing d, the area of an object (A), e.g. a capillary, is simply:

$$A = P_i \cdot d^2$$

where P_i is the number of points, formed by the intersection of grid lines, falling on the capillary. The total line length of the grid is:

$$L = 2P_t \cdot d$$

where P_t is the total number of points on the grid. The length of capillaries per unit area of tissue is then:

$$L_{cap} = \frac{\pi}{2} \cdot P_i / L$$

and capillary surface area is

$$S_{cap} = \bar{D}^2 \pi \cdot L_{cap} / 4$$

where $\bar{D}$ is the mean capillary diameter (Underwood 1970). Muscle volume is easily estimated by integrating the cross-sectional area from a set of parallel sections, allowing the length density (L_{cap} per unit volume) to be calculated. Allowing a correlation for the discrepancy between projected and actual capillary length, Schmid-Schönbein et al. (1977) calculated an error of around 5% for rat spinotrapezius, while data agreed to within 10% of direct estimates. These authors

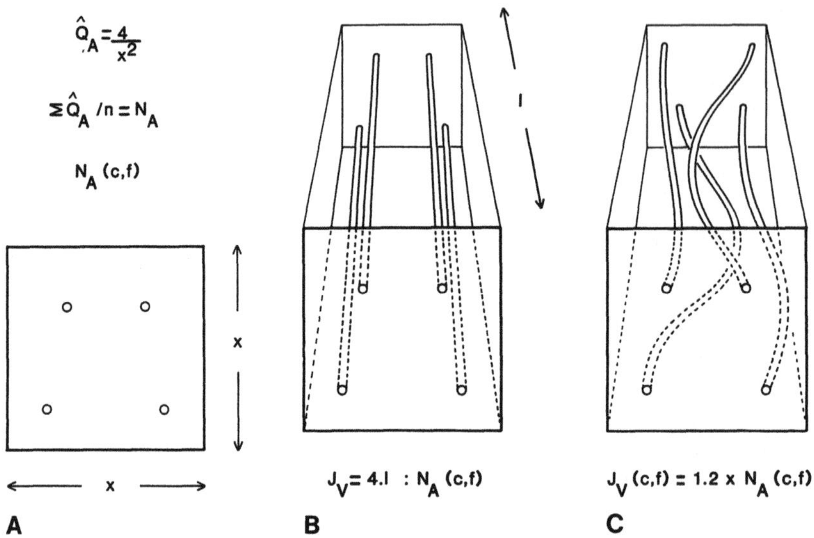

Fig. 5. Effect of tortuosity on estimates of capillary supply. *A* In cross-section, the density of capillaries is estimated by the number of capillary profiles per unit area of tissue:

$$\hat{Q}_A = \frac{4}{x^2}$$

If summed over n sections and the reference phase is muscle fibre area, then

$$\frac{\Sigma\hat{Q}_A}{n} = N_A(c,f)$$

B CD is strictly only a sufficient descriptor of capillary supply for a network of straight, unbranched vessels (as in the Krogh model) where the length per unit volume (length density; J_v) is simply the numerical count per unit area integrated over length, L:

$$J_v = \hat{Q}_A \cdot L$$

C Tortuous capillary paths increase the effective capillary length and surface; for example, a 20% increase in length would not be evident from single tissue cross-sections, as CD does not change, but:

$$J_v = 1.2 \times N_A(c,f)$$

and the proportionality constant therefore needs to be empirically derived

recognized some limitations of the method, e.g. the equation for length density assumes the angle of intersection between capillaries and grid lines of $0-\pi/2$ radius, which may be acceptable for larger vessels but not for the highly orientated capillaries.

Many other applications of stereological principles quote an early article by E.R. Weibel (1969), in which equations for length and surface density are given:

$$J_v(c,f) = 2 \cdot Q_A$$

$$S_v(c,f) = 4B_A/\pi$$

where Q_A is the number of capillary profiles per unit area of tissue, and B_A is the boundary length of capillaries per unit area. Despite their wide usage these relationships are strictly applicable only for sections taken with random orientation of isotropic (i.e. lacking directional order) structures: in contrast most data is obtained from sections in a fixed orientation (TS or LS) of highly ordered (anisotropic) capillary beds. Fortunately, a rather simple method has been shown to overcome these problems to a large extent (Mathieu et al. 1983; Egginton and Johnston 1983; Hoppeler 1984; Mattfeldt 1987). Length density, of fundamental importance due to its sensitivity to orientation, is related to numerical density:

$$J_v \propto Q_A(\Theta)$$

where $Q_A(\Theta)$ is the numerical density derived from a section whose normal (a projection at 90° to the section) forms an angle, Θ, with the axis of anisotropy. For TS sections, $\Theta = 0$ and for LS sections, $\Theta = \pi/2$ radians.

For perfect anisotropy, $J_v = Q_A$

For perfect isotropy, $J_v = 2 \cdot Q_A$ (Weibel 1980)

As the muscle capillary bed has a high (although unknown) degree of anisotropy, which is technically very difficult to determine in sectioned material, it is useful to evaluate a concentration parameter (K):

$$J_v = c(K, \Theta) \cdot Q_A(\Theta)$$

Using the assumption of a spherical-normal (Dimroth-Watson) distribution of orientation, it has been shown that:

$K = 0$ for an isotropic (random) distribution,
$K = +\infty$ for a component oriented parallel to the axis of anisotropy (linear or tubular distribution, e.g. a stack of pipes),
$K = -\infty$ for a component oriented parallel to the plane of anisotropy (planar or lamellar distribution, e.g. cartwheel spokes),

whereas for the partial anisotropy evident in most capillary beds $+\infty > K > 0$ (e.g. sheaf of corn). A simplified method of estimating K uses the ratio of numerical densities from TS and LS sections:

$$\frac{Q_A(0)}{Q_A(\pi/2)} = \frac{c(K, \pi/2)}{c(K, 0)} = K$$

from which $c(K, 0)$ can be calculated. Once this constant of proportionality is known, capillary length density may be calculated from random transverse sections:

$$J_v(c, f) = c(K, 0) \cdot Q_A(0)$$

i.e.

$$J_v(c, f) = c(K, 0) \cdot N_A(c, f)$$

This allows calculation of capillary volume, and surface densities:

$$V_v(c, f) = \bar{a}(c) \cdot J_v(c, f)$$

$$S_v(c, f) = \bar{b}(c) \cdot J_v(c, f)$$

where $\bar{a}(c)$ = mean capillary cross-sectional area and $\bar{b}(c)$ = mean capillary circumference (boundary length). In skeletal muscle fixed at resting length, $c(K,0)$ has a range of 0.016 – 0.12 (Egginton and Johnston 1983; Hoppeler 1984), suggesting that CD does indeed provide a reasonable estimate of capillary supply; although in tissue with more tortuous capillaries such as brain, $c(K,0)$ may be higher. In addition, comparison of tissue following experimental or pathological adaptation may be subject to error if the underlying control of capillary orientation is altered. Nevertheless, this approach provides nearly identical values of length and surface with true random (isotropic) sections, although J_v estimation is far more efficient (Mattfeldt and Mall 1984).

This approach relies on a mathematical model of a partially anisotropic system of lines in 3-D space, an approximation of the Fisher axial distribution, and is strictly applicable only if the degree of anisotropy is not too large (Weibel 1980). In skeletal muscle where the capillary bed is highly anisotropic this may cause problems, as there is a bimodal probability distribution of capillary orientation showing clusters parallel to the axis of muscle fibres and perpendicular to that axis. K may not change if sprouting or cross-connections occur evenly in both planes, although J_v would be increased, and cannot distinguish between sprouting in either plane. Specifically, as the uniaxial distribution is orientated parallel to the fibre longitudinal axis it may systematically underestimate lateral capillary anastomoses. Hence a biaxial model distribution would be more appropriate. Given that anastamoses are thought to be important in evening out pressure differences among capillaries, making for a more homogeneous perfusion, it is well not to have to rely on an assumed model distribution and utilize a deterministic rather than a probablistic approach, although both approaches provide qualitatively similar results (Atherton et al. 1982). Underwood (1970) describes the analysis of a system of lines separable into populations parallel ($||$) and perpendicular ($\perp$) to an orientation axis; a section plane parallel to this axis will intersect isometric, but not orientated lines:

$$L_v = 2(P_A)_{||}$$

where L_v = length density and P_A = point density (in Weibel's notation these are equivalent to J_v and Q_A respectively). In this case $P_A(\perp) = 0$. For partially orientated systems:

$$L_v = (P_A)_{\perp} + (P_A)_{||}$$

and the degree of orientation is simply the ratio of differences to sum of these counts:

$$\Omega_{1,3} = \frac{(P_A)\perp - (P_A)_{||}}{(P_A)\perp + (P_A)_{||}}$$

This approach was used to identify a longitudinal orientation of new capillaries in mouse EDL undergoing experimental hypertrophy (James 1981).

Alternatively, the problem of inappropriate models may be avoided by an analysis independent of assumed distributions by taking three sections at mutual right angles: the orthogonal triplet probe of Mattfeldt et al. (1985). If the first section has random orientation with respect to the tissue, the subsequent sections form a systematic random sample allowing application of the original (isotropic) formula for surface density with minimum error:

$$S_v(c,f) = \sum_{i=1}^{i=3} \left(\frac{2 \cdot P_l}{d \cdot P_t} \right)$$

using line intercept (P_l) and point count (P_i) analyses respectively, while estimates of volume density are unbiased for any orientation:

$$V_v(c,f) = \frac{\sum P_i}{\sum d^2 \cdot P_t}$$

Capillary volume density provides an estimate of the instantaneous blood volume within the capillary network, an index of the maximal convective delivery of O_2 and substrates to muscle; capillary surface density is an estimate of the total surface area available for blood-tissue exchange, an index of structural limitations to muscle O_2 and metabolite flux. These indices provide an anatomical parallel to those obtained by dye-dilution and tracer efflux experiments and provide a non-parametric estimate of length and surface, with excellent precision (Mattfeldt et al. 1985).

Recent available stereological methods for estimating volume and surface may be of particular use in tissue with a recognisable axis, such as muscle. If an arbitrary horizontal plane is chosen (e.g. a random streak along the length of a muscle), and a vertical plane perpendicular to this is selected for sectioning (≡LS), then the condition of isotropic uniform random (IUR) sampling (necessary for application of traditional stereological theory) may be achieved by means of the test system which overlies the section, rather than using IUR sections per se. This IUR line sampling technique requires a special cycloid curve test system (Baddeley et al. 1986). Unfortunately, length estimates are not available from this approach, but may be conveniently obtained by the ratio of $\hat{Q}_A$'s from the LS and an orthogonal (TS) sample, as above. Practical applications of these recent developments in stereology have been reviewed elsewhere (Braendgaard and Gundersen 1986; Mattfeldt 1987).

4 Linear Analyses

4.1 Introduction

One form of analysis that has attracted much less attention than the morphometric indices described in the previous section has been topology of the microvascular bed. Description of its geometric complexity and functional capacity has required development of analytical techniques, often adaptations of theoretical approaches used in geomorphology, which have found applications in neuroscience, pulmonary physiology and, only recently, microvascular research (Popel and Johnson 1986). Primarily this involves a description of branching patterns, but has developed to view the terminal vascular bed as a whole by applying network analysis techniques. The utility of such approaches in providing physiologically relevant data has received less attention, such that the methods outlined in this section represent a field whose potential has not yet been realized or, indeed, fully explored. Although much of the available theoretical work has concerned larger vascular elements (or even different structures), the basic principles are broadly applicable; a flexible analysis is needed because, despite the considerable body of literature concerning capillary supply to skeletal muscle, alternative approaches may be more appropriate for other organs. The arteriolar network cannot be described as a dichotomously branching tree such as is found in the pulmonary airways, the capillary bed in the brain may more closely resemble the branching patterns formed by hyphae or dendrites than vessels in anisotropic limb muscles, while capillary supply to the mesentery or similar 'planar' structures may quite closely follow the models derived from thin muscle preparations.

Many studies have used thin tissues which approximate planar (2-D) structures – mesentery, cheek pouch, spinotrapezius and cremaster muscles – and readily lend themselves to the concept of a tree-like arrangement of vessels, regarded as generally applicable to all nutritive vascular networks. Whether this holds for larger organs is unknown, although differences in size of supply vessels (and the subsequent network) appear to scale with tissue mass and metabolic capacity (Lund et al. 1987). Branching characteristics of the capillary bed, under a constant pressure head, should determine the proportion of blood flow to different segments. However, temporal variation in flow distribution observed by IVM indicates that another level of control is also operating, presumably involving arteriolar vasomotion. Indeed, the presence of actin in capillary endothelial cells at branch points suggests adjustment may even be possible at the level of an individual capillary. While anastomoses at all levels of the vascular tree permit flow along many different paths, the functional consequence is unclear. For example, Schmid-Schönbein and co-workers consider a complex tree organization to be necessary for relatively uniform perfusion of capillaries, i.e. similar flow velocities in all vessel orders, while Honig et al. (1977) conclude that intercapillary anastomoses do not improve flow distribution during exercise. On the other hand, Groom and co-workers suggest that such an arrangement makes the classical assumption about all capillaries being equivalent sources of oxygen supply untenable, leading to both spatial and temporal heterogeneity of perfusion. It may be that the relationship between the anatomical path length of individual capillaries, and RBC transit time, for instance, is not straightforward.

4.2 Branching Pattern

As with many aspects of morphometry the description of branching patterns began outside biology, as mathematical analysis of connected linear structures (or 'trees') has long been recognized (Cayley 1857). Many biological applications have paralleled Horton's (1945) study on the classification of stream orders in an analysis of drainage basin characteristics (see Popel and Johnson 1986), but considering a centrifugal rather than centripetal order of branch points. Asymmetric dichotomous branching trees can be described in a forward direction, as for airways (Weibel 1963), or retrograde as for rivers (Strahler 1952). The latter is a mod-

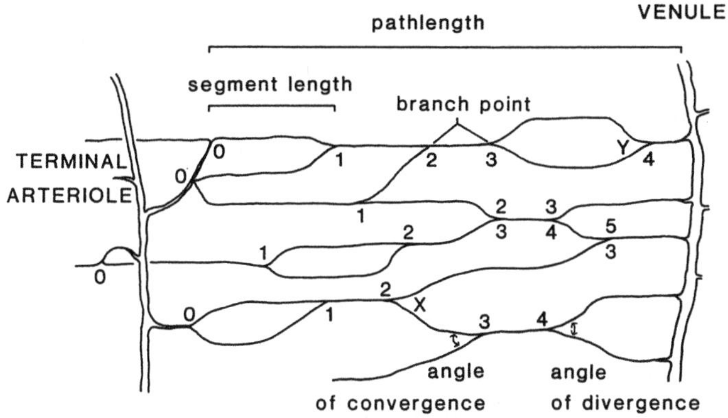

Fig. 6. A vascular network showing the parameters measured in lineal analyses and defining the terminology used in studying the capillary bed. *X* New (divergent) capillary; *Y* rejoined (convergent) capillary; *numbers* = branch point of order n (after Plyley et al. 1976). Note the ambiguity in ascribing node orders, dependent on which path length is followed

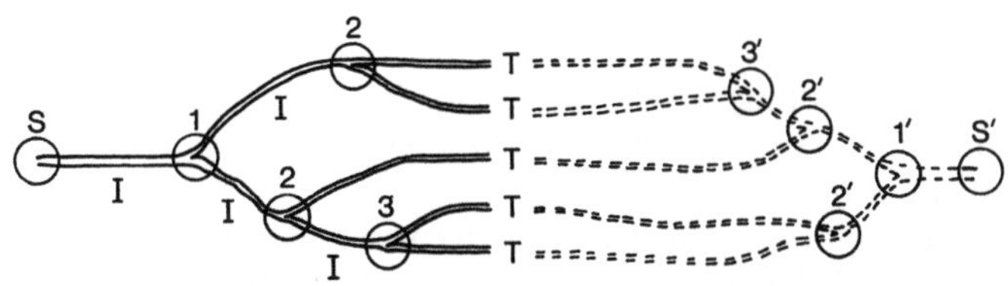

Fig. 7. Classification of topological trees, composed of intermediate (*I*) and terminal (*T*) segments. The node order is equal to the number of *I* segments between that bifurcation and the source (*S*). The degree of a tree is defined as the number of terminal segments; in this example it is 5 = [2, (1, 2)]. In a capillary bed, *S* may represent a terminal arteriole and *T* a capillary segment after the final division. The description of vascular division may be complemented by that of vascular confluence, although this is unlikely to be simply a mirror image of the former and often includes more segments, counting node order in this case from the sink (*S'*) or venular end of the capillary bundle

ification of Horton's analysis, and is more widely adopted. Here, joining of two vessels of order 1 form a vessel of order 2; when two vessels of unequal order form, the higher order vessel is continued. The number of vessels of order n is given as:

$$N(n) = R_B{}^{1-n} N(c)$$

where R_B is the branching order number; N(c) is the number of capillaries. Similar principles have been applied to branching patterns found in (botanical) trees (Leopold 1971), fungi (Edelstein 1982), dendritic fields (Berry et al. 1986), arteriolar and venous networks (Engelson et al. 1985a, b), and the capillary bed of tumours (Balding and McElwain 1985) and skeletal muscle (Plyley et al. 1976; Fig. 6).

Branching patterns are formed by segments linked via nodes, giving a particular connectivity. The number of links, or segments, associated with a node is termed the node degree (D); in the simplest branching pattern, a bifurcation, $D = 3$. Where $D > 3$ each segment should be ascribed a sign to denote direction of flow in or out of the node, although this adds a further complexity to the analysis and is not often adopted. Sequential appearance of nodes along a path length defines the node order (1, 2...n) which, with a mean value of D, will determine the extent of branching; e.g. where $D = 3$ the total number of branches of order n is 2^n (see Fig. 7).

4.3 Topological Trees

Geometry of the capillary bed may be characterized by topological tree analysis, using principles of graph theory to determine the in situ branching pattern of enclosed structures. This is a potentially powerful approach, as the mechanism generating constituent elements may be inferred from the statistical properties of the graph (Na and Rapoport 1970). Many well-established growth models are based on random processes that may be influenced by local restrictions: the null hypothesis states that linking of nodes is random and will give rise to a particular distribution of node degrees. Systematic departure from this distribution indicates operation of biases, or restrictions, the nature of which may help elucidate mechanisms controlling node association. Any derived hypothesis may then be tested by comparison of empirically determined distributions with those generated from an assumed bias.

Dendritic growth structures are widely found in nature, and their analysis has been simplified by a recent classification of branching patterns forming topological (ambilateral) tree structures (van Pelt and Verwer 1986). A tree can grow by elongation of segments, increasing the number of segments (and hence nodes), or a combination of both processes. The resultant shape of branching patterns will be governed by many interacting factors; the characteristics of observed branching patterns may indicate whether growth is a uniform or regular process, or whether it is adaptive in the sense of meeting changing needs (i.e. does supply meet demand or vice versa?). Modelling growth on the basis of empirical rules will, as a first approximation, concern only the topology of the pattern. Most trees are assumed to be binary such that branching replaces a segment with two consecutive segments, being connected to a third (source) segment by a node ($D = 3$). In practice, this is not a serious limitation, as higher node degrees ($D = 4$ or 5) in the capillary bed are rarely observed with intravital microscopy or histochemical thick sections of striated muscle or brain. A normalized branching probability may depend on its centrifugal (radial) order, which can be incorporated in the model by an exponential term based on the probability of terminal (end) branching and centrifugal order; e.g. in dendritic trees, branching probability decreases away from the source (van Pelt and Verwer 1986). Properties of growth models applied to neuronal trees and appropriate statistical procedures for testing them have recently been reviewed (Verwer and van Pelt 1986). An extension of branching pattern analysis to incorporate trichotomous nodes ($D = 4$) has been presented (Berry et al. 1986), while most authors adopt a simpler solution, assuming that trichotomies may be decomposed into a series of two dichotomies.

Given that diameter, length or a number of individual branches in each successive order follow an orderly progression (Horton's law) such that:

$$N_u = N_1 (R_b)^{1-u} \quad \text{or} \quad W_u = W_1 (R_w)^{u-1}$$

where $N_u(W_u)$ = number (width) of branch of order n,
$N_1(W_1)$ = number (width) of branch of order 1,
R_b = branching, or bifurcation, ratio,
R_w = width ratio;

then using mean branch (segment) length and cross-sectional area per order, one can estimate total length and volume within a vascular tree. Tree structure may also be characterized by the frequency distribution of nth-degree elements. Statistical analysis usually follows either segmental or terminal growth hypotheses, where the basic step is bifurcation of an intermediate or terminal segment respectively. It is implicitly assumed that each segment has the same probability of growth which describes river systems very well, although often growth will depend to a large extent on the position of the segment such that bifurcation probabilities should be assigned to each segment.

In contrast to where distance from the root, or source, is most important (above), Pape and Schopper (1987) describe growth at terminals which enter into mutual competition. This holds well for neuronal and botanical trees, and may be adapted for use with that part of capillary networks including division or confluence of vessels at nodes. This model was derived from analysis of leaf vein patterns based on self-similar structures (Pape et al. 1989), and has the potential to describe the capillary network as an integral part of the larger vascular tree using the theory of fractal dimensions, i.e. non-integer exponents. If the branching structure maintains its shape or pattern at successive orders, then the fractal dimension (D^s) is given by:

$$D^s = \log N / \log V^{-1}$$

where N is the number of segments of order n arising from segments of order n+1; V is the size ratio (length, area, volume) between the two orders. This "pigeon-hole" model uses a cascade of spreading arcs where deletion occurs as a result of competition, and the intersections form linear paths. It appears to fit such diverse structures, as the convolutions of mouse intestine and the branching systems of (botanical) trees. Of particular interest in any application to vascular networks is that the model starts with a random process to determine root distribution and growth. Variation in segments of the same order will be probabilistic, while the main growth process is strongly deterministic, leading to a high degree of ordering. This approach would appear to be justified in the study of capillary networks, given the ordered tree structure described by Plyley et al. (Fig. 6).

The concept of self-similarity is also implicit in other studies; e.g. Zhuang et al. (1985) corrected earlier data from pulmonary circulation where the size and number of small vessels, including capillaries, were obtained by extrapolating linear regression lines of logarithms of diameter, length or branch number of larger vessels against order number. The number of vessels of given length (L) in a unit volume of tissue is given as:

$$N_v = \frac{2N_A}{L}$$

where N_A is the number of vessels per unit area of section. Strictly speaking, this is valid only for planar images (infinitely thin sections) where vessels are of similar length. Considering the errors due to projection and section thickness, these authors obtained a corrected formula:

$$N_v = \frac{2}{L}\left(1+\frac{2h}{L}\right)^{-1} N_s$$

where h is section thickness and N_s is the total number of vessels cut by the slab. Interestingly, they found that the ratio of number of vessels in successive orders is approximately a constant, allowing the numerical density of order 1 vessels to be estimated from the formula:

$$N_{vi} = \left[\sum_{i=1}^{n} \frac{L_i}{2}\; 1+\frac{2h}{L_i}\; \frac{1}{B^{i-1}-1}\right]^{-1} \sum_{i=1}^{n} N_s$$

where B is the branching ratio, or the average ratio of N_{vi}/N_{vi+1}. While this analysis may be applicable for pulmonary or cerebral capillaries, its weakness lies in the assumption of isotropy; only then does the projected height of all capillary segments equal L/2, which in anisotropic tissue will vary from 0 to L depending on the orientation of section plane and tissue axis.

A fundamental problem of "modelling" the capillary bed is that one has to assume the 3-D geometry is pre-determined and/or that the influence of the surrounding tissue is predictable. For these reasons detailed analyses such as those applied to dendritic or hyphae branching patterns may be less useful for capillaries; differences may simply reflect whether the system is open or closed, i.e. developed within a constraining boundary or not. For closed systems such as bronchial passages or major vessels, (a) the number of branches of each order tend to form an inverse geometric series with branch order and (b) the mean length of branches per order tend to form a direct geometric series with branch order. In open systems, however, the mean length of branches per order tend to form a direct arithmetic series with branch order; the arithmetic relationship appears to result from a constant linear growth rate, whereas a geometric relationship may result from a progressively declining linear growth rate (Park 1985).

Topological tree analyses can only describe one-half of the vascular tree, i.e. centrifugal vascular division or centripetal vascular confluence, although it has the potential of including higher order vessels. Only network analysis seems to be able to cope with the concept of loops or arcades, although conceptually the arteriolar or venular systems may be regarded as opposites (Engelson et al. 1985a, b), so it may be worthwhile to attempt a composite topological tree analysis using centrifugal and centripetal components respectively. This offers some possibility of reconciling purely descriptive studies models. For example, Plyley et al. (1976) determined by empirical observation that in frog sartorius the frequency distribution of individual capillary a-v length was Gaussian, that of segment lengths was exponential, and number of segments per capillary length was Poisson. These data could be more readily incorporated into existing growth models if e.g. the cumulative mean segment length were to be computed (Park 1985). Just as model-based analyses utilizing assumptions about the growth process give rise to an expected branching pattern, so the descriptive analyses of in vivo capillary beds may indicate possible mechanisms generating the observed pattern. For example, if the frequency of nodes along a capillary path follow a Poisson distribution, the segment lengths will follow an exponential distribution, and the mean number of nodes will be proportional to the total capillary path length. Such observations have interesting implications which physiologists need to address. The correlation of number of segments per path with path length suggests that growth occurs by addition of extra segments (as suggested by Sarelius et al. 1981), rather than elongation of existing segments, implying that some upper limit exists for mean segment length. In addition, random node distribution in 1-D, a linear Poisson point process, implies that branching develops at random along the path length during growth and that control is not exerted at a local level. Care must be taken in drawing conclusions from such data, however, as it could equally be argued that any local metabolic signal for angiogenesis is randomly distributed along the muscle, or that the oxygen demand of the tissue is responsive to the available supply. More probably, it may reflect control at an intermediate level such that the mean number of nodes or segment length within, say, a capillary bundle supplied by one arteriole varies with metabolic demand of the enclosed tissue. Intravital microscopic observations have shown longer capillaries with fewer branch points and straighter paths around glycolytic than those around oxidative fibres (Dawson et al. 1987). If supply were regulated on a micro-scale, therefore, one might expect an asymmetric branching pattern in mixed muscles, with more nodes and shorter segments in oxidative regions. The question of cellular vs. tissue level of control will be dealt with further in later sections, but it is worth noting that the nature of tissue being supplied by the capillary bed is rarely incorporated into these models.

4.4 Optimized Branching

The methods outlined above may be developed further to predict microvascular branching patterns by elucidating the underlying growth processes, but are conceptually limited in ignoring the physiological consequences. Most analyses have, therefore, been purely descriptive, and many parameters of uncertain value have been measured. While the angle of divergence or convergence of elements of a binary tree may simply reflect local routes of least physical resistance along which endothelial cells migrate during angiogenesis, the consequences for RBC perfusion may also be very important (see below). Likewise, changes in vessel dimensions at successive orders will markedly affect rheological properties, e.g. a simple bifurcating network without anastomoses would quickly decrease the hydrostatic pressure gradient along the capillary path length, if total blood volume were maintained. An analysis of branching characteristics based on physiological 'cost' is now available that allows for unequal branching patterns and vessel dimensions. Theoretically, it can be shown that there are optimal values for branching

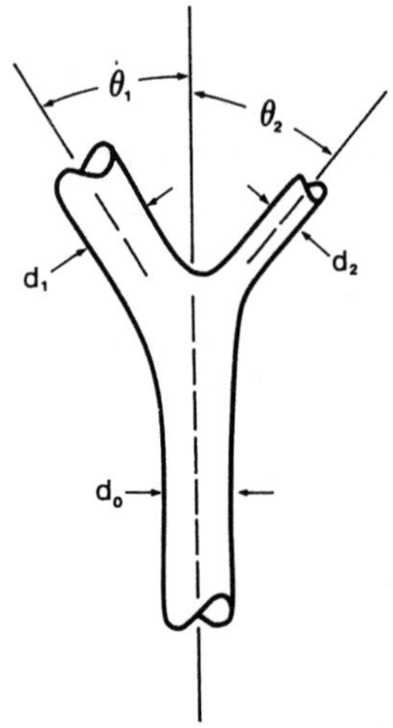

Fig. 8. Vascular networks typically consist of a succession of bifurcations where a parent vessel of diameter d_0 divides into two branches of diameters d_1 and d_2, shown schematically above. An important branching characteristic is the ratio d_2/d_1, where d_1 is always taken to be the larger of the two diameters. This bifurcation index thus reflects inequality of branching, and has a range of 0 to 1. A "symmetrical" bifurcation is one in which the two branch diameters are equal, and hence $d_2/d_1 = 1$, and a "side branch" is a bifurcation in which one branch is very much smaller than the other, hence $d_2/d_1 > 0$. Required measurements are the diameters d_0, d_1 and d_2, and angles Θ_1 and Θ_2 (From Zamir and Chee 1986)

angle and segment diameters, based on (1) hydrostatic pressure required to adequately perfuse a bifurcation, (2) drag forces exerted on vessel walls by flowing blood and (3) total lumen volume and surface at the bifurcation (Zamir 1976). In general the branching angle affects flow properties, while branch diameters determine the shear forces and work required to pump blood through a bifurcation. The geometrical characteristics of a node are easily determined (Fig. 8), while analysis is simplified because the dichotomous divisions appear to be mostly planar, allowing measured angles to be plotted directly against bifurcation indices. These plots show good agreement with calculated optimal values for human coronary arteries, although the departure from optimality in terms of physiological cost is considerably greater in rats (Zamir and Chee 1986). This approach may not be applicable to all vascular systems, e.g. major arterial branches are under anatomical constraints and act to damp the ventricular pressure pulse, and would not be expected to behave in a physiologically optimal manner. However, branching characteristics close to the predicted optimum in any network may reflect a lack of anatomical constraints, allowing freedom of branching. This may, for example, explain the otherwise puzzling phenomenon of random capillary branching described by Plyley et al. (1976). An analysis of the vascular supply to skeletal muscles of varying metabolic capacity, and heterogeneity of fibre type distribution, may help resolve whether capillary branching pattern is determined by tissue metabolic demand, or by cardiovascular constraints. In contrast, the more isotropic orientation of capillaries in brain would appear a more ideal situation for development of "optimal" (i.e. unrestricted) branching characteristics.

4.5 Networks

No single parameter can describe all geometrical properties of a tree structure, and several are required in order to compare branching patterns. However, the functional implications are poorly understood, partly because there is no consistent framework for analysis of the capillary bed. More serious is the inability to reconcile network structure with the role of individual vessels in oxygen transport, as affected by such parameters as red blood cell velocity, microvascular pressure, local haematocrit, etc. Some workers believe that such knowledge of capillary bed geometry is insufficient to permit development of a quantitative theory of blood flow in skeletal muscle. Arguing that a realistic description of connectivity of the capillary bed with arterioles and venules is required for developing a transport-based model of microvascular supply, Skalak and Schmid-Schönbein (1986) use a network analysis based on capillary bundles which may be divided into series and parallel resistance elements (Fig. 9). A reconstruction of the capillary network is made with the aid of normalized ratios of element numbers having well-defined anatomical meanings; in the simplified Krogh model these values would be:

$$A/P = V/P = 1$$

and

$$N/P = B_A/A = B_v/V = C/P = 0 \qquad \text{(see Fig. 9)}$$

The capillary models thus constructed ought then to be able to be added in series with similar models developed for arterioles and venules in order to construct a predictive model of vascular perfusion pressure and blood flow. One attraction of this approach is that by appropriate mathematical manipulation of the matrix of network connections it is possible to reduce considerably the sampling effort (Wetter et al. 1983). So far the physiological implications of such analyses have not been fully explored; e.g. in spinotrapezius muscle of hypertensive (SHR) rats there is a significant reduction in the ratios A/P, V/P and N/P, suggesting a more straight a-v path length (Skalak and Schmid-Schönbein 1986), but it is uncertain whether these differences have a direct influence on vascular resistance, as

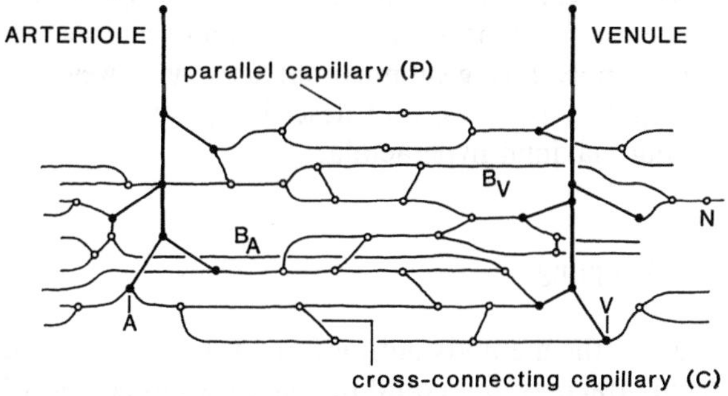

Fig. 9. A schematic capillary bundle may be analyzed in terms of series and parallel resistance elements – a transverse arteriole, collecting venule and the intervening capillary bed – and the following counts: capillaries parallel to muscle fibres (*P*), capillary-arteriole connections (*A*), capillary-venule connections (*V*), capillary nodes (*N*), segments linking neighbouring bundles at the arteriole (B_A) and venule (B_V) end, and anastomoses between neighbouring bundles laterally (*C*). In this network $P = 9$, $A = 8$, $V = 10$, $N = 23$, $B_A = 2$, $B_v = 2$, and $C = 0$ (After Skalak and Schmid-Schönbein 1986)

changes in capillary diameter, distensibility and segment length may act in a compensatory manner.

Adopting a similar rationale, Chen and Battarbee (1987) propose a simple mathematical approach to the description of vascular connectivity. Those parameters to be estimated include changes in vascular density, by vessel confluence or divergence, and modelling of microvascular networks require only the number of vessel segments (V), vessel loops (L) and bifurcations (B). In a tree-branched (open) network:

$$V = 2B + 1$$

while in a loop-branch (closed or arcade) network

$$V = 2B + 1 - L$$

This scheme allows various elaborations, including flow direction, branching angle, number of feeder vessels, etc., which appear to accurately describe networks previously analyzed by means of more laborious schematic diagrams. While such schemes may be of use in global models, analysis of local haemodynamics in the capillary bed requires not only information about tree structure, as described above, but also vessel orientation, diameter and tortuosity. Without physiological data (e.g. viscosity, perfusion pressure and PO_2 gradients) such models can only provide an approximation of tissue oxygen profiles. More sophisticated analyses are certainly required if the interrelationship (i.e. cooperativity) between neighbouring vessels and heterogeneous flow profiles are to be incorporated. Finally, if morphometric analysis of vascular networks is to be more than simply descriptive, an appropriate statistical treatment is required. Adopting the concept of minimizing physiological cost at vessel bifurcations, it may be possible to apply network optimization algorithms to models of the vascular tree. Such iterative models need to consider a set of functions at each node in a network as a function of some global parameter, such as tissue blood flow. For example, this may take the form of minimizing the fraction of mean blood flow perfusing each capillary segment, i.e. reducing inter-capillary heterogeneity. With the growing body of evidence for the importance of spatial and temporal heterogeneity of the microvascular supply during submaximal activity, however, such analyses may be restricted to capillary bundles, and describing the "efficiency" of the total capillary bed during maximal hyperaemia.

4.6 Sampling

Most of the methods outlined are at present applied to 2-D networks, e.g. camera lucida tracings, but could be readily adapted for 3-D networks measured directly with IVM of thin preparations. Circumventing the restricted depth of field in IVM by taking sections, either physical or optical, will introduce a bias proportional to section thickness. Smaller trees have a higher probability of being contained in the section although in flat tissue, where the Z-axis is of minor importance, this bias may be negligible. There remain, however, projection artefacts whereby the length of a line in 3-D will tend to be underestimated when viewed in 2-D and, due to superimposition of lines forming apparent nodes, will tend to overestimate segment number. Schmid-Schönbein et al. (1977) have explored these problems and derive total segment length from:

$$L = \frac{L_A}{h}$$

(where L_A is length per unit area and h is average muscle thickness) and incorporate a correction for overprojection. The effect of overprojection on even relatively planar networks such as dendritic fields has been shown to cause significant error in estimating the number of endings and branching nodes; e.g. segment length is usually presented as declining exponentially with radial distance, while corrected data show the relationship to be linear (Berry et al. 1981). Such bias may, of course, result in very different distributions of network parameters and thereby lead to erroneous conclusions regarding the underlying growth process. Correlating results from IVM preparations, which tend to have few inflow and outflow vessels, with more complex networks from larger organs should determine whether current theories of vascular dynamics are generally applicable. These limitations may be minimized by direct observation of 3-D networks, using the recently introduced confocal microscope technology (e.g. TSLM, Boyde 1985).

A serious, and seemingly intractable problem in most areas of morphometry is that of nomenclature. Although the definition of a capillary bed is generally taken to be the terminal vascular ramifications within a tissue concerned primarily with gaseous and metabolic exchange, in practice there exist many varied anatomical definitions. Dawson et al. (1987) consider capillaries to start at the point of first division after a terminal arteriole, which probably underestimates total capillary length as arteriolar smooth muscle is lost before this point; Eriksson and Myrhage (1972) use only those elements with no further branches, thereby effectively restricting analysis to latter segments; while Wiedeman (1984) considers the capillary bed to end at the first divergence, thereby foreshortening the capillary bed at the venular end.

Differences in definition may promote erroneous comparisons between different parts of the vascular bed, which is exacerbated by an indistinct functional division, e.g. oxygen diffusion may also occur across arteriolar walls, while much of the macromolecular transport takes place across large post-capillaries at the venular end of the network (Renkin 1977).

5 Planar Analyses

5.1 1-D Indices: Spatial Point Patterns

When a tissue is sampled in cross-section, normal to the axis of the capillary bed, it is clear that a description of capillary supply needs to take into account not only the density, but also the spatial heterogeneity. For statistcial purposes each capillary may be thought of as an event, or point occurrence, in 2-D space. The simplest point process is one in which points occur at random (homogeneous planar

Poisson process), where the intensity, λ, is equivalent to the number of events per unit area, N_A, in a subset of a Poisson distribution. Each event is located independently of all others and all are equally likely to occur at any point in the plane. In any random spatial pattern there will appear local aggregations of points, or *clumps*; this is not to be confused with non-random *clustering* of points which are a property of the underlying point process. The problem is to determine what growth process leads to local aggregation of capillaries, such as those in Fig. 4. Here a random process acts as a reference, or benchmark, playing a role in point process theory analogous to that of the normal distribution in the study of random variables (Diggle 1983).

The first-order properties of a spatial point process describes the local intensity. However, the observed heterogeneous distribution of capillaries creates varying local intensities which are difficult to analyse without invoking assumptions about the mechanism generating such a spatial distribution. Second-order properties describe variation in the relative frequency of pairs of events as a function of their position. Perhaps the most useful is the K-function, K(t), which describes the mean number of events within distance t of an arbitrary event, excluding that event (Ripley 1981). Clearly, for complete spatial randomness (CSR) t equates with the radius of a circular sampling zone, such that $K(t) = \lambda\pi t^2$; CSR may then be used as a benchmark for description of in situ patterns, which are best revealed graphically by plotting $[K(t) - \pi t^2]$ vs t (Ripley 1981). Analysis of point distributions, and formal tests of significance, involve comparison of experimental results with those generated according to known functions (Monte Carlo simulations), in order to test hypotheses regarding the nature of the underlying process.

It is important to distinguish between this approach and nearest-neighbour statistics (Sect. 5.2), which use only distances from each capillary to its nearest neighbour rather than all pair-wise distances. While the former approach may be adequate for testing hypotheses, particularly if combined with analysis of distances from arbitrary points in the tissue to the nearest capillary, it conveys less information about the underlying point process if a benchmark hypothesis is rejected. Truly random distribution of events or their spatial representation (i.e. CSR) appear to be quite rare. More often there is a minimum spacing between events, which may be explored using a 'hardcore' model, based on point distributions with an imposed constraint on interpoint distance (Diggle 1983). The degree of regularity imposed due to inhibitory interactions between neighbouring capillaries are likely to be complex and not amenable to such straightforward analysis; e.g. any zone of inhibition around a capillary must be asymmetric, having both a radial (across a fibre) and a lateral (around a fibre) component (Fig. 10). One great advantage of the point process approach is that two distinguishable types of event may be handled in the analysis of bivariate spatial point patterns (Diggle 1986). This has a potential application in revealing cooperativity and/or autoregulation of relative capillary perfusion using fluorochromes (Egginton et al. 1987a). If two compounds are sequentially infused that fluoresce at different wavelengths (e.g. Lissamine-rhodamine B200 and fluoroscein-isothiocyanate; Vetterlein et al. 1982), the pattern of perfused capillaries may be determined at time t, again at time t_1, and the degree to which the second distribution is dependent on the first determined. A practical application of this approach has recently been presented

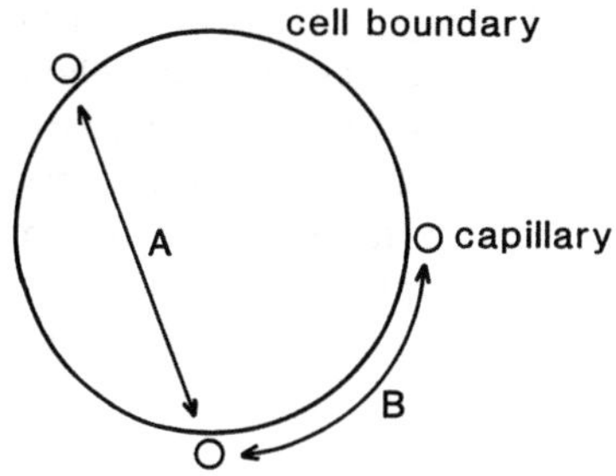

Fig. 10. Imposed constraint on interpoint distances applied to capillaries interspersed between cells has two major components: *A* simple physical constraint; this will be proportional to cross-sectional area of cells, with a maximal value of $2\sqrt{(A/\pi)}$. *B* Complex physiological constraint influenced by composition of the extracellular matrix, cooperativity of adjacent capillaries, and local capillary density; this will be related to cell perimeter, with a maximal value of $(\pi \cdot d)/2$. Unless CD is low, therefore, cell size will be the primary determinant of diffusion distances

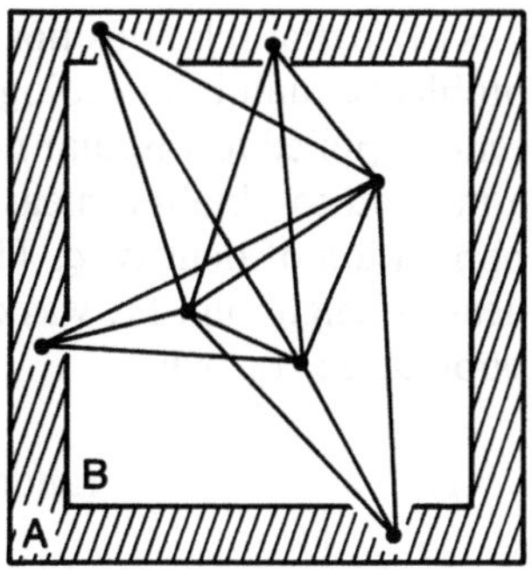

Fig. 11. Interpoint distance (IPD) calculation. The counting frame (*B*) is constructed within a sampling region (*A*). All pair-wise distances are measured except those between points lying in the guard area (*hatched*) (After Appleyard et al. 1985)

in a general form (Appleyard et al. 1985), where interpoint distance (IPD) is quantified using a quadrat analysis to delineate sample area and edge effects corrected by means of a guard area (Fig. 11). Distances are calculated between each point in the sample area and all other points, including those in the guard area, but distances between points in the guard area are ignored. The distance between points, t, is plotted against L(t), where:

$$L(t) = \sqrt{(2A/\pi \cdot m \cdot n)} \times \sqrt{n(t)}$$

(A = sample area; m = number of points in the sample area; n = number of points in the guard area).

L(t), therefore, is an index proportional to the square root of the number of pairs of points closer together than distance t. For a random distribution L(t) = t, producing a straight line passing through the origin; for clustered and dispersed patterns the lines lie above and below this line respectively.

In the present context it is important to recognize that point process statistics use the assumption, implicit though not stated in many studies, that capillaries may be regarded as point sources. The various probability plots may be used to test for aggregations or repulsions, equivalent to local concentrations of angiogenesis or zones where capillary growth is inhibited, when compared with known distributions. However, the form of the distribution may be misleading, as the basic statistical comparison is derived from the concept of complete spatial randomness. As a 'random' distribution is impossible due to the presence of muscle fibres, this approach is inappropriate in detailed analyses such as oxygen transport calculations at a cellular or individual capillary level.

5.2 Neighbourhood Analyses

5.2.1 Nearest Neighbour

As mentioned earlier, a frequent use of spatial information is to perform some estimate of linear separation such as mean intercapillary distance (ICD), which is taken to be twice the mean radius of the Kroghian tissue cylinder (R_K), and the most common approach is some form of nearest-neighbour analysis. Again, analysis of spatial relationships originally developed in response to non-physiological problems, mainly in ecological and astronomical studies, and was only quite recently applied to vascular networks. For example, Clark and Evans (1954) devised a method to obtain a measure of the spacing of individual plants in a grassland population of density, ϱ. The observed mean distance from each individual to its nearest neighbour, $\bar{r}_A$, was compared to the expected mean distance in an infinite population, r_E, where:

$$\bar{r}_E = 2\frac{1}{\sqrt{\varrho}} \quad \text{for a random distribution.}$$

The ratio $R = \bar{r}_A/\bar{r}_E$ then provides a measure of the degree to which the observed distribution deviates from random expectation. R has a calculated range of 0–2.149 for a distribution with maximal aggregation and separation respectively. Although the authors adopted a biased sampling rule, using a quadrat and measuring nearest-neighbour distances including all points surrounding the counting frame, they also suggested a circular counting frame as an improvement. Loats et al. (1978) applied this concept to estimates of capillary spacing, while Kayar et al. (1982a) added the statistical refinement of using a random point in the tissue rather than over an individual capillary (Sect. 3.7). From this distribution, which is useful in describing both the minimal and maximal intramuscular diffusion distances, Krogh's radius (≡ICD/2) can be derived by an iterative process.

Nearest-neighbour analysis is a specific application of point process statistics, and hence subject to similar limitations. In particular, certain assumptions are required concerning the probability of capillary distribution, which may be physiologically or anatomically unrealistic. While cumulative probability plots are sensitive tools with which to detect changes in distribution, they are open to false interpretation in equating calculated and observed patterns. For example, that a distribtution of capillary spacing *on average* follows a curve computed for a known distribution is of comparative value, and may adequately describe a given geometric arrangement and subsequent perturbations; however, it is likely to provide little information about the underlying stimuli or functional consequences.

5.2.2 General Neighbourhood

For analysis of capillary distribution alone it is sufficient to compare data using tests based on pattern recognition techniques. However, more information may be obtained from graph-theory methods applied to a set of points, forming a conceptual link between the LS (network) and TS (point pattern) analyses outlined above (Sects. 3 and 4). Spatial distribution patterns are connected at different cellular levels; for single capillaries, measurements may be performed to either distinguish between different types of vessels (perfused or unperfused) or to analyse the geometric relationship between adjacent capillaries. Analysis of geometric regularities tend to concentrate on nearest-neighbour characteristics (above), but this may hide essential cooperativity within the capillary bed which also requires non-nearest adjacent capillaries to be examined. If they are neighbours, nearest or otherwise, the geometric arrangement of capillaries may be assessed using a similar approach to that outlined for LS networks, but with capillary profiles representing nodes.

5.2.3 Syntactic Structure Analysis

A mathematical analysis of the geometric arrangement of structural units such as vessels has been proposed (Lu and Fu 1978). In transverse sections, a network of neighbouring structures may be constructed where centres of gravity of vessels constitute vertices (points) and are connected by edges (lines). A number of parameters may then be estimated using the principles of graph theory, e.g. number of edges, cyclomatic number, distribution of N-stars or N-simplices, number of elementary collapses (see Fig. 12; Kayser et al. 1986):

$$N_c = N_e - N_v + 1$$

where N_c is the cyclomatic number and N_e, N_v are number of edges and vertices, respectively.

Unfortunately, the utility of these indices in biological structures has not been systematically investigated beyond pattern recognition, although some equivalence with other analyses are evident, e.g. edges will equate with ICD, and N_c may be derived from a Delaunay tessellation (Sect. 5.3).

5.2.4 Minimum Spanning Tree (MST)

Related to the above analysis is the minimum spanning tree. Edges (equivalent to segments in LS networks) are drawn to the nearest neighbour, producing an undirected simple (connected linear) graph where segment lengths are minimized along all spanning trees of the graph. The degree of order in a set of points may

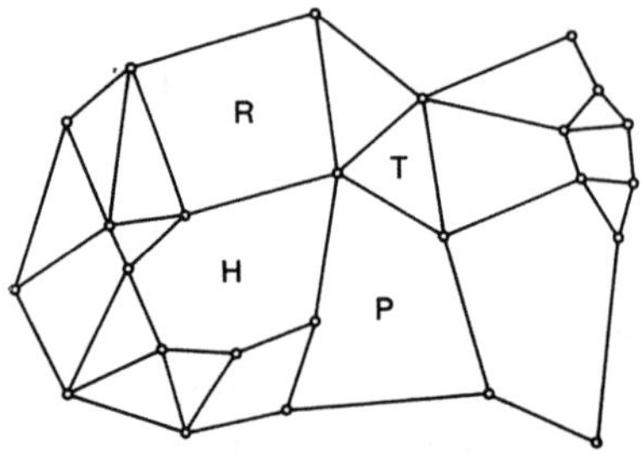

Fig. 12. Minimum spanning tree (MST) is constructed by edges connecting vertices. Two characteristics can be observed. The number of different enclosed geometric shapes (N-simplices) and number of surrounding points (N-stars): N-simplices are formed by 25 vertices (*points*) and 44 edges (*lines*), giving a cyclomatic number of 20. The number of 3-, 4-, 5- and 6-simplices (i.e. *triangles*, *rectangles*, *pentagons* and *hexagons*) is 11, 6, 2 and 11 respectively. N-stars represent the number of neighbouring vertices: 2, 11, 9 and 3 vertices having 2, 3, 4 and 5 neighbours respectively (After Kayser et al. 1986)

be assessed from inspection of the graph, and distributions compared by, e.g. plotting the standard deviation vs mean segment length. The principal advantages of such an approach are that (1) fixation and staining artefacts may be less important than in other morphometric analyses, as small errors in vertices coordinates have little effect on the network, and (2) neighbourhood analysis of discrete structures appears to be a very sensitive tool with which to detect regular or irregular patterns at various levels of organization (Dussert et al. 1987).

5.2.5 O'Callaghan's Definition

Different concepts of neighbourhood of a point have been used in analyses of spatial point patterns, including the number of shared neighbours of points (e.g. SF). However, definitions based on fixed distances – lengths (or 'depth') of minimum spanning tree edges or n-nearest neighbours of a point – may not identify clustering of simple patterns. This has led to the development of an alternative definition incorporating constraints to exclude points both greater than a set distance from the central point, and those "behind" others in a given direction (O'Callaghan 1975). The effect of a distance and direction constraint is to control overall size of neighbourhoods and limit the total number of neighbours, respectively. This approach, then, preserves the notion of proximity by the use of local computations (see also Sect. 5.3), and makes neighbourhood size dependent on local point density. One advantage of this method is that distances between points may be much less affected by boundaries, compared with k-nearest-neighbour distances which tend to increase for boundary points with local curvature, although this effect may be limited by use of an appropriate sampling protocol (Fig. 3).

5.2.6 *Alternative Strategies*

Any distribution pattern of capillaries classified solely in terms of the distance to nearest neighbour is incomplete, as it ignores other spatial relationships within the tissue, possibly indicative of higher order levels of control such as capillary bundles. It is possible to extend analysis to distinguish between 2nd, 3rd ... nth-nearest neighbour most easily by the concentric circle method. However, this concept has only really been explored in linear graphs such as the Delaunay tessellation (Sibson 1980) and the MST (Kayser et al. 1986). It would appear that cell distribution in normal tissue is qualitatively similar to random points in a plane, forming convex polygons that tend towards circularity with increasing value of n; whether or not this also holds for the interspersed vessels is unknown. Deviation from a given pattern may be used as a novel diagnostic method to assess disturbance of normal distributions using texture analysis (Kayser et al. 1986), and may be useful as a link between spatial distributions of capillaries and higher-order vessels such as arterioles and venules.

5.3 2-D Indices of Neighbourhood

We have seen that analyses based on point patterns are most useful as global parameters, and that linkages revealed in linear graphs may offer certain advantages. Although these indices may be used to ascertain underlying control processes during growth, they are difficult to incorporate into physiological models such as oxygen transport to tissue without applying many assumptions. For example, estimates of diffusion resistance tend to be based on a regular pattern of identical capillaries, and distances between adjacent (usually nearest) capillaries. Although oxygen and other small molecular metabolites tend to obey the 1-D diffusion equation, the in situ realization must be a 2-D integration of all linear path lengths. For modelling of intramuscular diffusion distances, therefore, we would like to integrate all local ICDs to provide an index representing the zone of influence of individual capillaries.

The need for a realistic spatial representation of the area of tissue thus 'supplied' by one capillary can be illustrated by the widely used Krogh cylinder model of capillary O_2 supply. Here the close packing of cylinders leaves a void proportional to the cylinder radius, R_K, which the Krogh-Erlang formula represents as a zone of anoxia (Fig. 13a). The fallacy of this situation is readily appreciated in sparse capillary networks, where a cell may be partially or wholly contained in this space; not only does it lack direct capillary supply, it is apparently starved of oxygen. Perhaps a more realistic approximation of the in vivo situation is given in Fig. 13b, where close packing of capillary centres will define a zone of minimum PO_2. Despite serious limitations, the Krogh model is still the basis of most analyses of tissue oxygen supply. However, to be more useful at a local level it requires a space-filling, non-overlapping representation of the zone of capillary influence.

Most approaches outlined so far are strictly applicable only if the objects can be regarded as a point, which may be unrealistic in some cases. Here we use an

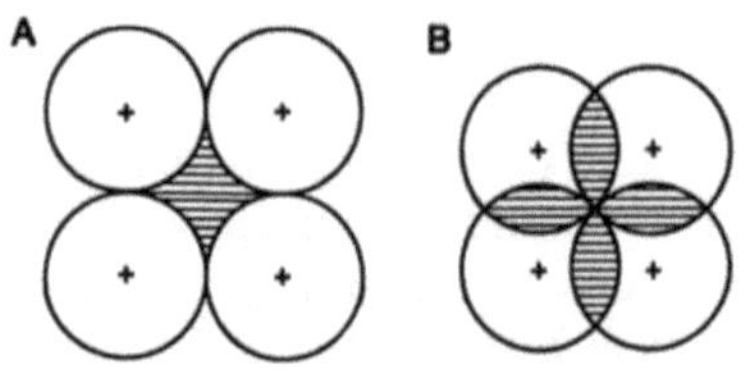

Fig. 13. Krogh's cylinder model of tissue oxygenation. *A* Literal interpretation with close packing of tissue cylinders showing a void (*anoxic*) space equivalent to $(4-\pi)(r^2)$, or 27% of cylinder cross-section (*hatched*). In tissue with a sparse capillary network, i.e. low CD, cells may be partially or wholly contained in this region, apparently lacking O_2 supply. *B* Close packing of capillary centres (*crosses*) produces zones of overlap with equivalent (minimal) PO_2. In reality, capillaries are not regularly spaced, and the area of tissue supplied with O_2 does not resemble a circular field of influence

area or volume distribution associated with a point: rather than obtain a 0-D index of density or 1-D index of separation, we can determine a 2- or 3-D estimate of the tissue mass supplied by an individual capillary. Note that this is essentially a decomposition of such global parameters as Krogh's radius, maximal hypothetical diffusion distance, etc. into the component area (or volume) around individual capillaries, and is best represented by a tessellation of convex polygons.

5.3.1 Voronoi Tessellation (V_t)

2-D space may be divided into bounded and connected subsets (referred to as tiles or cells in the mathematical literature) of the plane R^2. A tessellation is an aggregrate of tiles that fit together to cover space without overlapping. Three familiar examples of regular tessellations are equilateral triangles, squares, and regular hexagons (Fig. 15) whose characteristic parameters are the number of sides or vertices, perimeter, surface area and, in 3-D, volume. Given that most analyses of capillary distribution use these physiologically unrealistic models, with the consequence that each capillary is surrounded by (supplies) the same area of tissue, we are interested in utilizing more realistic non-regular convex polygons with random or non-random origins. A Voronoi tessellation is generated by a finite set of distinct points, each of which acquires a tile comprising that part of the plane closer to it than any other. In 1-D the construction is trivial with each tile running halfway to the next point, or off to infinity from both end points. Drawing the construction in 2-D reveals a more subtle relationship between irregularly spaced points and the associated tiles. Formally:

$$T_n = [x : d(x,n) < d(x,m) \quad \text{for all } m \neq n]$$

where d is linear distance (Sibson 1980). The tile, T_n, is an intersection of many open half-planes, with the mth half-plane being bounded by the perpendicular bisection of the line joining adjacent points n and m (p_n, p_m). T_n is therefore an open convex polygon which may be bounded or extend to infinity, which may be regarded as the 'territory' of p_n. Clearly, any point lies either in the interior of one tile, in the common side ot two tiles, or at a common vertex of > 3 tiles (Figs. 14, 15).

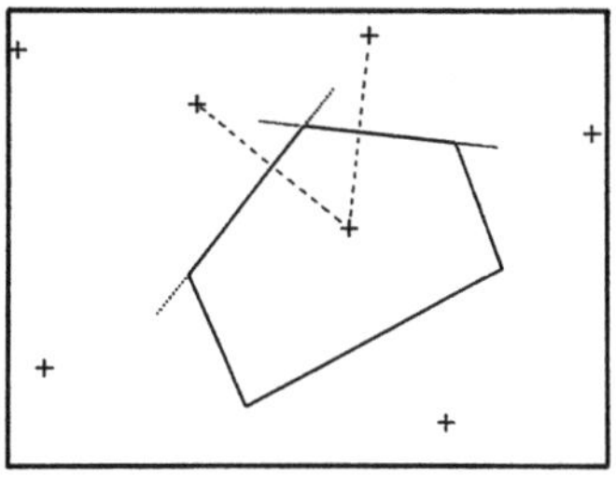

Fig. 14. Construction of a Voronoi tile. In the case of capillary domains, boundaries of the convex polygon (*solid lines*) are formed by perpendicular bisectors (*dotted*) of lines (*dashed*) joining adjacent capillaries (*crosses*)

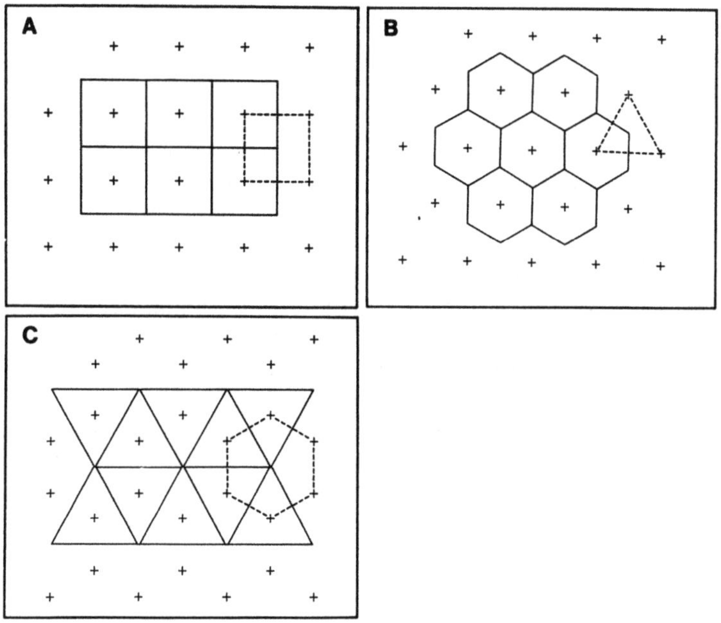

Fig. 15. Examples of Voronoi tessellations constructed using commonly modelled regular patterns of capillary distributions. *A* For a square array the area around each capillary is likewise square. *B* For capillaries distributed in an equilateral triangle array there are six neighbours, and the associated area is therefore a regular hexagon. *C* For capillaries in a hexagonal distribution it follows that the associated area is the complementary, triangular shape

Voronoi tessellations have been adopted for many practical and theoretical studies (Voronoi 1908; Miles and Maillardet 1982). An efficient computer algorithm was developed by Green and Sibson (1978) using ANSI Fortran, while specific application to analysis of capillary supply has been developed by L. J. C. Hoofd (Fortran-77; Hoofd et al. 1985) and H. F. Ross ('C' language; unpublished). This construction is known by various other names (Dirichlet or Thiessen polygons), reflecting its widespread utility for statistics/data analysis, particularly in geography (Boots 1986), and appears to have been rediscovered on a number of occasions. Biological applications have been more recent, most commonly in the context of ecological spacing (Cormack 1979), e.g. with respect to competitive behaviour between conspecific nest builders: the strength of aggressive tendency of a territory owner decreases monotonically with distance from the nest, and the pressure between identical neighbours balances on the perpendicular bisector or boundary line. In this way a habitat is partitioned into polygons. There are clear parallels with the physiological role of V_t associated with capillaries in terms of oxygen delivery, where boundaries define the minimum PO_2 of the tissue between adjacent capillaries of identical transport capacity.

5.3.2 Capillary Domains (Dom)

For capillary distributions that may be equated with a planar point process (e.g. muscle, liver or brain), the 2-D Voronoi tessellation may be useful, although in more linear tissue (e.g. intestinal or uterine wall) a definition of neighbourhood based on the MST might be more appropriate: i.e. analyses based on packing density or configurations respectively (O'Callaghan 1975; Kayser et al. 1986). In V_t each capillary is associated with a tile containing all the points in the tissue that are closer to it than any other capillary, and acts as the source or generating point for the tile, which is referred to as the domain of that capillary. The boundary of each domain is formed by the perpendicular bisectors of lines forming a capillary with all adjacent capillaries, producing irregular convex polygons (Fig. 14).

Hoofd et al. (1985) introduced the analysis of capillary domains in myocardium as a means of assessing capillary spacing, a procedure that also proved useful with sparse networks in skeletal muscle (Egginton et al. 1988a). For capillaries relatively far apart (long ICD) the domains will be larger than for those lying close together, and the distribution of individual domain areas will therefore reflect the heterogeneity of capillary supply (Fig. 16). Radii of circles with an equivalent area to the domain in cross-section provides a direct estimate of Krogh's radius:

$$R_K = \sqrt{\left[\frac{\bar{a}(\text{Dom})}{\pi}\right]}$$

where $\bar{a}$(Dom) = mean domain area.

As the domain may be thought of as an integration of 1-D distances between adjacent capillaries in 2-D, it is encouraging to note that analysis of ICD distribution in striated muscle by concentric circles (Turek and Rakusan 1981), triangulation (Renkin et al. 1981) and domains give rise to a similar, log-normal distribution (Fig. 17).

Many advantages accrue from such a local-based approach due to the extra information available from distribution analyses of the non-integer index a(Dom); moreover, the greater flexibility offered by 2-D permits interaction of capillaries and surrounding cell types to be examined, as well as providing a morphological framework within which to develop mathematical models of O_2 diffusion. Evaluation of the capillary supply is primarily derived from overlapping of domains and muscle fibres, and three main indices can be calculated:

1. Fibre : Domain Ratio (FDR); the number of whole or partial muscle fibre cross-sections that overlap each domain. This represents the number of muscle fibres

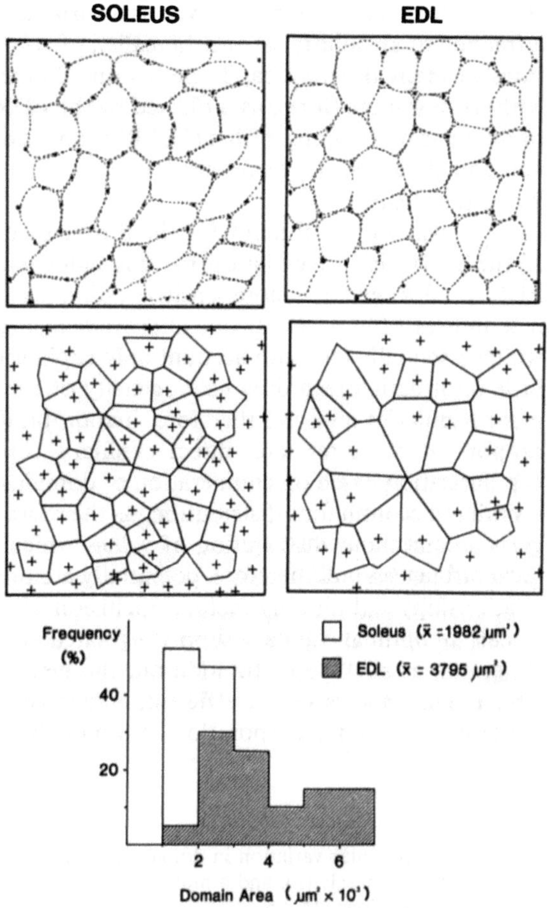

Fig. 16. Example of capillary domains applied to oxidative (soleus) and mixed (EDL) rat skeletal muscles. Fibre outlines (*dashed*) are shown in the *upper panel* along with capillary locations (*dots*) identified by alkaline phosphatase staining of cryostat sections. These samples were chosen from regions where mean fibre areas are similar in the two muscles, thereby avoiding a scaling affect. The *middle panels* show the calculated capillary domains. Note that in the oxidative muscle, capillaries (*crosses*) are located toward the centre of most domains, whereas the influence of glycolytic fibres in EDL is seen to produce highly asymmetric domains. Clearly, in this case, adopting a circular zone of influence, as in the Krogh model, is inappropriate. The histogram *at the bottom* shows that domains from an oxidative tissue are both smaller, and less variable than those from mixed tissue

in the vicinity of (probably supplied by) that capillary which, unlike sharing factor (SF; Sect. 3), accounts for non-contiguous capillaries. In mammalian muscle FDR approximates SF, but will be greater in sparse capillary networks such as in fast muscle of fishes.

2. Domain : Fibre Ratio (DFR); the number of whole or partial domains overlapping each muscle fibre cross-section. This is equivalent to the number of contiguous capillaries around a fibre (CAF; Sect. 3) only where CD is high, such as mammalian cardiac or slow muscle, but is significantly greater where the capillary supply is sparse or very heterogeneous.

3. Local Capillary : Fibre Ratio (LCFR); as a domain usually overlaps more than one fibre, the sum of the fractional areas of domains overlapping each fibre cross-section is calculated (Fig. 18). Note that this non-integer index is different from the overall C : F obtained using integer counts of capillaries and fibres in a section, and, when plotted against individual fibre area, shows that some fibres interact with only a fraction of a 'complete' capillary, while for most other fibres the

equivalent capillary supply is given as a variable number of whole capillaries plus a fraction of an additional one. In reality, of course, a fibre will interact with fractions of many different capillaries as one domain always overlaps (is in contact with) two or more fibres, as indicated by the FDR. The mean cumulative fraction of domains overlapping a fibre (LCFR) is approximated by the ratio of mean fibre and domain areas, and represents the supply of an average fibre in terms of 'capillary equivalents'.

It is also an useful index for the degree of heterogeneity in capillary supply, as the coefficient of variation for this index tends to be greater than for either FDR or DFR, giving a more sensitive estimate of intramuscular variability. These new indices offer a significant improvement over those described in Section 3, which essentially take into account only capillaries adjacent to a cell and provide little information regarding heterogeneity of capillary spacing, as they ignore remote capillaries and are, therefore, inappropriate for tissue where a significant proportion of fibres lack direct capillary contact. Morphometric indices of capillarization based on enumeration of domains and fibres (FDR, DFR) provide a similar discontinuous distribution, as the more usual anatomical indices; LCFR gives a continuous distribution of values which is likely to be more sensitive to local influences and, therefore, potentially a more informative index of functional relationships and limiting factors. Inclusion of remote capillaries in the analysis reveals an optimal capillary supply regulated at both the tissue and cellular levels (Egginton et al. 1988a). In addition, the new indices are readily computerized, which makes analysis more efficient, amenable to elaboration and, therefore, potentially useful for incorporation into models of tissue oxygen supply.

Fig. 17. Intramuscular variation in domain area along a transect over transverse sections of a purely oxidative muscle (*1*), and a muscle showing regional differentiation in aerobic capacity (*2*). Cryostat sections, 14 µm
1) Soleus. Both size and heterogeneity are similar all across the muscle, irrespective of fibre type composition. *Inset*: serial sections stained for m. ATPase to highlight the relatively homogeneous composition. Darker staining fibres are fast oxidative (FO).
2) Tibialis anterior (TA). Both size and heterogeneity are seen to vary across the muscle, reflecting the transition from predominantly large glycolytic fibres with low CD in the cortex (*sample A*) to the oxidative core (*sample D*). *Inset*: SDH sections showing the clear differentiation into regions of low and high oxidative capacity

	n	Mean	SD	Mean log	Variance log
Soleus					
A	219	316.6	100.8	5.712	0.088
B	251	269.1	87.7	5.545	0.099
C	261	265.4	90.0	5.525	0.115
D	193	354.6	104.9	5.826	0.095
TA					
A	88	3062	969.3	7.981	0.091
B	193	1369	682.7	7.105	0.238
C	248	1045	375.3	6.892	0.119
D	256	1039	391.5	6.883	0.123

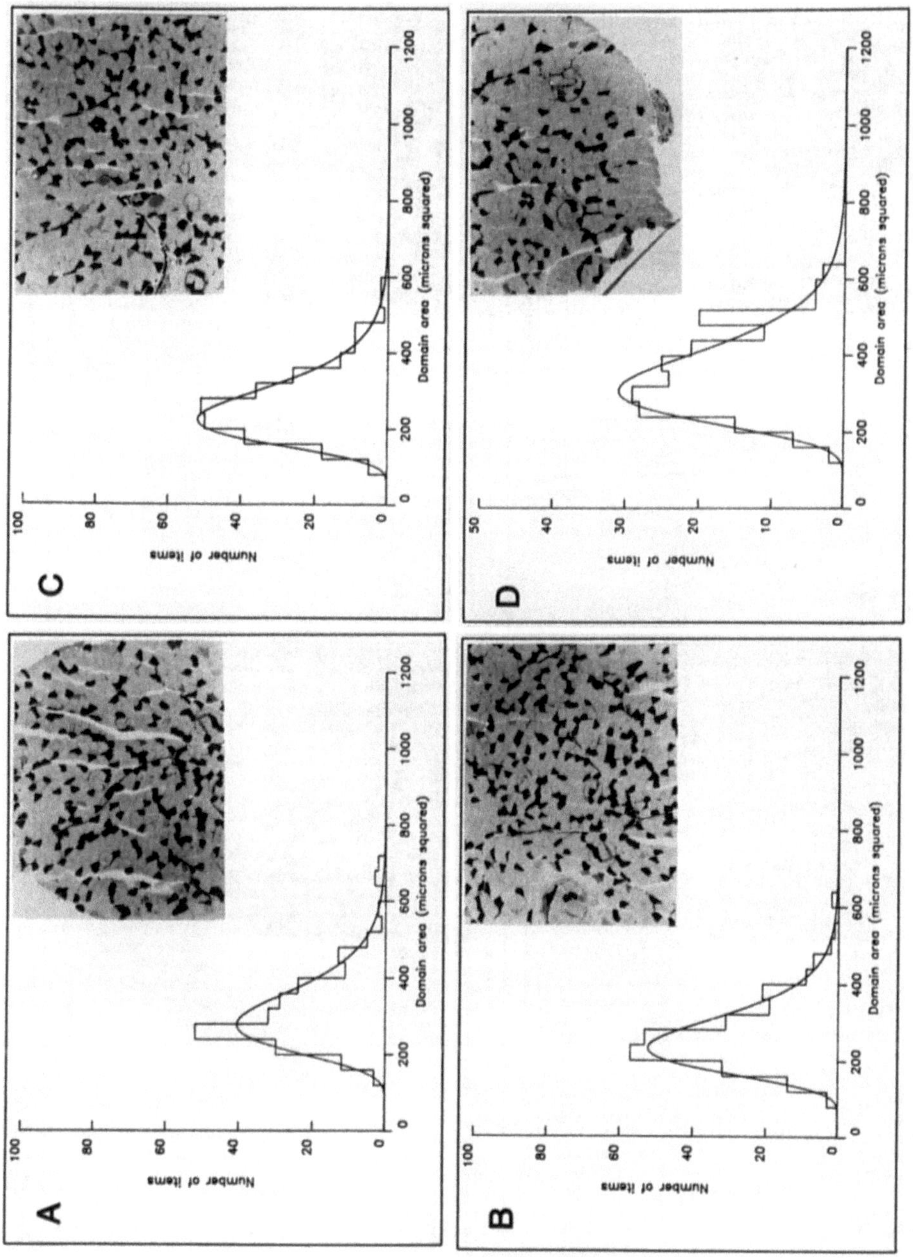

A
B
C
D
Number of items
Domain area (microns squared)
0
20
40
60
80
100
10
30
50
200
400
600
800
1000
1200

Fig. 17A–D (I)

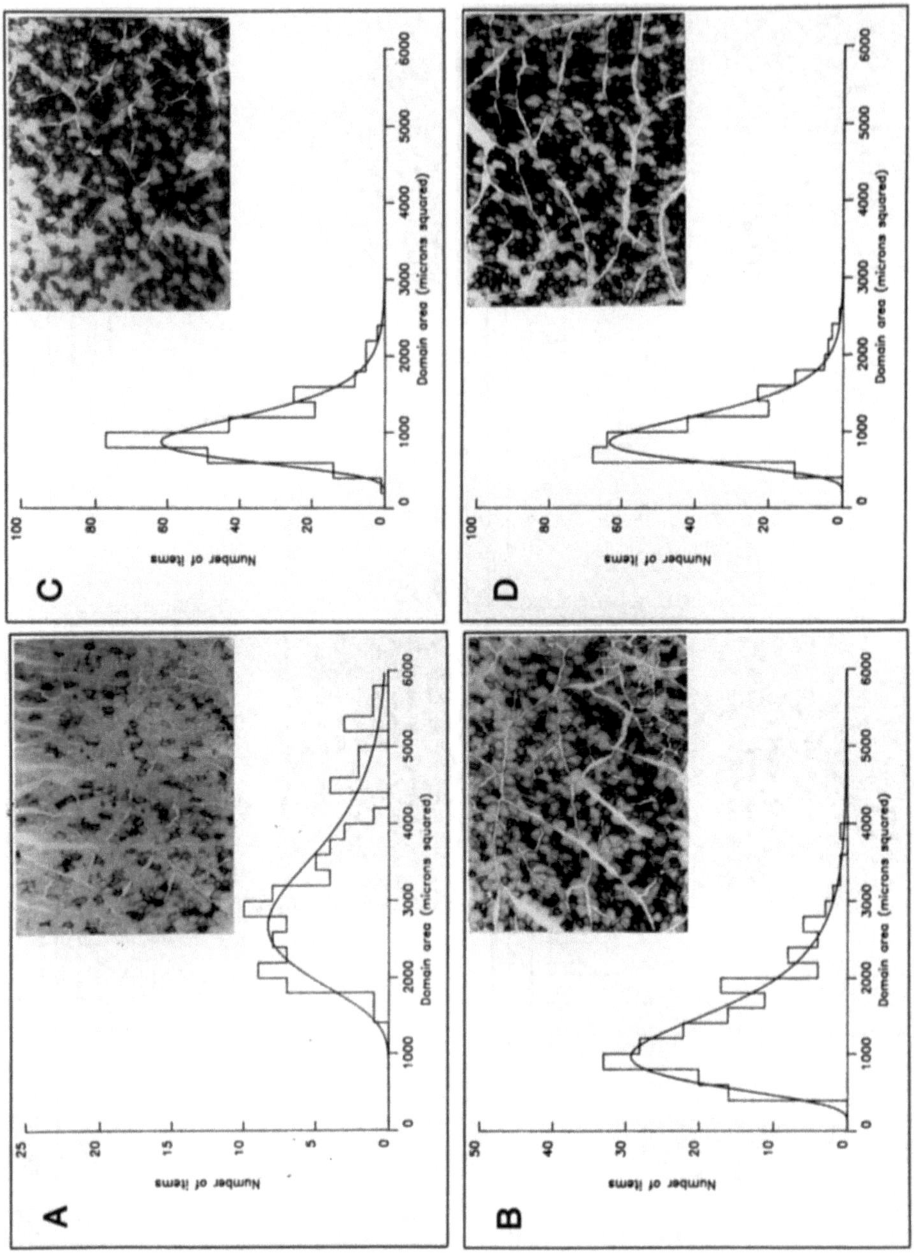
A
B
C
D
Number of items
Domain area (microns squared)
0
1000
2000
3000
4000
5000
6000

Fig. 17A–D (2)

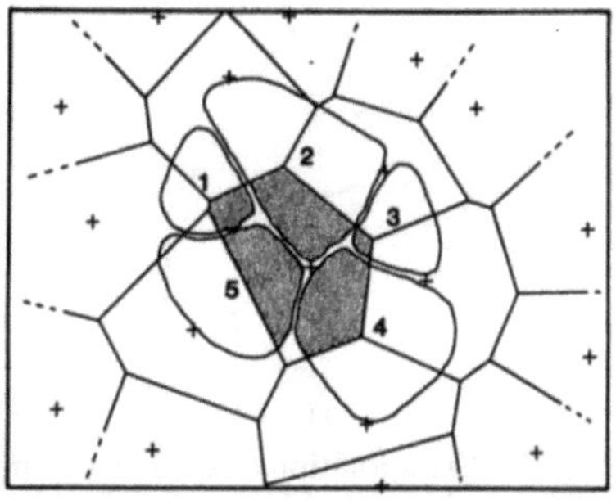

Fig. 18. Interaction of local capillary supply and individual cells. The central domain overlaps five cells (*1*–*5*) in proportion to its proximity, e.g. cell *3* is likely to derive most of its supply from the contiguous capillaries. Transport capacity of the central capillary will then be divided among adjacent cells in proportion to the fraction of domain area overlapping the cell (*stipple*) if all cells have equal metabolic demand. In this example, cell 3 would receive < 5% of the potential capillary supply while cells *2*, *4* and *5* would each receive > 20%. Summing the fractions of domains overlapping individual cells will provide an index of potential local capillary supply (LCFR), as equivalent number of capillaries (± fractions). If all capillaries are assumed to have similar transport capacity, this index may quantitatively reflect the available supply of O_2 or other foodstuffs. Note that cell *1* lacks direct capillary contact but potentially receives a modest proportion of total supply from four sub-adjacent capillaries. This can only be realistically evaluated using planar analysis of capillary distribution

In highly anisotropic tissue such as skeletal muscle, domains can be constructed using capillary or surface centres as the point of generation with very little error. One could be more objective in, say, first calculating the lumen centre of gravity, but as capillary cross-section usually accounts for < 2% of domain area the improved accuracy is minimal and probably dependent more on tissue preparation than on analytical refinement. In more isotropic tissue such as brain or cardiac muscle, not all capillaries are sampled in cross-section, giving rise to oblique profiles. Although linear distance to nearest neighbours may be accurately estimated, the error increases in proportion to distance of nth neighbours. Just as a domain can be constructed maximizing the distance from points, it can also be applied to distances between surfaces (a linear series of connected points), which may minimize error if boundaries of oblique capillary profiles are used. In the case of unusually large capillaries (e.g. icefish muscle; Fitch et al. 1984), estimates of diffusive flux into surrounding tissue will be improved by the use of capillary surface in this manner, although the required increase in computational effort necessitates development of a fast algorithm for routine usage.

If we accept that spatial point representation of capillary centres provides minimal error, the relative ease of computation may be utilized to provide additional information regarding capillary supply on a tissue-wide basis, and at the level of individual fibres. Capillary density is approximated by the inverse of mean domain area. Also, stereological theory states that an area associated with any point in 2-D is equivalent to the volume associated with unit length in 3-D. Hence, domain area, A(Dom), may be extrapolated to volume of tissue supplied per vessel, V(Dom), on multiplying by capillary segment length, or per capillary bed by muscle length (Fig. 19); reverting to use of equivalent Krogh's radius:

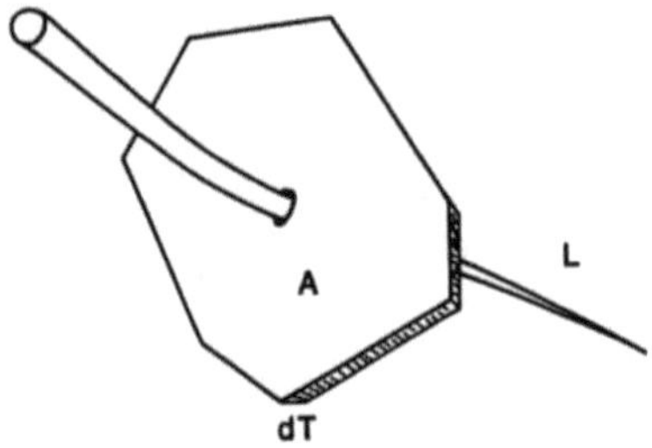

Fig. 19. Extrapolation of capillary domains into 3-D. A domain of area *A* and section thickness δT has volume $A \cdot \delta T$ associated with the source or generating point, a capillary of length *L*. In a regular, linear network the volume of tissue supplied by one capillary would be $A \cdot \delta T \cdot L$, following the Kroghian cylinder model, although realistically we need to analyze non-parallel networks. Length density is a global index that accounts for both changes in CD and tortuosity over the whole muscle: J_v is the length of capillaries associated with unit volume of tissue, while the reciprocal represents $A \cdot \delta T$ associated with δL. As $\delta T = \delta L$, then $1/J_v =$ mean domain area

$$V(\mathrm{Dom}) = \pi \cdot R_K{}^2 \cdot L$$

Clearly such a treatment is only valid for tissue conforming to the Krogh model in its purest form, i.e. a parallel linear capillary bed, and will deviate from reality in direct proportion to tortuosity of the vascular network.

Capillary length density (Sect. 3) is the length of capillary per unit volume of tissue, with the reciprocal then providing an independent check on the above relationship. The difference between these global parameters (A(Dom) and $1/J_v(c, f)$) will represent the error due to partial isotropy of the system; individual estimates of A(Dom) may then provide novel information regarding local, or at least regional, variation in capillary orientation. The only data set available for comparison at present is from slow and fast skeletal muscle of conger eel; $J_v(c, f) = 6.25$ and $0.22\ \mathrm{cm}^{-2} \times 10^{-4}$ respectively (Egginton and Johnston 1983), while corresponding values for A(Dom) are 1470 and 46530 μm^2 (Egginton et al. 1988a). In fast muscle the difference between the reciprocal of length density and mean domain area is therefore only 2%, i.e. corresponding to the calculated deviation from anisotropy of 1.6%. In slow muscle the error is greater (8%), although still acceptable for comparative purposes, possibly due to inter-animal variation, as different samples were used in these two studies.

5.3.3 Delaunay Tessellation (D_t)

The Voronoi tessellation is therefore very useful in describing planar neighbourhood conditions, inherently dealing with the distance and direction constraints required by linear approaches (e.g. O'Callaghan 1975) and providing a sound basis for numerous elaborations and applications (a few of which are outlined below). In addition, there is one derivative which permits a more objective (and possibly more efficient) estimate of ICD than that achieved by lineal methods.

Previously described methods of calculating ICD suffer from an inherent uncertainty as to which capillaries are contiguous. Nearest neighbours are strictly

Fig. 20. Delaunay tessellation (D_t). Given the primary (Voronoi, V_t) tessellation of points in 2-D space, the unambiguous complementary tessellation links centres of adjacent (contiguous) tiles. V_t is shown as *solid lines*, D_t as *dashed lines*. Note that in the majority of cases V_t produces three contiguous tiles at a vertex, such that D_t represents a space-filling pattern of non-overlapping triangles. Where vertices are formed by four contiguous tiles, D_t includes a square array

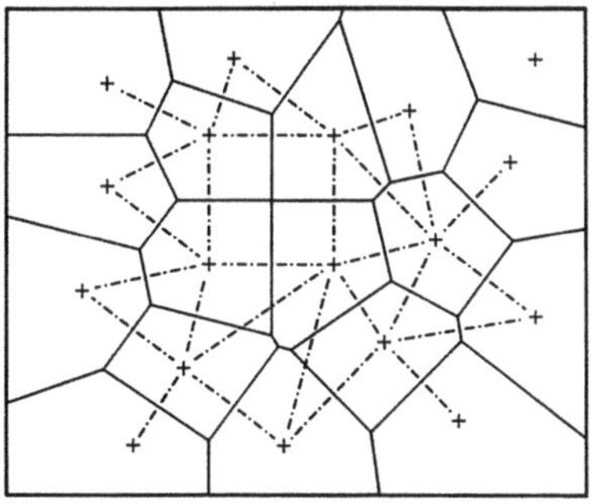

valid only for estimating minimal diffusion distances between capillaries in tissue of homogeneous $\dot{V}O_2$, while an objective scheme is necessary in order to account for non-nearest neighbours. In this construction, the line connecting pairs of points which share a common tile edge (i.e. domain boundary) is termed a contiguity; tiles with a common edge, and the corresponding points from which they are generated, are therefore contiguous. In general, three contiguous domains meet at vertices (although more are possible) such that lines joining contiguous polygon centres form a space-filling tessellation of triangles, the distribution of which is clearly linked to that of the primary tessellation: the perpendicular bisectors of triangle edges form domain boundaries, and domain vertices form their circumcentres. This geometric structure is well defined mathematically, and its properties intimately associated with those of the complimentary tessellation, V_t; e.g. the internal angle distribution is known for a number of defined spatial point patterns, and both V_t and D_t can be readily extended to 3-D (Miles 1970). It will be readily seen that contiguous capillaries form a planar graph, the Delaunay tessellation (Fig. 20). Given the coordinates of points (capillaries) and the domain boundaries we can therefore unambiguously identify the interpoint distance (equivalent to ICD) between adjacent capillaries by means of D_t, paralleling the linear graph approach (Sect. 4) where capillaries represent nodes and ICD links, and resembling the non-overlapping triangulation method of Renkin et al. (1981). Sequential construction of D_t around a capillary leads to a rosette of ICDs (Fig. 21), which, when repeated for all capillaries in a sample, will allow distribution analysis of the true ICD. This approach has a number of advantages over previous attempts to obtain such information. First, distances are measured, not derived, and limits are natural, not imposed. Second, estimation of all distances does not limit analyses to nearest or k-nearest neighbour, although such information may be readily obtained.

5.3.4 Sampling

Tests of random dispersion based on frequencies within quadrats are strongly influenced by the size of the sampling frame. Measurement of spacing between individuals may, in theory, operate without quadrats and therefore eliminate this problem, although in practice some delineation of sample area is required, proba-

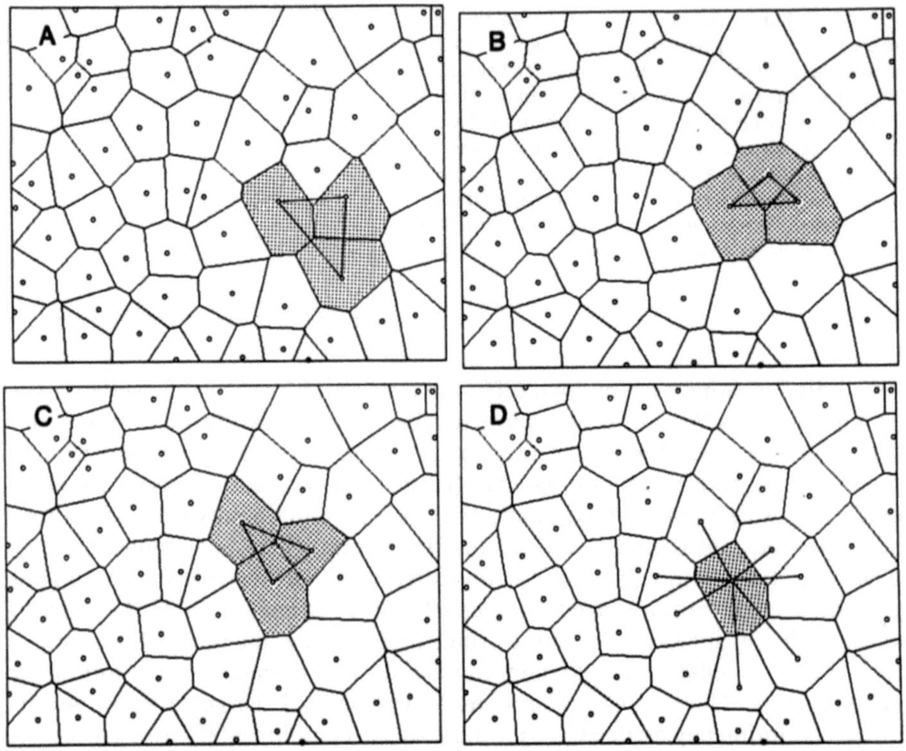

Fig. 21. Use of D_t for defining neighbourhood conditions in the analysis of linear distances (ICD) in fish slow muscle. For a given capillary (*filled circle*) a Delaunay tessellation is constructed sequentially around its domain (*A–C*), providing a rosette of distances to adjacent capillaries (i.e. those capillaries acting as generating points for contiguous domains). Note that in *panel A* this approach obviates the need for both a distance and direction constraint, including both nearest and non-nearest neighbours in an unambiguous manner. Constructing a rosette similar to that in *panel D* for all capillaries in the sample provides an efficient way of performing distribution analyses. Furthermore, this construction may be useful in syntactic structure analysis where higher order patterns are to be investigated, e.g. using sub-adjacent domains as the primary tessellation

bly by a self-limiting algorithm. A large sample is desirable, as the degree of spatial randomness observed depends on the imposed boundaries, since a set of points may appear randomly distributed within a small area but occur non-randomly with respect to a larger containing space. Point sampling bias may be removed by adopting the type of counting frame described in Section 2, similar to that used in estimating IPD (Appleyard et al. 1985) or ICD by triangulation (Renkin et al. 1981); this will coincidentally remove bias for all distances contained within the sample frame. However, if we wish to maintain the concept of neighbourhood (i.e. linear interpoint distances integrated over 2-D) a sample must be unbiased in the area, rather than the number distribution. For planar analysis of neighbourhood conditions, quadrat analysis is complicated in that a point may lie outside the frame, while its tile lies partly within. The simplest algorithm leads

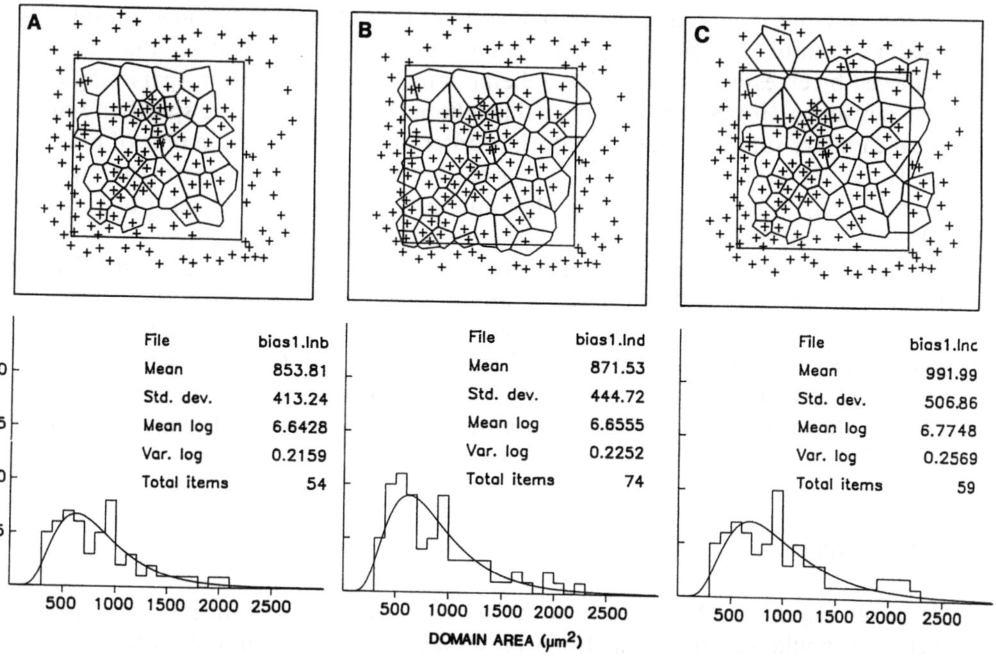

Fig. 22. Sampling bias in planar analyses. The potential error associated with erroneous counting rules adopted for capillary domains will increase directly with the heterogeneity of capillary spacing. In this example from rat EHP (a mixed fast muscle), two biased (though superficially attractive) and an unbiased sampling regime are illustrated. *A* Using a quadrat to sample only those domains entirely within the frame will tend to exclude larger domains, which have a probability of being transected by one or more frame edges proportional to their size. *B* Sampling domains associated with all capillaries contained within the quadrat reduces, but does not eliminate, this error. Note that this regime will provide an unbiased estimate of capillary *number* across a tissue (i.e. CD), but a biased sample of the distribution of *area* associated with the capillaries. *C* Domains are sampled in an unbiased manner if their boundaries lie entirely within the quadrat, or do not transect the 'forbidden' edges of the counting frame. Note that as the sampling bias is reduced the sample heterogeneity is increased: it is therefore quite easy to arrive at erroneous conclusions about the degree of order (variance) in a system if the sampling regime is initially biased

to exclusion of incomplete tiles (located at the periphery of the sampling frame and identified by coincidence of a domain edge and frame boundary) from the analysis (Fig. 22). This will introduce a bias toward smaller domains and give a weighted mean, underestimating both $\bar{a}$(Dom) and the frequency of larger domains. The bias will vary inversely with both homogeneity and density of the capillary supply, although in the examples shown this amounts to $< 5\%$ of $\bar{a}$(Dom) for both muscles.

This problem can be solved by applying an unbiased counting rule to the domains themselves, or adapting an associated point (Jensen and Sundberg 1986) such as centre of gravity or constant proportion (e.g. if $< 5\%$ of a domain intersects a frame, it is excluded) sampling regime. The first condition is relatively easy to apply with convex polygons, excluding those constructed from capillaries in the guard zone but intersecting the forbidden lines (Fig. 22). This provides an unbiased and more efficient sampling regime for distribution analysis than that achieved by excessive erosion of peripheral domains, particularly in sparse capillary networks. However, for associated neighbourhood analysis, i.e. specificity of local capillary supply, a more appropriate criterion is an unbiased sample of muscle fibres. The same approach can be used, but in this case the guard zone will be much larger, as it must include all domains overlapping peripheral fibre profiles.

Both Kroghian tissue radii and domain areas show a logarithmic-normal distribution; therefore, the heterogeneity can be represented by the logarithmic standard deviation $\sigma \log_{10} x$, more usually denoted by the abbreviation LogSD. This can be calculated from log-transformed variates using standard parametric statistics. Unlike the normal SD, the LogSD is dimensionless and independent of the magnitude of the mean, making it a useful comparative statistic (Diem and Lentner 1970).

5.3.5 Applications

Perfused Capillaries. Morphometric indices describing the anatomical interaction between capillaries and muscle fibres presumably represent a limiting factor only during the relatively infrequent bouts of maximal activity. During sub-maximal work the functional relationship will depend on additional factors such as oxygen consumption, blood flow and haematocrit. Domain analysis may therefore provide the geometric requirements for modelling of oxygen transport if used in conjunction with histological sections reflecting temporal differences in capillary perfusion (Egginton et al. 1988b). Increased blood flow to working muscle is often considered in terms of sequential recruitment of unperfused (non-patent) capillaries, although IVM observations suggest rather that this is due to a reduced intermittency of perfusion as a result of vasomotion. Whatever the cause, the effect is to produce an instantaneous spatial heterogeneity in capillary supply which ought to decrease in inverse proportion to demand (i.e. CD of patent capillaries increase and apparent ICD decrease). Little is known about perfusion of capillaries supplying muscles of differing aerobic capacity. In a muscle with regional differentiation of glycolytic and oxidative fibres, tibialis anterior, minimum activation caused a large reduction in heterogeneity of domains around perfused capillaries, which was not reduced further with increased activity (Fig. 23). It was therefore concluded that increased capillary perfusion is quantitatively independent of the degree of muscle activity (Egginton et al. 1988b). However, total blood

Fig. 23. Temporal heterogeneity of local capillary supply at rest. The anatomical capillary supply is represented by ALP-stained vessels in the oxidative core (*A*) and of glycolytic cortex (*B*) rat tibialis anterior. The lower CD in (*B*) results in a proportional increase in mean domain area. The corresponding functional capillary supply at rest is represented by fluorescence-labelled vessels (*C,D*) following timed infusion of a fluorochrome. Note that this instantaneous picture of perfused capillary spacing is qualitatively similar in both regions of the muscle, suggesting a lack of fibre type specificity in resting blood flow (Egginton et al. 1988b)

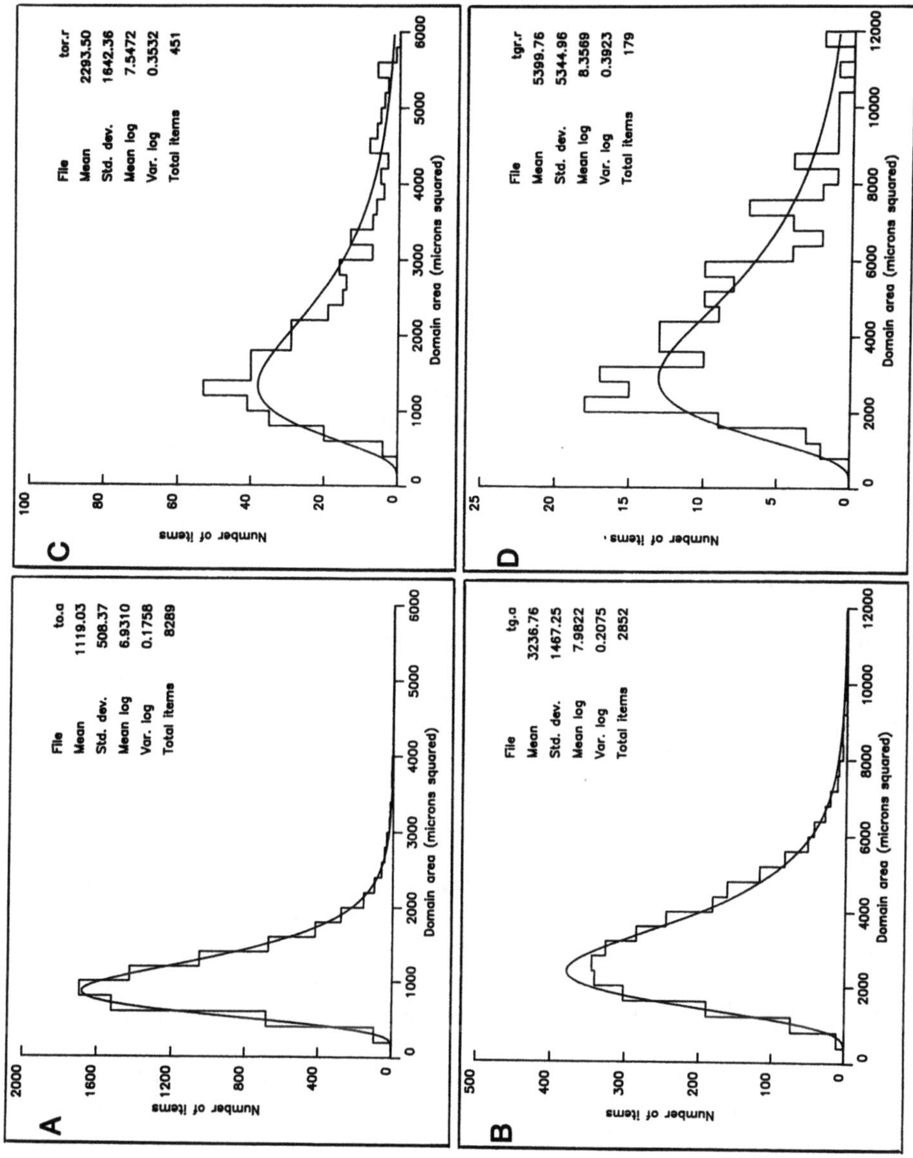
A
Number of items
Domain area (microns squared)
File to.a
Mean 1119.03
Std. dev. 508.37
Mean log 6.9310
Var. log 0.1758
Total items 8289
B
Number of items
Domain area (microns squared)
File tg.a
Mean 3236.76
Std. dev. 1467.25
Mean log 7.9822
Var. log 0.2075
Total items 2852
C
Number of items
Domain area (microns squared)
File tor.r
Mean 2293.50
Std. dev. 1642.36
Mean log 7.5472
Var. log 0.3532
Total items 451
D
Number of items
Domain area (microns squared)
File tgr.r
Mean 5399.76
Std. dev. 5344.96
Mean log 8.3569
Var. log 0.3923
Total items 179

Fig. 23 A–D

flow increased more in oxidative than in glycolytic regions; this difference cannot be accounted for by recruitment of 'unperfused' capillaries, and must result from a significant increase in velocity (Egginton et al. 1988). The clustering of perfused capillaries supports the hypothesis that capillaries are arranged in functional units, each fed by a common arteriole. Capillary units have been reported in a number of tissues and it has been proposed that they function on the fundamental control and distribution elements of the microcirculation (Eriksson and Myrhage 1972). Lund et al.'s (1987) intravital observations of hamster TA showed such units to consist of approximately 15 capillaries, each supported by a common arteriole and drained by a common venule. It remains to be demonstrated whether the perfusion pattern is influenced at levels other than vasoactive arterioles. Some indication of this may be gleaned from the relationship of bivariate spatial point patterns, but proximate factors will have to be empirically determined. Domain analysis may help to distinguish between the two main candidates at the cellular level: fibre size and fibre type.

Fibre Size. Much of the data from traditional morphometric indices (Sect. 3) suggest that capillary supply may be primarily regulated by physical separation of individual vessels, imposed by normal cellular hypertrophy. Important changes in capillarization of skeletal muscle occur as a function of fibre area; an increase in $\bar{a}(f)$ appears to be compensated by an increase in C : F and CAF (Loats et al. 1978; Aquin and Banchero 1981), limiting the maximal distance of tissue from a capillary to within the 50-μm radius assumed adequate for oxygenation (Akmal et al. 1978). However, given the limitations of these indices, the interaction of capillaries with cells of different size is not readily apparent, particularly where cells have little or no direct contact.

Mean values of the corresponding planar indices, DFR and FDR, give comparable results only where fibre size and domain area are of similar magnitude; CAF and SF would be more informative if fibre size were to be taken into account. From our studies it seems likely that relatively distant capillaries may contribute to the metabolic activity of a fibre, albeit to a lesser extent than adjacent capillaries, in proportion to the extent of domain overlap. The importance of distant capillaries is reflected in a high correlation between the local capillary to fibre ratio (LCFR) and fibre area in skeletal muscle, which suggests that the oxygen/substrate supply and/or metabolite removal of working skeletal muscle fibres is not restricted to capillaries contiguous with the fibre surface, but extends to remote capillaries whose domains show some overlap with fibres with which they have no direct contact (Egginton et al. 1988a).

In the example shown (Fig. 24), CAF shows rather a coarse correlation with $\bar{a}(f)$, whereas LCFR shows a much tighter distribution; it is interesting to note that the slope is not significantly different between animals, although the fibres in cold-acclimated fish have undergone significant hypertrophy. A positive correlation between LCFR and fibre area suggests that supply is increased directly with fibre area, i.e. the tissue-wide integration of capillary supply is homologous, irrespective of mean fibre area. A similar conclusion has been reached in a study of rat cardiac tissue, where it was calculated that an increase in the number of capillaries surrounding a fibre with fibre area is potentially an adequate mechanism

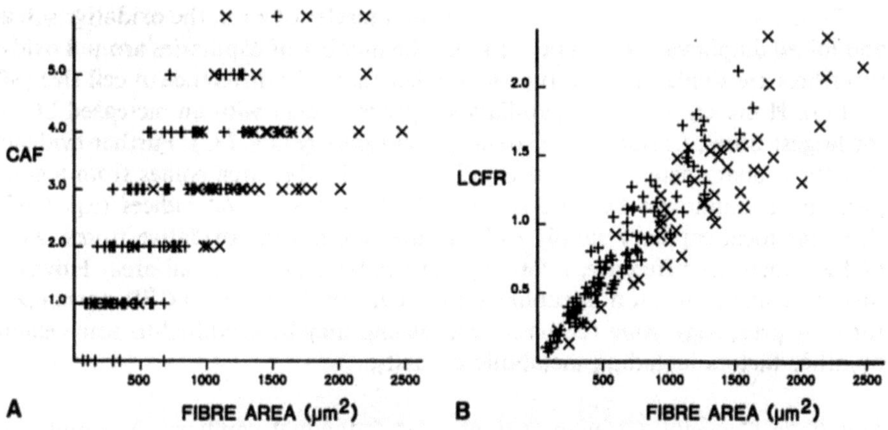

Fig. 24. Effect of cell size on local capillary supply. Examples from slow muscle of striped bass acclimated to 25 °C(+) or 5 °C(x); 0.5 μm resin sections, toluidine blue stain. There is a large scatter in number of capillaries around a fibre (CAF), while the fractional capillary supply (LCFR) shows a better correlation. Note that the effect of normal muscle growth (intra-group variation) and experimentally induced fibre hypertrophy (inter-group variation) produces a similar regression line, passing through the origin

for maintenance of intracellular PO_2 levels (Rakusan et al. 1984). In skeletal muscle, then, increasing fibre area is matched by capillary proliferation, resulting in a relatively constant average LCFR per unit area of fibre (CFD).

In high CD networks where capillarization is limited by cell size, the point process describing capillary distribution is likely to be paralleled by a point process associated with the intervening cells, e.g. centre of gravity. This may, therefore, lead to complementary tessellations representing supply and demand, and provide an extension to the MST approach in detection of abnormal distributions.

Fibre Type. It is generally assumed that the size of the capillary bed increases directly with the level of oxidative metabolism in a tissue. While in most cases this is undoubtedly true on a global scale, the degree to which this is reflected on a local level is uncertain, being dependent on the chosen index. The improvements offered by planar analysis in modelling capillary supply (inclusion of contiguous and remote vessels without duplication, and fibre area) offers the potential to differentiate between the effects of fibre size and phenotype with a resolution not previously possible. In a survey of muscle types, the index of heterogeneity, LogSD, shows a decrease in the sequence mammalian fast > mammalian slow > fish skeletal muscle, while an inverse relationship is seen with the correlation coefficient of LCFR and fibre area. This may reflect differences in muscle composition, as EDL has three (FG, FOG, SO), soleus has two (FOG, SO) and fish muscle consists of only one fibre type (SO or FG). The qualitative distribution of capillaries (heterogeneity), therefore, appears to be independent of aerobic capacity per se, while it is quantitatively related to the level of oxidative metabolism by CD (Egginton et al. 1987a). The homologous spatial distribution of capillaries between muscles must, therefore, be relatively independent of the capacity for oxidative metabolism.

Table 1 shows a comparison between two muscles of a rat, the oxidative soleus and mixed diaphragm. It can be seen that the number of capillaries around oxidative fibres are similar, irrespective of whether there is a difference in cell size (SO vs. FO). However, scaling of capillary supply is evident with an increased LCFR for largest fibres, irrespective of oxidative capacity (FO v. FG). Further evidence that the primary determinant of capillarization is fibre area comes from a comparison of diaphragm FG and soleus SO fibres: traditional indices (e.g. CAF) show the local capillary supply to be greater around the oxidative fibres, while LCFR shows no difference, reflecting the similar cross-sectional area. However, there is a small variation in capillary fibre density (CFD), i.e. LCFR normalized for fibre area, suggesting the basic relationship may be modified to some extent by other factors including metabolic capacity.

Metabolic Capacity. Given partitioning of individual capillary O_2 supply between surrounding fibres of different oxidative capacity (above), it is now possible to test assumptions regarding the equivalence of O_2 transport capacity for all capillaries. Respirometry studies of tissue slices provide an estimate of $\dot{V}O_{2max}$ and, from muscles composed of a single fibre type, the relative capacity for O_2 demand. The magnitude will clearly vary among species; e.g. in fishes, slow muscle respiration is some 2.5 times that of fast muscle for sluggish species like eel (Egginton 1982), while in the active and highly aerobic tunas this differential may be as great as 6.2 (Gordon 1968). We can then construct a vector diagram for each capillary, with O_2 flux given as the product of relative $\dot{V}O_2$ (e.g. a ratio of 4:1 for red:white) and fraction of the domain overlapping that fibre. This is conceptually similar to the approach of Gray and Renkin (1978), although is less laborious (making inter-tissue surveys feasible) and, being a continuous variable, of higher resolution. This will provide only the first stage in an analysis of local control: a narrow distribution of compound values per capillary would only indicate regulation of O_2 transport capacity and/or O_2 demand.

However, it is known that within a given motor unit or fibre type there is a significant heterogeneity in enzymatic and structural correlates of oxidative metabolism (Nemeth et al. 1979; Egginton and Johnston 1982); hence fibre typing may be too coarse a criterion with which to assess the specificity of capillary supply at a local level. Hoppeler and Lindstedt (1985) have recently demonstrated a relatively constant rate of maximal O_2 uptake per unit volume of mitochondria in mammals, irrespective of the overall aerobic capacity of the muscle; 2–3 ml O_2/g wet, wt. mitochondria/h. If capillary supply is indeed matched at a local level with maximal tissue O_2 demand, then the higher volume density of mitochondria in small domains (usually overlapping oxidative fibres) should equate with the same total O_2 demand of a low $V_V(\text{mit, f})$ in larger domains, i.e.

$$V_V(\text{mit, f}) \times a(\text{Dom}) = V(\text{mit, dom}) = \text{a constant}$$

This autoregulation of capillary spacing and $\dot{V}O_{2max}$ of intervening tissue is implicit in some of the 1-D analyses; we are now able to explore this relationship in two or three dimensions.

Table 1. Effect of fibre type and size on local capillary supply[a]

	Diaphragm							Soleus	
N(f)	189							130	
N(c)	278							307	
CD, mm^{-2}	936.2							802.7	
C:F	1.47							2.36	
ā(f), μm^2	1457							2510	
Fibre type	SO	P[b] < >	FO	P < >	FG	P < >	SO	P < >	FO
Numerical fraction	0.222	–	0.497	–	0.280	–	0.600	–	0.400
ā(f), μm^2	918 ± 36.9	NS	981 ± 33.7	***	2729 ± 69.7	NS	2723 ± 79.3	***	2190 ± 54.6
Areal fraction	0.140	–	0.375	–	0.485	–	0.643	–	0.357
CAF	3.52 ± 0.178	NS	3.39 ± 0.120	*	3.87 ± 0.214	***	5.19 ± 0.193	NS	4.92 ± 0.162
FCD, mm^{-2}	4065 ± 255	NS	3666 ± 149	***	1451 ± 79.7	***	1998 ± 84.7	*	2285 ± 79.6
LCFR	1.05 ± 0.057	NS	1.07 ± 0.042	***	2.24 ± 0.078	NS	2.31 ± 0.074	***	1.93 ± 0.054
CFD, mm^{-2}	1145 ± 47.3	NS	1100 ± 27.3	***	825 ± 24.1	NS	856 ± 16.4	NS	894 ± 24.1

[a] Values given as mean ± SEM. Data collected from one rat and one representative field per muscle; serial cryostat sections stained for ALP, m·ATPase and SDH to localize capillaries and identify fibre type on the basis of phenotype and oxidative capacity, respectively.
[b] Level of significance is shown for adjacent columns: * $P<0.05$; *** $P<0.001$; NS, not significant.

Weighted Tessellations. Linear separation of capillaries has, for many years, been utilized in modelling of O_2 transport, since mean ICD is equivalent to twice Krogh's radius, R_K. ICD of adjacent capillaries based on contiguous domains (i.e. D_t) will only equate with minimal or nearest neighbour estimates with small fibres, high CD and a homogeneous planar distribution. However, this approach has the advantage of being able to readily scale ICD according to mean $\dot{V}O_2$ of the intervening tissue (e.g. metabolic fibre type), thereby producing an alternative vector diagram of O_2 flux for individual capillaries.

It is important to reiterate that most of this chapter, and literature on capillarization in general, deals with anatomical indices that are thought to reflect capillary supply, with the implicit assumption that all vessels have an equal transport function. Using this approach, correlation between structure and function is essentially limited to describing the capacity of a system; in this case tissue $\dot{V}O_{2max}$. Vector diagrams for perfused and all capillaries are a step toward accounting for local variations (temporal and spatial heterogeneity) in the magnitude of O_2 supply and/or demand. An alternate approach is to construct domains in a manner reflecting differential O_2 transport. Weighting may be applied to vessels such that the domain boundary becomes more distant from a capillary in proportion to the relative O_2 supply, e.g. flow rate. If flow is less in neighbouring capillaries, a domain will expand to include tissue not originally supplied by that capillary showing hyperaemia; if flow is increased equally in adjacent capillaries the common boundary will be unaltered, as relative rather than absolute values are important in describing the zone of influence of an individual capillary (Fig. 25). Of course, if domain size were unaltered following increased flow, there

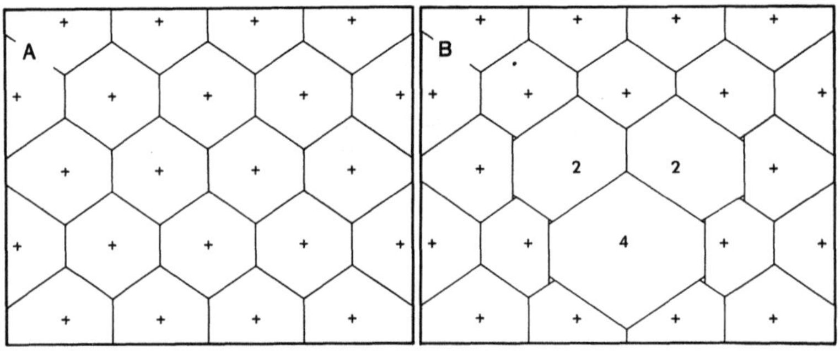

Fig. 25. Weighted capillary domains. *A* Domains are normally constructed from the distribution of capillaries with equal weighting. A regular diamond array of capillaries with equivalent transport capacity, supplying tissue of homogeneous metabolic capacity, will produce a hexagonal compressed tessellation. *B* A crude positive weighting may be applied in proportion to the flow rate through individual capillaries, or a negative weighting proportional to the mean O_2 demand of the enclosed tissue. For example, let *crosses* represent capillaries with unit transport capacity, and *2 & 4* represent the location of capillaries with double and quadruple capacity respectively. The zone of influence (domain) of these capillaries increases accordingly, reducing the area of tissue supplied by contiguous capillaries of lower transport capacity. A similar picture would be obtained where the scaling factors represent the differential demand of tissue covered by a domain (half and one-quarter of other tissue respectively) supplied by capillaries of equal transport capacity. Clearly, if weighted domains are to be representative of in situ conditions, account has to be taken of variations in both supply and demand

must be an increased O_2 flux to tissue. Clearly the difficulty here is determining the appropriate weighting; factors which will affect this are well known from thin or isolated preparations (RBC transit time; local haematocrit, etc.), and this approach may well be limited to such cases, or restricted to analytical simulations.

Scaling for tissue O_2 demand is probably more useful; if an approximately equal supply is assumed for all capillaries, then the simplest solution is to adopt a negative weighting relative to the highest tissue $\dot{V}O_2$. In the example given (Fig. 25), a weighting of 1 may signify a domain overlapping oxidative fibres, while the capillary with a weighting of 4 is supplying glycolytic fibres with one-fourth the O_2 demand of oxidative fibres; hence, a capillary with similar O_2 supply capacity can accommodate a greater tissue volume. In this way domain area will be weighted according to the average oxidative capacity of the tissue surrounding each capillary. More helpful would be to locally weight demand within a domain according to the oxidative capacity of fibres with which it intersects. As the Voronoi polygon is constructed of a set of points associated with a given source (capillary), weighting, using a lattice framework, can be ascribed to all such points to define a weighted distance between any location in the polygon and the capillary. Such a scheme may produce domains with straight, convex or concave edges (Boots 1986) and will answer Krogh's original plea for the influence of surrounding tissue to be incorporated into models of tissue oxygenation, essentially giving a complex realisation of the 'cylinder' boundaries.
Using the 1-D diffusion equation:

$$PO_2 = (\dot{V}O_2/2 \cdot KO_2)(x - x_0)^2$$

where $\dot{V}O_2$ = rate of O_2 consumption (i.e. flux),
KO_2 = Krogh's coefficient of diffusion $(D \cdot \alpha)$,
x = any linear distance from a capillary,
x_0 = linear distance to end (minimal) PO_2,
and having defined the domain boundary (a linear series of points, x_0), then O_2 flux lines may be generated:

$$\dot{V}O_2 = KO_2(d^2PO_2/dx^2)$$

Previously this has only been applied to idealized cylinders, although recently this has been extended to include more realistic distributions (Fig. 26). That domain boundaries are seen to closely mirror zones of zero O_2 flux in cardiac tissue does indeed indicate that these anatomical limits do have some functional relevance. Clearly a more stringent test of their validity will be a comparison of flux lines in tissue of heterogeneous O_2 demand, and variable supply, with boundaries of weighted domains.

It has been shown that domains are asymmetric, i.e. the capillary is located toward one edge. The above analysis of diffusion shows higher PO_2s will occur towards the shortest boundary, while the larger distances to the opposite edge of the domain leads to lower PO_2s than predicted by the Krogh model (Hoofd et al. 1987). Given that the Krogh cylinder defines the limits for most mathematical models of oxygen transport, these more realistic boundaries may help answer some fundamental questions, such as whether all capillaries are equivalent sources of O_2 and the effect of spatial heterogeneity in flow rate. Capillary do-

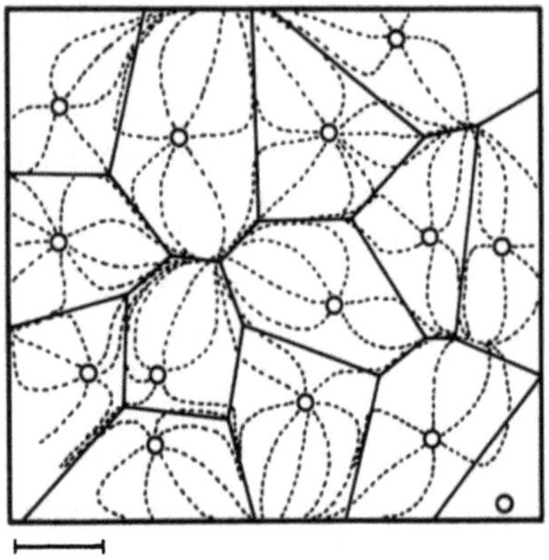

Fig. 26. Calculated structural and mathematical views of capillary O_2 supply. In this example from rat left ventricular wall (camera lucida tracing of wax-embedded, haematoxylin and eosin stained section) the capillaries (*open circles*) are assumed to possess equivalent transport capacity (equal size, flow rate and PO_2) supplying tissue with a homogeneous O_2 consumption. All flux lines (*dashed*) end at the same end point, the location of minimal PO_2. Note that this closely matches the domain boundaries (*solid lines*). Hence, under these particular conditions and likely to represent maximal hyperaemia in highly oxidative tissue, capillary domains provide an anatomical representation of the irregular, local Kroghian tissue 'cylinder' (Courtesy of L. J. C. Hoofd)

mains also provide a unique opportunity to test postulated control mechanisms; for example, it is assumed that the decrease in PO_2 from arterioles to venules will be compensated for by an increased CD. If local capillary PO_2 and tissue oxidative capacity can be determined, the efficiency of reduced capillary spacing may be realistically assessed. We may then be in a position to reconcile physiological and anatomical realizations of 3-D capillary distributions.

6 The Role of Capillaries in Oxygen Transport to Tissue

From this series of books it will be evident that there are several possible modes of O_2 transport, each of which may be important at different levels of the respiratory cascade. For example, stirring may be important at the environmental interface, while diffusion is paramount at the mitochondrial surface. Blood vessels are mainly concerned with the convective delivery of substrates to cells, or removal of metabolites, at which point simple or facilitated diffusion predominates. Oxygen delivery is particularly important in organs such as the brain or working muscle, where a continuous and adequate supply is required to fuel mitochondrial respiration and energy production. The vascular supply of vertebrate striated muscle is plastic and clearly matched with functional demand of the tissue, e.g. capillary density (CD) is determined by muscle phenotype and/or activity level (Hudlická 1984; Hudlická and Tyler 1986). Although it is unclear what factors control functional capacity of the microcirculation, the role of capillary supply has been mainly examined in terms of oxygen delivery, rather than removal of metabolites and heat. However, while capillary density is known to covary with levels of oxidative enzymes, these parameters simply define the morphological and biochemical capacity (potential) for maximal work. There are few studies that address the question of dynamic correlation between the functional capillary supply and level of tissue oxygenation. Furthermore, it seems likely that the role in purely oxidative or glycolytic muscle is different, and may account for many of the modifications acquired during phylogenetic development of functionally homologous tissue. Acute increases in tissue oxygen demand are thought to be met by capillary recruitment, while chronic increases in aerobic capacity are met by capillary proliferation, which optimizes intramuscular diffusion. Indeed capillary supply often varies directly with mitochondrial content (Egginton and Johnston 1983), but capillary proliferation does not parallel an increased mitochondrial content in small diameter fibres (Egginton and Johnston 1984), nor when oxidative metabolism of slow muscle is increased following training (Müller 1976). Conversely, increased activity in fast muscles with limited blood supply leads to an increase in oxidative capacity without changing capillarization (Hudlická et al. 1988). Furthermore, following induced activity, capillary blood flow increases more around fast glycolytic (FG) than slow oxidative (SO) fibres (Dawson et al. 1987). These data demand a reappraisal of the role of capillaries in fast skeletal muscle, and support the proposals of Schmidt-Nielsen and Pennycuik (1961) that, during activity, the delivery of substrates and removal of lactic acid and other metabolites are important functions of the microcirculation. The capillary density of glycolytic muscle is low and is directly related to lactate production (Hudlická et al. 1987). Tissue PO_2 at maximal activity may, therefore, be a limiting factor in determining the size of capillary bed in oxidative muscle, while metabolite concentration may act similarly in fast muscle, although it is unclear whether PO_2 retains a regulatory function. For example, a reduction in intracellular diffusion distances may ameliorate limitations in O_2 supply (Egginton et al. 1987c), while differential oxygen solubilities through a tissue will markedly affect O_2 flux densities (Longmuir 1981).

Caution must be exercised in inferring functional capacity of any system from morphometric data alone. For example, muscle capillarization has been correlated with metabolic fibre type (Romanul 1965; Hudlická et al. 1987), body size (Krogh 1919; Schmidt-Nielsen and Pennycuik 1961; Weibel et al. 1981) and whole animal $\dot{V}O_2$ (Hoppeler et al. 1981), and can be induced by experimental hypertrophy (James 1981) and chronic electrical stimulation (Brown et al. 1976). However, the distribution of newly transformed oxidative fibres following hormone treatment did not correspond to regions of neovascularization, nor was oxidative enzyme activity always correlated with capillary supply (Capó and Sillau 1983; Egginton et al. 1987a, b). It seems likely that there is an integrated reserve capacity at different levels of organization within the respiratory cascade, an important factor in determining the functional consequences of morphometric estimates of capillary supply. Direct effects of a change in O_2 demand may, therefore, only be manifest in microvascular adaptation if the initial situation was limiting with respect to O_2 transport.

The structure of capillaries is characteristic of exchange systems – short transport distances and large surface area – which may be quantified to provide an estimate of transport capacity. The diffusing capacity (physiological conductance) for O_2 can be related to the capillary dimensions and the physiochemical properties of the tissue medium through which diffusion occurs, by a modification of Fick's law of diffusion (Wangensteen and Weibel 1982). Diffusing capacity, D, depends on Krogh's diffusion coefficient and the barrier to diffusion made up of compartments in series: erythrocytes, plasma, capillary wall, interstitial space, plasmalemma, sarcoplasm and mitochondrial membranes, but no systematic quantitative study has been made of O_2 path lengths between different muscle types and species. Morphometric analyses can be used to determine the resistance to oxygen diffusion from capillary lumen to mitochondrial cristae, allowing its incorporation in models of oxygen transport to tissue. The resistance to gaseous diffusion may be determined by surface area of the capillary lumen and the external endothelial cell boundary, while translational distances across the capillary wall (barrier thickness) provide a sensitive index of both active and passive transport capacity. The capillary wall is thought to present a significant barrier to transendothelial flux of oxygen in an isolated rete (Rasio and Goresky 1979), but so far whether this also applies to capillaries of muscle and interstitial spaces is only conjecture. Morphometric estimates of D may also provide information about maximal O_2 delivery, although this will need to be compared with direct estimates of maximal O_2 consumption ($\dot{V}O_2$ max) measured by in vivo a-v differences. However, if diffusion is limiting, greater demand (maximal flux) should be reflected by changes in local structural relationships, while global parameters (such as CD) will scale according to the magnitude of metabolic demand.

Developments in computational techniques has facilitated mathematical modelling of physiological processes, although progress has not been equal in all disciplines. For example, in the field of pulmonary gas transport it is recognized as essential to interpret physiological measurements in the light of theoretical principles, while close co-operation between experimentalists and theoreticians studying gas exchange within tissue is less common. This is reflected in many experimental studies neglecting theory and modelling completely, while theoretical studies of O_2 supply are often based on unrealistic assumptions. That modelling of O_2 supply to tissue is still in its infancy is illustrated by the fact that the most frequently used model is that of August Krogh, developed in 1919. Central to this first attempt at morphometric quantification was the suggestion that a capillary network can only make efficient use of the available blood volume if the number of patent capillaries, those open to flow, varies with time. The observation of a tenfold increase in apparent density of perfused capillaries in contracting relative to resting muscles led to the concept of capillary recruitment (Krogh 1919), which remains an implicit basis for many studies on capillarization. In this scheme only a fraction of the total number of capillaries are perfused at rest, an average of around 50% being quoted even in the highly oxidative myocardium (Henquell and Honig 1976; Weiss and Conway 1985). Increased tissue demand, $\dot{V}O_2$, stimulates an increased supply, $\dot{Q}O_2$ which is achieved by 'recruitment' of previously unperfused vessels. This increase in apparent CD will reduce intramuscular diffusion distances, thereby enhancing O_2 flux. The Krogh-Erlang formula permits a quantitative analysis of global oxygen supply based on a cylinder of tissue served by a straight, unbranching capillary at the centre. This identifies the major determinants of tissue oxygenation in a radial manner:

$$P_x = P_c - \frac{\dot{M}O_2}{KO_2}\left(\frac{R_k^2}{2}\ln\frac{x}{r} - \frac{x^2 - r^2}{4}\right)$$

where P_x is the PO_2 at distance x from capillary centre, P_c the capillary PO_2, $\dot{M}O_2$ the tissue O_2 consumption, KO_2 Krogh's diffusion coefficient (the product of diffusivity, D, and solubility, α), R_k tissue cylinder radius, r capillary radius. The model requires at least eight basic assumptions about the uniformity of capillary supply, blood flow, O_2 diffusion, tissue respiration rates, etc., that are rarely found in vivo (Longmuir 1981; Kreuzer 1982). Despite this, it does provide good predictions of muscle PO_2 in some cases (Akmal et al. 1978), although it also predicts that 20% of the myocardium in 'resting' hearts may be hypoxic (Bourdeau-Martini 1974)! Some workers believe, therefore, that in contrast to Krogh's 'centrifugal' model, a 'centripetal' model (in which a tissue cylinder is supplied with O_2 from the periphery, e.g. Hill 1928) is more realistic, particularly in contracting muscle (Groom et al. 1984). Nevertheless the simplicity of Krogh's model ensures its popularity, given that quantitative information about O_2 supply may be obtained from routine histological preparations – counts of capillaries viewed in transversely sectioned material – since R_k, the boundary with no radial O_2 flux, can be derived from estimates of CD or ICD. It is not difficult, therefore, to see why CD has come to be regarded as an index of fundamental importance. However, O_2 flux depends not only on the distance from a source (capillary), but also on the area of tissue supplied by that capillary and the topological relationships with neighbouring capillaries.

The inhomogeneity of oxygen delivery to and uptake from tissue results in temporal and spatial heterogeneity in local PO_2. In mammals, systemic blood PO_2 is relatively constant, so that all tissue would receive similarly oxygenated blood. In lower vertebrates there is variation in oxygen delivery which reflects different circulatory systems; e.g. exercise in eels causes no hyperaemia, but increases the O_2-saturation of blood (Davie and Forster 1980), thereby presumably increasing capillary PO_2. In addition, changes in the physical capillary bed may be accentuated or rendered unnecessary by the various metabolic and physiological modifiers of oxygen delivery discussed elsewhere in this book. Although it is clear that changes in the hierarchy of resistance to oxygen delivery above the capillary bed, from respiratory surfaces to the terminal vasculature, may have a profound effect on its functional capacity, much less is understood about how changes at a lower level of organisation may affect oxygen flux. While morphometric analyses may determine the anatomical potential, constraints imposed by physiology need also to be considered. A more complete description of oxygen supply will require a modification of Krogh's approach to not only allow for heterogeneities in perfusion distribution and blood flow but also factors affecting oxygen-haemoglobin association (such as tissue myoglobin content) and mitochondrial respiration.

Above all, the capillary bed requires a high degree of flexibility in functional capacity, but how the architecture contributes to this remains uncertain. For example, plasma skimming occurs in fast flowing blood, with a preferential flow into capillaries branching close to arterioles, while the haematocrit may increase in distal capillaries where blood flow has slowed. This potential heterogeneity in O_2-carrying capacity may result in either steep or shallow longitudinal gradients in tissue PO_2, depending on the organization of the vascular tree. While mean tissue PO_2 fluctuates little around a characteristic set point in tissue such as brain and liver, it may vary greatly in low CD networks such as skeletal muscle, and between neighbouring cells. Either the function of a set point is less important in muscle, or else local oscillations in PO_2 are tolerated better. The underlying control mechanisms may have important consequences for the present topic; e.g. temporal heterogeneity in spacing of perfused capillaries significantly increases the proportion of tissue with low PO_2, while mean values are little changed, suggesting that the goal of describing an 'optimal' system in morphometric terms may be unattainable. We therefore still have some way to go before a model of oxygen transport to tissue can be developed which will account for local variations in both physical and physiological characteristics.

Acknowledgements. This review is based upon work supported by the N.E.R.C. and the Wellcome Trust. Particular thanks are due to Professor Olga Hudlická, whose interest and support were a great encouragement. Hamish Ross provided excellent computer support and many useful discussions, as did Drs Margaret Brown, Hans Hoppeler and Zdenek Turek.

References

Aherne WA, Dunhill MS (1982) Morphometry. Arnold, London, pp 205

Akmal K, Bruley DF, Banchero N, Artigue R, Maloney W (1978) Multi-capillary model for oxygen transport to skeletal muscle. Adv Exp Med Biol 94:139–147

Andersen P (1975) Capillary density in skeletal muscle of man. Acta Physiol Scand 95:203–205

Andersen P, Henriksson J (1977) Capillary supply of the quadriceps femoris muscle of man: adaptive response to exercise. J Physiol 270:677–690

Appleyard ST, Witkowski JA, Ripley BD, Shotton DM, Dubowitz V (1985) A novel procedure for pattern analysis of features present on freeze-fractured plasma membranes. J Cell Sci 74:105–117

Aquin L, Banchero N (1981) The cytoarchitecture and capillary supply in the skeletal muscle of growing dogs. J Anat 132:341–356

Asmussen G, Kiessling A (1974) Charakterisierung von besonderen Muskelfasergruppen in der Skelettmuskulatur des Frosches durch ihre motorische Innervation und ihre Gefäßversorgung. Acta Anat 90:226–242

Atherton GW, Cabric M, James NT (1982) Stereological analyses of capillaries in muscles of dystrophic mice. Virchows Arch 397A:347–354

Baddeley AJ, Gundersen HJG, Cruz-Orive LM (1986) Estimation of surface area from vertical sections. J Microsc 142:259–276

Balding D, McElwain DLS (1985) A mathematical model of tumour-induced capillary growth. J Theor Biol 114:53–73

Bär T (1978) Morphometric evaluation of capillaries in different laminae of rat cerebral cortex by automatic image analysis: changes during development and aging. Adv Neurol 20:1–9

Berry M, McConnell P, Sievers J (1981) Dendritic growth and the control of neuronal form. In: Hunt RK (ed) Current Topics in Developmental Biology, vol 15. Academic Press, London

Berry M, Sadler M, Flinn R (1986) Vertex analysis of neural free structures containing trichotomous nodes. J Neurosci Methods 18:167–177

Boots BN (1986) Voronoi (Thiessen) Polygons. Catmog 45, GeoBooks, Norwich UK
Bourdeau-Martini J, Odoroff CL, Honig CR (1974) Dual effect of oxygen on magnitude and uniformity of coronary intercapillary distance. Am J Physiol 226:800–810
Boyde A (1985) Stereoscopic images in confocal (tandem scanning) microscopy. Science 230:1270–1272
Braendgaard H, Gundersen HJG (1986) The impact of recent stereologic advances on quantitative studies of the nervous system. J Neurosci Methods 18:39–78
Brown MD, Egginton S (1988) Capillary density and fine structure in rabbit papillary mscles after a high dose of norepinephrine. Microvasc Res 36:1–12
Brown MD, Cotter MA, Hudlická O, Vrbová G (1976) The effects of different patterns of muscle activity on capillary density, mechanical properties and structure of slow and fast rabbit muscles. Pflügers Arch 361:241–250
Capó LA, Sillau AH (1983) The effect of hyperthyroidism on capillarity and oxidative capacity in rat soleus and gastrocnemius muscle. J Physiol 342:1–14
Casley-Smith JR, Green HS, Harris JL, Wadey PJ (1975) The quantitative morphology of skeletal muscle capillaries in relation to permeability. Microvasc Res 10:43–64
Cayley A (1857) On the analytical form called trees. Phil Mag 18:374–378
Chen IIH, Battarbee HD (1987) The quantitative estimation of microvessels in microvascular networks. J Theor Biol 127:315–320
Clark PJ, Evans FC (1954) Distance to nearest neighbour as a measure of spatial relationships in populations. Ecology 35:445–453
Cormack RM (1979) Spatial aspects of competition between individuals. In: Cormack RM, Ord JK (eds) Spatial and Temporal Analysis in Ecology. Int Co-op Publ House, Maryland, pp 151–212
Cotter MA, Hudlická O, Vrbová G (1973) Growth of capillaries during long-term activity in skeletal muscle. Bibl Anat 11:395–398
Davie PS, Forster ME (1980) Cardiovascular responses to swimming in eels. Comp Biochem Physiol 67A:367–373
Dawson JM, Tyler KR, Hudlická O (1987) A comparison of the microcirculation in rat fast glycolytic and slow oxidative muscles at rest and during contractions. Microvasc Res 33:167–182
Diem K, Lentner C (eds) (1970) Scientific tables. JR Geigy, SA, Basel, pp 165–166
Diggle PJ (1983) Statistical analysis of spatial point patterns. Academic Press, London
Diggle PJ (1986) Displaced amacrine cells in the retina of a rabbit: analysis of a bivariate spatial point pattern. J Neurosci Methods 18:115–125
Dussert C, Rasigni M, Palmari J, Rasigni G, Llebaria A, Marty F (1987) Minimal spanning tree analysis of biological structures. J Theor Biol 125:317–323
Edelstein L (1982) The propagation of fungal colonies: a model for tissue growth. J Theor Biol 98:679–701
Egginton S (1982) Structural and functional differentiation of teleost skeletal muscle. PhD Thesis. University of St Andrews, Scotland
Egginton S, Johnston IA (1982) Suitability of measured parameters and minimum sample sizes required to quantify capillary supply to fish muscle. Acta Stereol 1(2):309–319
Egginton S, Johnston IA (1983) An estimate of capillary anisotropy and determination of surface and volume densities of capillaries in skeletal muscles of the conger eel (*Conger conger* L.). Q J Exp Physiol 68:603–617
Egginton S, Johnston IA (1984) Effects of acclimation temperature on routine metabolism, muscle mitochondrial volume density and capillary supply in the elver (*Anguilla anguilla* L.). J Therm Biol 9:165–170
Egginton S, Hargreaves D, Hudlická O (1988b) Changes in capillary perfusion induced by different patterns of activity in rat skeletal muscle. J Physiol 399:24 p
Egginton S, Ross HF, Sidell BD (1987c) Morphometric analysis of intracellular diffusion distances. Acta Stereologica 6:449–454
Egginton S, Hargreaves D, Hudlická O (1987a) Is capillary perfusion in skeletal muscle linked to a specific pattern of activity? Int J Microcirc Clin Exp 6:295
Egginton S, Turek Z, Hoofd LJC (1988a) Differing patterns of capillary distribution in fish and mammalian skeletal muscle. Respir Physiol 74:383–396

Egginton S, Turek Z, Hoofd LJC (1987b) Morphometric analysis of sparse capillary networks. In: Silver IA, Silver A (eds) Oxygen transport to tissue IX. Adv Exp Med Biol 215, Plenum Press, New York, pp 1–12
Engelson ET, Skalak TC, Schmid-Schönbein GW (1985a) The microvasculature in skeletal muscle. I. Arteriolar network in rat spinotrapezius muscle. Microvasc Res 30:29–44
Engelson ET, Schmid-Schönbein GW, Zweifach BW (1985b) The microvasculature in skeletal muscle. III. Venous network anatomy in normotensive and spontaneously hypertensive rats. Int J Microcirc Clin Exp 4:229–248
Eriksson E, Myrhage R (1972) A repeating modular organization of the microcirculation of the cat mesentery. Microvasc Res 4:62–76
Fitch NA, Johnston IA, Wood RE (1984) Skeletal muscle capillary supply in a fish that lacks respiratory pigments. Respir Physiol 57:201–211
Flood PR (1979) The vascular supply of three fibre types in the parietal trunk muscle of the Atlantic hagfish (*Myxine glutinosa* L.) Microvasc Res 17:55–70
Gordon MS (1968) Oxygen consumption of red and white muscles from tuna fishes. Science 159:87–90
Gray SD, Renkin EM (1978) Microvascular supply in relation to fibre metabolic type in mixed skeletal muscles of rabbits. Microvasc Res 16:406–425
Green PJ, Sibson R (1978) Computing Dirichlet tessellations in the plane. Comput J 21:168–173
Groom AC, Ellis CG, Potter RF (1984) Microvascular architecture and red cell perfusion in skeletal muscle. Prog Appl Microcirc 5:64–83
Gundersen HJG (1977) Notes on the estimation of the numerical density arbitrary particles: the edge effect. J Microsc 111:219–223
Henquell L, Honig CR (1976) Intercapillary distances and capillary reserve in right and left ventricles; significance for control of tissue PO_2. Microvasc Res 12:35–41
Heroux O, St. Pierre J (1957) Effect of cold acclimation on vascularization of ears, heart, liver and muscles of white rats. Am J Physiol 188:163–168
Hill AV (1928) The diffusion of oxygen and lactic acid through tissues. Proc R Soc Lond 104:39–96
Homer LD (1984) Moments of distributions of distances to the nearest capillary in tissue. Microvasc Res 27:114–116
Honig CR, Feldstein ML, Frierson JL (1977) Capillary lengths, anastomoses, and estimated capillary transit times in skeletal muscle. Am J Physiol 233:H122–129
Hoofd L, Turek Z, Kubat K, Ringnalda BEM, Kazda S (1985) Variability of intercapillary distance estimated on histological sections of rat heart. Adv Exp Med Biol 191:239–247
Hoofd L, Turek Z, Rakusan K (1987) Local capillary-myocyte ratio in the rat myocardium: its effect on the calculated PO_2 profiles. Fed Proc 46:348
Hoppeler H (1984) Morphometry of skeletal muscle capillaries. Prog Appl Microcirc 5:33–43
Hoppeler H, Lindstedt SL (1985) Malleability of skeletal muscle tissue in overcoming limitations: structural elements. J Exp Biol 115:355–364
Hoppeler H, Mathieu O, Weibel ER, Krauer R, Lindstedt SL, Taylor CR (1981) Design of the mammalian respiratory system. VIII. Capillaries in skeletal muscles. Respir Physiol 44:129–150
Horton RE (1945) Erosional development of streams and their drainage basins; hydrophysical approach to quantitative morphology. Bull Geol Soc Am 56:275–370
Hudlická O (1984) Development of the microcirculation: capillary growth and adaptation. In: Renkin EM, Michel CC (eds) Handbook of Physiology. The cardiovascular system, vol IV. Am Physiol Soc, Bethesda, Md, pp 165–216
Hudlická O (1985) Development and adaptability of microvasculature in skeletal muscle. J Exp Biol 115:215–228
Hudlická O, Tyler KR (1986) Angiogenesis. The growth of the vascular system. Academic Press, London, pp 221
Hudlická O, Hoppeler H, Uhlmann E (1987) Relationship between the size of the capillary bed and oxidative capacity in various cat skeletal muscles. Pflügers Arch 410:369–375
Hudlická O, Egginton S, Brown MD (1988) Capillary diffusion distances – their importance for cardiac and skeletal muscle performance. News Physiol Sci 3:134–138

James NT (1981) A stereological analysis of capillaries in normal and hypertrophic muscle. J Morphol 168:43–49
Jensen EB, Sundberg R (1986) Generalized associated point methods for sampling planar objects. J Microsc 144:55–70
Jensen EB, Gundersen HJG, Østerby R (1978) Determination of membrane thickness distribution from orthogonal intercepts. J Microvasc 115:19–33
Johnston IA (1982) Quantitative analyses of ultrastructure and vascularization of the slow muscle fibres of the anchovy. Tissue Cell 14:319–328
Kayar SR, Banchero N (1983) Distribution of capillaries and diffusion distances in guinea pig myocardium. Pflügers Arch 396:350–352
Kayar SR, Archer PG, Lechner AJ, Banchero N (1982a) Evaluation of the concentric-circles method for estimating capillary-tissue diffusion distances. Microvasc Res 24:342–353
Kayar SR, Archer PG, Lechner AJ, Banchero N (1982b) The closest-individual method in the analysis of the distribution of capillaries. Microvasc Res 24:326–341
Kayar SR, Lechner AJ, Banchero N (1982c) The distribution of diffusion distances in the gastrocnemius muscle of various mammals during maturation. Pflügers Arch 394:124–129
Kayser K, Shaver M, Modlinger F, Postl K, Moyers JJ (1986) Neighbourhood analysis of low magnification structures (glands) in healthy, adenomatous, and carcinomatous colon mucosa. Pathol Res Pract 181:153–158
Kreuzer F (1982) Oxygen supply to tissues: the Krogh model and its assumptions. Experientia 38:1415–1426
Krogh A (1919) The number and distribution of capillaries in muscles with calculations of the oxygen pressure head necessary for supplying the tissue. J Physiol 52:409–415
Kryvi H, Flood PR, Galyaer D (1980) The ultrastructure and vascular supply of the different fibres types in the axial muscle of the sturgeon *Acipenser stellatus* Pallas. Cell Tissue Res 212:117–126
Laursen H, Diemer NH (1980) Capillary size, density and ultrastructure in brain of rats with urease-induced hyperammonaemia. Acta Neurol Scandinav 62:103–115
Leopold LB (1971) Trees and streams: the efficiency of branching patterns. J Theor Biol 31:339
Loats JT, Sillau AH, Banchero N (1978) How to quantify skeletal muscle capillarity. Adv Exp Med Biol 94:41–48
Longmuir IS (1981) Channels of oxygen transport from blood to mitochondria. Adv Physiol Sci 25:19–22
Lu SY, Fu KS (1978) A syntactic approach to texture analysis. Comput Graph Image Process 7:303–330
Lund N, Damon DH, Damon DN, Duling BR (1987) Capillary grouping in hamster tibials anterior muscles: flow patterns, and physiological significance. Int J Microcirc Clin Exp 5:359–372
Mathieu O, Cruz-Orive LM, Hoppeler H, Weibel ER (1983) Estimating length density and quantifying anisotropy in skeletal muscle capillaries. J Microsc 131:131–146
Mattfeldt T, Mall G (1984) Estimation of length and surface of anisotropic capillaries. J Microsc 135:181–190
Mattfeldt T, Möbius H-J, Mall G (1985) Orthogonal triplet probes: an efficient method for unbiased estimation of length and surface of objects with unknown orientation in space. J Microsc 139:279–289
Maxwell LC, White TP, Faulkner JA (1980) Oxidative capacity, blood flow, and capillarity of skeletal muscles. J Appl Physiol 49:627–633
Miles RE (1970) On the homogeneous planar Poisson point process. Math Biosci 6:85–127
Miles RE, Maillardet RJ (1982) The basic structures of Voronoi and generalized Voronoi polygons. J Appl Probab 19A:97–111
Mosse PRL (1978) The distribution of capillaries in the somatic musculature of two vertebrate types with particular reference to teleost fish. Cell Tissue Res 187:281–303
Müller W (1976) Subsarcolemmal mitochondria and capillarization of soleus muscle fibers in young rats subjected to an endurance training. Cell Tissue Res 174:367–389
Myrhage R (1978) Capillary supply of the muscle fibre population – hindlimb muscles of the cat. Acta Physiol Scand 103:19–30
Na HS, Rapoport A (1970) Distribution of nodes of a tree by degree. Math Biosci 6:313–329

Nemeth P, Hofer HW, Pette D (1979) Metabolic heterogeneity of muscle fibres classified by myosin ATPase. Histochemistry 63:191–201
O'Callaghan JF (1975) An alternative definition for "neighbourhood of a point". IEEE Trans Comput C24:1121–1125
Pape H, Schopper JR (1987) Growth models, regarding the influence of competition on the formation of dendritic structures. Acta Stereol 6:343–348
Pape H, Riepe L, Schopper JR (1989) Theory of self-similar network structures in sedimentary and ingeneous rocks and their investigation with microscopical and physical methods. J Microsc (in press)
Park D (1985) Does Horton's law of branch length apply to open branching systems? J Theor Biol 112:299–313
Plyley D, Groom AC (1975) Geometrical distribution of capillaries in mammalian striated muscle. Am J Physiol 228:1376–1383
Plyley MJ, Sutherland GJ, Groom AC (1976) Geometry of the capillary network in skeletal muscle. Microvasc Res 11:161–173
Popel AS, Johnson PC (1986) Microvascular networks: experimental and theoretical studies. Karger, Basel, 223 pp
Porter GA, Bankston PW (1987) Myocardial capillaries in the fetal and the neonatal rat: a morphometric analysis of the maturing myocardial capillary bed. Am J Anat 179:108–115
Potter RF, Groom AC (1983) Capillary diameter and geometry in cardiac and skeletal muscle studied by means of corrosion casts. Microvasc Res 25:68–84
Rakusan K, Hoofd L, Turek Z (1984) The effect of cell-size and capillary spacing on myocardial oxygen supply. Adv Exp Med Biol 169:463–475
Rakusan K, Korecky B, Sarkar K, Turek Z (1986) Merits and pitfalls in morphological assessment of cardiac growth. Fed Proc 45:2580–2584
Ranvier L (1874) Note sur les vaisseaux sanguilns et la circulation dans les muscles rouges. Arch Physiol Norm Pathol 1:446–450
Rasio EA, Goresky CA (1979) Capillary limitation of oxygen distribution in the isolated rete mirabile of the eel (*Anguilla anguilla*). Circ Res 44:498–503
Renkin EM (1977) Multiple pathways of capillary permeability. Circ Res 41:735–743
Renkin EM, Gray SD, Dodd LR, Lia BD (1981) Heterogeneity of capillary distribution and capillary circulation in mammalian skeletal muscles. In: Bachrach AV, Matzemm MM (eds) Underwater Physiology, Vol 7. Undersea Medical Society, Bethesda, pp 464–474
Romanul FCA (1965) Capillary supply and metabolism of muscle fibers. Arch Neurol 12:497–509
Safranyos RGA, Ellis CG, Tyml K, Groom AC (1983) Heterogeneity of capillary diameters in skeletal muscle of the frog. Microvasc Res 26:151–156
Sarelius IH, Damon DN, Duling BR (1981) Microvascular adaptations during maturation of striated muscle. Am J Physiol 241:H317–324
Schmidt-Nielsen K, Pennycuik P (1961) Capillary density in mammals in relation to body size and oxygen consumption. Am J Physiol 200:746–750
Schmidt-Schönbein H (1981) Blood rheology and oxygen transport to tissues. Adv Physiol Sci 25:279–289
Schmid-Schönbein GW, Zweifach BW, Kovalcheck S (1977) The application of stereological principles to morphometry of the microcirculation in different tissues. Microvasc Res 14:303–317
Sibson R (1980) The Dirichlet tessellation as an aid in data analysis. Scand J Stat 7:14–20
Sillau AH, Banchero N (1977) Effect of maturation on capillary density, fiber size and composition in rat skeletal muscle. Proc Soc Exp Med Biol 154:461–466
Skalak R, Chien S (1981) Capillary flow: history, experiments and theory. Biorheology 18:307–330
Skalak TC, Schmid-Schönbein GW (1986) The microvasculature in skeletal muscle. IV. A model of the capillary network. Microvasc Res 32:333–347
Spälteholz W (1888) Die Verteilung der Blutgefäße im Muskel. Abh Sachs Ges Wiss Math Phys 14:507–535
Strahler AN (1952) Hyposometric (area-altitude) analysis of erosional topography. Bull Geol Soc Am 63:1117–1142

Sullivan SM, Pittman RN (1987) Relationship between mitochondrial volume density and capillarity in hamster muscle. Am J Physiol 255:H149–155
Turek Z, Rakusan K (1981) Lognormal distribution of intercapillary distance in normal and hypertrophic rat heart as estimated by the method of concentric circles: its effect on tissue oxygenation. Pflügers Arch 391:17–21
Turek Z, Hoofd L, Rakusan K (1986) Myocardial capillaries and tissue oxygenation. Can J Cardiol 2:98–103
Turek Z, Hoofd L, Rakusan K (1987) A comparison of the methods for assessment of the heterogeneity of myocardial capillary spacing. Adv Exp Med Biol 215:13–19
Underwood EE (1970) Quantitative stereology. Addison-Wesley, Reading, MA, 251 pp
Valdivia E (1958) Total capillary bed in striated muscle of guinea pigs native to the Peruvian mountains. Am J Physiol 194:585–589
van Pelt J, Verwer RWH (1986) Topological properties of binary trees grown with order-dependent branching probabilities. Bull Math Biol 48:197–211
Verwer RWH, van Pelt J (1986) Descriptive and comparative analysis of geometrical properties of neuronal tree structures. J Neurosci Methods 18:179–206
Vetterlein F, dal Ri H, Schmidt G (1982) Capillary density in rat myocardium during timed plasma staining. Am J Physiol 242:H133–141
Voronoi G (1908) Nouvelles applications des paramètres continus à la théorie des formes quantatiques. Deuxième mémoire: recherches sur les parallelloedres primitifs. J Reine Angew Math 134:198–287
Wangensteen D, Weibel ER (1982) Morphometric evaluation of chorioallantoic oxygen transport in the chick embryo. Respir Physiol 47:1–20
Weibel ER (1963) Morphometry of the human lung. Springer, Berlin Göttingen Heidelberg
Weibel ER (1969) Stereological principles for morphometry in electron microscopic cytology. Int Rev Cytol 26:235–302
Weibel ER (1980) Stereological methods, Vol 2: Theoretical foundations. Academic Press, London
Weibel ER, Knight BW (1964) A morphometric study on the thickness of the pulmonary air-blood barrier. J Cell Biol 21:367–384
Weibel ER, Taylor CR, Gehr CR, Hoppeler H, Mathieu O, Maloiy GMO (1981) Design of the mammalian respiratory system. IX. Functional and structural limits for oxygen flow. Respir Physiol 44:151–164
Weiss HR, Conway RS (1985) Morphometric study of the total and perfused arteriolar and capillary network of the rabbit left ventricle. Cardiovasc Res 19:343–354
Wetter T, Hoffmann D, Schmid-Schönbein H (1983) Analysis of network flow distribution: computational aid to minimize experimental expenditure. Microvasc Res 26:221–237
Wiedeman MP (1984) Architecture. In: Renkin EM, Michel CC (eds) Handbook of Physiology, Section 2, Microcirculation, vol IV, Part 1. Am Physiol Soc, Bethesda
Zamir M (1976) Optimality principles in arterial branching. J Theor Biol 62:227–251
Zamir M, Chee H (1986) Branching characteristics of human coronary arteries. Can J Physiol Pharmacol 64:661–668
Zhuang FY, Yen MRT, Fung YC, Sobin SS (1985) How many pulmonary alveoli are supplied by a single arteriole and drained by a single venule? Microvasc Res 29:18–31

Hemoglobin Physiology in Vertebrate Animals: a Cautionary Approach to Adaptationist Thinking

R. M. G. Wells[1]

Contents

"Some organisms use less oxygen than others under conditions that seem to entail the same effort. Can we say that those organisms that consume less oxygen are more efficient and better adapted? Certainly not, because as long as they are alive, they have all met the requirements for an uninterrupted ontogeny. Comparisons about efficiency belong to the realm of the observer's descriptions; they are not directly related to what happens in the individual histories of conservation of adaptation." (Maturana and Varela 1987)

1 Introduction

The apparent "goodness of fit" of animals to environments brings into focus immediate adaptations to local conditions. This interaction of organism and environment has led us into a particular way of thinking about the role of hemoglobin in oxygen transport, in which the near omnipotence of natural selection is acknowledged in the design of the protein for function under particular circumstances (e.g., Powers 1980; Perutz 1983). The neo-Darwinian concept of adaptation implies pre-existent problems to which animals are fitted by a dynamical process, the process being "adaptation" (Lewontin 1978). In this view, there is a tendency to perceive natural selection as operating with few constraints, and non-adaptive features, such as those imposed by phyletic inertia, structural and ontogenetic constraints, play only a minor role in shaping the physiological traits of animals.

[1] Department of Zoology, University of Auckland, Auckland, New Zealand

Advances in Comparative and Environmental Physiology
Vol. 6. Edited by R. G. Boutilier

Adaptation is a central concept in biology; therefore, one must be careful about its intended meaning in any given discussion. "Adaptation" refers to both the process by which natural selection adjusts the frequency of genes coding for traits that affect fitness, as well as the traits affecting fitness themselves. Physiologists frequently refer to adaptation as both a specific property, trait, or feature of animals, and as a process of adjustment to local conditions. Nevertheless, all definitions of adaptation in the evolutionary sense are expressed in terms of selective processes (Brookfield 1982; Krimbas 1984), allowing all organism-environment interactions to be viewed against a background of traits that have been optimized by natural selection (cf. Bock 1980; Brady 1982). The mere existence of good fit between an animal and its environment is, however, insufficient for inferring the action of natural selection.

Much of what seems to be adaptation in the field of comparative respiratory physiology may be epiphenomenal, and no assumptions can be made that selection has shaped it or maintains it. Nor can the intermediate steps by which an ancestral trait is transformed (historical origins) be often assumed. Current utility of a trait may be interpreted in the neo-Darwinian mode as pre-adaptive. One real danger in adaptationist thinking among panselectionist physiologists is the tendency to atomize animals into sets of traits or features, each of which are separately optimized for its particular role by natural selection. If this is so, and traits such as hemoglobin-oxygen binding are optimized, then the integration of the whole respiratory system, or interaction between the parts, counts for nothing. Consider the possibility of a genetically engineered hemoglobin-optimizing oxygen loading and unloading for the purpose of creating faster racehorses. The speed gained would count for little in reproductive fitness if the horse died of a heart attack after a hard race! Adaptationists may counter with a view of constrained optimization; a trade off between competing systems, with the result being the best of possible worlds. Nonetheless, animals are integrated wholes, not decomposable into independent and separately optimized parts that selection has shaped into adaptations.

Neo-Darwinism acknowledges the great power of natural selection, and the idea that some physiological traits may not be adaptive might appear a radical departure from orthodoxy. Yet, Darwin himself emphasized, "I am convinced that natural selection has been the main, but not the exclusive means of modification." (*Origin* 1872, p. 395). Gould (1982), Lewontin (1978) and other evolutionary biologists seem to agree that natural selection is not the only process that can give rise to new traits or adjust the frequency of existing traits in a population. Others are sexual selection, genetic drift, learning, founder effect, meiotic drive or unusually high mutation rates.

The purpose of this review is to examine the role of hemoglobin in oxygen transport against the framework of adaptationist thinking, and to illustrate some contingent pitfalls. Further, a different kind of argument is frequently required where epiphenomena are concerned. Perhaps we should abandon the idea that all physiological features are adaptive, because it encourages us to invent adaptive stories. Non-adaptive and non-selectionist reasons may in some cases help us better to explain the presence and evolution of physiological traits. Failure to do this may lead to our inability to understand patterns because we view secondary adaptation as reasons for the patterns themselves. As we shall see, just because things seem non-adaptive, it does not render them unintelligible.

2 Universal Patterns of Organization in Hb Structure

The molecular biology of hemoglobin has proved to be a rich and productive field of enquiry. One of the fundamental points to emerge has been that the structure of hemoglobin exists through universal requirements expected under the basic organization of protein structure. The constraints in tertiary and quaternary Hb structure appear conserved in vertebrate evolution, with the residues responsible for ligand binding, R-T transformations when allosteric factors are present, and Bohr proton binding, remarkably constant between species (Perutz 1983). We may therefore agree that the structure of heme and the tetrameric pattern of organization in the Hb molecule cannot find explanation through environmental conditions or through adaptive radiation.

Nevertheless, intra- and interspecific variability is considerable in Hb_{O_2} affinity, the degree of cooperativity and the dissociation constants of tetramers, and in the pH and phosphate sensitivities of the oxygen equilibrium reaction (Weber 1982; Weber and Wells 1989), due to substitutions of a few key residues (Perutz 1983). Accordingly, it will not be helpful to seek explanations as to whether natural selection has been directional, disruptive or stabilizing, to view Hb as a lump of putty, remodelled under the pressures of natural selection, nor for animals with unusual traits to be regarded as "living fossils", or atavisms eking out precarious existences alongside the ruthlessly selected and efficient molecular systems operating in "higher" animals.

3 Pitfalls for Molecular Adaptationists

Beyond the descriptions of observable phenomena, molecular biologists have promoted a genetically deterministic way of thinking about animals as collections of traits, each selected for optimised function (e.g. in hemoglobin: Brunori 1975; Powers 1980; Perutz 1983). Monod (1972) uses the term "teleonomy" to encompass his view of proteins as agents endowed with purpose, whereby their complexity and efficiency are guided by natural selection towards stereospecific properties permitting concerted performance in allosteric interactions. Teleonomy is useful in the proposition of a hypothesis regarding structure and function of traits, but can be counterproductive if it leads one to accept teleonomical speculations as historical facts. "End-directed" thinking, wherein evidence of purpose or design (-er) is sought in nature, is useful only as an explanatory analogy (see O'Grady 1984, 1985). This manner of thinking may be unwarranted, because organisms are not the product of their genes, and genes do not determine form. One must, however, be prepared to qualify this remark by saying that there is mounting evidence that genes are being identified that do in fact account for morphogenesis. Few neo-Darwinists would ever admit that selection acts at the level of genes, which is precisely what the molecular biologists imply. Hubbard (1982) harshly criticizes those molecular biologists who view genes in the same way as a child who concludes that the on-off knob on a T.V. set "causes" the picture. It is thus essential to appreciate that the difference between genotype and phenotype is simply ontogeny, and we shall explore this difference and its implications in Section 9.

4 Multiplicity and Polymorphism: The Adaptationist's Approach of Different Genes for Different Scenes

A feature of nearly all animals is the synthesis of multiple hemoglobins during the lifetime of an individual. Much of this information has arisen from electrophoretic and chromatographic studies of hemolysates, whereby Hb components are separated according to surface charge or size. Great care must be taken when analyzing these data, however, because the physio-chemical environment of the treated isolates differs from that inside the erythrocyte, with the consequence that it is often difficult to exclude method artifacts. Conversely, a number of the key residues invoked in the reactions of Hb reside internally and may thus appear electrophoretically "silent". The same kind of problem is likely to arise where heme or ligand concentration dependence influences the dissociation constants of the monomer-dimer-tetramer equilibrium (e.g. fish: Ikeda-Saito et al. 1983; reptiles, Matsuura et al. 1987; mammals, Brittain 1987 a). These technical problems render both the detection of multiple globins, and matching of their properties with those presumed in vivo, a difficult exercise.

A key question often asked is how much observed globin-chain variability is due to natural selection, and how much is due to purely random factors? Natural selection is not always directional, however, and may be disruptive or stabilizing. Powers (1972, 1980) provides an adaptationist argument for us with sympatric catostomid fishes. Active species living in well-oxygenated water are generally characterized with a marked Bohr shift that exploits an arterio-venous pH difference to provide a feedback control of oxygen release for carbon dioxide gain. *Catostomus clarkii* possesses a hemoglobin with no Bohr effect. The adaptationist argument is that, living in fast-flowing water, the fish has nowhere to hide and repay an oxygen debt after severe exercise, and so accumulated lactate might cause acidosis and ultimately, asphyxiation. Collections of similar fables have become fashionable in the new disciplines of "molecular ecology" and "molecular physiology", and share the common base of adaptation and selection (viz. Powers 1974, 1980). A physiologist would have difficulty in accepting the catostomid story because he or she would want to know, in adddition to the oxygen-binding properties of whole blood, the blood gas values and acid-base status before and after exercise. Furthermore, a physiologist knows very well that the blood contributes almost nothing to the buffer capacity of fish, and that adaptationist arguments need to include the buffering ability of skeletal muscle (cf. Heisler 1984; Dobson et al. 1986). The fundamental flaw in the adaptationist arguments for "molecular ecology" is, however, that traits have been atomized to the point where natural selection is unable to pick out hemoglobin genes from an integrated and complex system of respiratory control. One cannot automatically assume that variation in Hb is the result of natural selection. Selection acting on one trait may, of course, cause changes in other traits if the other traits are genetically linked to the first.

On the other hand, a genetic basis for multiplicity is evident in many cases where no adaptation or selection is implied. For example, 77 genera of Amazonian fish examined by Fyhn et al. (1979) showed a mean of four electrophoretic components per phenotype, but no clear correlation between Hb heterogeneity and habitat preference or mode of life. Further, the heterogeneity of components from two species of flatfish show little intraspecific variation in oxygen-binding parameters (Weber and de Wilde 1976). A Hb variant that has arisen through processes other than natural selection may currently be under stabilizing selection (i.e. individuals lacking the trait will exhibit reduced fitness). One cannot automatically assume that Hbs that have not involved natural selection in early evolution are not currently under selection.

The discovery of polymorphic hemoglobins with distinctive functional properties is obviously an important observation for molecular adaptationists. Unfortunately, too many physiologists have failed to clearly distinguish polymorphisms from other forms of hemoglobin multiplicity (e.g. Giles and Randall 1980).

Evidence has been gathered for the polymorphism of sheep allohemoglobins Hb*A* and Hb*B*, which reveal differences in their oxygen-binding characteristics, and, in conditions of severe anaemia or hypoxia, Hb*A* is replaced by Hb*C* (Bauer and Jung 1975; Dodgson and Holland 1983; Maginniss et al. 1986). Can natural selection scrutinize these genes? Bauer and Jung (1975) correctly point out that at sea level there is no selective dominance of one or other allele, and conclude that the *AB*-polymorphism is pre-adaptive and without selection. The replacement of Hb*A* by Hb*C* with hypoxia is therefore non-selective in the Darwinian sense, and an example of phenotypic plasticity.

Hemoglobin polymorphism is also common in rodent populations (Stratton and Duffy 1976) with evidence for different oxygen-binding characteristics arising from β-chain variants in the mouse (Newton and Peters 1983). The frequency appears to be maintained by natural selection, but the authors astutely point out that differences in oxygen binding are not the only properties important in determining the differential survival of a particular phenotype under environmental stress. The possibility is raised that a closely linked gene may have been selected and the polymorphism is manifested through a "hitch-hiking" effect.

The classic case of Darwinian selection maintaining a balanced hemoglobin polymorphism arises from an intensively studied animal, the human. Almost every introductory course in biology draws attention to the correlation between geographic areas where malaria is endemic, and the distribution of the hemoglobinopathic diseases sickle cell and thalassemia (see Dickerson and Geis 1983). Thalassemic disease had not been reported in Polynesian populations because these people came from an area historically free of malaria (Carrell 1980). Remarkably, recent gene-mapping studies have revealed that more than 16% of Polynesians show the α-thalassemia trait (Mickleson et al. 1985; Trent et al. 1985).

How can adaptationists explain such a high frequency in the absence of a mechanism for selection? An hypothesis rescue might involve explanations of genetic drift and the founder effect, but historical and geographic evidence is inadequate to sustain the conjecture (see Trent et al. 1985). For the present, neither malarial selection nor other processes are adequate to explain the observed pattern.

The Root effect in the hemoglobin of the frog, *Xenopus*, is a prime example of the problems that may arise from considerations at different levels of integrative organization. In the case of certain hemoglobins from teleost fish, the normal acid Bohr effect is absent, and the reversal in oxygen affinity that is usually observed at ≈pH 6.5 is absent. Affinity continues to fall with lowered pH, and cooperativity becomes diminished, and may even register as negative values (Brittain 1987b). Originally described as a decrease in oxygen-carrying capacity at low pH (Root 1931), it is now considered an enhanced version of the Bohr effect (Brittain 1987b), responsible for secretion of oxygen into the eye of certain teleosts (Ingermann and Terwilliger 1982). The earlier adaptive story about the role of the Root hemoglobin in filling the swim bladder explains why a Root effect was not reported in higher vertebrates.

Consequently, the finding that the structural basis for the Root effect lay in a serine residue at position β(F9)93 led to the proposal that Root effects should also be found in hemoglobins of other animals showing the same structural feature – like the frog, *Xenopus*, lungfish and some sharks (Perutz and Brunori 1982). Indeed, it has been confirmed that in the presence of appropriate concentrations of phosphate cofactors, *Xenopus* hemoglobin manifests a Root effect (Brunori et al. 1987). Problems in interpretation have arisen because physiologists have never demonstrated a Root effect in whole blood from *Xenopus* (Jokumsen and Weber 1980; Bridges et al. 1985; Boutilier and Shelton 1986). What is the explanation? The conditions of high phosphate concentrations under which the molecular biologists are able to demonstrate a Root effect do not exist in the living frog. That the variation at the serine position is currently not subject to selection does not mean that past selection cannot account for it. Neither does it demonstrate that selection has occurred; one simply cannot say. We must therefore be cautious about extrapolations.

5 Purpose and Design for Optimizing Strategies

The comparative approach invites us to make predictions about the performance of physiological systems. From the extensive base of a half-century of accumulated literature on the oxygen-binding properties of blood, is it possible to prescribe optimized oxygen-binding properties with the best P_{50} and n values for particular roles? This ought to be possible if natural selection has screened all Hbs and moulded them for optimal utility.

Consider optimization of blood in quick fishes such as tuna. These highly aerobic, fast-swimming fish possess complex hemoglobin systems characterized by subunit heterogeneity, biphasic equilibrium curves, multiple Bohr groups and unusual temperature insensitivity, with one of the purified components showing negative cooperativity at low pH and temperature (Ikeda-Saito et al. 1983). The traits are important in confirming molecular mechanisms and possibly some observations of whole blood, but they cannot predict higher integrative phenomena operating at the organismal level.

The blood of the tuna, *Thunnus alalunga*, is described by Cech et al. (1984) as having a large Bohr effect ($\varnothing = -1.17$), negligible Root effect, high affinity ($P_{50} \approx 8$ mmHg) and hyperbolic oxygen equilibrium curve ($n = 1.1$). The authors conclude that "... blood gas transport has evolved in response to ... conditions encountered by this species." This poses a difficulty for an adaptationist's interpretation of the characteristics of tuna blood from *Euthynnus affinis*, which has half the Bohr effect ($\varnothing = -0.59$), a substantial Root effect, low affinity ($P_{50} \approx 21$ mmHg) and a sigmoid equilibrium curve ($n = 1.72$) (Jones et al. 1986). The adaptationist is bound to invoke both traits as selected optimization. Hypothesis rescuers will sit uncomfortably with this and, in the first instance, appeal to technical explanations. For instance, differences in all of the equilibrium properties might be explained if, in the former study, P_{100} was defined for alkaline blood at higher oxygen pressures. If this is the case, then both sets of observations are correct, but the adaptive stories must nevertheless be discarded.

Alternatively, the prime contention that blood traits are always adaptive may be mistaken. Thus, in the latter study, the predictions of dorsal and ventral aortic pH based on the magnitude of the Bohr effect might not only depend on how the authors view and define a Root effect, but depend on the notion that the equilibrium traits are functionally optimized. Jones et al. (1986) further commented "... finding a hyperbolic blood curve in an extremely active and highly aerobic animal like tuna is quite unexpected." How does the optimization argument hold for large, active, fast-swimming fish like the mako shark, whose blood has no Bohr or Root effect, a relatively high affinity for oxygen and low hemoglobin concentration (Wells and Davie 1985; Dobson et al. 1986)? Clearly, it does not hold without special pleading that the primitive (but highly successful) elasmobranchs have not received the scrutiny of selective forces meted out to tunas. This line of reasoning simply will not do, and one must therefore consider the possibilities that (1) environment has not influenced hemoglobin traits, or (2) the traits may not be adaptive with respect to their current utility (see Sect. 7).

The dominant social paradigm which has become translocated into science is one of maximization represented as optimization. But animals do not have to be optimized, merely adequate. The concept of constrained optimality (i.e. the best, given the constraints) takes cognizance of the fact that parts cannot be optimized. Constrained optimality (or "sufficing"; Routley 1984) is an accepted biological concept with much potential for physiologists, but retains the same cautions as adaptive methodology.

6 Non-Standard Oxygen-Binding Curves: Observation Shaped by Theory

For many years, physiologists have employed Hill's (1910) equation to describe cooperative oxygen binding in whole blood. Because it was assumed that this model for cooperative binding was an adequate descriptor of oxygen-binding reactions between approximately 5% and 95% saturation, the accumulated data for more than 60 years showed regular oxygen-binding curves tided up by regression techniques or, perhaps, with the aid of a flexicurve rule.

The stereochemical model described in terms of the Monod-Wyman two-state structure of hemoglobin provided a further basis for interpretation of the phenomenological descriptions of oxygen-binding reactions for a range of animal hemoglobins (Perutz 1983). With a sound molecular basis for the binding reaction, physiologists have interpreted cooperativity traits of whole blood in terms of the paradigm.

Hemoglobin physiologists were therefore sceptical when Lutz et al. (1973, 1974) reported that, for bird blood, the cooperativity constant, n, was not constant throughout the saturation range, but increased at high saturations. A number of workers, myself included, were unable to accept these observations because it appeared to violate theoretical predictions, and so we sought methodological explanations (see Lutz's 1980 review).

In recent years, however, non-standard equilibrium curves have been reported by independent groups using different techniques and different animals. These include examples from fish (Barnikol and Burkhard 1979); amphibians (Lykkeboe and Johansen 1978; Maginniss et al. 1980; Wells and Weber 1985), reptiles (Lapennas and Lutz 1982; Lutz and Bentley 1985), birds (Lapennas and Reeves 1983 a, b; Maginniss 1985; Johansen et al. 1987) and mammals (Wells 1979; Wells and Brittain 1981).

Several explanations have been advanced to explain these phenomena. n-Values rising from 1.5 at low saturation to 6.5 at high saturation are reported in chick blood (Lapennas and Reeves 1983 a), and values >4.0 in turtles, frogs and birds (Lykkeboe and Johansen 1978; Lapennas and Lutz 1982; Lapennas and Reeves 1983 b), and the authors suggested that oxygenation-dependent polymerization of tetrameric hemoglobin might account for high n-values, although no evidence for polymerization was found in the frog (Wells and Weber 1985). At low saturation, decreased n-values have been explained by lysis, causing a hump in the curve of bird blood (Lapennas and Reeves 1983 b), or by tetramer-dimer dissociation in embryonic mammals (Brittain 1987 a). An alternative explanation for sparrow blood is that the isohemoglobins with different oxygen-binding charactertistics combine to produce non-standard curves (Maginniss et al. 1980). It will be nevertheless apparent that no single set of explanations is adequate for all cases.

A number of authors have enquired as to the purpose of heterogeneous oxygen binding. Lapennas and Reeves (1983 b) comment on the physiological significance of rising n-values with saturation, and note that, in fowl, oxygen loading is favoured at high saturation, and unloading favoured at low saturation. Note that their phenomenological explanations appear adequate in the absence of inventive adaptive stories.

Adaptationist explanations for similar phenomena in frogs involve multiple loci for gas exchange, incomplete separation of pulmonary and systemic blood circulation, or diving behaviour (Maginniss et al. 1980). Lutz and Bentley (1985) point out that blood properties in diving turtles are generally similar to those of non-diving reptiles, and contrast with enhanced Bohr factors found in diving mammals (Butler and Jones 1983). The authors further point out that the polyphasic component of turtle blood may be advantageous in securing oxygen during a dive at low internal P_{O_2} from inflated lungs. The turtle can extract oxygen to very low P_{O_2} and, with lungs inflated during a dive, may exploit the low saturation hump of the curve without the disadvantage that would be entailed by the large Bohr factor characterized by diving mammals. The curve is steeper at low saturations, and *n* approaches 1, thus tending to preserve arterial oxygen content (Lapennas and Lutz 1982). These speculations are coupled with a greater degree of integrative physiology than is usual. This holistic approach clearly forecloses a number of alternative explanations, but problems in interpreting the affinity constants clearly remain. Lutz and Bentley (1985) report such huge differences in P_{50} between green turtles ($P_{50} \approx 29$ mmHg) and loggerhead turtles ($P_{50} \approx 47$ mmHg), both good divers, as to render purposeful explanations of the fine structure of the reaction very doubtful.

Polyphasic oxygen equilibria demonstrate the problem of describing oxygen-binding reactions by simple parameters, as well as in offering functional explanations.

7 The Environment-Organism Problem: Current Utilities that are Non-Adaptive

I have argued that for hemoglobin, traits of quaternary organization and features of the hemoglobin-oxygen equilibrium do not find their origins in adapatation to particular features of the environment. Of all the adaptationist's claims for environment-organism selection of hemoglobin traits, the strongest cases appear for modulation of whole-blood oxygen-binding traits in response to oxygen and temperature factors. The allosteric behaviour of erythrocyte phosphate factors in response to environmental challenges has often been cited in support of adaptationist arguments (see reviews of Grigg 1974; Johansen and Weber 1976; Wood 1980; Weber and Wells 1989). In the next section we turn to address the problem of intraspecific variation in blood oxygen traits and, in particular, the putative role of allosteric phosphate interactions. I shall argue that many of the traits frequently reflect current utility, and that they were never shaped directly by natural selection for the purpose they now serve.

Physiologists have a tendency to use the term "adaptation" to refer not only to traits which have been selected genetically in the Darwinian sense, but also to ontogenetic features which indicate *phenotypic plasticity* and are non-genetic and non-adaptive traits. (Of course, the *capacity* to display a phenomenon may be genetic. Selection can act on all of the traits in a population, regardless of their heritability. Only if the traits are heritable can there be a response to selection in the next generation.) Physiological explanations of such phenotypic plasticity should address current utilities, and not become tangled with adaptationist proposals in the Darwinian mode. Many so-called physiological adaptations have not been selected for Darwinian fitness, and are not heritable. Other traits that have arisen through a non-adaptive process may, however, currently be undergoing stabilizing selection.

8 Fine Tuning to Meet Specific Environmental Challenge or Non-Adaptive OEC?

Adaptive changes in oxygen affinity plausibly appear to compensate for low levels of environmental oxygen, and many animals encountering periodically low oxygen conditions have the necessary machinery to deal with the problem. What I question is whether this machinery is adaptive and currently under selection, and that maybe the animals are simply making a response to the perturbation that is typical of their class, perhaps better developed, perhaps not. Too infrequently have "control" observations been carried out on similar animals from a different habitat, or with different lifestyles for comparison. Such observations are a powerful tool and have the potential to strengthen the adaptationist theory according to the falsifiability criterion of Popper (1972).

With these ideas in mind, Gordon Grigg (Sydney University) and I visited Antarctica to study the responses of the fish *Pagothenia borchgrevinki* to extended hypoxia. Living in near-constant physical conditions at sub-zero temperatures and in water of exceptionally high oxygen content, these fish have among the lowest routine rate of oxygen consumption for any vertebrate (Wells 1987). To our surprise, these fish showed a marked reduction in blood-oxygen affinity following an

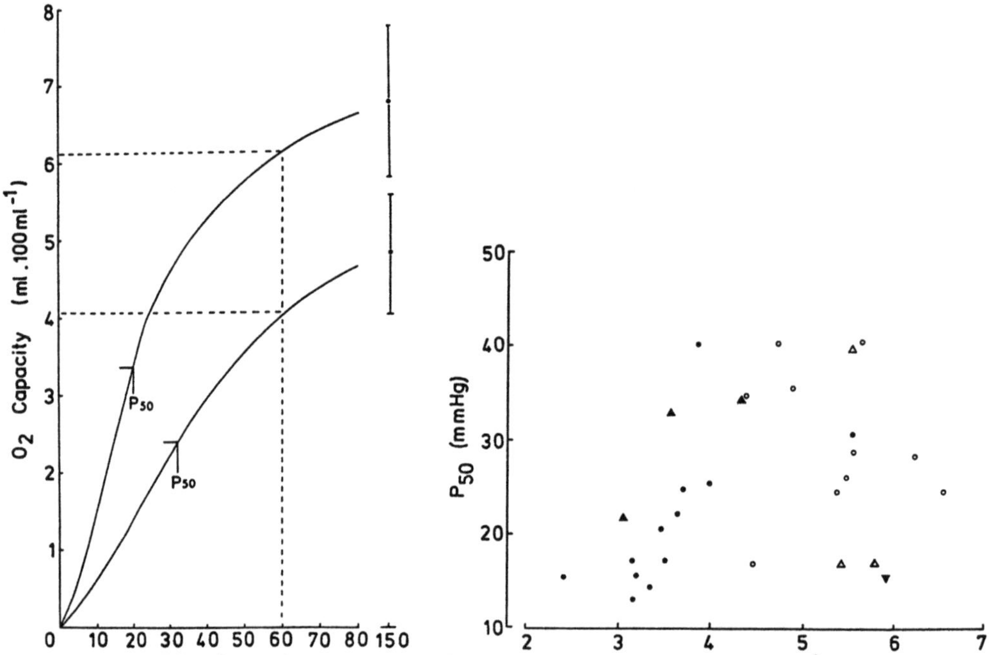

Fig. 1. Hypoxic acclimation in the antarctic fish *Pagothenia borchgrevinki*. *A* Oxygen binding curves for hypoxic (*left*) vs controls (*right*) showing both the increased capacity (±SD) and affinity in the former, and *B* erythrocyte ATP and whole blood oxygen affinity (P_{50}) for *hypoxic* (●) and *control* (○) groups. Fish with ice-damaged gills are shown with corresponding symbols (▲, △)

11 – 14-day exposure to an oxygen pressure of 50 – 60 mmHg, something that they would never be exposed to in nature. The extent of the response (Fig. 1) is similar to that seen in species for which adaptive stories have been told (see Weber 1982; Weber and Wells 1989).

Birds flying at high altitude provide another example for adaptationist thinking. Perutz (1986) states: "The bar-headed goose can reach these heights thanks to a haemoglobin with high oxygen-affinity that has been generated by a chance mutation...". The greylag goose, with a low affinity, is relegated to the plains all year, and the traits in blood-oxygen binding are clearly seen as resulting from selective pressures (Petschow et al. 1977; Black and Tenney 1980; Perutz 1983). Unfortunately, the adaptationist claim rests entirely on laboratory responses to hypoxia in only one of the species (Black and Tenney 1980), and more importantly, on hearsay evidence of migratory behaviour. Banding studies clearly demonstrate that the bar-headed goose has a tendency to remain in Central Asia, while it is the greylag goose that migrates over the Himalayas (McClure 1974)! Furthermore, the non-migratory house sparrow has all the physiological traits necessary to fly over the Himalayas under conditions which render prostrate a mouse of similar size and metabolic weight, and indistinguishable blood-oxygen binding properties (Tucker 1968). Detailed investigation of sparrow blood confirms its efficacy for high altitude, but the adaptive features for flight are pulmonary and cardiovascular (Maginniss 1985). The adaptive significance of blood traits for high flying are therefore doubtful.

The discovery by Chanutin and Curnish (1967) of the powerful modulating effect of 2,3-diphospholycerate (DPG) on the oxygen affinity of mammalian hemoglobin prompted comparative physiologists to look for other cofactors in the remaining vertebrate classes. A general phylogenetic picture emerged confirming the primary role of organic phosphates in controlling blood-oxygen affinity: DPG in mammals, inositol pentaphosphate in birds, ATP in reptiles and amphibians, and ATP or GTP in fish (Eaton 1974; Johansen and Weber 1976). In general, red cell trinucleotide levels are determined by the rate of oxidative phosphorylation, and DPG levels by glycolysis. Thus, in low environmental oxygen when body fluids become acidic, phosphoglycerate mutase promotes glycolysis so that DPG tends to accumulate, and ATP or GTP tend to disperse. The result is that under hypoxic challenge, mammals with red cell DPG show a decrease in blood-oxygen affinity, which tends to preserve tissue oxygen delivery, and fish with trinucleotides increase oxygen affinity, which optimizes oxygen loading. However, the distribution of phosphate cofactors in vertebrates does not strictly follow the pattern outlined (Bartlett 1978, 1980, 1982; Hazard and Hutchinson 1982; Kim et al. 1981). The contention that current utility rather than adaptation should be the manner with which to view allosteric regulation is strengthened by new evidence from the cane toad *Bufo marinus.*

The red cells of the cane toad are unusual in that they contain significant amounts of both ATP and DPG. This poses a problem for adaptationists wishing to make predictions about the blood respiratory responses of the toad to hypoxia. Does ATP decrease in tandem with a rise in DPG, and what happens to oxygen affinity? To answer these questions, we kept a group of toads at 30 °C continuously at a P_{O_2} of 31 – 34 mmHg for 11 days, against a control group in air. The

Table 1. Whole blood parameters from *Bufo marinus* acclimated to hypoxia at 30 °C (Data are means ± SD for $N = 5$)

	ATP µmol/g Hb	DPG µmol/g Hb	P_{50} mmHg	pH	Hb (g/l)	Hematocrit (%)	Lactate (mmol/l)	Weight (g)
Hypoxia	8.33 ± 0.43	35.1 ± 2.5	59.8 ± 5.6	7.72 ± 0.04	73.5 ± 8.3	32.6 ± 4.6	3.25 ± 0.42	88.1 ± 21.3
Control	8.40 ± 1.14	28.3 ± 3.1	56.7 ± 3.1	7.94 ± 0.01	75.1 ± 12.9	31.7 ± 4.1	2.42 ± 0.82	113.5 ± 17.3
	N.S.	$P < 0.5$	N.S.	N.S.	N.S.	N.S.	N.S.	N.S.

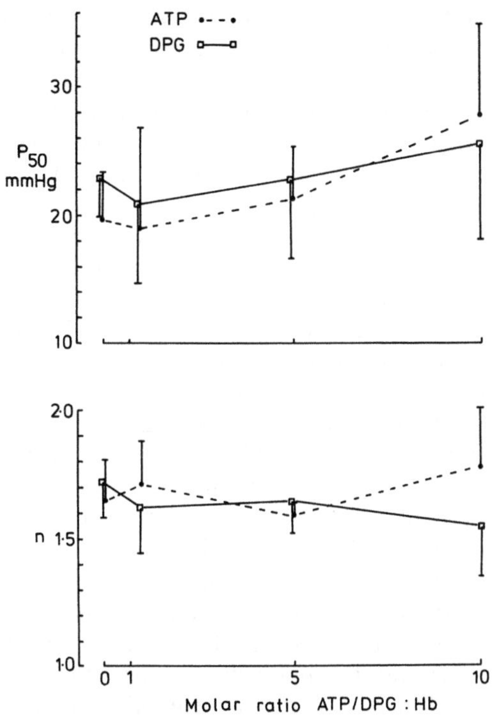

Fig. 2. Oxygen affinity (P_{50}) and cooperativity constant (n) of toad hemoglobin (*Bufo marinus*) at different concentrations of ATP and DPG, showing possible influence of phosphates only at the highest concentrations

blood was sampled and ATP, DPG, oxygen affinity, and hematological variables were recorded (Table 1). The decrease in ATP was not significant ($P > 0.5$) and the increase in DPG significant ($P < 0.05$). Despite the apparent mammalian-type response of elevated DPG, the affinity increase was not significant ($P > 0.05$). These data might be interpreted as suggesting that toad hemoglobin is intensitive to DPG. Both ATP and DPG affect the isolated major Hb fraction (Fig. 2), but relatively high concentrations ($\approx$ ten-fold excess) are required to appreciably alter the Hb affinity constants. In the cane toad, an increase in DPG resulting from hypoxic challenge does not influence affinity by allostery. This conclusion poses no difficulty in dealing with a phenomenological approach, but clearly presents a problem for adaptationists.

9 Hemoglobin Ontogeny and the Functionalist/Structuralist Dichotomy

In mammalian species so far investigated, embryonic hemoglobins are associated with the nucleated erythrocytes derived from the embryonic yolk sac, and later on in development a transition to fetal globins may occur (Brittain and Wells 1983). The hemoglobin system develops over time, according to specific ground rules of synthesis in the yolk sac, the liver and, finally, in the marrow. Similar observations of developmental transitions to adult hemoglobin have been made for other vertebrate classes (see Weber and Wells 1989). Early embryonic hemoglobins show low cooperativity, phosphate and pH insensitivity, and high affinity for oxygen (Brittain and Wells 1983; Weber et al. 1987). This remarkable conservatism of hemoglobin traits exemplifies the historical constraints of development on future evolutionary change and adaptations. These relations and dynamic constraints are inconsistent with the panselectionist's paradigm, and were enunciated long ago by Von Baer (1828) in his fundamental laws of embryology: viz. that the early stages of development are conservative and restrictive of later development. Development cannot be pulled apart, piece by piece, in evolution through selection.

A second example of ontogenetic constraints during development is the salamander. Here, the oxygen affinity of blood decreases in the growing larvae in an allometric fashion (Burggren et al. 1987), not as any consequence of environmental input, but directly reflecting the declining energetic status of larger individuals.

"Functionalism" is a doctrine promoting the atomization of parts of a system (see Hughes and Lambert 1984), and invites questions such as "What is the *function* of embryonic hemoglobin, how is it *selected*, and what is it *adapted* for?" Perhaps hemoglobin has no function in the sense that a physicist cannot ask the question "What is the *function* of an electron?" An alternative approach attempts to describe the traits of complex and unified systems in a "structuralist" mode, whereby not the elements or parts, but the interrelations between them are emphasized; thus, possible category mistakes may be avoided. For example, the structuralist's approach recognizes that phosphorylation and glycolysis are, first and foremost, part of a metabolic framework, rather than a collection of separate biochemical pathways sequestered by the red cell in its mission of efficient oxygen transport. Metabolic processes are descriptions of events and do not provide explanations about hemoglobin function. Science is explanatory, not descriptive, and metabolism therefore cannot be considered out of context.

Given the historic theme to hemoglobin ontogeny, with underlying organizational rules restricting development, how then does change proceed given such constraints? One possibility is illustrated by the neotenous salamander, where a change in the timing of development results in adult hemoglobin traits of low cooperativity and Bohr effect, and high affinity (Weber et al. 1985).

A number of the so-called physiological adaptations to hypoxia may also be viewed in terms of the plastic phenotype that has moulded organisms during ontogeny. Atavisms – or reversions to previous evolutionary states – appear to transcend adaptive boundaries and highlight nature's imperfections. For example, *Sphenodon* reptiles (Rhynchocephalia) have low cooperativity and small Bohr factors, and relatively high affinity, though none of these traits are typical of lizards (Wells et al. 1983). Humans, too, have retained the genetic ability to produce non-cooperative embryonic hemoglobins, or mutant hemoglobinopathies during development (Lehmann and Huntsman 1974). The thalassemic diseases are also becoming increasingly difficult to interpret in the functionalist mode (Sect. 4), and perhaps the structuralist mode of thinking might enable us to interpret intelligibly the presence of apparently deleterious genes.

10 Conclusions

The respiratory system begins with a set of constraints. Some of the "spaces" on Gould and Lewontin's (1979) spandrels are used for ingenious effects, which reflect current utility, e.g. allosteric phosphate cofactors and the Root effect. We are urged not to focus exclusively on immediate adaptation to local conditions (which may be epiphenomena) and ignore the constraints (noses never evolved to support spectacles). The hemoglobin molecule is not so plastic that it must always be moulded under the omnipotence of natural selection; instead, its traits have frequently been co-opted rather than adapted. Acclimatory variation in hemoglobin can undergo selection, but no evolutionary change in Hb will result from this selection; i.e., there will be no "response to selection". The evolution of hemoglobin has not proceeded by gradualism with the summation of small changes, adapting more finely to the environment. Such thinking leads to the idea of complete adaptation (cf. Gould 1986); that every species is adapted to its environment is tautologous to existence (Krimbas 1984).

We must reject optimizing selection on the hemoglobin part of the respiratory system. The concept of sufficient design with best compromises among competing demands for loading and unloading respiratory gases may be more appealing, and one cannot do better than cite the example of the Antarctic icefish. These channichthyids are quite unique in having no hemoglobin, yet, despite this, they have well-regulated systems for oxygen uptake, which are comparable with red-blooded nototheneiids living in the same extreme environment (Hemmingsen et al. 1969; Hemmingsen and Douglas 1970). Their complex and unified structural systems include low metabolic rate, comparatively large cardiac chambers and blood volume, and extensive cutaneous gas exchange. Selection cannot have acted independently on the separate parts of the respiratory system of the icefish. A functionalist explanation is unsatisfactory because the loss of hemoglobin does not seem intelligible when such radical restructuring is required for the remainder of the respiratory system.

A recurrent phenomenon is that Hb traits, having evolved under one selective regime, or by means other than selection, enhance organisms' fitness under another selective regime. Traits that serve in a context other than that in which they originated are termed "pre-adaptations". This concept is clearly established in the literature (see Feder 1988).

The oddities and apparent imperfections in nature are useful in formulating non-adaptive hypotheses. Among those we have mentioned are: allosteric control in the toad, hypoxia in antarctic fish, non-standard binding curves in turtles and birds, and conservative binding properties in embryos, salamanders and the *Sphenodon*.

In discussing evolution, it becomes extremely important to specify which assumptions are being made. If we wish to draw inferences about effects on fitness, mode of origin, selection, etc., it becomes exceedingly important that the assumptions be verified, which is not typically done by otherwise well-meaning comparative physiologists and biochemists.

True knowledge, the cornerstone of scientific method, cannot be reached by interpreting phenomena in terms of final causes (i.e. purpose). How, for example, would one go about proving the non-existence of purpose? We no longer need to be guided by Monod's vision of Maxwell's demon, but are better served by adopting new ways of thinking about the traits of proteins. The proposed holistic approach does not make life easy for scientific investigators but seems instead to close doors to research. Slower but potentially more useful gains may be made if we view adaptation not as discrete traits, but as mergent properties of an organized system of development.

Acknowledgement. I am most grateful for constructive disscussion with Dr. M.E. Feder.

References

Barnikol R, Burkhard O (1979) The fine structure of O_2 Hb binding in animals: *Salmo irideus.* In: Gilles R (ed) Animals and environmental fitness. Pergamon, Oxford, pp 47–48
Bartlett GR (1978) Phosphate compounds in reptilian and avian red blood cells; developmental changes. Comp Biochem Physiol 61A:191–202
Bartlett GR (1980) Phosphate compounds in vertebrate red blood cells. Am Zool 20:103–114
Bartlett GR (1982) Developmental changes of phosphates in red cells of the emu and the rhea. Comp Biochem Physiol 73A:129–134
Bauer C, Jung H-D (1975) A comparison of respiratory properties of sheep haemoglobin A and B. J Comp Physiol 102:167–172
Black CP, Tenney SM (1980) Oxygen transport during progressive hypoxia in high-altitude and sea-level waterfowl. Respir Physiol 39:217–239
Bock WJ (1980) The definition and recognition of biological adaptation. Am Zool 20:217–227
Boutilier RG, Shelton G (1986) Respiratory properties of blood from voluntarily and forcibly submerged *Xenopus laevis.* J Exp Biol 121:285–300
Brady RH (1982) Dogma and doubt. Biol J Linn Soc 17:79–96
Bridges CR, Pelster B, Scheid P (1985) Oxygen binding in blood of *Xenopus laevis* (Amphibia) and evidence against Root effect. Respir Physiol 61:125–136
Brittain T (1987a) A possible role for protein dissociation in the functioning of embryonic haemoglobins. J Exp Biol 127:443–447
Brittain T (1987b) The Root effect. Comp Biochem Physiol 86B:473–481
Brittain T, Wells RMG (1983) Review: oxygen transport in early mammalian development: the molecular physiology of embryonic hemoglobins. In: Johnson M (ed) Mammalian development, vol 5. Elsevier, Amsterdam, Chapt 4, pp 135–154
Brookfield JFY (1982) Adaptation and functional explanation in biology. Evolut Theory 5:281–290
Brunori M (1975) Molecular adaptation to physiological requirements: the hemoglobin system of trout. In: Horecker BL, Stadtman ER (eds) Current topics in cellular regulation, vol 9. Academic Press, London, pp 1–39
Brunori M, Bellelli A, Giardina B, Condo S, Perutz MF (1987) Is there a Root effect in *Xenopus* hemoglobin? FEBS Lett 221:161–166
Burggren WW, Dupré RK, Wood SC (1987) Allometry of red cell oxygen binding and hemotology in larvae of the salamander, *Ambystoma tigrinum.* Respir Physiol 70:73–84
Butler PJ, Jones DR (1983) The comparative physiology of diving in vertebrates. In: Lowenstein D (ed) Advances in comparative physiology and biochemistry. Academic Press, New York, pp 180–364
Carrell RW (1980) Hemoglobinopathies in Oceania. Hemoglobin 4:427–429
Cech JJ, Laurs RM, Graham JB (1984) Temperature-induced changes in blood gas equilibria in the albacore, *Thunnus alalunga,* a warm-bodied tuna. J Exp Biol 109:21–34
Chanut in A, Curnish RR (1967) Effect of organic and inorganic phosphates on the oxygen equilibrium of human erythrocytes. Arch Biochem Biophys 121:96–102

Dickerson RE, Geis I (1983) Hemoglobin: structure, function, evolution, and pathology. Benjamin Cummings, Menlo Park CA
Dobson GP, Wood SC, Daxboeck C, Perry SF (1986) Intracellular buffering and oxygen transport in the pacific blue marlin (*Makaira nigricans*): adaptations to high speed swimming. Physiol Zool 59:150–156
Dodgson SJ, Holland RAB (1983) The reaction kinetics of four sheep haemoglobins with identical α-chains. Respir Physiol 53:31–45
Eaton JW (1974) Oxygen affinity and environmental adaptation. Ann NY Acad Sci 241:491–504
Feder ME (1988) The analysis of physiological diversity. In: Feder ME, Bennet AF, Burggren WW, Huey RB (eds) New directions in ecological physiology. Cambridge University Press, Cambridge
Fyhn UEH, Fyhn HJ, Davis BJ, Poers DA, Fink WL, Garlick RL (1979) Hemoglobin heterogeneity in Amazonian fishes. Comp Biochem Physiol 62A:39–66
Giles MA, Randall DJ (1980) Oxygenation characteristics of the polymorphic hemoglobins of coho salmon (*Oncorhynchus kisutch*) at different developmental stages. Comp Biochem Physiol 65A:265–271
Gould SJ (1982) Darwinism and the expansion of evolutionary theory. Science 216:380–387
Gould SJ (1986) Archetype and adaptation. Nat Hist 10:16–28
Gould SJ, Lewontin RC (1979) The spandrels of San Marco and the Panglossian paradigm: a critique of the adaptationist programme. Proc R Soc Lond B 205:581–598
Grigg GC (1974) Respiratory function of blood in fishes. In: Florkin M, Scheer BT (eds) Chemical zoology, vol VIII. Academic Press, London, pp 332–368
Hazard ES, Hutchison VH (1982) Distribution of acid-soluble phosphates in the erythrocytes of selected species of amphibians. Comp Biochem Physiol 73A:111–124
Heisler N (1984) Acid-base regulation in fishes. In: Randall DJ, Hoar WS (eds) Fish physiology XA. Academic Press, London, pp 315–401
Hemmingsen EA, Douglas EL (1970) Respiratory characteristics of the hemoglobin-free fish *Chaenocephalus aceratus*. Comp Biochem Physiol 33:733–744
Hemmingsen EA, Douglas EL, Grigg GC (1969) Oxygen consumption in an Antarctic hemoglobin-free fish, *Pagetopsis macropterus*, and in three species of *Notothenia*. Comp Biochem Physiol 29:467–470
Hill AV (1910) The possible effects of the aggregation of haemoglobin on its dissociation curve. J Physiol Lond 40:iv–vii
Hubbard R (1982) The theory and practice of genetic reductionism – from Mendel's laws to genetic engineering. In: Rose S (ed) Towards a liberatory biology. Allison and Busby, New York, pp 85–102
Hughes AJ, Lambert DM (1984) Functionalism, structuralism, and "ways of seeing". J Theor Biol 111:787–800
Ikeda-Saito M, Yonetani T, Gibson QH (1983) Oxygen equilibrium studies on hemoglobin from the bluefin tuna (*Thunnus thynnus*). J Mol Biol 168:673–686
Ingermann RL, Terwilliger RC (1982) Presence and possible function of Root effect hemoglobins in fishes lacking functional swim bladders. J Exp Zool 220:171–177
Johansen K, Weber RE (1976) On the adaptability of haemoglobin function to environmental conditions. In: Davies PS (ed) Perspectives in experimental biology, vol 1. Pergamon Press, Oxford, pp 219–234
Johansen K, Berger M, Bicudo JEPW, Ruschi A, de Almeida PJ (1987) Respiratory properties of blood and myoglobin in hummingbirds. Physiol Zool 60:269–278
Jokumsen A, Weber RE (1980) Haemoglobin-oxygen binding properties in the blood of *Xenopus laevis*, with special reference to the influences of aestivation and of temperature and salinity acclimation. J Exp Biol 86:19–37
Jones DR, Brill RW, Mense DC (1986) The influence of blood gas properties on gas tensions and pH of ventral and dorsal aortic blood in free-swimming tuna, *Euthynnus affinis*. J Exp Biol 120:201–213
Kim HD, Zeidler RB, Sallis JD, Nichol SC, Isaacks RE (1981) Adenosine triphosphate-deficient erythrocytes of the egg-laying mammal, echidna (*Tachyglossus aculeatus*). Science 213:1517–1519

Krimbas CB (1984) On adaptation, neo-Darwinian tautology, and population fitness. Evol Biol 17:1–57
Lapennas GN, Lutz PL (1982) Oxygen affinity of sea turtle blood. Respir Physiol 48:59–74
Lapennas GN, Reeves RB (1983a) Oxygen affinity and equilibrium curve shape in blood of chicken embryos. Respir Physiol 52:13–26
Lapennas GN, Reeves RB (1983b) Oxygen affinity of blood of adult domestic chicken and red jungle fowl. Respir Physiol 52:27–39
Lehmann H, Huntsman RG (1974) Man's haemoglobins. North-Holland, Amsterdam
Lewontin RC (1978) Adaptation. Sci Am 239:156–169
Lutz P (1980) On the oxygen affinity of bird blood. Am Zool 20:187–198
Lutz PL, Bentley TB (1985) Respiratory physiology of diving in the sea turtle. Copeia 1985(3):671–679
Lutz PL, Longmuir IS, Schmidt-Nielsen K (1973) Dissociation curve of bird blood and effect of red cell oxygen consumption. Respir Physiol 17:269–275
Lutz PL, Longmuir IS, Schmidt-Nielsen K (1974) Oxygen affinity of bird blood. Respir Physiol 20:325–330
Lykkeboe G, Johansen K (1978) An O_2-Hb "paradox" in frog blood? (n-values exceeding 4.0). Respir Physiol 35:119–127
Maginniss LA (1985) Red cell organic phosphates and Bohr effects in house sparrow blood. Respir Physiol 59:93–103
Maginniss LA, Song YK, Reeves RB (1980) Oxygen equilibria of ectotherm blood containing multiple hemoglobins. Respir Physiol 42:329–343
Maginniss LA, Olszowka AJ, Reeves RB (1986) Oxygen equilibrium curve shape and allohemoglobin interaction in sheep whole blood. Am J Physiol 250:R298–305
Matsuura MSA, Ogo SH, Focesi A (1987) Dimer-tetramer transition in hemoglobins from *Liophis miliaris* – I. Effect of organic polyphosphates. Comp Biochem Physiol 86A:683–687
Maturana HR, Varela FJ (1987) The tree of knowledge: the biological roots of understanding. Shambhala Publications, Boston
McClure HE (1974) Migration and survival of the birds of South East Asia. SEATO Medical Project, Publ US Army, Bangkok
Mickleson KNP, Dixon M, Hill PJ, Black R, Eales M, Rutherford J, Yakas J, Trent RJ (1985) Influence of α-thalassaemia on haematological parameters in Polynesian neonates. NZ Med J 98:1036–1038
Monod J (1972) Chance and necessity. Collins London
Newton MF, Peters J (1983) Physiological variation of mouse haemoglobins. Proc R Soc Lond B 218:443–453
O'Grady RT (1984) Evolutionary theory and teleology. J Theror Biol 107:563–578
O'Grady RT (1985) Historical processes, evolutionary explanations, and problems with teleology. Can J Zool 64:1010–1020
Perutz MF (1983) Species adaptation in a protein molecule. Mol Biol Evol 1:1–28
Perutz MF (1986) A new view of Darwinism. New Scientist, 2 Oct 1986:36–38
Perutz MF, Brunori M (1982) Stereochemistry of cooperative effects in fish and amphibian hemoglobins. Nature 291:682–684
Petschow D, Wurdinger I, Baumann R, Duhm J, Braunitzer G, Bauer C (1977) Causes of high blood-oxygen affinity of animals living at high altitude. J Appl Physiol 42:139–143
Popper KR (1972) The logic of scientific discovery. Hutchinson, London
Powers DA (1972) Hemoglobin adaptation for fast and slow water habitats in sympatric catostomid fishes. Science 177:360–362
Powers DA (1974) Structure, function, and molecular ecology of fish hemoglobins. Ann NY Acad Sci 241:472–489
Powers DA (1980) Molecular ecology of teleost fish hemoglobins: strategies for adapting to changing environments. Am Zool 20:139–162
Root RW (1931) The respiratory function of the blood of marine fishes. Biol Bull 61:427–456
Routley R (1984) Maximizing, *satisficing, satisizing*: the difference in real and rational behaviour under rival paradigms. Discussion papers in environmental philosophy no 10. Australian National University, Canberra

Stratton LP, Duffy LK (1976) Hemoglobin polymorphism in *Microtus pennsylvanicus*. Comp Biochem Physiol 54B:413–415

Trent RJ, Mickleson KNP, Wilkinson T (1985) Alpha globin gene rearrangements in Polynesians are not associated with malaria. Am J Hematol 18:431–433

Tucker VA (1968) Respiratory physiology of house sparrows in relation to high-altitude flight. J Exp Biol 48:55–66

Von Baer KE (1828) Entwicklungsgeschichte der Tiere. Bornträger, Königsberg

Weber RE (1982) Intraspecific adaptation of hemoglobin function in fish to oxygen availability. In: Addink ADF, Spronk N (eds) Exogenous and endogenous influences on metabolic and neural control. Pergamon, Oxford, pp 87–102

Weber RE, de Wilde JAM (1976) Multiple haemoglobins in plaice and flounder and their functional properties. Comp Biochem Physiol 54B:433–437

Weber RE, Wells RMG (1989) Hemoglobin structure and function. In: Wood SC (ed) Comparative pulmonary physiology: current concepts. Lung biology in health and disease, vol 39. Marcel Dekker, New York, pp 279–310

Weber RE, Wells RMG, Rossetti JE (1985) Adaptations to neoteny in the salamander, *Necturus maculosus*. Blood respiratory properties and interactive effects of pH, temperature and ATP on hemoglobin oxygenation. Comp Biochem Physiol 80A:495–501

Weber RE, Kleinschmidt T, Braunitzer G (1987) Embryonic pig hemoglobins Gower I ($\zeta_2 \varepsilon_2$), Gower II ($\alpha_2 \varepsilon_2$), Heide I ($\zeta_2 \gamma_2$) and Heide II ($\alpha_2 \gamma_2$): oxygen-binding functions related to structure and embryonic oxygen supply. Respir Physiol 69:347–357

Wells RMG (1979) Haemoglobin-oxygen affinity in developing erythroid cells of the mouse. J Comp Physiol 129:333–338

Wells RMG (1987) Respiration of Antarctic fishes from McMurdo Sound. Comp Biochem Physiol 88A:417–424

Wells RMG, Brittain T (1981) Transition to cooperative oxygen-binding by embryonic mice. J Exp Biol 90:351–355

Wells RMG, Davie PS (1985) Oxygen binding by the blood and hematological effects of capture stress in two big game fish: mako shark and striped marlin. Comp Biochem Physiol 81A:643–646

Wells RMG, Weber RE (1985) Fixed acid and carbon dioxide effects as functions of hemoglobin-oxygen saturation and intra-erythrocytic pH in the blood of the frog, *Rana temporaria*. Pflügers Arch Eur J Physiol 403:7–12

Wells RMG, Weber RE (1985) Fixed acid and carbon dioxide effects as functions of hemoglobin-oxygen saturation and intra-erythrocytic pH in the blood of the frog, *Rana temporaria*. Plügers ARch Eur J Physiol 403:7–12

Wood SC (1980) Adaptation of red blood cell function to hypoxia and temperature in ectothermic vertebrates. Am Zool 20:163–172

Chapter 6

Kinetics of Oxygen and Carbon Dioxide Exchange

E. R. Swenson[1]

Contents

1 Introduction

The aim of this chapter is to compare the kinetics of gas exchange among the vertebrates and examine whether the rates of diffusion and chemical reaction pose any limitation to O_2 and CO_2 exchange. These kinetics have been extensively studied in mammals, and the reader seeking an overview of the literature may find

[1] Pulmonary and Critical Care Medicine, VA Medical Center, University of Washington, USA

Advances in Comparative and Environmental Physiology
Vol. 6. Edited by R. G. Boutilier

it in a number of recent reviews (Klocke 1987; Meldon 1987; Kreuzer and Hoofd 1987; Baumann et al. 1987; Wagner 1977; Förster 1987). Surprisingly there has been little attention paid to non-mammalian vertebrates, despite considerable interest otherwise in their blood-gas equilibrium chemistry and cardiorespiratory structure-function relationships.

1.1 Temporal Constraints on Gas Exchange

High rates of O_2 uptake and CO_2 elimination are necessary to sustain vertebrate metabolism, and are probably only exceeded in magnitude by H_2O transfer between the animal and its environment. Moreover, these rates rise above basal levels for hours at a time during normal activity and may, with extreme exertion, increase by an order of magnitude. Fundamentally, the basic constraint to gas exchange is the temporal limitation imposed by finite rates of convection, diffusion and chemical reactions of carbon dioxide and oxygen. Since diffusion is only effective over limited distances, convective processes of gas transport (perfusion and ventilation) and the necessary chemical reactions with blood evolved of necessity to permit respiration in large animals. The design of vertebrate gas exchangers allows remarkably efficient performance under the time constraints of these processes. Although the rates of ventilation and perfusion are not usually considered under the rubric of gas exchange kinetics, strictly speaking, they can alter kinetics of gas transfer, particularly in extreme exercise or severe hypoxia and acidosis. A discussion of the influence of ventilation and perfusion in limiting gas exchange is outside the scope of this review, and the reader is referred to the excellent theoretical treatment of these aspects by Piiper and his colleagues (Piiper 1982). Directly germane to the subject of gas exchange kinetics, however, is the determining role of blood flow on capillary transit time; for it is against this that all diffusion and chemical reaction rates must ultimately be compared in any analysis of physiological or pathological gas exchange.

As a general description for the transfer of any gas across a perfused surface, the following equation serves to highlight many of the important factors that determine the time course of equilibration between blood and air, water or tissue (Wagner 1977; Piiper and Scheid 1980):

$$Pc(t) = Pm + (Pi - Pm) \times e^{-Ad\beta mt / \times \beta bVcMW} , \quad (1)$$

where Pc, Pm and Pi are gas tensions in the capillary, medium and afferent (incoming) blood; A,d, and × are the area, diffusion coefficient and thickness of the membrane barrier, Vc is capillary blood volume, βm and βb are the solubilities of gas in membrane and blood, t is time and MW is the molecular weight of the gas. In this equation, the capillary transit time equal to Vc/Q, where Q is equals the blood flow of the region under consideration. Increasing the value of any factor within the numerator of the exponential accelerates gas equilibration, while the opposite is true for factors within the denominator. It is useful here to raise several points of definition. While end-capillary equilibration represents efficient gas exchange, in quantitative terms it is not always true that failure of equilibration between blood and tissue or respired gas is a cause for reduced gas uptake or elimination. For instance, increases in both Q and βb will slow equilibration, but generally result in more gas transfer even if equilibration does not occur within capillary transit.

A comparison between the kinetics of inert gas and CO_2/O_2 exchange helps to underscore one of the most important determinants in gas transfer equilibration. Inert gases (which dissolve in blood without undergoing any chemical reaction) equilibrate much more rapidly than do carbon dioxide or oxygen, both of which require a considerable fraction of transit time. In the human lung, Wagner (1977) has calculated an inert gas (with equivalent molecular weight to CO_2 or O_2) requires only 15–20 ms to reach 99% equilibration, while O_2 and CO_2 need 430 and 210 ms respectively. Thus, for inert gases, full equilibration can be assumed, since it is unlikely any capillary transit times are this short (vide infra). The principle difference in the time courses of an inert gas and CO_2 or O_2 lies in the value of $\beta m/\beta b$. For an inert gas this is almost unity, but for CO_2 and O_2, $\beta m/\beta b$ is one to two orders of magnitude less. The importance of this ratio can be appreciated intuitively as follows: βm determines the rate at which a gas diffuses into or out of blood across a surface or membrane, since its diffusing capacity is proportional to βm. The rate at which the partial pressure of the gas in blood changes is dependent upon its solubility in blood. Thus, the more soluble a gas, the slower will be its rate of partial pressure change for a given amount of gas transferred. The far greater solubility of O_2 and CO_2 in blood relative to the membrane arises from reversible chemical binding and reaction with hemoglobin and other proteins in red cells and plasma. In an analogous fashion, Gray (1971) has shown that the rate of approach to equilibrium in any solution upon the addition of CO_2 is strongly dependent on the buffering capacity (i.e. the capacitance).

Although Eq. (1) rightly emphasizes the significance of a large blood-gas capacitance in delaying O_2 and CO_2 equilibration, it ignores the considerable differences by which they attain high solubilities and, more importantly, it fails to account for factors in blood which additionally delay O_2 and CO_2 equilibration. These include non-linear blood dissociation curves, red cell and plasma diffusion resistances, divergent plasma/red cell gas solubilities and finite chemical reaction rates. These factors influence the time courses of O_2 and CO_2 exchange quite differently, as will be taken up in detail in the course of this review.

Gas exchange occurs mainly across the capillary, although it has been possible to detect or infer that some small fraction of gas transfer occurs in pre- and post-capillary vessels (Conhaim and Staub 1980; Feisal et al. 1963). Contact of blood with a gas-exchanging surface should be long enough to permit equilibration (with some reserve) and yet not compromise the adequacy of total cardiac output. Capillary transit times in vertebrates range from a few hundred milliseconds to many seconds (Table 1), depending on the tissue, species, temperature, metabolic rate, blood flow, pressure and viscosity, and capillary density, number and length. In general, as metabolism rises and blood flow increases, transit times fall but are prevented from decreasing too precipitously by capillary recruitment (Honig and Odoroff 1981; Knopp and Bassingthwaighte 1969; Randall and Daxboeck 1984; Malvin 1988).

Lindstedt (1984) found capillary transit times (in seconds) scale disportionately to body mass (in kilograms) at rest in mammalian lungs to the power of 0.20.

Table 1. Representative capillary transit times in various vertebrate species and organs

Species	Organ	Condition	Transit time (seconds)	Reference
Human	Lung	Rest	0.7	Johnson et al. (1968)
			2.8	Lindstedt (1984)
		Exercise	0.3	Johnson et al. (1968)
			0.2	Groebe and Thews (1987)
			0.6	Lindstedt (1984)
		Rest-hypoxia	0.3	Groebe and Thews (1987)
Cow	Lung	Rest	4.6	Lindstedt (1984)
		Exercise	0.9	Lindstedt (1984)
Dog	Lung	Rest	1.8	Capen et al. (1987)
		Exercise	0.3	Karas et al. (1987)
	Muscle	Rest	0.9	Honig et al. (1982)
		Exercise	0.13	Honig and Gayeski (1987)
	Heart	Exercise	0.6	Honig and Gayeski (1987)
Fox	Lung	Exercise	0.12	Longworth et al. (1989)
Mouse	Lung	Rest	0.6	Lindstedt (1984)
		Exercise	0.12	Lindstedt (1984)
Frog	Lung	Rest	1.3	Capen et al. (1987)
	Skin	Rest	1.0	Malvin (1988)
Fish	Gill	Rest	1.0	Hughes et al. (1981)
Bird	Lung	Rest	0.8	Powell and Mazzone (1983)
		Exercise	0.5	Powell et al. (1985)

Thus, transit time can vary from tenths of a second in the smallest mammals to several seconds in the largest. This is a consequence of the fact that capillary blood volume scales linearly with body mass, but cardiac output scales with body mass to the power of 0.75 (Weibel 1984). As Table 1 highlights, there are very few data on non-mammalian species, and it is unknown if these same considerations apply, especially with regard to the vastly different environmental habitats and physiology that characterize non-mammalian vertebrates. Whether a trend toward shorter transit times is present in smaller animals of the other vertebrate classes is unknown.

1.2 Methods of Measuring Gas Exchange Kinetics

All present methods that measure kinetic events during gas exchange suffer from necessary simplifications that depart from the physiological situation. While newer techniques may approach more closely the conditions prevailing in capillaries, important artifacts may still remain. A brief review of these techniques will serve to give some perspective and underscore limitations in our present knowledge.

1.2.1 Steady-State Techniques

The rapid-reaction continuous-flow technique developed by Hartridge and Roughton (1923) provided the first kinetic data for processes with time constants less than a second, and has been used extensively since then in blood and other fluids. It involves the rapid mixing of two solutions and flow away from the point of mixing down well-calibrated tubes. Because the flow and distance down the tube are known, the time elapsed at any distance can be calculated. Measuring or sampling devices interposed at the end of various size tubes allow multiple time points in a reaction to be studied. As powerful as this method has been, measurements must often be taken under very non-physiological conditions. Low solubilities of carbon dioxide and oxygen in solution require high and often rapidly varying gas tensions during the course of a reaction, which do not occur in vivo. Furthermore, experiments measuring CO_2 and O_2 transfer between red cells and extracellular fluid must be done at low hematocrits to avoid problems of high viscosity. Lastly, turbulent flow in relatively large diameter tubes does not mimic the physiological conditions of slower flow through small capillaries. The very important issues of red cell shape, motion and depth of unstirred layers surrounding the red cell will be examined in the section on O_2 kinetics. Other steady-state methods involving flux across simple membranes (Donaldson and Quinn 1974) or the isolated lung (Klocke 1978; Enns and Hill 1983) have been used. These have an advantage of permitting analysis of a gas-fluid interface, but are limited to conditions in which chemical reaction rates are far slower than diffusion.

1.2.2 Non-Steady State Techniques

The development of rapidly responding instruments such as spectrophotometers and, more recently, of pH electrodes and mass spectrometers has generated a number of non-steady-state methods to assess gas exchange kinetics. The stopped-flow rapid-reaction apparatus of Gibson (1954), like the earlier Hartridge-Roughton instrument, rapidly mixes two solutions of different composition, which then proceed to an observation point where flow is then suddenly stopped and the reaction continues to completion. The advantages of this technique are that much smaller volumes of reactants are needed and that a complete kinetic description is obtained on a single sample. It suffers from some of the same problems described for the continuous-flow rapid-reaction method. Gas exchange kinetics in the lung can be studied with plethysmographic techniques, as Feisal et al. (1963) have done for CO_2 and, more recently, for O_2 by Klocke (1988). Instantaneous changes in lung volumes measure gas transfer between alveoli and blood in an extremely accurate fashion; however, the experiments can only involve transport of a single gas. Recently, a variety of two-phase systems which seek to model capillary gas exchange in a more realistic manner have been employed. Older work by Luckner (1939) on CO_2 reaction kinetics in blood layers and by Kreuzer and Yahr (1960) on O_2 kinetics involved fairly thick layers (100 μm). These have been extended to much thinner films for O_2 by Thews (1959); Frech et al. (1968), Sinha (1969), and Reeves and Heidelberger (1987). Ono and Tazawa

(1975) developed an in situ microphotometric method for measuring O_2 exchange rates in single stationary cells within a capillary. Artificial capillaries developed by Heineken et al. (1976) have been further improved by Boland et al. (1987). As yet, CO_2 kinetics have not been investigated with such very thin blood film or capillary methods.

1.2.3 Mathematical Models

Models of gas exchange in the lung and other tissues have proven very useful in the interpretation of data obtained both in vitro and in vivo. The ability to quickly and extensively alter conditions in such models provides valuable confirmation of experimental data and, more significantly, highlights possible deficiencies in data and general concepts. A number of models of gas exchange in the mammalian lung (Wagner and West 1972; Hill et al. 1973b; Hlastala 1973; Swenson and Maren 1978; Crandall and Bidani 1981; Piiper and Scheid 1981; Wieth et al. 1982) and tissues (Forster and Crandall 1975; Moll 1968/1969; Kutchai 1970; Gutierrez 1986a; Vandergriff and Olson 1984c) have explored the effects of altering chemical reaction and diffusion rates on gas transport. Gas-exchange models in other vertebrates exist for fish gills (Scheid et al. 1986; Malte and Weber 1985; Cameron and Polhemus 1974), avian lung (Scheid 1978) and amphibian skin (Piiper and Scheid 1975). Accuracy of data used in models is a crucial issue, and the necessity for precise determination of these rates under physiological conditions cannot be overstated.

2 Oxygen Exchange

Rate-limiting steps in oxygen exchange have been variably ascribed to both diffusion resistances and chemical reaction rates. The source for much of this uncertainty lies with the experimental methods described above. While the speeds at which these processes occur are fairly rapid and do not require enzymes or membrane transport carriers (as is necessary for CO_2 exchange and the Bohr-Haldane effects), they happen both simultaneously and sequentially, so that the overall process of oxygen exchange is slowed. Oxygen uptake and release from blood are primarily hindered by diffusion resistance, with only a minor limitation due to chemical reaction with hemoglobin. Experimental and model work show that O_2 transfer and equilibration are generally complete in mammals under resting conditions when blood leaves the capillary, although a substantial fraction of transit time is drawn upon in all species and tissues. It is far from certain, however, that end-capillary PO_2 equilibration occurs in non-mammalian vertebrates, since, in most species, anatomic and physical factors would appear to be disadvantageous.

2.1 Oxygen Diffusion

In its path from environmental air or water to mitochondria, oxygen diffuses over several anatomically distinct distances that constitute varying degrees of resistance to oxygen exchange. The diffusion of O_2 (or any gas) through a medium is related to the product of its diffusion coefficient (d) and solubility (β). This measure of diffusivity, K (Krogh's constant of diffusion), for O_2 and CO_2 in different media is presented in Table 2. Despite the fact that these diffusion distances are usually small, diffusion across these surfaces and in and out of red cells is slow enough, relative to the rates of Hb-O_2 reaction and consumption by the mitochondria, that diffusive equilibrium probably does not occur at any point in the capillary or within tissue.

2.1.1 Diffusion in Respired Air or Water

In air breathing vertebrates, the kinetics of O_2 diffusion in the gas phase are generally not considered a limitation to uptake by the lung, because the diffusivity of O_2 in air is about 200000 times greater than in water or other physiological media. The enormous differences in diffusivity are such that even when one considers that the diffusion distances involved in the gas phase (500–2000 μm) exceed those in blood and respiratory epithelia by orders of magnitude, complete equilibration should require only a few milliseconds. Despite this apparent rapidity of diffusional equilibration, evidence for a diffusion resistance to O_2 exchange

Table 2. Krogh's diffusion constants (K) for O_2 and CO_2 in various media and tissues

	O_2	CO_2	KCO_2/KO_2
	(nmol/cm s torr)		
Air (0 °C)	10.45[a]	8.16[a]	0.72
Water (37 °C)	4.5×10^{-5}[b]	8.5×10^{-4}[c]	17
Plasma (37 °C) (albumin = 6.5 g/dl)	3.1×10^{-5}[d]	8.0×10^{-5}[e]	25
Red cell (37 °C) (Hb = 33 g/dl)	1.4×10^{-5}[f]	3.1×10^{-4}[c]	22
Alveolar-capillary membrane	2.2×10^{-5}[g]		
Muscle			
Mammalian (37 °C)	2.1×10^{-5}[b]	2.9×10^{-4}[h]	14
Frog (20 °C)	1.4×10^{-5}[g]		

[a] Dejours (1981).
[b] Kawashiro et al. (1975).
[c] Gros and Moll (1971).
[d] Goldstick et al. (1975).
[e] Tanashita et al. (1985).
[f] Clark et al. (1985).
[g] Bartels (1971).
[h] Kawashiro and Scheid (1976).

in the gas phase in mammalian lung has been adduced from a number of studies, in which a concentration gradient (or stratification) for oxygen in the alveolar space has been observed. As reviewed by Piiper and Scheid (1987), the subject of intrapulmonary gas mixing is exceedingly complex, and no single experimental approach has yet dealt adequately with the confounding influences of V/Q heterogeneity, heterogeneities in ventilation to gas volume and diffusing capacity to perfusion, and convective-diffusive interactions within the airways that may contribute to apparent gas stratification.

Since oxygen is not the ideal gas to measure gas-phase diffusion resistance, the best experimental approach utilizes retention and excretion characteristics of inert gases with differing molecular weights and nearly equal blood-gas partition coefficients. However, even here the data are difficult to reconcile. Robertson et al. (1986) showed very carefully that the retentions of higher molecular weight gases were slightly greater than lower molecular weight gases in the dog, findings consistent with diffusion limitation in the alveolar gas phase. Since the mammalian acinar length increases by a power of 0.2 with body mass (Weibel 1984), animals of smaller size should then show less stratified inhomogeneity. However, Troug et al. (1979) showed a significant molecular weight dependence on inert gas elimination in rats, which Hlastala et al. (1981) concluded was consistent with an alveolar gas-phase diffusion limitation of enough magnitude to impair O_2 transfer in the lung. While the data remain consistent with some degree of stratified gas-phase inhomogeneity, and some models (Scheid et al. 1981; Paia and Engel 1985) predict its presence, the magnitude is small and the mechanism unclear. Most investigators consider that any real diffusive limitation to O_2 exchange in mammalian lungs resides past the alveolar gas phase.

In the avian lung, gas-phase diffusion in the air capillaries which radiate from the parabronchial lumen is not thought to be a limiting factor in gas exchange at rest in model calculations (Crank and Gallagher 1978; Scheid 1978). Experimental work by Burger et al. (1979) and Powell and Wagner (1982) support this. With exercise, models predict some gas-phase stratification, but confirmatory physiological data are lacking. As in the mammalian lung, the more important diffusive resistances are present in the blood-gas barrier and blood. Gas exchange across the avian eggshell is unique in being almost entirely diffusion-dependent (Rahn and Paganelli 1982). The length of eggshell pores (300 μm) is such that gas-phase diffusive resistance is marked. Ar et al. (1980) showed that the pressure of a gas in a chick embryo air cell could be varied nearly fourfold, simply by altering its diffusivity with substitution of SF_6 or He for N_2 in air. Despite the rather simple non-alveolated lung in amphibia and reptilia, data in artificially ventilated tegu lizards show no molecular weight dependence in inert gas excretion (Hlastala et al. 1985). However, since breathing can be very periodic in these animals, gas phase diffusion limitation may develop during apnea (Donnelly and Woolcock 1978; Glass and Wood 1983). It is doubtful whether, in air, cutaneous gas exchange in reptiles and amphibians is limited by diffusion in gas overlying skin, since the thickness of the epidermis is many cells deep, and even the motion of "still air" around an animal probably causes effective convective mixing (Feder and Burggren 1985). The situation for laminar water flow over the gills of fish and amphibians is appreciably different. Since diffusion of gases in water is orders of magnitude slower than in air, the wide 10–200-μm intralamellar distances in gills are predicted to impose a large diffusion resistance. For example, Piiper et al. (1986) calculated that it may account for more than 50% of the limitation to O_2 uptake in the shark. In skin-breathing aquatic organisms, locomotion or movements of body parts may help to abolish diffusion limitation in water layers adjacent to the skin (Feder and Burggren 1985).

2.1.2 Diffusion in the Gas-Blood Membrane Barrier

The respiratory epithelium, interstitial space and capillary endothelium comprise the gas-blood membrane barrier. The depth of this barrier varies considerably, depending upon the type of gas-exchange organ and therefore may or may not pose a significant resistance to gas transfer. Roughton and Forster (1957) proposed that the resistance to O_2 uptake in the mammalian lung (L) could be separated into additive membrane (m) and red cell (rbc) components:

$$RL = Rm + Rrbc \tag{2}$$

or in terms of conductances

$$1/DL = 1/Dm + 1/\theta \cdot Vc \ , \tag{3}$$

where R is resistance and D is diffusing capacity. Red cell resistance was defined as $1/\theta \cdot Vc$ where θ is the red cell-specific conductance in ml/min mmHg ml of whole blood and Vc is capillary blood volume. θ can be measured in vitro and used to solve for the other components. In their classic paper, Roughton and Forster attributed half of the resistance to the membrane and the other to the red cell. Implicit in Eq. (3) is the assumption that diffusion of O_2 beyond the membrane is very rapid, since the second term is purely one describing the resistance posed by intracellular O_2 diffusion and the chemical reaction of hemoglobin with O_2. Thus, Dm is in reality the diffusing capacity of the combined membrane and plasma interface. Measurements by morphometry (Weibel 1984) or in isolated saline-perfused dog lungs, using dithionite to eliminate any diffusion resistance within the capillary (Burns and Sheppard 1979), give far greater values than calculated by the Roughton and Forster equation. It is becoming increasingly clear that underestimation of θ, if not the validity of the concept itself (vide infra), caused by substantial unstirred layers around red cells in rapid reaction measurements and the in vivo effect of unstirred layers combine to falsely incriminate the membrane as a major resistance. The significance of unstirred plasma surrounding red cells in vitro and in vivo is considered in the next section. Calculations by Wagner and West (1972) and Hill et al. (1973b) show the resistance of the human alveolar-capillary membrane, whose thickness is 0.6 μm (Weibel 1984), must be increased four- to tenfold before there is any failure of end-capillary PO_2 equilibration.

The avian membrane barrier is even thinner and would not be expected to confer any appreciable resistance (Powell and Mazzone 1983). The branchial membrane barrier in fish varies in thickness, depending upon the species (Hughes 1984). In benthic fish, the barrier is thick (2.5–11 μm) and in elasmobranchs may account for almost half the diffusion resistance between intralamellar water and blood (Piiper et al. 1986). Most pelagic species have thinner barriers (0.5–2 μm) that should pose less resistance (Randall and Daxboeck 1984). Cutaneous gas exchange in amphibia, reptiles and fish is clearly diffusion-limited, as skin depths (20–50 μm) are at least tenfold greater than other gas exchanging organs. This has been studied in amphibians using inert gases by Feder et al. (1988), who showed a clear dependence on molecular weight, in accord with others in the field (Gatz 1982; Malvin 1988). Although skin poses a large diffusion resistance to O_2 exchange, this limitation is not so profound as to obscure an important perfusion dependence (Feder and Burggren 1985) that can be utilized in hypoxia to increase O_2 uptake (Malvin and Hlastala 1986).

2.1.3 Diffusion Through Blood

The diffusion of O_2 through plasma, red cell membrane and red cell interior, and chemical reaction with hemoglobin, contribute to further resistance in O_2 exchange. The first analysis of rapid-reaction data implicated the red cell membrane as a significant barrier to O_2 diffusion (Nicolson and Roughton 1951), but recent experimental and model data demonstrate otherwise. The rates of steady-state O_2 diffusion in packed red cells and concentrated Hb solutions are almost identical (Kreuzer and Yahr 1960; Kutchai and Staub 1969; Kutchai 1970; Stein et al. 1971). Such studies suggest the membrane does not impede O_2 transfer, but membrane O_2 diffusivity has not been studied directly. Indirect measurements yield a calculated membrane O_2 diffusivity of about 10^{-6} cm^2/s (Fischkoff and Vanderkooi 1975; Holland et al. 1985), which would only account for 5% of the total resistance in blood to O_2 uptake or release (Huxley and Kutchai 1981). Supporting the minor role of the membrane in limiting O_2 diffusion are observations that marked changes in membrane fluidity have virtually no effect on O_2 exchange rates (Rotman et al. 1980; Hook et al. 1988).

It is becoming increasingly clear that O_2 diffusion in plasma and the red cell interior (once thought not rate-limiting) is a large component in overall O_2 diffusional resistance. A major assumption in early rapid-reaction measurements with flowing RBC suspensions was that the fluid around cells was well mixed and no extracellular gradients in gas tension existed. Therefore a resistance was attributed to the red cell membrane and interior (Forster 1964) to explain the 10–20-fold slower oxygenation of red cells compared to equally concentrated hemoglobin solutions. What causes the membrane to appear less O_2 permeable in flowing suspensions is the presence of an unmixed fluid layer adjacent to the cell surface. This was first postulated by Sha'afi et al. (1967), who concluded that a 5 μm unstirred layer was responsible for the delay in shrinkage of red cells exposed to a hypertonic solution. Koyama and Mochizuki (1969) reached the same conclusion on the basis that red cell O_2 uptake was greater when higher Reynolds numbers were generated in a rapid-reaction apparatus. Further work by Gad-el-Hak et al. (1977), Rice (1980), and Huxley and Kutchai (1983) underscored the complexities of turbulent flow and inefficient mixing in cell suspensions and the effects of cell shape and size on the extent to which unstirred layers develop. Reduction in O_2 diffusivity of the extracellular medium with the addition of albumin

(Huxley and Kutchai 1981; Yamaguchi et al. 1985) or synthetic polymer (Coin and Olson 1979) markedly slows red cell O_2 uptake in a rapid-reaction apparatus. In contrast, elimination of any O_2 gradient external to the cell by the addition of dithionite, which chemically consumes O_2 and negates the effect of an unstirred layer on O_2 release, markedly speeds up the rate of deoxygenation in cell suspensions (Coin and Olson 1979; Yamaguchi et al. 1985). Model calculations suggest the thickness of the unstirred layer ranges from 1 – 5 μm, depending on the nature of the experiment. Vandergriff and Olson (1984a) have successfully simulated their work on O_2 uptake and release by RBCs in stopped flow measurements by modelling the red cell as a cylinder about which an unstirred layer develops. More recently, Hook et al. (1988) modelled the red cell as a sphere and came to the same conclusions regarding the effects of an unstirred layer. Williams and Kutchai (1986) monitored the time course of quenching by red cell membrane-bound rhodamine isothiocyanate of iodide and found that the data were best fitted to a model in which unstirred layers of 5 – 7 μm were present around the red cells. These data offer the best evidence to date for unstirred layers, since, unlike O_2, the iodide need not diffuse through the red cell to reach its binding site.

The variable and time-dependent depth of these unstirred layers in vitro have led to significant problems in the calculations of θ, the specific conductance of red cells to O_2 exchange. Important misconceptions in the interpretation of diffusing capacity measurements arise from the falsely low and relatively constant values of θ (0.7 to 2.7 ml/min mmHg ml whole blood) that were generated in earlier studies (reviewed in Forster 1987). Recent experiments by Yamaguchi et al. (1985) show roughly twofold higher values for θ during the initial course of red cell O_2 uptake or release, which decline thereafter and are sensitive to changes in the external medium. The range of θ encompassed by these studies is unsatisfactory because, depending on the value used in calculations of diffusing capacity, one can either assign a majority of resistance to the alveolar capillary membrane (high θ) or to the red cells (low θ), when in fact the real hindrance is located in the unstirred plasma layer. Heidelberger and Reeves (1988) have virtually eliminated these layers in their thin-blood film method. They showed that θ varies over a wide range (1 – 25 ml/min mmHg ml whole blood), depending upon a number of intracellular factors (masked by large unstirred layers in other methods) such as hemoglobin saturation and O_2 affinity (P 50), red cell thickness, partial pressure driving gradient and the direction of O_2 flux (release vs uptake). The overwhelming conclusion to be drawn from these experiments is that the θ concept as conventionally taught may need to be abandoned.

The existence and effect of extracellular unstirred layers in vitro is well established. The importance of this phenomenon for in vivo capillary gas exchange is less clear at present, but low flow, lack of turbulence and red cell deformation in the capillary could easily create significant unstirred layers. This is suggested by recent studies in the isolated rabbit lung using red cell and hemoglobin solutions of equal O_2 carrying capacity, in which the measured O_2 diffusing capacity was almost 40% greater with hemoglobin solutions (Geiser and Betticher 1989). Where the RBC and capillary endothelium are in intimate contact it is difficult to estimate the depth of interposed plasma, but layers as deep as 1 – 2 μm have been noted (Middleman 1972). Some amount of O_2 also must traverse the plasma between red cells during gas exchange. These distances may be many μm (and more in anemia) and should behave as unstirred layers (Federspiel 1989), since there is no mixing of the plasma between cells (Aroesty and Gross 1970). Federspiel and Sarelius (1984) and Homer et al. (1981) examined the role of cell spacing to O_2 flux in simulations of O_2 uptake in muscle, and concluded that capillary blood could not be considered homogeneous for O_2 supply when cells are separated by more than four cell lengths at rest and one length in maximal exercise. Recent model simulations by Federspiel (1989) in the lung show the importance of the particulate (two phase) nature of blood in limiting O_2 uptake and give support to the findings of Geiser and Betticher (1989). Calculations by Gutierrez (1986a) and Homer et al. (1981) suggest that, in resting normoxia, O_2 release rate from red cells (chemical reaction and diffusion) is fast enough to maintain near-equilibrium between red cells and plasma, but under conditions of low O_2 delivery (hypoxia/anemia) O_2 is depleted faster from the plasma layer than from red cells during capillary transit. This disequilibrium is dissipated as blood leaves the capillary and gas transfer ceases. The possible consequences of intracapillary O_2 disequilibration in assessment of tissue metabolism will be taken up later.

Diffusion of oxygen within the red cell constitutes a further resistance to overall oxygen exchange. The increased viscosity of red cell cytoplasm reduces the diffusivity of O_2 by about two-thirds (Kreuzer and Hoofd 1987). Although Hb diffusivity is at least 100-fold less than free O_2, its high concentration provides some facilitated O_2 diffusion (Kreuzer and Hoofd 1987). Moll (1968/1969) has calculated that Hb diffusion may speed complete saturation by 30% and more in hypoxia. Kutchai (1970) made similar findings and showed that the importance of Hb diffusion increases with increasing cell diameter. Holland and Forster (1966) showed that smaller vertebrate red cells have faster O_2 uptake than do larger cells and, in part, attributed this to the greater intracellular diffusion distances in larger cells. Vandergriff and Olson (1984b) found the same relation between red cell size and rate of O_2 uptake, but ascribed this to the fact that larger cells develop a deeper unstirred layer. Yamaguchi et al. (1987) analyzed temperature and pH dependence of O_2 kinetics in red cells and solutions of Hb from stopped-flow reaction measurements and concluded that, with minimal unstirred layers, the main limitation to O_2 transfer would be intracellular O_2 diffusion. Hook et al. (1988) modelled intracellular O_2 exchange and reached similar conclusions that O_2 diffusion would account for roughly 60% of the limitation to gas transfer and Hb diffusion about 20%.

Lastly, the possibility that intracellular convective mixing may speed O_2 exchange has been proposed, since it is evident that red cells are deformed within the capillary (Bagge et al. 1980). Miyamoto and Moll (1972) have demonstrated that cells in a rapid-reaction apparatus remain biconcave discs, but are deformed quite extensively in their passage through a lung capillary (Miyamoto and Moll 1971). Most models do not incorporate this potentially relevant factor. The only data suggesting an importance to intracellular convective mixing is that of Zander and Schmid-Schönbein (1973), who showed an increased O_2 release by erythrocytes when subjected to a deforming shear stress. The tank-tread motion of red cell membranes during capillary passage (Fischer and Schmid-Schönbein 1977) may cause convective mixing of the intracellular contents during gas exchange. Clark et al. (1981) calculated that red cell membrane motion has little influence on free O_2 diffusivity, but may enhance HbO_2 diffusion considerably. Overall O_2 exchange rates may be improved by 20%. This is clearly an area deserving further study, especially with techniques able to measure O_2 exchange rates with and without cell motion.

2.1.4 Diffusion in Tissues

Much larger distances are involved in O_2 diffusion in tissues than those present across gills and lungs. In his classical treatment of tissue O_2 exchange, Krogh (1919) showed an exponential decrease in PO_2 as distance from the capillary wall increased. In these simulations capillary blood was homogeneous and in constant equilibrium, and the principle diffusion resistance was the tissue extracellular and intracellular volume. Many investigators have attempted to measure tissue O_2 gradients and define the sites of diffusion resistance (i.e. where gradients are largest). The data are divergent for several reasons, including different tissues (muscle vs tissues lacking myoglobin), the presence or absence of CO_2 and/or erythrocytes in the perfusate, and differing metabolic conditions.

Data suggesting steep tissue or intracellular PO_2 gradients comes principally from work in mammalian heart in which overall delivery of oxygen is limited by the use of blood-free perfusates (Hassinen et al. 1981). Contrasting data showing almost uniform tissue PO_2 in exercising dog muscle have been reported by Gayeski and Honig (1986). In these and other experiments by this group, the muscle is perfused with blood and the intracellular spatial distribution of PO_2 is determined by cryo-spectrophotometric measurements of myoglobin in very small volumes of muscle ($<30\ \mu m^3$) frozen rapidly in situ. This sophisticated technique has not been independently validated, although a theoretical study by Clark and Clark (1983) suggests that such a method can capture a spatial distribution of chemical reactions in tissue (e.g. $Mb\text{-}O_2$ equilibria). It will be important to verify this technique, since it has led to other controversies such as the concept that there are no hypoxic areas of sufficient magnitude in maximally working muscle to cause anerobic lactate production (Connett et al. 1984). These results lead the authors to conclude that within muscle the greatest PO_2 gradient lies outside the myocyte in the capillary-interstitial space, largely as a consequence of facilitated oxygen diffusion by myoglobin (Kreuzer and Hoofd 1987) and to the above-mentioned diffusion limitations in blood which may contribute 50% of the total resistance (Hellums 1977; Clark et al. 1985; Federspiel and Popel 1986). The remaining resistance must be in the extra-sarcolemmal space.

The importance of myoglobin-facilitated O_2 diffusion has been shown by Cole (1983) in canine skeletal muscle and by Bailey and Driedzic (1986) in fish heart. Myoglobin-containing skeletal or cardiac muscle demonstrated reductions in O_2 consumption if myoglobin was reversibly inactivated with H_2O_2, while the heart of a fish without myoglobin was unaffected. Federspiel (1986) and Covell and Jacquez (1987) have calculated that a myoglobin-rich fibre can sustain maximal O_2 consumption with a relatively flat O_2 pressure drop from the periphery to the core. It is doubtful, however, whether this is the case for tissues lacking myoglobin, although some of these tissues are quite capable of very high rates of O_2 consumption, such as the salt-secreting glands of certain fishes and birds (Swenson and Maren 1984).

2.2 Hemoglobin-Oxygen Reactions

2.2.1 Hb-O_2 Kinetics in Solution

The reaction of O_2 and hemoglobin is very complex inasmuch as there are four binding sites for O_2 and differences exist between the alpha and beta subunits. Furthermore, the rates are affected by other ligands binding elsewhere in the molecule (Baumann et al. 1987). In spite of these problems, one can gain a reasonable description of Hb-O_2 kinetics by treating the reaction as a monomeric process ($Hb + O_2 \rightarrow HbO_2$), except at the extremes of saturation. In solution, initial rates of hemoglobin oxygenation and deoxygenation are very rapid, with half-times of only a few milliseconds (Bauer et al. 1973). In the only comparative study of vertebrate hemoglobin reaction rates in solution, Holland and Forster (1966) showed no significant differences between amphibian and mammalian hemoglobins at equal temperatures. Temperature and allosteric modifiers of Hb such as pH and 2,3-DPG alter the rates several-fold over a very wide range (Bauer et al. 1973; Lawson et al. 1965), but, in physiological terms, the influence of these changes is probably minimal, since these rates are already very fast. Clearly the rates of oxygen binding and release from hemoglobin are very rapid and much faster than diffusion through the plasma, red cell and membrane. Direct measurement of their rate-limiting impact is impossible, because diffusion limitation will always be present. Model analyses of O_2 exchange show reaction kinetics play only a small role in slowing overall O_2 transfer. Hook et al. (1988) calculated that reaction kinetics account for only 15%–25% of the total resistance to O_2 exchange between plasma and red cells. Groebe and Thews (1987) calculated only 5% for exchange between alveolar gas and red cells. Thus, while reaction kinetics are not instantaneous and can be shown to alter the time course of equilibration, their contribution to oxygen-exchange resistance is quite small, even in those organs in which diffusion distances are the least.

2.2.2 Hb-O_2 Kinetics in Red Cells

It has been observed ever since the pioneering studies of Roughton and colleagues that solutions of hemoglobin bind or give up oxygen much faster than red cells in vitro. As discussed above, the kinetics of red cell oxygenation and deoxygenation are quite complex, since oxygen exchange requires combined diffusion and chemical reaction in a non-homogeneous medium, but it is well established that diffusion through plasma and the red cell poses the greatest limitation. The data in Table 3 show that rates in mammalian red cells for complete saturation or desaturation are far slower than in Hb solutions as measured with rapid-reaction

Table 3. Half-times of O_2 uptake and release by red cells of various representative vertebrates

Species	Temperature (°C)	MCV (μm^3)	Deoxygenation (ms)	Oxygenation (ms)
In vitro rapid reaction technique				
Artificial RBC	25	0.004		4[a]
Sheep	37	33		41[b]
Human	37	88	45[c]	
	25		50[d,e]	23[d,e]
				70[a]
Chicken	41	104	66[k]	
Frog	25	680		290[b]
Salamander	25	10800		680[b]
Turtle	10	327	350[l]	
Eel	25	1000		80[f]
Other techniques				
Chicken	25		110[g]	55[g]
Eel	20		1550[h]	250[h]
Carp	20		650[h]	160[h]
Human	37			40[i]
	37		15[j]	7.5[j]

[a] Vandergriff and Olson (1984b).
[b] Holland and Forster (1966).
[c] Bauer et al. (1973).
[d] Yamaguchi et al. (1985).
[e] Hook et al. (1988).
[f] Forster and Steen (1969).
[g] Tazawa et al. (1976).
[h] Hills et al. (1982).
[i] Thews (1959).
[j] Reeves and Heidelberger (1987).
[k] Nguyen-Phu et al. (1986).
[l] Yamagucki et al. (1989).

techniques. Red cells from several non-mammalian species display similar kinetics, which suggests diffusion limitation occurs with these cells as well. Vandergriff and Olson (1984b) studied various morphological and physiological factors affecting O_2 uptake and release by cells in stopped-flow methods and found differences in effects of external unstirred layers, HbO_2 affinity, 2,3-DPG, and heme content between O_2 uptake and release. High dithionite in O_2-release studies largely abolish the impact of unstirred plasma layers making O_2-release rates dependent upon the intracellular chemical mileau and intracellular diffusion distance (i.e. cell size), in contrast to O_2 uptake data which show little dependence on anything but the unstirred layer depth which also varies in proportion to cell size. In general, O_2 uptake by cells is about two- to threefold faster than deoxygenation under equivalent conditions, due to the nonlinear nature of the Hb dissociation curve (Wagner 1977). Since conditions in vivo are not well simulated in vitro rapid-reaction methods, the interplay of all of the factors above on O_2 exchange kinetics is uncertain.

Several recent attempts to study O_2 kinetics under conditions more closely simulating physiological capillary gas exchange have yielded data supporting the importance of unstirred layers. Boland et al. (1987) fashioned artificial capillaries of about 30 μm in diameter (considerably larger than true capillaries) and showed the half-time for full saturation of a human hemoglobin solution at 37 °C was 50 ms, and that of a red cell suspension was 250 ms. In the deoxygenation reaction, the half-time for a hemoglobin solution was 500 ms and for cells almost 1000 ms. The slower reaction times of hemoglobin solutions and red cells compared to data from rapid reaction techniques probably lies in the "capillary" wall depth (70 μm) and the unstirred layer of plasma around flowing red cells. These authors calculate that only 50% of the total resistance can be ascribed to the thick silicone wall. With this system, Lemon et al. (1987) examined the roles of pH, 2,3-DPG, PO_2, temperature and Hb concentration on rates of O_2 uptake and release and found results similar to those of Vandergriff and Olson (1984b). Although these artificial capillaries represent an important new and sophisticated method, the problems of substantial non-physiological unstirred layers and large intraluminal diameters remain.

Tazawa et al. (1976) measured the rates of oxygenation and deoxygenation in stationary red cells contained in capillaries of the chorioallantoic membrane of a chicken embryo, using a microphotocolorimetric method for detecting Hb saturation in a single red cell. In these experiments only one side of the membrane was exposed to gas, potentially reducing the kinetics by 50%. The half-time of oxygen saturation was 110 ms and twice as long for oxygen desaturation. Since these measurements were made at room temperature, they would be expected to be even less at the higher physiologic temperatures of birds. In contrast to the artificial capillaries, conditions for gas exchange in natural capillaries are significantly better, probably as a result of reduced unstirred layers. The fact that half-times for O_2 uptake and release are still tenfold greater than those in hemoglobin solutions confirms the remaining rate limitations of diffusion in vivo. Using the same technique, Hills et al. (1982) found that in secondary lamellar capillaries of gills which were ventilated with air instead of water (reducing the normal resistance present with water) the half-times for oxygenation were 320–500 ms and 1350–3100 ms for deoxygenation. The considerable difference between these rates and those in the chicken stem from the larger size of fish red cells and the thicker membrane barrier. Unfortunately, they could not study rates in free hemoglobin solutions to assess more directly the role of chemical reaction rates and diffusion limitation. Recently, Reeves and Heidelberger (1987) reported kinetic data on human blood, using a thin film technique, and found half-times of 7 ms for O_2 saturation and 15 ms for O_2 desaturation at 37 °C in a film of red cells 1.6 μm in depth, the average red cell thickness in a human lung capillary (Weibel 1984). Their technique probably minimizes any external plasma layers and the fast half-times under such conditions (intact red cells separated from a gas phase by a very porous membrane with virtually no resistance to gas diffusion) affirm the concept that external plasma layers pose the major resistance to O_2 exchange. There appears to be no significant impact of the intracellular milieu on reaction and diffusion rates, since the kinetics were essentially the same as those in lysates of equal depth. Lack of red cell flow and deformation and almost no unstirred plasma make for more optimal conditions for gas exchange than probably exist in vivo.

2.3 O_2 Exchange Kinetics in Vivo

Mammalian red cell studies in vitro and virtually all model simulations suggest that, under resting normoxic conditions, rates of O_2 diffusion and chemical reaction in blood should not prevent end-capillary equilibration in most lungs, because transit times are sufficiently long. While the same probably holds true for birds as well, it is uncertain that such is the case for gill and skin gas exchange, since transit times do not appear to be proportionately longer in relation to the lower temperatures and greater diffusion distances. It is of interest that in the human lung, for which the data are most abundant, calculated rates of O_2 equilibration in the lung capillary (reviewed in Wagner 1977) are remarkably close to those derived from measurements in blood. The coincidence may be fortuitous, but more likely it suggests that the effect of resistance in the alveolar capillary membrane (not a factor in blood studies) and effects of diffusion resistance in stagnant plasma layers (ignored in all calculations) are roughly of equal magnitude. Both the membrane and plasma are layers of low oxygen solubility, which become depleted of O_2 in the face of uptake by red cells and thus pose the major resistances to oxygen exchange.

Situations unfavorable to complete end-capillary PO_2 equilibration include severe hypoxia (diminished PO_2 driving gradient and higher blood O_2 solubility), large erythrocytes (greater unstirred layers), low temperature (reduced rates of diffusion and chemical reaction), high cardiac outputs (shorter transit times) and thicker membrane barriers separating water or air from blood. The extent to which any of these prevent equilibration ultimately depends on transit time. Unfortunately, transit times under these conditions are sorely lacking. Heterogeneity of capillary transit times within a tissue adds further quantitative uncertainty. In man, diffusion resistance may be sufficient to prevent end-capillary equilibration during severe exercise, as transit times fall as low as 200 ms (Groebe and Thews 1987). Even if there is failure of end-capillary O_2 equilibration, the actual amount of oxygen this represents may be minimal, at least during O_2 uptake, owing to the shape of the Hb dissociation curve, which results in much greater O_2 loading earlier in capillary transit. The situation during oxygen unloading in tissues however may be different. Gutierrez (1986a) has modelled the rate of O_2 release by red cells into the tissues, using a capillary transit time of 0.5 s and showed that significant failure of end-capillary PO_2 equilibration can occur with hypoxemia and anemia. He did not simulate states of increased metabolism or higher O_2 consumption in the face of reduced delivery, but on the basis of his calculations even greater disequilibrium and possible impairment of O_2 uptake would be expected. The importance of such calculations is twofold. It calls into question the validity of equating venous and tissue O_2 tensions (Tenney 1974), particularly in conditions different from a normal resting state. Secondly, kinetic limitations in oxygen exchange may cause overestimations in the extent of heterogeneity attributed to other factors important in tissue O_2 uptake. Many models (Ellsworth et al. 1988; Federspiel and Popel 1986) do not calculate the effect of finite kinetics but assume instantaneous equilibration. Kinetic limitation has not been addressed as a factor in the observation that venous PO_2 rises in maximally exercising skeletal muscle (Connett et al. 1984) and in diseases associated with low O_2 delivery and loss of functional capillary bed, such that, in the face of high cardiac output, transit times may be extremely short (Gutierrez 1986b).

Simultaneous CO_2 exchange will further slow O_2 equilibration, since the fraction of O_2 exchange linked to CO_2 exchange (the Bohr effect) must await its completion. This has not been studied directly; however, calculations (Hill et al. 1973b; Hlastala 1973) predict that O_2 equilibration in the lungs will only be delayed some 10–20 ms further. This has not been analyzed for tissue O_2 exchange where O_2 uptake may be more dependent on the Bohr effect (see Sect. 3.5 for further discussion of the kinetic implications of O_2-CO_2 interactions).

There is little direct physiological evidence for O_2 uptake limitation on the basis of O_2 exchange kinetics. In the skin and mammalian lung, equilibration failure between ambient or alveolar gas and arterial blood may be considered as presumptive evidence for rate-limiting exchange kinetics relative to capillary transit time if ventilation-perfusion heterogeneities are factored out. In lungless amphibia, there is obvious failure of arterial PO_2 to reach ambient PO_2 (Feder and Burggren 1985). In some mammals, maximal exercise at sea level and high altitude (Dempsey et al. 1984; Wagner et al. 1986) lead to large differences between calculated alveolar and measured arterial PO_2 ($AaDO_2$). Substantial diffusion limitation had not been predicted to occur at sea level but had been calculated by West and Wagner (1980) to develop at high altitude. Wagner et al. (1986, 1988) used the multiple inert gas elimination technique to measure what fraction of the $AaDO_2$ was caused by ventilation-perfusion inequality and calculated that diffusion limitation accounts for the majority of the alveolar-arterial PO_2 difference in humans and horses exercising at sea level, and in humans during lesser levels of exercise in hypoxia. Since it is clear that chemical reaction rates in O_2 exchange play little role in limiting O_2 transfer, it is fair to attribute end-capillary disequilibrium to diffusion limitation. When arterial O_2 desaturation occurs with maximal exercise, breathing supplemental oxygen further extends exercise performance (Dempsey et al. 1984) and underscores the possible importance of both tissue and pulmonary diffusion limitation in defining the limits to work output in some mammals. Crucial factors limiting pulmonary end capillary disequilibrium and arterial O_2 desaturation appear to be a high mixed venous pO_2 (decreased mean blood O_2 solubility) and relative alveolar hyperventilation (raised alveolar pO_2 and higher driving gradient). An excellent example of this is the fox in whom there is no significant end capillary disequilibrium despite a very short transit time (0.12) during maximal exercise (Longworth et al. 1989).

While it is possible that diffusion limitation might occur in birds and fish in hypoxia and/or at high rates of metabolism, the cross- and counter-current design of blood and gas flows in bird lungs and fish gills prevents the type of analysis cited above (Scheid et al. 1986). Likewise, in amphibians and reptiles variable large intrapulmonary shunts (Glass and Wood 1983) cause such considerable arterial O_2 desaturation that any contribution of diffusion limitation would be difficult to detect. If in fact diffusion limitation occurs in birds in flight at high altitude, it surely must be small, since Black and Tenney (1980) found almost no difference between inspired and arterial PO_2 in the resting bar-headed goose at a simulated altitude of 11800 m ($PIO_2 = 23$ mmHg). The situation in gills is quite complicated, but the combination of lower temperature, diffusion resistance in water and greater gill epithelial thickness could act to promote significant diffusion limitation and failure of end-capillary equilibration. However, under hypoxic or metabolic stress, diffusion resistances in water and gill epithelium will fall as increased ventilation diminishes the extent to which O_2 is depleted from intralamellar water and as increased perfusion of the gills thins the branchial epithelium from distension of lamellar capillaries (Randall and Daxboeck 1984). This effective increase in diffusing capacity may sufficiently compensate or even correct for decrease in capillary transit time or change in venous PO_2 that otherwise would cause failure in end-capillary equilibration.

3 Carbon Dioxide Exchange

Carbon dioxide exchange is a multi-step process whose speed is determined more by rates of chemical reaction and transport events and less by diffusion (Fig. 1). They include (1) the diffusion of CO_2 and O_2 through tissues, plasma, red cells and the respiratory membrane barrier; (2) the rate of release or consumption of Bohr protons involved in the reversible binding of O_2 and CO_2 (as carbamate) to

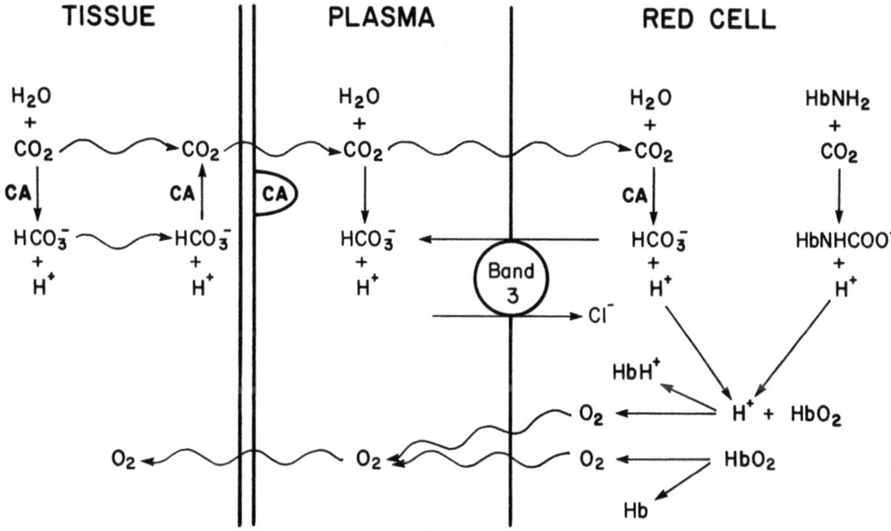

Fig. 1. Reactions of carbon dioxide and oxygen in blood during tissue gas exchange. At the lungs, gills and skin, reactions proceed in the opposite direction. Symbols: *CA* carbonic anhydrase either unbound in the tissue or red cell cytoplasm or membrane-bound, as is the case for the capillary endothelium with enzyme activity available to plasma; *Band 3* the anion exchanging protein within the red cell membrane; $\rightsquigarrow$, diffusion; $\rightarrow$, chemical reaction (including the binding and translocation of anions by band 3 protein; *Hb*, hemoglobin and HbNHCOO, carbamate. The linkage between CO_2 and O_2 exchange (Bohr/Haldane effect) is illustrated by the proton binding and release reactions of hemoglobin and CO_2. In the lungs, gills and skin (not shown) CA within the respiratory epithelium may facilitate CO_2 diffusion in similar fashion as shown in the figure for tissues. In water-breathers (gills and skin) there may be a small fraction of CO_2 excreted as HCO_3^- ion.

hemoglobin as it undergoes changes in saturation; (3) the rates of CO_2 hydration and dehydration in the red cell catalyzed by the great excess of intracellular carbonic anhydrase (CA), and in the plasma by the much smaller and variable amount of CA within the luminal facing membrane of the capillary vascular endothelium; (4) the rate of red cell membrane exchange of HCO_3 and Cl (the Hamburger shift) mediated by the anion-exchanging capacity of band 3 protein and (5) the rate of water movement across the red cell membrane. It is important to include the rates of O_2 diffusion in consideration of CO_2 exchange kinetics, because roughly 50% of CO_2 transfer in vivo is linked to O_2 transfer (the Haldane effect).

Since CO_2 is roughly 20 times more diffusible than O_2, it has been naively taught that CO_2 equilibration must be faster than O_2 equilibration by the same factor. However, as is evident in Eq. 1, equilibration is not only determined by diffusion (diffusing capacity or conductance) but by a convective conductance defined by the product ($Q \times \beta b$). A much steeper blood dissociation curve and higher membrane solubility for CO_2 than O_2 limit the apparent advantage to only about twofold, as discussed in the introduction (1.1). The magnitude to which certain critical rate-limiting CO_2 reactions reduce this even further has been the subject of much intense and somewhat controversial investigation. While direct physiological evidence is minimal, it is generally assumed that in vivo CO_2 equilibration is not only slower than for O_2 but may not occur within capillary transit even at rest. Since O_2 uptake and release by blood are principally determined by O_2 diffusivity and not rate-limited by hemoglobin reaction, the term diffusion limitation has been used to define the kinetic limitation in O_2 exchange. However, failure of CO_2 equilibration should best be described as an equilibration limitation (Piiper et al. 1980) or chemical reaction limitation, since these more accurately identify the major rate limiting aspect of CO_2 exchange.

3.1 Carbon Dioxide Diffusion

The transfer of carbon dioxide across body compartments and into the environment occurs solely by passive diffusion of molecular CO_2, except in those animals inhabiting water, which accomplish a small fraction of CO_2 elimination by active extrusion of ionic HCO_3 linked to NaCl or NH_3 metabolism (Perry 1986; Smith 1988; Wright et al. 1988). As Table 2 indicates, CO_2 diffusion through a variety of biological media is far more rapid than O_2 diffusion and should never be rate-limiting during the course of CO_2 transfer. Two exceptions to this generalization may be raised. The first is that during the earliest phase of CO_2 uptake or elimination, intra-erythrocytic catalyzed CO_2-HCO_3-introconversion may in fact be faster than diffusion; however, as discussed in Sect. 3.2.1, this phase has a very short duration. The other is that in tissues (e.g. kidney) which transport large amounts of H^+ or HCO_3^-, tissue PCO_2 may rise substantially above that in blood (Bidani et al. 1984) due to extra non-metabolic production of CO_2 (from intracellular buffering) and to a venular to arteriolar diffusional shunt of CO_2, particularly in states of low blood flow, due to close approximation of these vessels in some organs (Piiper et al. 1984).

In those tissues that contain carbonic anhydrase, the flux of CO_2 is enhanced by simultaneous and quantitative HCO_3 diffusion (Kreuzer and Hoofd 1987). The enzyme rapidly catalyzes the formation or consumption of HCO_3 to diffuse along with CO_2 down its concentration gradient. This was proposed originally by Enns (1967) and has been extensively studied in a number of tissues and artificial membranes. It is beyond the scope of this review to detail the physical chemistry of this phenomenon (see Kreuzer and Hoofd 1987; Meldon et al. 1982); however, a few salient points will serve to gain a physiological perspective. The degree of "facilitated diffusion" depends upon the pH, the absolute PCO_2 and PCO_2 gradient, and the diffusivity and quantity of buffering substances. In general, facilitated diffusion is greater with lower PCO_2 and smaller gradients, higher pH and greater mobility of the buffer compounds necessary to supply and consume hydrogen ion for HCO_3 formation and consumption. In a certain sense, the enzyme subserves the same function at the cell and membrane level for tissues as red cell CA does for the organism as a whole, i.e. increasing CO_2 transport by promoting rapid interconversion of CO_2 into one of its carrier forms. It may enhance CO_2 diffusion by as little as 10% to as much as 200%. Its significance in particular tissues will be taken up in the section on tissue carbonic anhydrase (3.2.2).

3.1.1 Diffusion in Respired Air or Water

In air-breathing vertebrates, the kinetics of CO_2 diffusion in the gas phase are very close but slightly slower than for O_2, due to carbon dioxide's greater molecular weight. It is very doubtful, as is also true for O_2, that CO_2 diffusion through the gas phase presents any important resistance to CO_2 elimination by mammalian and avian lungs. CO_2 elimination via long pores in the avian eggshell is clearly diffusion-limited (Ar et al. 1980). In the simple, less structured lungs of amphibians and reptiles, which may remain apneic for long periods of time, gas phase diffusional limitation may be present, but this has never been investigated. Cutaneous CO_2 elimination into air is probably not hindered by gas phase diffusion. While the greater distance over which gases must diffuse in the intra-lamelar water of gills poses a large diffusion limitation for O_2, it is probably only a minor limitation for CO_2, since its diffusivity is roughly 20-fold greater. A similar argument can be made as well for cutaneous CO_2 elimination into water.

3.1.2 Diffusion in the Gas-Blood Membrane Barrier

Diffusion of CO_2 through the respiratory epithelium, as through any aqueous medium, is far faster than O_2. Thus, in those gas-exchanging organs where O_2 diffusion has been either shown or calculated to cause little resistance to O_2 uptake (lungs of most species), it is fair to assume that any resistance attributable to CO_2 diffusion should be vanishingly small. Direct measurements of human CO_2 diffusing capacity (DL) with 13-carbon labeled CO_2 give values of

200 ml/min mmHg (Hyde et al. 1968; Piiper et al. 1980). The ratio of CO_2 to O_2 diffusing capacity by this method is only 4:1, considerably less than a predicted 20:1 ratio based on the aqueous diffusion coefficients. This suggests a crucial blood-phase chemical reaction limitation to CO_2 movement out of the blood into the alveoli under normal conditions, with the alveolar capillary membrane accounting for, at most, about 10% of the total resistance to lung CO_2 exchange (Piiper et al. 1980 and Schuster et al. 1988). In support of this, Hyde et al. (1968) used acetazolamide, an inhibitor of carbonic anhydrase, to further amplify the chemical reaction limitation in the blood and found that the CO_2 diffusing capacity dropped to 40 ml/min·mmHg.

Schuster (1985) used 18-oxygen labeled CO_2 and found values ranging from 860 to 1100 ml/min mmHg. These much higher figures approach the true membrane diffusing capacity for CO_2, since red cell and lung water (to a lesser extent) provide an infinite sink for the oxygen label, as it is rapidly transfered to water through carbonic anhydrase-mediated catalysis. This provides a pathway for the disappearence of the oxygen label, which is not available to the carbon label, which must depend upon other slower chemical reactions for its disappearence. Schuster (1987) points out that even his data can only be taken as a lower limit to the membrane diffusing capacity, because he obtained greater values with doubly oxygen labeled CO_2 than with singly labeled CO_2. These findings imply partial chemical reaction and diffusion limitation in blood and lung tissue, since the diffusion kinetics of various isotopes of CO_2 should not differ. Model simulations by Wagner and West (1972) and Hill et al. (1973b) show that even eightfold reductions in Dm for carbon dioxide do not prevent end-capillary CO_2 equilibration in the human lung. Experimentally, Effros et al. (1981b) could only establish true end-capillary PCO_2 disequilibrium on the basis of diffusion limitation in the isolated saline perfused lung, when exceedingly high and unphysiological perfusion rates were used. They concluded that pulmonary blood flow in vivo would have to be at least an order of magnitude greater before diffusion of CO_2 through the alveolar-capillary membrane could limit CO_2 exchange. Klocke (1987) calculated a half-time for CO_2 diffusion from blood to alveolar gas of 2.4 ms, a process far too rapid to hinder CO_2 exchange in the mammalian lung.

In those gas exchangers where significant O_2 diffusion limitation through the membrane is present (namely gills and skin) it is possible that a CO_2 diffusion resistance, albeit smaller, may exist, but this has not been well studied. The classical twenty-to-one difference in diffusivity of CO_2 over O_2 in purely aqueous media may be reduced in tissues, since gas solubilities in lipid, protein and other tissue constituents can vary from those in water. Enns and Hill (1983), for instance, found that the diffusing capacity of dog lung tissue for CO_2 and CO differed only a factor of nine rather than 24.5, as predicted by Graham's law. Cameron and Polhemus (1974) modelled CO_2 exchange in a typical salmonid fish, whose epithelial thickness is about 3μ, and found that CO_2 efflux was highly sensitive to variations in the rate of CO_2 diffusion across the gill epithelium. Greater diffusion resistance for CO_2 might be expected in skin, since the thicknesses involved are vastly increased.

3.1.3 Diffusion Through Blood

The diffusion of CO_2 within blood includes its movement through plasma and the red cell membrane and cytoplasm. The red cell membrane is very likely quite permeable to CO_2, although direct measurements have not been done. Forster (1969) calculated a permeability of 0.58 cm/s based on data of NH_3 permeability into red cells and differences in solubility and molecular weight between the two gases. This is remarkably close to the value of 0.35 cm/sec obtained by Gutknecht et al. (1977) in lipid bilayers. Gros and Moll (1971) showed no difference between the diffusion of CO_2 in packed red cell suspensions and equally concentrated hemoglobin solutions, supporting the concept that the membrane does not limit CO_2 diffusion through blood. A recent controversy has arisen over the work of Silverman et al. (1976), who calculated a red cell permeability for CO_2 of 0.0076 cm/s from measurements of oxygen-labeled CO_2 exchange in red cells. These authors were careful to conclude that this measurement in actual fact had to include the diffusion resistance not only of the membrane but of the cell interior and any unstirred layers surrounding the cell. This value has been somewhat unfairly cited by some reviewers to contradict other published values, despite clarification of this point by these investigators (Silverman et al. 1981). Taken together, the data show a very high red cell membrane permeability which should have no rate-limiting consequences for CO_2 exchange. Measurements of carbon dioxide exchange in vitro with red cells whose membrane fluidity has been altered (such as was done for O_2 exchange by Rotman et al. 1980 and Hook et al. 1988) would be of great interest as an independent test of this conclusion.

Diffusion of CO_2 within the red cell cytoplasm is also very rapid. Measurements of CO_2 diffusivity vary, depending upon the PCO_2 gradients employed and the hemoglobin concentration studied. Gros and Moll (1971) showed that at high PCO_2, CO_2 diffusion in red cells and hemoglobin solutions is reduced about 60% from that in water, due to the higher viscosity. However, at lower and more relevant PCO_2 ranges, CO_2 diffusion may be enhanced by the presence of carbonic anhydrase such that, when compared to plasma, diffusion inside the erythrocyte may only be decreased about 10%–30% (Tanashita et al. 1985). Any slight actual decrement in CO_2 diffusivity within the red cell is likely of no importance in CO_2 exchange, because even this reduced rate is extremely fast and the cellular distances are quite small. Klocke (1987) calculated that a step change in CO_2 at the red cell membrane would be dissipated throughout the red cell, with a half-time of 0.3 ms.

CO_2 diffusion in plasma is rapid and nearly equal to that in water (Tanashita et al. 1985). In comparison to other chemical reactions involved in carbon dioxide exchange, it probably cannot be rate-limiting to CO_2 transfer in vivo. This contrasts with O_2 whose diffusivity is some 20-fold less. However, the effect of unstirred plasma layers in vitro may have similar consequences for quantitative analysis of CO_2 exchange kinetics in red cells, as discussed for oxygen; namely, that measurement of events with kinetics near to those of diffusion may be in error if possible diffusion limitation is ignored. This is of particular concern for the determination of intracellular activity of erythrocyte carbonic anhydrase (Sect. 3.2.1), but may also apply to measurements of the Haldane effect and carbamate reactions. As has been done for oxygen, it would be of great interest to study the rates of CO_2 uptake or release by red cells under conditions in which CO_2 diffusivity is reduced (increased plasma viscosity; Huxley and Kutchai 1981 and Coin and Olson 1979), where depth of external unstirred plasma layers vary (varying red cell size; Vandergriff and Olson 1984b) and under conditions where unstirred layers are virtually eliminated (thin naked blood film technique; Reeves and Heidelberger 1987). Such studies would help to clarify the true impact of unstirred plasma layers in CO_2 exchange.

3.2 Hydration-Dehydration Reactions of Carbon Dioxide

3.2.1 Erythrocyte Carbonic Anhydrase

The greatest amount (85%) of CO_2 added to blood by tissues and liberated into air or water is transported in the circulation as bicarbonate, with the remainder carried as dissolved CO_2 (10%) and chemically bound hemoglobin and protein carbamate (5%). This pattern of CO_2 exchange in mammals (Klocke 1987) does not differ significantly throughout the vertebrates, except possibly in those few fish lacking red cells and hemoglobin (Dejours 1981). The very slow chemical rates of CO_2 hydration and HCO_3 dehydration are incapable of supporting any level of normal CO_2 exchange, a fact readily appreciated as far back as the early 1900s. Since the bulk of carbon dioxide is carried as bicarbonate, and capillary transit times are in the order of seconds, catalysis of these reactions is necessary to maintain physiological levels of CO_2 elimination. While animals can tolerate complete loss of red cell CA activity by pharmacological inhibition (Maren 1967), genetic manipulation (Lewis et al. 1988), and some species such as the Antarctic icefish and small lungless salamanders have little or no CA in blood (Toews et al. 1978), most animals have very high enzyme levels in the red cell cytoplasm. The degree of catalysis is substantial enough to take these rates from potentially rate-limiting to among the fastest in carbon dioxide exchange. The ability of the few vertebrates lacking blood or red cell carbonic anhydrase to survive probably involves some combination of small size, low metabolic rate and life at very low temperatures.

The catalytic potentiation of CO_2 hydration by carbonic anhydrase within vertebrate red cells varies considerably (Table 4). This variation stems from different activities and concentrations of several isozymes found throughout the vertebrates, and from variable inhibition by intracellular chloride (Maren et al. 1980). Despite 10- to 15-fold differences in catalytic potential, there is no evidence that CA activity is ever rate-limiting in CO_2 elimination. In mammals, it is necessary to achieve enzyme inhibition of more than 95% – 98% before the pattern of CO_2 transport is in any way altered and greater than 99.5% – 99.9% before CO_2 output will not be maintained by compensatory mechanisms (Swenson and Maren 1978; Wistrand 1981). Although compensations, such as hyperventilation, increased cardiac output and tissue CO_2 retention enable animals to tolerate total CA inhibition at rest (Cain and Otis 1961; Swenson and Maren 1978) by increasing contributions of dissolved CO_2 and carbamino-CO_2 to expired CO_2, ad-

Table 4. Representative values for red cell carbonic anhydrase activity and catalytic potentiation in various vertebrates

Species	Enzyme activity[a]	Catalytic potentiation[b]	Reference
Hagfish	1000	1850	Carlsson et al. (1980)
Dogfish	2000	4000	Maren et al. (1980)
Goosefish	25000	26500	Maren et al. (1980)
Trout	24000	14700	Maren et al. (1980)
Turtle	6650	5600	Hall and Schaer (1979)
Frog	6600	12600	Bundy and Cheng (1976)
Chicken	20000	3400	Bernstein and Schaer (1972)
Human	23000	17500	Maren (1967)

[a] Enzyme activity = enzyme units/ml blood at 0 °C (Maren et al. 1980).
[b] Catalytic potentiation = enzymic rate/uncatalyzed rate.

equate CO_2 output may not be maintained under conditions of extreme exertion (Swenson and Maren 1978) or cardiopulmonary dysfunction (Swenson 1984). These concepts are further substantiated by the apparent lack of effect on CO_2 elimination or exercise tolerance in humans lacking the high activity form of red cell CA (CA II), in whom the overall catalytic potential is reduced by over 90% (Sly et al. 1985). Similar calculations hold for species and classes of vertebrates with lesser red cell CA activities, such as the spiny dogfish, in whom we showed almost 50% inhibition had no effect on the pattern of CO_2 excretion, and calculated that greater than 90% inhibition would be necessary before rates of CO_2-HCO_3 interconversion would be rate-limiting (Swenson and Maren 1987). These quantitative considerations should give rest to the concept first proposed by Larimer and Schmidt-Nielsen (1960) and later advanced by Magid (1967) and Dodgson and Forster (1983) that differences in mammals of up to fourfold in red cell catalysis in relation to body mass or red cell size have any physiological significance. Likewise, reports showing 5% – 30% increases in red cell CA activity with physical training (Little 1980; Tokuda et al. 1984) or anemia (Mondrup and Anker 1976) are probably trivial and represent epiphenomena. Such variations in activity have also been detected in euryhaline fish in response to ambient temperature or salinity changes, but these are small and may relate more importantly to osmoregulation than to gas exchange (Houston and McCarty 1978).

The very huge reserve of catalytic activity present in vertebrate red cells serves to accelerate the interconversion of CO_2 and HCO_3 to a par with that of intracellular CO_2 diffusion. In mammals, the catalysis is so great that hydration and dehydration may indeed outpace CO_2 diffusion. Calculations of human red cell CA catalysis based on assay in dilute hemolysates yield values of 17000 to 23000. However, measurements in intact red cells have generated figures considerably lower, depending upon the particular assay conditions (Forster et al. 1984). Some of the discrepancy stems from assumed or measured rates of the uncatalyzed reaction at 37 °C, which vary twofold depending upon the technique used (Garg and Maren 1972). Factors acting to additionally decrease calculated catalytic values in red cells are the accumulation of end products, intracellular and extracellular diffusion gradients for CO_2, failure to mimic the erythrocyte intracellular milieu in hemolysates and decreased specific activity of CA in high concentrations. This last point has been disproved by Kernohan et al. (1962) and Donaldson and Quinn (1974), who showed the kinetics for carbonic anhydrase are constant over a wide range of enzyme concentration up to and including that found in red cells.

Recently, measurements of 18-oxygen labeled carbon dioxide exchange by mass spectrometry in red cell suspensions (Silverman et al. 1976; Itada and Forster 1977) have reduced much of the discrepancy. Measurements are performed at chemical equilibrium, and intracellular water serves as an infinite sink for 18-O, thus eliminating the problem of end product accumulation. Details of the method are beyond the scope of this review, but it appears that the rate of red cell CO_2 uptake in vitro may not be entirely dependent on intracellular catalysis, as maintained by Forster et al. (1984). Silverman et al. (1981) cite evidence for partial diffusion control from the necessity to inhibit greater than 80% of intracellular enzyme activity before rates of isotopic exchange between labeled CO_2 and H_2O are affected. In fact, this is also evident in the experiments of Barinov et al. (1986) and Dodgson and Forster (1983), although not emphasized by them. Furthermore, the failure to show a twofold difference between uptake rates of singly and doubly labeled CO_2, which would be predicted if intracellular catalysis were actually the rate limiting step (Tu et al. 1978), argues against a true chemical reaction limitation. It is interesting to compare 18-oxygen exchange data in vivo (Schuster 1987) and in vitro (Silverman et al. 1981); together they show evidence for partial diffusion and chemical reaction limitation. Hoffman and Henkens (1987) used 13C-NMR to measure intracellular CO_2 diffusion and hydration kinetics at 37 °C, and found the half-times of these processes are 1.5 and 0.26 ms respectively. This lends more credence to the idea that CA catalysis is not solely limiting in relation to CO_2 diffusion. What appears to promote intracellular hydration rates slightly faster than diffusion is the high intracellular hemoglobin concentration, which, on the one hand, reduces CO_2 diffusion (Tanashita et al. 1985) while, on the other, it increases CA activity through buffer enhancement of proton transfer to the enzyme's active site (Silverman et al. 1978). An additional factor that slows CO_2 diffusion is the unstirred layer effect, although its magnitude is unknown. To place matters in some physiological perspective, the issue of which process limits the other is only of academic interest, since within 50 ms of the onset of CO_2 exchange, rates at which HCO_3 can be transported across the red cell membrane become the physiologically important rate-limiting event in all vertebrates (Crandall and Bidani 1981).

In mammalian red cells, there appears to be no evidence that carbonic anhydrase is associated with the plasma membrane or has access to extracellular fluid, either by quantitative analysis of enzyme activity contained in red cell ghosts (Randall and Maren 1972) or by functional assay in intact cells (Effros et al. 1988). The possibility that metabolically active nonmammalian vertebrate red cells have membrane-associated enzyme has been noted for the trout, where it could subserve ionic and volume regulation (Smeda and Houston 1979). A role for red cell membrane-associated CA in gas exchange (such as facilitated diffusion) is difficult to conceive, since the membrane is so permeable to CO_2. Large molecular weight inhibitors of the enzyme have been found in plasma in a number of animals (Booth 1938; Rispens et al. 1985; Hill 1986; Haswell and

Randall 1976). These are poorly characterized substances, whose presence is sporadic even within species of the same vertebrate class. The function is entirely unknown and they appear not to penetrate the red cell membrane. Rispens et al. (1985) speculate that they may neutralize any CA liberated with intravascular hemolysis, but the need for such a capacity is not obvious, particularly in light of the more universal finding of carbonic anhydrase bound to the luminal membrane of the capillary endothelium (Lutjen-Drecoll et al. 1985). Hill (1986) cautions against extrapolating from results in artificially perfused organs if an endogenous CA inhibitor is present in that species.

3.2.2 Tissue Carbonic Anhydrase

In addition to its role in fluid, electrolyte and acid-base regulation (Maren 1988), carbonic anhydrase enhances the diffusion of CO_2. This will effectively lower the PCO_2 gradient, down which a given quantity of CO_2 must diffuse; the net result being a lower tissue PCO_2 and higher pH for the same metabolism and perfusion. The unique role of CA bound to the luminal membrane of capillary endothelia with access to plasma will be taken up later in this section.

In the mammalian lung, the presence of carbonic anhydrase is well established from histochemical (Lonnerholm 1982; Sugai et al. 1981) and chemical enzymic assays of whole lung, various cell subtypes and membrane fractions (Ryan et al. 1982; Whitney and Briggle 1982; Henry et al. 1986). The only comparative studies are by Fain and Rosen (1973) on amphibian and reptilian lungs, Dimberg et al. (1981), Lacy (1983) and Rahim et al. (1988) on fish gill and Rosen and Friedley (1973) on frog skin.

Possible functions of lung CA include augmented CO_2 excretion across the alveolar-capillary membrane. Enns and Hill (1983) have shown that CO_2 diffusion across the lung is enhanced by enzyme in the membrane barrier. They found that this facilitated diffusion of carbon dioxide increases the membrane-diffusing capacity and labeled CO_2 uptake by 40%. Klocke (1978) and Crandall and O'Brasky (1978) observed that inhibition of lung tissue CA reduces CO_2 excretion in saline perfused lungs by 20%–40%. Increased CO_2 excretion appears to be dependent upon enzyme within the alveolar-capillary membrane (cytoplasmic CA of the capillary endothelium and alveolar epithelium) and not bound to membranes facing either the alveolar air space or capillary blood, since (1) Heming et al. (1986) showed that CO_2 excretion in isolated lungs is only effected if inhibitors are able to pass out of the vascular space and (2) Effros et al. (1981 a) found no change in CO_2 uptake when either carbonic anhydrase or CA inhibitors were added to alveolar fluid in saline-filled isolated lungs. Whether tissue CA plays such a large role in CO_2 excretion in normal blood-perfused lungs has not been investigated, owing to difficulties in selectively inhibiting lung tissue. Conditions in the isolated saline-perfused lung (larger PCO_2 gradients across the lung and reduced buffering capacity of the perfusate) appear to enhance any contribution of facilitated CO_2 diffusion to total CO_2 output. It is unlikely in vivo that lung tissue CA augments CO_2 elimination by any quantitative amount, since erythrocytes have 100 times the enzymic activity of plasma and/or lung tissue

(Bidani et al. 1983; Henry et al. 1986) and ten times the buffering capacity. The confinement of CA within red cells gives it immediate proximity to hemoglobin, the only buffer for the CO_2-HCO_3 reactions of a sufficient amount (in part due to its oxylabile character) to sustain physiological CO_2 elimination. This is supported by model simulations of lung-gas exchange by Crandall and Bidani (1981) and Mochizuki et al. (1987) in humans, which suggest lung tissue CA can maximally account for less than 10% of normal CO_2 output. At rest, the loss of this activity would require very minimal compensation, but it might be more important in heavy exercise.

While this degree of augmentation probably has little impact in mammalian lungs, whose diffusing capacity is large, it may take on more significance in gills and skin whose diffusing capacities are lower and where conditions are possibly more favorable for facilitated CO_2 diffusion. Surprisingly, this is not borne out by studies in a number of vertebrates. Perry et al. (1982) showed that CO_2 excretion in a spontaneously ventilating blood-perfused trout preparation was correlated with hematocrit, and inferred that gill CA does not function in CO_2 exchange. Swenson and Maren (1987) were able to selectively inhibit gill CA in intact elasmobranch sharks and saw no fall in CO_2 excretion or change in blood gases, despite the above prediction and earlier claims that gill CA had a role in CO_2 excretion in fish (Haswell et al. 1980). Henry et al. (1988) more recently came to similar conclusions about lack of any function for teleost gill carbonic anhydrase in gas exchange. The only situation in which it has been possible to demonstrate a participation of gill CA in CO_2 excretion has been in the saline-perfused isolated shark-head preparation (Swenson and Evans 1984), whose rate of CO_2 excretion is quite low compared to the living animal and has the same problems attending it as discussed above for the isolated perfused lung. There have been no investigations of cutaneous CO_2 excretion with CA inhibition. Perhaps failure to demonstrate a role for CA in branchial gas exchange should not be expected, since enzyme appears to be most intimately associated with the chloride cells (Lacy 1983; Rahim et al. 1988) responsible for salt and acid base homeo-stasis, in which it is clear that CA is involved (Swenson and Maren 1987; Heisler, this volume). Recently, Wright et al. (1986) found that CA is present within the mucus layer of the secondary lamellae in trout and has access to water passing over the gills. There is sufficient activity to rapidly acidify intralamellar water, as CO_2 is eliminated for the purpose of increasing ammonia excretion by ionic trapping (Wright et al. 1988).

Other functions for lung CA have been proposed. These include fetal airway/alveolar fluid secretion (Adamson and Waxman 1976; Davis et al. 1989). The maintenance of an acidic and poorly buffered alveolar fluid may be important in optimal surfactant function (Wildeboer-Venema 1984) in the newborn as well as in the adult. CA may accelerate the dynamics of lung tissue CO_2 capacitance and minimize alveolar PCO_2 changes during the respiratory cycle (Fahri et al. 1976). Swenson et al. (1986) documented that deterioration in ventilation-perfusion matching occurs with CA inhibition. They speculate that lung tissue CA serves to rapidly catalyze changes in local PCO_2 caused by temporal variations in regional perfusion or ventilation to changes in local pH. Both perfusion and ventilation are pH sensitive; thus, a primary change in one is countered by an appropriate directional change in the other to maintain V/Q equality.

A host of other vertebrate tissues not directly involved with H^+ or HCO_3^- transfers contain CA (Maren 1967), whose function is perhaps to augment CO_2 diffusion. Several CA isoenzymes have been discovered in mammalian skeletal muscle (see review by Gros and Dodgson 1988) and, despite original predictions that it was absent in muscle where it might prove disadvantageous (Roughton 1935), the weight of recent evidence suggests it enhances carbon dioxide output from muscle and in some cases improves performance. This was first proposed by Zborowska-Sluis et al. (1974) and then shown by Swenson (1984) in normally perfused muscle, and by Barclay (1987) and Geers and Gros (1988) in isolated nonperfused muscles. In addition, muscle CA associated with the sarcoplasmic reticulum may accelerate the CO_2-HCO_3 reaction sufficiently to provide for rapid formation and consumption of protons entering and leaving the sarcoplasmic reticulum in exchange for calcium (Bruns et al. 1986). Carbonic anhydrase has been found in amphibian (Scheid and Siffert 1985) and fish skeletal muscle (Sanyal et al. 1984).

The enzyme appears to be ubiquitous in vertebrate skeletal muscle, but the data regarding the presence of CA in myocardial muscle are conflicting. Zborowska-Sluis et al. (1975) were unable to detect a significant change in CO_2 space of dog heart with CA inhibition, as is easily detected in skeletal muscle. Moynihan (1977) found small amounts in the heart but could not rule out small contaminating amounts of red cells. More recently, Gay et al. (1981) detected trace amounts of CA in hearts of avian chicks by histochemical and activity measurements, as have Shiels et al. (1984) and Gros and Dodgson (1988) in mammalian heart membrane fractions. Swenson et al. (1988a) used positron emission tomography and found specific uptake into dog myocardium of 11-C labeled acetazolamide. The actual role for myocardial muscle CA has not been demonstrated, but facilitated CO_2 diffusion remains a reasonable possibility, considering the high metabolic rate of cardiac muscle. Its low concentration may be a function of the small size of cardiac myocytes and the very rich vascularity of the heart. Equally important, the enzyme may subserve acid-base regulation in the myocardium, since it has been shown that cardiac muscle has a considerable buffering capacity (Strome et al. 1976), but the effect of CA inhibitors has never been examined.

Other tissues with known high metabolic rates or very poor perfusion also appear to utilize CA to facilitate CO_2 diffusion; these include the avian and elasmobranch salt glands (Nechay et al. 1960; Swenson and Maren 1984) and the mammalian and fish lens (Friedland and Maren 1981). The enzyme is found in various mitochondria, especially liver and kidney (Dodgson et al. 1980), where it may be important in generating HCO_3^- from the inward flux of cytoplasmic CO_2 during high rates of gluconeogenesis and ureogenesis (Balboni and Lehninger 1986; Gros and Dodgson 1988). Considering the small size of these organelles and the absence of CA in other tissue mitochondria (e.g. the heart), it is doubtful that the CA facilitates CO_2 diffusion to any appreciable or physiologically important extent.

Carbonic anhydrase is bound to the plasma-facing membranes of the vascular endothelium of many, but not all, organs (O'Brasky and Crandall 1980). The concept that plasma pH would never reach equilibrium during capillary gas exchange because plasma lacks CA was first posed by Roughton (1935) and subsequently reiterated by Sirs (1970), Hill et al. (1973b) and Forster and Crandall (1975) in computer modeling or in in vitro blood experiments. However, in vivo data of Hill et al. (1977) and Bidani and Crandall (1978a) did not demonstrate any apparent disequilibrium. Effros et al. (1978) were able to show there was sufficient CA activity in lung tissue to accelerate CO_2 reactions 100-fold and thus eliminate any delay in plasma pH equilibration during gas exchange in the pulmonary capillary on this basis. Crandall and O'Brasky (1978), Klocke (1980), Hanson et al. (1981) and Nioka et al. (1988) have confirmed these data. The need for plasma pH equilibration in quantitative CO_2 output by lungs can be neglected, since selective inhibition of luminal membrane-bound CA by drugs linked to high molecular weight dextrans fails to reduce CO_2 output (Heming et al. 1986). The only situation where plasma catalysis appears to enhance CO_2 elimination is in swim bladders of air-breathing teleosts (Randall et al. 1978).

Despite adequate catalysis available to plasma in its transit of the lung capillary (Effros et al. 1980; Bidani et al. 1983), there still remain small measurable post-lung capillary pH disequilibria (Ponte and Purves 1980; Rispens et al. 1980; Chakrabarti et al. 1983) which are not abolished by adding additional CA to plasma (Ponte and Purves 1980). Delayed pH equilibration in plasma following gas exchange has been invoked as a potential factor in the control of ventilation, particularly in exercise (Filley and Heineken 1976) and as cause for positive alveolar to arterial PCO_2 differences during rebreathing and CO_2 inhalation (see review by Piiper 1986). Lewis and Hill (1980) infused large amounts of CA into dogs and showed no significant changes in the pattern of ventilation, with exercise suggesting no role for slow equilibration due to insufficient plasma catalysis in ventilatory control. Bidani and Crandall (1978b) were unable to account for any positive A-a PCO_2 differences due to delayed plasma pH equilibration except under conditions of no CO_2 transfer. It should be noted that plasma pH disequilibrium in the presence of adequate plasma catalysis has only been reported in situations where blood is the perfusing medium, suggesting that other aspects of CO_2 exchange may possibly be responsible. There have been no measurements of this phenomenon in the blood of non-mammalian species. Further discussion of the basis for delayed pH and CO_2 equilibration will follow in a later section on capillary CO_2 equilibration in vivo (Sect. 3.6). Lastly, Lutjen-Drecoll et al. (1985) have proposed that capillary endothelial CA may subserve the function and maintenance of fenestrae, particularly in tissues with high rates of fluid and solute transport. Diaz et al. (1982) provide evidence that membrane-bound CA forms channels in the lipid bilayer, which could participate in ionic exchanges.

3.3 Carbamate Reactions

The binding of CO_2 to proteins, most notably hemoglobin, at amino groups provides another means by which CO_2 is transported in the blood (Klocke 1987). In quantitative terms, carbamates contribute only about 10% to CO_2 eliminated in mammalian lungs (Bauer and Schroeder 1972) and up to 15%–20% in exercising muscles in which higher temperatures and lower hemoglobin O_2 saturations favor more carbamate contribution (Hlastala et al. 1977). These considerations only apply tó mammals with a hemoglobin to organic phosphate ratio around one. In ruminants with low 2,3-DPG, carbamates take on more importance (Hlastala et al. 1977). The carbamate contribution in birds is neglible because of the high concentration of inositol phosphate in avian red cells and its strong affinity for hemoglobin (Maginniss 1985). This inverse relationship between direct CO_2 binding to hemoglobin and organic phosphate stems from competition for common binding sites (Klocke 1987). The importance of carbamate to carbon dioxide exchange has been less well studied in the lower vertebrates, but carbamates play some role even in the jawless cyclostomes (see Bauer 1980 for review).

The kinetics of CO_2 binding to proteins (Gros et al. 1976) and to hemoglobin in solution in the absence of any O_2 saturation change (Forster et al. 1968) are very fast, with half-times almost that of CO_2 diffusion and catalyzed CO_2 hydration. It should be noted that very little CO_2 is fixed to hemoglobin in the absence of saturation changes and to other non-heme-containing proteins (Dejours 1981). Thus it is principally O_2 exchange-linked changes in carbamino-hemoglobin that are of physiological interest, and this chemistry is considerably slower. Klocke (1973) studied the rate of the Haldane effect in the presence of sufficient CA inhibition to evaluate the physiologically significant carbamate kinetics. In a continuous-flow apparatus, he found a half-time of 140 ms such that almost 500–600 ms are required for completion. The reasons for the discrepancy between results in Hb solutions undergoing no change in O_2 saturation, and red cells undergoing a large change in saturation, are not entirely obvious. Klocke (1987) argues against a rate-limiting speed to the conformational changes occurring, with O_2 binding, since none of the other ligand-induced conformational changes in hemoglobin are slow. It is more likely that a combination of unstirred layer effects and multiple sequential steps act to slow the overall process. Even if carbamate kinetics are not rapid enough to bring the reaction to completion within capillary transit, the impact on CO_2 exchange will be minimal, since carbamate contributes only a small fraction of total CO_2 output. However, the relative slowness of this process may partly contribute to slow post capillary pH and PCO_2 equilibration (see Sect. 3.2.2 and Sect. 3.6).

3.4 Bicarbonate-Chloride Exchange

Rapid red cell transmembrane anion exchange confers upon plasma a very high CO_2-carrying capacity and permits more of the buffering capacity of hemoglobin to be utilized in CO_2 exchange. Exchange of HCO_3 for Cl prevents the otherwise earlier depletion of intracellular HCO_3 during CO_2 elimination from the blood and, conversely, premature intracellular buildup of HCO_3 during CO_2 uptake in the peripheral tissues, which would occur if no red cell plasma exchange took place (Klocke 1987). Anion exchange is mediated by the now well characterized band 3 protein, of which in the mammalian red cell membrane there are about one million molecules (Wieth 1979). The contribution of red cell anion exchange to total CO_2 exchange is almost one-third (Wieth et al. 1982) and of all the reactions involved in CO_2 exchange it is the slowest.

In mammalian blood at 37 °C, HCO_3/Cl exchange has a half-time of about 100 ms (Chow et al. 1976; Klocke 1976) and varies with extracellular pH and temperature (Obaid and Crandall 1979) in rapid-reaction apparatus experiments. The effect of unstirred layers around the erythrocytes in these experiments is probably minimal since the transmembrane movement of anions is far slower than their diffusion in an aqueous medium such as plasma (Klocke and Flasterstein 1982). The permeability of the mammalian red cell membrane for bicarbonate is calculated to be in the range of $2-5\times10^{-4}$ cm/s in these experiments, agreeing well with other methods such as 13-C NMR (Chapman et al. 1986) and 18-O exchange (Silverman et al. 1976; Itada and Forster 1977). By itself, the rate of anion exchange should not be cause for any end-capillary disequilibrium at rest, since the process would be 90% complete in 400–500 ms and 99% complete in 700–800 ms (Wieth et al. 1982). However, this analysis ignores the fact that CO_2 exchange is not the result of any one single process but is a sequential and interrelated set of events.

Crandall and Bidani (1981) studied the possible rate-limiting influence of anion exchange in model simulations of pulmonary gas exchange and observed that increasing the effective erythrocyte membrane HCO_3 permeability increased CO_2 exchange by 8% at rest and 14% in exercise. Decreasing HCO_3 permeability reduced CO_2 output by 30%. In the isolated rat lung, perfusion with red cells pretreated with DIDS, an irreversible inhibitor of band 3 protein, reduced CO_2 output 16% at a half normal hematocrit and caused marked downstream pH changes consistent with failure of end-capillary equilibration (Crandall et al. 1981). Equivalent reductions in CO_2 elimination during single breaths were noted in intact dogs when an appropriately timed pulmonary artery infusion of a potent and fast-acting reversible inhibitor, DNDS, was given (Brahm et al. 1987). Further support for a critical rate-limiting role of anion exchange in CO_2 transfer is found in further model calculations by Bidani and Crandall (1982) on the effects of hematocrit changes, particularly in anemia. They showed a marked sensitivity of CO_2 output to the hematocrit in the absence of any compensatory responses. A large fraction of the drop in CO_2 output with anemia can be attributed to the loss of surface area available for anion exchange. This may help to explain the rather profound degree of hyperventilation and increased cardiac output which occur with salicylates, which are known inhibitors of anion exchange at levels causing toxicity in mammals. In like fashion, the ventilatory and hemodynamic effects of anemia may be the result of changes in CO_2 transport and kinetics. This is suggested by failure of cardiac output in severely anemic dogs to return to normal even when the animals are placed in a hyperbaric chamber at three atmospheres of 100% O_2: a level of ambient oxygen sufficient to provide tissue-oxygen demands solely by that dissolved in plasma (Dedichen 1967).

Anion exchange capacity has been demonstrated in nearly all vertebrates studied, and in general the kinetics do not appear to differ from those in mammals when temperature and pH are taken into consideration (Cameron 1978; Obaid et al. 1979; Romano and Passow 1984; and review by Hoffman 1986). In nucleated red cells, anion exchange appears to subserve intracellular pH and volume regulation as well as gas exchange (Hoffman 1986). While it seems red cell HCO_3^-/Cl^- exchange activity and high anion permeability have been well conserved in vertebrate evolution, there is recent evidence that the most primitive vertebrates such as the hagfish and lamprey have less than 0.1% the exchange capacity of other fish (Ellory et al. 1987), despite respectable intracellular CA activity (Carlsson et al. 1980). The physiological consequence for these species is that plasma plays no quantitative role in CO_2 transport since all HCO_3 formed in CO_2 uptake by tissue blood is carried wholly within the red cell (Tufts and Boutilier 1989).

3.5 Bohr, Haldane and Root Effects

The interactions of CO_2 and O_2 in blood are important at many levels in vertebrate gas exchange. To varying degrees, each gas in the course of its own transfer enhances quantitative transfer of the other (Tyuma 1984) but is delayed in its final equilibration (Hlastala 1973). The kinetics of these several effects are complex, because they necessarily incorporate the kinetics of both processes. The linked nature of CO_2 and O_2 exchange is probably a fundamental factor in failure of end-capillary CO_2 equilibration and may even contribute to a lesser extent to end-capillary O_2 disequilibria under extreme conditions.

The kinetics of the Bohr effect in red cell hemolysates are extremely rapid, reflecting the very fast catalyzed generation of protons from CO_2, the rapid binding of both CO_2 and hydrogen ions to critical amino acids and release of O_2 from the heme subunit (Gray 1970). In a rapid-reaction apparatus, the half-time is only several ms in human hemoglobin (Craw et al. 1963). However, in red cell suspensions, the half-time is 120 ms at 37 °C (Forster and Steen 1968; Maren and Swenson 1980). The large difference between cells and hemoglobin solutions lies, of course, in the rate-limiting aspects of O_2 and possibly CO_2 diffusion in unstirred layers. Maren and Swenson (1980) examined the kinetics of the Bohr effect in a series of representative vertebrates and found that these did not differ substantially. The half-times of the Bohr effect in poikilothermic vertebrates at 16 °C were roughly twice as long, findings no different from human red cells at the same temperature. Extension of these in vitro results to in vivo capillary conditions is not easy. The effect of unstirred layers acting to retard the process in vivo is likely less but, since normal physiological PCO_2 gradients are much less than those studied in vitro, these kinetics may be overestimated. The situation is further complicated by the fact that the Bohr effect data cited above (and Haldane and Root data cited below) have been measured only under conditions in which either PCO_2 or PO_2, but not both, were changed. Since both gases are transferred in normal capillary gas exchange, this interaction may be quite important and potentially a decelerating factor, as has been shown in a single study by Tazawa and Mochizuki (1976) in chicken blood. In their microphotometer technique of single red cells in capillaries of the chorioallantoic membrane, they found that the overall rate was at least two to three times slower with both gases in flux than when the PO_2 was held constant.

Klocke (1973) studied the kinetics of the Haldane effect in human red cells and found a time course similar to that for the Bohr effect. This should not be surprising, since the two effects share the same fundamental chemistry and are thermodynamically indistinguishable (Wyman 1964). The half-time at 37 °C is

140 ms and is nearly independent of external pH or 2,3-DPG. Since the contributions of bicarbonate and carbamate to the Haldane effect vary considerably with these factors (Klocke 1973), but the overall rate is constant, it is reasonable to assume that neither intracellular CO_2 hydration nor carbamate kinetics are rate-limiting. As in the Bohr effect, it is probably anion exchange kinetics which are responsible. The question of whether the kinetics of the Haldane effect as studied in vitro can be applied in vivo becomes a more critical issue for CO_2 than does the Bohr effect for O_2, because a much larger fraction of CO_2 exchange is normally linked to O_2 exchange than vice versa. In the above experiments, the rate at which hemoglobin became fully saturated was greatly accelerated (compared to physiological O_2 uptake) by mixing desaturated red cells with a solution of almost 100% O_2 to give a half-time of about 20 ms. Thus, despite the problem of unstirred layers in the in vitro measurements, the rate of the Haldane effect may have been enhanced. In the lung, where at least 50% of CO_2 output is dependent upon O_2 uptake (Hill et al. 1973b), a large fraction of CO_2 exchange must await the completion of oxygen binding to hemoglobin, and this may prevent CO_2 equilibration within capillary transit. There have been no systematic studies of the Haldane effect kinetics in other vertebrates. Our results on the Bohr effect in other vertebrate classes (Maren and Swenson 1980) would suggest that Haldane effect kinetics should not be significantly different.

The Root effect is another manifestation of the heterotropic effect of protons on the binding of oxygen to hemoglobin that appears to be a feature in many fish (Brittain 1987). Although some controversy remains, most investigators consider the Root effect to be an exaggerated alkaline Bohr effect, occurring in the absence of the normal acid Bohr effect that, in most vertebrate hemoglobins, causes O_2 affinity to actually increase below a pH of 6.5. Many species of fish utilize the Root effect to concentrate oxygen within the swim bladder (Berg and Steen 1968) and the eye (Wittenberg and Wittenberg 1962). In both tissues, this process can be blocked by CA inhibitors (Fange 1966; Fairbanks et al. 1969) that slow lactate release by gas gland cells in the swim bladder (Kutchai 1971) and also slow the rate of intracellular red cell acidification or alkalinization (Forster and Steen 1969). The Root effect half-time in eel red cells in the direction of O_2 release (Root off shift) at 24 °C in vitro varies depending on the acidifying conditions (CO_2 or lactic acid) but occurs with half-times of 100–400 ms, well under the many seconds transit time of the very long rete capillaries (Forster and Steen 1969). The rate of O_2 uptake (Root on shift), in contrast, is twenty times slower in intact cells but not hemolysates. Whether such rate differences between the on and off shifts exist in vivo may be questioned, since the conditions under which Forster and Steen (1969) studied the Root on shift were extremely unphysiological. The extracellular pH was 9, which made a large fraction of the shift dependent upon uncatalyzed extracellular CO_2 hydration, in marked contrast to normal conditions in which the PCO_2 falls much more rapidly as CO_2 diffuses out from efferent to afferent capillaries of the rete.

3.6 CO_2 Exchange Kinetics in Vitro and in Vivo

Measurements of overall CO_2 exchange kinetics in mammalian blood, using in vitro rapid-reaction techniques, show that both pH and PCO_2, as measured in extracellular fluid, take many seconds to attain complete equilibrium (Constantine et al. 1965; Forster and Crandall 1975). The slow rate of complete equilibration in these experiments is adequately explained by the uncatalyzed CO_2 hydration and HCO_3 dehydration reactions in plasma, for when CA is added the half-time for pH equilibration falls markedly (Hill et al. 1977, Holland and Forster 1975, Forster and Crandall 1975). These findings were taken as evidence that blood might remain in a state of disequilibrium with regard to CO_2 throughout the circulation and were corroborated by mathematical simulations of Sirs (1970); Hill et al. (1973b); Forster and Crandall (1975) and Bidani et al. (1978). However, discovery of carbonic anhydrase in lung and tissue capillaries has now turned attention to other causes for measured in vivo end-capillary CO_2 disequilibria.

Analysis of CO_2 exchange kinetics in vivo has been even more difficult than for O_2. In addition to those factors discussed for O_2 which confound the interpretation of measured arterial partial pressures in relation to alveolar or ambient O_2 (V/Q mismatching, shunt and counter- or cross current blood-gas flow arrangements), there is the problem that measured blood PCO_2, HCO_3 and pH are inaccurate for precise quantitation. Since kinetic limitation for CO_2 is not a diffusional defect, but rather chemical reaction limitation, blood gases (measured ex vivo at equilibrium) can not reflect true end-capillary status. Thus, while molecular CO_2 reaches complete equilibration between capillary blood and tissue or respired medium (i.e. PCO_2 equilibration), the CO_2 content of blood will be at a lower level (tissue output) or greater level (lung, gill or skin output) than the expected equilibrium value. As blood exits the capillary, there will be an internal re-equilibration of the $CO_2/H/HCO_3$ system in blood, causing the PCO_2 to fall or rise respectively. Recent calculations of Crandall and Bidani (1981) and Zock et al. (1981), which incorporate vascular CA activity available to plasma and experimental data from animals with intravascular pH catheters (Sect. 3.2.2), show there are other factors responsible for end-capillary CO_2 disequilibration. These include (1) the finite speed of O_2 exchange upon which 50% of CO_2 exchange is dependent (i.e. the Haldane effect and oxylabile carbamate reactions) and (2) the finite speed of red cell-plasma anion exchange, which confers on plasma a large capacitance for CO_2 and a quantitative role in CO_2 exchange that its otherwise poor buffering capacity and little CA activity would not permit. The predicted magnitude of end-capillary CO_2 disequilibration in any particular animal and tissue will depend on (1) the extent of vascular CA activity available to plasma, (2) kinetics of red cell anion exchange, (3) hematocrit, (4) plasma buffer capacity, (5) presence or absence of a plasma CA inhibitor and (6) capillary transit time.

The presence of slow and incomplete equilibration of in vivo capillary CO_2 exchange has been difficult to assess experimentally for several reasons. The first, already alluded to above, is the lack of good techniques to measure end-capillary PCO_2 and pH. The second is that even standard ex vivo measurements may not be sensitive enough to qualitatively track critical in vivo disequilibria. This is because CO_2 has such a high solubility in blood that its transport and excretion are accomplished over very small a-v PCO_2 gradients (2–6 mmHg). Physiologically significant changes of even 1 mmHg, which border on the measurement error of even the best CO_2 electrodes, may pass unrecognized. Lastly, effects of ventilation-perfusion or metabolism-perfusion mismatching and other non-kinetic factors will likely be as responsible for measured PCO_2 differences between blood and tissue or respired medium as chemical reaction limitation. Various investigators have attempted to predict the magnitude of these PCO_2 gradients, using data largely taken from in vitro studies and often relying on a single lumped parameter to describe CO_2 kinetics. In the human lung, Hlastala (1973) and Hill et al. (1973 b) calculated a 1 mmHg alveolar to arterial PCO_2 gradient. In the placenta where the capillary transit is greater than 1 s, Hill et al. (1973 a) showed no gradient between maternal and fetal blood. The more recent work of Piiper et al. (1980) and Crandall and Bidani (1981) has estimated that the alveolar to arterial PCO_2 gradient in the human lung may only be 0.2 to 0.5 mmHg. However, Piiper et al. (1980) and Johnson and Ramanathan (1985) calculate that, in exercise, failure of end-capillary equilibration may worsen considerably and cause the PCO_2 difference to rise 6–7 mmHg. In simulations of moderate to severe lung disease, Johnson and Ramanathan (1985) estimate equally large arterial to alveolar PCO_2 differences. Failure of end-capillary CO_2 equilibration will have as its consequence either higher tissue CO_2 levels (respiratory acidosis) or greater levels of alveolar ventilation than would be necessary if equilibrium were achieved in capillary transit.

The issue of end-capillary PCO_2 and pH disequilibration in vivo has other important physiological implications. Traditional assumptions in mammalian physiology that alveolar PCO_2 is equal to arterial PCO_2 and that blood PCO_2 is constant after it leaves a site of gas exchange are erroneous. Thus, calculations of the physiological dead space and alveolar PO_2 and assumptions about the arterial PCO_2 and pH at sites in the circulation important in the control of ventilation will be only semi-quantitative. The magnitude of error, while minimal at rest, may increase significantly in importance in the interpretation of gas exchange and acid-base status in disease and during high metabolic stress as capillary transit times decrease. Furthermore any drugs that inhibit carbonic anhydrase or red cell anion exchange will only worsen the extent of end-capillary CO_2 disequilibration. The presence and consequences of CO_2 chemical reaction limitation in non-mammalian species is a virtually unexplored area, and it is hoped that similar experimental methods and mathematical models will be employed in a variety of vertebrates.

References

Adamson TM, Waxman BP (1976) Carbonate dehydratase (carbonic anhydrase) and the fetal lung. In: Lung liquids. Ciba Found Symp 38:221–234

Ar A, Visschedijk AHJ, Rahn H, Piiper J (1980) Carbon dioxide in the chick embryo towards end of development: effects of He and SF_6 in breathing mixture. Respir Physio 40:293–307

Aroesty J, Gross JF (1970) Convection and diffusion in the microcirculation. Microvasc Res 2:247–253

Bagge V, Branemark PI, Karlsson R (1980) Three-dimensional observations of red blood cell deformation in capillaries. Blood Cells 6:231–240

Bailey JR, Driedzic WR (1986) Function of myoglobin in oxygen consumption by isolated perfused fish hearts. Am J Physiol 20:R1144–R1150

Balboni E, Lehninger AL (1986) Entry and exit pathways of CO_2 in rat liver mitochondria respiring in a bicarbonate buffer system. J Biol Chem 261:3563–3570

Barclay JK (1987) Carbonic anhydrase III inhibition in normocapnic and hypercapnic contracting mouse soleus. Can J Physiol Pharmacol 65:100–104

Barinov I, Dodgson SJ, Itada N, Lin L (1986) Graded inhibition of intracellular carbonic anhydrase activity in human red blood cells. J Physiol 371:159P

Bartels H (1971) Table 14 p. 22 of Biological handbooks: respiration and circulation. FASEB Bethesda

Bauer C (1980) Phylogeny and physiological significance of oxygen-linked CO_2 binding in hemoglobin. In: Bauer C, Gros H, Bartels H (eds) Biophysics and physiology of carbon dioxide. Springer, Berlin Heidelberg New York, pp 84–88

Bauer C, Schröder E (1972) Carbamino compounds of hemoglobin in human adult and foetal blood. J Physiol 227:457–471

Bauer C, Klocke RA, Kamp D, Forster RE (1973) Effect of 2,3-diphosphoglycerate and H^+ on the reaction of O_2 and hemoglobin. Am J Physiol 224:838–847

Baumann R, Bartels H, Bauer C (1987) Blood-oxygen transport. In: Fahri LE, Tenney SM (eds) Handbook of physiology: the respiratory system, Vol IV. American Physiological Society, Bethesda, pp 147–172

Berg T, Steen JB (1968) The mechanism of oxygen concentration in the swim-bladder of the eel. J Physiol 195:631–638

Bernstein RS, Schraer R (1972) Purification and properties of an avian carbonic anhydrase from the erythrocytes of *Gallus domesticus*. J Biol Chem 25:1300–1322

Bidani A, Crandall ED (1978a) Slow postcapillary pH changes in blood in anesthetized animals. J Appl Physiol 45:674–680

Bidani A, Crandall ED (1978b) Analysis of PCO_2 differences during rebreathing due to slow pH equilibration in blood. J Appl Physiol 45:666–673

Bidani A, Crandall ED (1982) Analysis of the effects of hematocrit on pulmonary CO_2 transfer. J Appl Physiol 53:413–418

Bidani A, Crandall ED, Forster RE (1978) Analysis of postcapillary pH changes in blood in vivo after gas exchange. J Appl Physiol 44:770–781

Bidani A, Mathew SJ, Crandall ED (1983) Pulmonary vascular carbonic anhydrase activity. J Appl Physiol 55:75–83

Bidani A, Crandall ED, Dubose TD (1984) Analysis of the determinants of renal cortical PCP_2. Am J Physiol 247:F466–F474

Black CP, Tenney SM (1980) Oxygen transport during progressive hypoxia in high-altitude and sea-level water fowl. Respir Physiol 39:217–239

Boland EJ, Nair PK, Lemon DD, Olson JS, Hellums JD (1987) An in vitro capillary system for studies on microcirculatory O_2 transport. J Appl Physiol 62:791–797

Booth VH (1938) The carbonic anhydrase inhibitor in serum. J Physiol 91:474–489

Brahm J, Gronlund J, Hlastala MP, Ohlsson J, Swenson ER (1987) Effects of in vivo inhibition of red cell anion exchange and Haldane effect on carbon dioxide output (V_{CO2}) in the dog lung. J Physiol 390:245P

Brittain T (1987) Minireview: the Root effect. Comp Biochem Physiol 86B:473–481

Bruns W, Dermietizel R, Gros G (1986) Carbonic anhydrase in the sarcoplasmic reticulum of rabbit skeletal muscle. J Physiol 371:351–364

Bundy HF, Cheng B (1976) Amphibian carbonic anhydrase: purification and partial characterization of the enzyme from erythrocytes of *Rana catesbeiana*. Comp Biochem Physiol 55B:265–271

Burger RE, Meyer M, Graf W, Scheid P (1979) Gas exchange in the parabronchial lung of birds: Experiments in unidirectionally ventilated ducks. Respir Physiol 36:19–37

Burns B, Shepard RH (1979) DLO_2 in excised lungs perfused with blood containing sodium dithionite. J Appl Physiol 46:100–110

Cain SM, Otis AB (1961) Carbon dioxide transport in anesthetized dogs during inhibition of carbonic anhydrase. J Appl Physiol 16:1023–1028
Cameron JN (1978) Chloride shift in fish blood. J Exp Zool 206:289–295
Cameron JN, Polhemus JA (1974) Theory of CO_2 exchange in trout gills. J Exp Biol 60:183–194
Capen RL, Latham LP, Wagner WW Jr (1987) Comparison of direct and indirect measurements of pulmonary capillary transit times. J Appl Physiol 62:1150–1154
Carlsson U, Kjellstrom B, Antonsson B (1980) Purification and properties of cyclostome carbonic anhydrase from erythrocytes of hagfish. Biochem Biophys Acta 612:160–170
Chakrabarti MK, Cobbe S, Loh L, Poole-Wilson PA (1983) Measurement of pulmonary venous and arterial pH oscillations in dogs using catheter tip pH electrodes. J Physiol London 336:61–71
Chapman BE, Kirk K, Kuchel PW (1986) Bicarbonate exchange kinetics at equilibrium across the erythrocyte membrane by ^{13}C NMR. Biochem Biophys Res Commun 136:266–272
Chow EI, Crandall ED, Forster RE (1976) Kinetics of bicarbonate-chloride exchange across the human red blood cell membrane. J Gen Physiol 68:633–652
Clark A, Clark PAA (1983) Capture of spatially homogeneous chemical reactions in tissue by freezing. Biophys J 42:45–59
Clark A, Cokelet GR, Federspiel WJ (1981) Erythrocyte motion and oxygen transport. Bibl Anat 20:385–388
Clark A Jr, Federspiel WJ, Clark PAA, Cokelet GR (1985) Oxygen delivery from red cells. Biophys J 47:171–181
Coin JT, Olson JS (1979) The rate of oxygen uptake by human red blood cells. J Biol Chem 254:1178–1190
Cole RP (1983) Skeletal muscle function in hypoxia: effect of alteration of intracellular myoglobin. Respir Physiol 53:1–12
Conhaim RL, Staub NC (1980) Reflection spectrophotometric measurement of O_2 uptake in pulmonary arterioles of cats. J Appl Physiol 48:848–856
Connett RJ, Gayeski TEJ, Honig CR (1984) Lactate accumulation in fully aerobic, working, dog gracilis muscle. Am J Physiol 246:H120–H128
Constantine HP, Craw MR, Forster RE (1965) Rate of the reaction of carbon dioxide with human red blood cells. Am J Physiol 208:801–811
Covell DG, Jacquez JA (1987) Does myoglobin contribute significantly to diffusion of oxygen in red skeletal muscle? Am J Physiol 21:R341–R347
Crandall ED, Bidani A (1981) Effects of red blood cell HCO_3^-/Cl^- exchange kinetics on lung CO_2 transfer: theory. J Appl Physiol 50:265–271
Crandall ED, O'Brasky JE (1978) Direct evidence for participation of rat lung carbonic anhydrase in CO_2 reactions. J Clin Invest 62:618–622
Crandall ED, Mathew SJ, Fleischer RS, Winter HI, Bidani A (1981) Effects of inhibition of RBC HCO_3^-/Cl^- exchange on CO_2 excretion and downstream pH disequilibrium in isolated rat lungs. J Clin Invest 68:853–862
Crank WD, Gallagher RR (1978) Theory of gas exchange in the avian parabronchus. Respir Physiol 35:9–25
Craw MR, Constantine HP, Morello JA, Forster RE (1963) Rate of the Bohr shift in human red cell suspensions. J Appl Physiol 18:317–324
Davis TA, Gause G, Perks AM, Kuck H, Cassin S (1989) The effects of acetazolamide on fetal lung liquid secretion. FASEB J 3:A1140
Dedichen H, Raace D, Schenk WG (1967) Hemodilution and concomitant hyperbaric oxygenation. Effects on cardiovascular function. J Thorac Cardiovasc Surg 53:341–348
Dejours P (1981) Principles of comparative physiology. Elsevier/North Holland, Amsterdam
Dempsey JA, Hanson PG, Henderson KS (1984) Exercise-induced arterial hypoxemia in healthy human subjects at sea level. J Physiol 355:161–175
Diaz E, Sandblom JP, Wistrand PJ (1982) Selectivity properties of channels induced by a reconstituted membrane-bound carbonic anhydrase. Acta Physiol Scand 116:461–463
Dimberg K, Hoglund LB, Knutsson PG, Ridderstrale Y (1981) Histochemical localization of carbonic anhydrase in gill lamellae from young salmon (*Salmo salar* L.) adapted to fresh and salt water. Acta Physiol Scand 112:218–220

Dodgson SJ, Forster RE (1983) Carbonic anhydrase activity of intact erythrocytes from seven mammals. J Appl Physiol 55:1292–1298
Dodgson SJ, Forster RE, Storey BT, Mela L (1980) Mitochondrial carbonic anhydrase. Proc Natl Acad Sci 77:5562–5566
Donaldson TL, Quinn JA (1974) Kinetic constants determined from membrane transport measurements: Carbonic anhydrase activity in high concentrations. Proc Natl Acad Sci 61:4995–4999
Donnelly PM, Woolcock AJ (1978) Stratification of inspired air in the elongated lungs of the carpet python, *Morellia spilotes variegata*. Respir Physiol 35:301–315
Effros RM, Chang RSY, Silverman P (1978) Acceleration of plasma bicarbonate conversion to carbon dioxide by pulmonary carbonic anhydrase. Science 199:427–429
Effros RM, Shapiro L, Silverman P (1980) Carbonic anhydrase activity of rabbit lungs. J Appl Physiol 49:589–600
Effros RM, Mason G, Silverman P (1981 a) Asymmetric distribution of carbonic anhydrase in the alveolar-capillary barrier. J Appl Physiol 51:190–193
Effros RM, Mason G, Silverman P (1981 b) Role of perfusion and diffusion in $^{14}CO_2$ exchange in the rabbit lung. J Appl Physiol 51:1136–1144
Effros RM, Taki K, Dodek P, Edwards J, Husczuk A, Silverman P, Hukkanen J (1988) Exchange of labeled bicarbonate and carbon dioxide with erythrocytes suspended in an elutriator. J Appl Physiol 64:569–576
Ellory JC, Wolowyk MW, Young JD (1987) Hagfish (*Eptatretus stouti*) erythrocytes show minimal chloride transport activity. J Exp Biol 129:377–383
Ellsworth ML, Popel AS, Pittman RN (1988) Assessment and impact of heterogeneities of convective oxygen transport parameters in capillaries of striated muscle: experimental and theoretical. Microvasc Res 35:341–372
Enns T (1967) Facilitation by carbonic anhydrase of carbon dioxide transport. Science 155:44–46
Enns T, Hill EP (1983) CO_2 diffusion capacity in isolated dog lung lobes and the role of carbonic anhydrase. J Appl Physiol 54:483–490
Fain W, Rosen S (1973) Carbonic anhydrase activity in amphibian and reptilian lung: a histochemical and biochemical analysis. Histochem J 5:519–528
Fairbanks MB, Hoffert JR, Fromm PO (1969) The dependence of the oxygen concentrating mechanism of the teleost eye (*Salmo gairdneri*) on the enzyme carbonic anhydrase. J Gen Physiol 54:202–211
Fange R (1966) Physiology of the swimbladder. Physiol Rev 46:299–322
Farhi L, Plewes JL, Olszowka AJ (1976) Lung carbonate dehydratase (carbonic anhydrase), CO_2 stores and CO_2 transport. In: Lung liquids. Ciba Found Symp 38:235–249
Feder ME, Burggren WW (1985) Cutaneous gas exchange in vertebrates: Design, patterns, control and implications. Biol Rev 60:1–45
Feder ME, Full RJ, Piiper J (1988) Elimination kinetics of acetylene and Freon 22 in resting and active lungless salamanders. Respir Physiol 72:229–240
Federspiel WJ (1986) A model study of intracellular oxygen gradients in a myoglobin-containing skeletal muscle fiber. Biophys J 49:857–868
Federspiel WJ (1989) Pulmonary diffusing capacity: implications of two phase blood flow in capillaries. Respir Physiol 77:119–134
Federspiel WJ, Popel AS (1986) A theoretical analysis of the effect of the particulate nature of blood on oxygen release in capillaries. Microvasc Res 32:164–189
Federspiel WJ, Sarelius IH (1984) An examination of the contribution of red cell spacing to the uniformity of oxygen flux at the capillary wall. Microvasc Res 27:273–285
Feisal KA, Sackner MA, DuBois AB (1963) Comparison between time available and the time required for CO_2 equilibration in the lung. J Clin Invest 42:24–28
Filley GF, Heineken FG (1976) A blood gas disequilibrium theory. Br J Dis Chest 70:223–245
Fischer T, Schmid-Schönbein H (1977) Tank tread motion of red cell membranes in viscometric flow: behavior of intracellular and extracellular markers. Blood Cells 3:351–360
Fischkoff S, Vanderkooi JM (1975) Oxygen diffusion in biological and artificial membranes determined by the fluorochrome pyrene. J Gen Physiol 65:663–676

Forster RE (1964) Rate of gas uptake by red cells. In: Fenn WO, Rahn H (eds) Handbook of physiology: respiration, Vol I. American Physiological Society, Washington, DC, pp 827–837
Forster RE (1987) Diffusion of gases across the alveolar membrane. In: Fahri LE, Tenney SM (eds) Handbook of physiology: the respiratory system, Vol IV. American Physiological Society, Bethesda, pp 71–88
Forster RE, Crandall ED (1975) Time course of exchanges between red cells and extracellular fluid during CO_2 uptake. J Appl Physiol 38:710–718
Forster RE, Steen JB (1968) Rate-limiting processes in the Bohr shift in human red cells. J Physiol 196:541–562
Forster RE, Steen JB (1969) The rate of the "Root shift" in eel red cells and eel hemoglobin solutions. J Physiol 204:259–282
Forster RE, Constantine HP, Craw MR, Rotman HH, Klocke RA (1968) Reaction of CO_2 with human hemoglobin solution. J Biol Chem 243:3317–3326
Forster RE, Dodgson SJ, Storey BT, Lin L (1984) Measurement of carbonic anhydrase activity inside cells and subcellular organelles. Ann NY Acad Sci 429:415–429
Frech WE, Schuitehinrichs D, Vogel HR, Thews G (1968) Modelluntersuchung zum Austausch der Atemgase. I. Die O_2-Aufnahmezeiten des Erythrocyten unter den Bedingungen des Lungen-Kapillarblutes. Pflügers Arch 301:292–301
Friedland BR, Maren TH (1981) The relation between carbonic anhydrase activity and ion transport in elasmobranch and rabbit lens. Exp Eye Res 33:545–561
Gad-el-Hak M, Morton JB, Kutchai H (1977) Turbulent flow of red cells in dilute suspensions: effect on kinetics of O_2 uptake. Biophys J 18:289–295
Garg LC, Maren TH (1972) The rates of hydration of carbon dioxide and dehydration of carbonic acid at 37 °C. Biochem Biophys Acta 261:70–76
Gatz RN (1982) Diffusion-limited gas exchange across amphibian skin. Fed Proc 41:2137–2139
Gay CV, Schraer H, Sharkey DJ, Rieder E (1981) Carbonic anhydrase in developing avian heart, blood, and chorioallantoic membrane. Comp Biochem Physiol 70A:173–177
Gayeski TEJ, Honig CR (1986) O_2 gradients from sarcolemma to cell interior in red muscle at maximal VO_2. Am J Physiol 251:H789–H799
Geers C, Gros G (1988) Carbonic anhydrase inhibition affects contraction of directly stimulated rat soleus. Life Sci 42:37–45
Geiser J, Betticher DC (1989) Gas transfer in isolated lungs with red cell suspension or hemoglobin solution. Respir Physiol 77:31–40
Gibson QH (1954) Stopped-flow apparatus for study of rapid reactions. Disc Faraday Soc 17:137–139
Glass ML, Wood SC (1983) Gas exchange and control of breathing in reptiles. Physiol Rev 63:232–260
Goldstick TK, Ciuryla VT, Zuckerman L (1975) Diffusion of oxygen in plasma and blood. Adv Exp Med Biol 75:183–190
Gray BA (1971) The rate of approach to equilibrium in uncatalyzed CO_2 hydration reactions: the theoretical effect of buffering capacity. Respir Physiol 11:223–234
Gray RK (1970) The kinetics of the alkaline Bohr effect of human hemoglobin. J Biol Chem 245:2914–2921
Groebe K, Thews G (1987) Time courses of erythrocytic oxygenation in capillaries of the lung: Lower and upper bounds on red cell transit times. Adv Exp Med Biol 215:165–169
Gros G, Dodgson S (1988) Velocity of CO_2 exchange in muscle and liver. Annu Rev Physiol 50:669–694
Gros G, Forster RE, Lin L (1976) The carbamate reaction of glycylglycine, plasma and tissue extracts evaluated by a pH stopped flow apparatus. J Biol Chem 251:4398–4407
Gros G, Moll W (1971) The diffusion of carbon dioxide in erythrocytes and hemoglobin solutions. Pflügers Arch 324:249–266
Gutierrez G (1986a) The rate of oxygen release and its effect on capillary O_2 tension: a mathematical analysis. Respir Physiol 63:79–96
Gutierrez G (1986b) Peripheral delivery and utilization of oxygen. In: Dantzer DR (ed) Cardiopulmonary critical care. Grune and Stratton, Orlando, pp 169–210

Gutknecht J, Bisson MA, Tosteson DC (1977) Diffusion of carbon dioxide through lipid bilayer membranes: effects of carbonic anhydrase, bicarbonate, and unstirred layers. J Gen Physiol 69:779–794
Hall GE, Schraer R (1979) Purification and partial characterization of high and low activity carbonic anhydrase isoenzymes from Malaclemys terrapin centrata. Comp Biochem Physiol 64B:561–565
Hanson MA, Nye PCG, Torrance RW (1981) Studies on the localization of pulmonary carbonic anhydrase in the cat. J Physiol 319:93–109
Hartridge H, Roughton FJW (1923) A method of measuring the velocity of very rapid chemical reactions. Proc R Soc Lond Ser A 104:395–415
Hassinen IE, Hiltunen JK, Takala TES (1981) Reflectance spectrophotometric monitoring of the isolated perfused heart as a method of measuring the oxidation-reduction state of cytochromes and oxygenation on myoglobin. Cardiovasc Res 15:86–93
Haswell MS, Randall DJ (1976) Carbonic anhydrase inhibitor in trout plasma. Respir Physiol 28:17–27
Haswell MS, Randall DJ, Perry SF (1980) Fish gill carbonic anhydrase: acid-base regulation or salt transport? Am J Physiol 238:R240–R245
Heidelberger E, Reeves RB (1988) Failure of specific conductance (O) to describe oxygen uptake by whole blood. FASEB J 2:A928
Heineken FG, Predecke RK, Filley GG (1976) Oxygen exchange in silicone rubber capillaries. J Bioeng 2:195–204
Hellums JK (1977) The resistance of oxygen transport in the capillaries relative to that in the surrounding tissue. Microvasc Res 13:131–136
Heming TA, Geers C, Gros G, Bidani A, Crandall ED (1986) Effects of dextran-bound inhibitors on carbonic anhydrase activity in isolated rat lungs. J Appl Physiol 61:1849–1856
Henry RP, Cameron JN (1988) The distribution of branchial carbonic anhydrase and effects of gill and erythrocyte carbonic anhydrase inhibition in the catfish, *Ictalurus punctatus*. J Exp Biol 134:201–208
Henry RP, Dodgson SJ, Forster RE, Storey BT (1986) Rat lung carbonic anhydrase: activity, localization, isozymes. J Appl Physiol 60:638–645
Hill EP (1986) Inhibition of carbonic anhydrase by plasma of dogs and rabbits. J Appl Physiol 60:191–197
Hill EP, Power GG, Longo LD (1973a) A mathematical model of carbon dioxide transfer in the placenta and its interaction with oxygen. Am J Physiol 224:283–299
Hill EP, Power GG, Longo LD (1973b) Mathematical simulation of pulmonary O_2 and CO_2 exchange. Am J Physiol 224:909–917
Hill EP, Power GG, Gilbert RK (1977) Rate of pH changes in blood plasma in vitro and in vivo. J Appl Physiol 42:928–934
Hills BA, Hughes GM, Koyama T (1982) Oxygenation and deoxygenation kinetics of red cells in isolated lamellae of fish gills. J Exp Biol 98:269–275
Hlastala MP (1973) Significance of the Bohr and Haldane effects in the pulmonary capillary. Respir Physiol 17:81–92
Hlastala MP, Woodson RD, Wranne B (1977) Influence of temperature on hemoglobin-ligand interaction in whole blood. J Appl Physiol 43:545–550
Hlastala MP, Scheid P, Piiper J (1981) Interpretation of inert gas retention and excretion in the presence of stratified inhomogeneity. Respir Physiol 46:247–259
Hlastala MP, Standaert TA, Pierson DJ, Luchtel DL (1985) The matching of ventilation and perfusion in the lung of the Tegu lizard, *Tupinambis nigropunctatus*. Respir Physiol 60:277–294
Hoffmann EK (1986) Anion transport systems in the plasma membrane of vertebrate cells. Biochim Biophys Acta 864:1–31
Holland RAB, Forster RE (1966) The effect of size of red cells on the kinetics of their oxygen uptake. J Gen Physiol 49:727–742
Holland RAB, Forster RE (1975) Effect of temperature on rat of CO_2 uptake by human red cell suspensions. Am J Physiol 228:1589–1596
Holland RAB, Shibata H, Scheid P, Piiper J (1985) Kinetics of O_2 uptake and release by red cells in stopped-flow apparatus: effects of unstirred layer. Respir Physiol 59:71–91

Homer LD, Weathersby PK, Kiesow LA (1981) Oxygen gradients between red blood cells in the microcirculation. Microvasc Res 22:308–320
Honig CR, Gayeski TEJ (1987) Comparison of intracellular PO_2 and conditions of blood-tissue O_2 transport in heart and working red skeletal muscle. Adv Exp Med Biol 215:309–321
Honig CR, Odoroff CL (1981) Calculated dispersion of capillary transit times: significance for oxygen exchange. Am J Physiol 240:H199–H208
Honig CR, Odoroff CL, Frierson JL (1980) Capillary recruitment in exercise: rate, extent, uniformity, and relation to blood flow. Am J Physiol 238:H31–42
Honig CR, Odoroff CL, Frierson JL (1982) Active and passive capillary control in red muscle at rest and in exercise. Am J Physiol 243:H196–H206
Hook C, Yamaguchi K, Scheid P, Piiper J (1988) Oxygen transfer of red blood cells: experimental data and model analysis. Respir Physiol 72:65–82
Houston AH, McCarty LS (1978) Carbonic anhydrase (acetazolamide-sensitive esterase) activity in the blood, gill and kidney of the thermally acclimated rainbow trout, *Salmo gairdneri*. J Exp Biol 73:15–27
Hughes GM (1984) General anatomy of the gills. In: Hoar WS, Randall DJ (eds) Fish physiology, vol 10, Part A. Academic Press, New York, London, pp 1–72
Hughes GM, Horimoto M, Kikuchi Y, Kakiuchi U (1981) Blood-flow velocity in microvessels of the gill filaments of the goldfish (*Carassius auratusl*). J Exp Biol 90:327–331
Huxley VH, Kutchai H (1981) The effect of the red cell membrane diffusion boundary layer on the rate of oxygen uptake by human erythrocyte. J Physiol 316:75–83
Huxley VH, Kutchai H (1983) Effect of diffusion boundary layers on the initial uptake of O_2 by red cells. Theory versus experiment. Microvasc Res 26:89–107
Hyde RW, Puy RJM, Raub WF, Forster RE (1968) Rate of disappearance of labeled carbon dioxide from the lungs of humans during breath holding: a method for studying the dynamics of pulmonary CO_2 exchange. J Clin Invest 47:1535–1552
Itada N, Forster RE (1977) Carbonic anhydrase activity in intact red blood cells measured with ^{18}O exchange. J Bio Chem 252:3881–3890
Johnson RL, Ramanathan M (1985) Buffer Equilibria in the Lungs. In: Seldin DW, Giebisch G (eds) The kidney: physiology and pathophysiology. Raven, New York, pp 149–171
Johnson RL, Spicer WS, Bishop JM, Forster RE (1968) Pulmonary capillary blood volumes, flow and diffusing capacity during exercise. J Appl Physiol 15:893–902
Karas RH, Taylor CR, Jones JH, Lindstedt SL, Reeves RB, Weibel RR (1987) Adaptive variation in the mammalian respiratory system in relation to energetic demand. VII. Flow of oxygen across the pulmonary gas exchange. Respir Physiol 69:101–115
Kawashiro T, Scheid P (1976) Measurement of Krogh's diffusion constant of CO_2 in respiring muscle at various CO_2 levels: evidence of facilitated diffusion. Pflügers Arch 362:127–133
Kawashiro T, Nusse W, Scheid P (1975) Determination of diffusivity of oxygen and carbon dioxide in respiring tissue: results in rat skeletal muscle. Pflügers Arch 359:231–251
Kernohan JC, Forrest WW, Roughton JW (1962) The activity of concentrated solutions of carbonic anhydrase. Biochem Biophys Acta 65:31–41
Klocke RA (1973) Mechanism and kinetics of the Haldane effect in human erythrocytes. J Appl Physiol 35:673–681
Klocke RA (1976) Rat of bicarbonat-chloride exchange in human red cells at 37 °C. J Appl Physiol 40:707–714
Klocke RA (1978) Catalysis of CO_2 reactions by lung carbonic anhydrase. J Appl Physiol 44:882–888
Klocke RA (1980) Equilibrium of CO_2 reactions in the pulmonary capillary. J Appl Physiol 48:972–976
Klocke RA (1987) Carbon dioxide transport. In: Fahri LE, Tenney SM (eds) Handbook of physiology: the respiratory system, vol IV. American Physiological Society, Bethesda, pp 173–197
Klocke RA (1987) Oxygen transfer from the red cell to the mitochondrion. In: Brion-Brown CW, Ayres SM (eds) New horizons in oxygen transport and utilization. Society Critical Care Medicine, Baltimore, pp 239–265
Klocke RA, Flasterstein F (1982) Kinetics of erythrocyte penetration by aliphatic acids. J Appl Physiol 53:1138–1143

Knopp TJ, Bassingthwaighte JB (1969) Effect of flow on transpulmonary circulatory transport functions. J Appl Physiol 27:36–43
Koyama T, Mochizuki M (1969) A study of the relationship between oxygenation velocity of the red blood cell and flow velocity in a rapid-flow method. Jpn J Physiol 19:534–546
Kreuzer F, Hoofd L (1987) Facilitated diffusion of oxygen and carbon dioxide. In: Fahri LE, Tenney SM (eds) Handbook of physiology: the respiratory system, vol IV. American Physiological Society, Bethesda, pp 89–111
Kreuzer F, Yahr WZ (1960) Influence of red cell membrane on diffusion of oxygen. J Appl Physiol 15:1117–1122
Krogh A (1919) Number and distribution of capillaries in muscles with calculations of oxygen pressure head necessary for supplying the tissue. J Physiol 52:409–421
Kutchai H (1970) Numerical study of oxygen uptake by layers of hemoglobin solution. Respir Physiol 10:273–284
Kutchai H (1971) Short communication: role of carbonic anhydrase in lactate secretion by the swimbladder. Comp Biochem Physiol 39A:357–359
Kutchai H (1975) Role of the red cell membrane in oxygen uptake. Respir Physiol 24:121–132
Kutchai H, Staub NC (1969) Steady-state, hemoglobin-facilitated O_2 transport in human erythrocytes. J Gen Physiol 53:576–587
Lacy ER (1983) Histochemical and biochemical studies of carbonic anhydrase activity in the opercular epithelium of the euryhaline teleost, *Fundulus heteroclitus*. Am J Anat 166:19–39
Larimer JL, Schmidt-Nielsen K (1960) A comparison of blood carbonic anhydrase of various mammals. Comp Biochem Physiol 1:19–23
Lawson WH, Holland RAB, Forster RE (1965) Effect of temperature on deoxygenation rate of human red cells. J Appl Physiol 20:912–918
Lemon DD, Nair PK, Boland J, Olson JS, Hellums JD (1987) Physiological factors affecting O_2 transport by hemoglobin in an in vitro capillary system. J Appl Physiol 62:798–806
Lewis SE, Erickson Barnett LB, Venta PJ, Tashian RE (1988) N-Ethyl-N-nitrosourea-induced null mutation at the mouse car-w locus: an animal model for human carbonic anhydrase II deficiency syndrome. Proc Natl Acad Sci 85:1962–1966
Lewis SM, Hill EP (1980) Effect of plasma carbonic anhydrase on ventilation in exercising dogs. J Appl Physiol 49:708–714
Lindstedt SL (1984) Pulmonary transit time and diffusing capacity in mammals. Am J Physiol 246:R384–R388
Little NG (1980) Erythrocyte carbonic anhydrase activity and physical work capacity. J Sports Med 20:239–243
Longworth KE, Jones JH, Bicudo JEPW, Taylor CR, Weibel ER (1989) High rate of O_2 consumption in exercising foxes: large pO_2 difference drives diffusion across the lung. Respir Physiol 77:263–276
Lonnerholm G (1982) Pulmonary carbonic anhydrase in the human, monkey and rat. J Appl Physiol 52:352–356
Luckner H (1939) Über die Geschwindigkeit des Austausches der Atemgase im Blut. Pflügers Arch 241:753–778
Lutjen-Drecoll E, Eichhorn M, Barany EH (1985) Carbonic anhydrase in epithelia and fenestrated juxtaepithelial capillaries of *Macaca fascicularis*. Acta Physiol Scand 124:295–307
Magid E (1967) Activity of carbonic anhydrase in mammalian blood in relation to body size. Comp Biochem Physiol 21:357–360
Maginniss LA (1985) Red cell organic phosphates and Bohr effects in house-sparrow blood. Respir Physiol 59:93–103
Malte H, Weber RE (1985) A mathematical model for gas exchange in the fish gill based on non-linear blood gas equilibrium curves. Respir Physiol 62:359–374
Malvin GM (1988) Microvascular regulation of cutaneous gas exchange in amphibians. Am Zool 28:999–1007
Malvin GM, Hlastala MP (1986) Regulation of cutaneous gas exchange by environmental O_2 and CO_2 in the frog. Respir Physiol 65:99–111
Maren TH (1967) Carbonic anhydrase: chemistry, physiology and inhibition. Physiol Rev 47:595–781

Maren TH (1988) The kinetics of HCO_3^- synthesis related to fluid secretion, pH control and CO_2 elimination. Annu Rev Physiol 50:695–717
Maren TH, Swenson ER (1980) A comparative study of the kinetics of the Bohr effect in vertebrates. J Physiol 303:535–547
Maren TH, Friedland BR, Rittmaster RS (1980) Kinetic properties of primitive vertebrate carbonic anhydrases. Comp Biochem Physiol 65 B:69–74
Meldon JH (1987) Blood-gas equilibria, kinetics and transport. Chem Eng Sci 42:199–211
Meldon JH, Stroeve P, Gregoire CE (1982) Facilitated transport of carbon dioxide: a review. Chem Eng Commun 16:263–300
Middleman S (1972) Transport phenomena in the cardiovascular system. Wiley-Interscience, New York
Miyamoto Y, Moll W (1971) Measurements of dimensions and pathway of red cells in rapidly frozen lungs in situ. Respir Physiol 12:140–156
Miyamoto Y, Moll W (1972) The diameter of red blood cells when flowing through a rapid-reaction apparatus. Respir Physiol 16:259–266
Mochizuki M, Shibuya I, Uchida K, Kagawa T (1987) A method of estimation contact time of red blood cells through lung capillary from O_2 and CO_2 concentration in rebreathing air in man. Jpn J Physiol 37:283–301
Moll W (1968/69) The influence of hemoglobin diffusion on oxygen uptake and release by red cells. Respir Physiol 6:1–15
Mondrup M, Anker N (1976) Carbonic anhydrase isoenzymes B in erythrocytes of subjects with different types of anemia. Clin Chim Acta 69:463–469
Moynihan (1977) Carbonic anhydrase activity in mammalian skeletal and cardiac muscle. Biochem J 168:565–569
Nechay BR, Larimer JL, Maren TH (1960) Effects of drugs and physiologic alterations on nasal salt excretion in sea gulls. J Pharmacol Exp Ther 130:401–410
Nguyen-Phu D, Yamaguchi K, Scheid P, Piiper J (1986) Kinetics of oxygen uptake and release by red blood cells of chicken and ducks. J Exp Biol 125:15–27
Nicolson P, Roughton FJW (1951) A theoretical study of the influence of diffusion and chemical reaction velocity on the rate of exchange of carbon monoxide and oxygen between the red corpuscle and surrounding fluid. Proc R Soc Lond 138:241–268
Nioka S, Henry R, Forster RE (1988) Total CA activity in isolated perfused guinea pig lung by 18-O exchange method. J Appl Physiol 65:2236–2244
Obaid AL, Crandall ED (1979) HCO_3^-/Cl^- exchange across the human erythrocyte membrane. Effects of pH and temperature. J Membr Biol 50:23–41
Obaid AL, Critz AM, Crandall ED (1979) Kinetics of bicarbonate/chloride exchange in dogfish erythrocytes. Am J Physiol 237:R 132–R 138
O'Brasky JE, Crandall ED (1980) Organ and species differences in tissue vascular carbonic anhydrase activity. J Appl Physiol 49:211–217
Ono T, Tazawa H (1975) Microphotometric method for measuring the oxygenation and deoxygenation rate in a single blood cell. Jpn J Physiol 25:93–107
Paiva M, Engel LA (1985) Model analysis of intra-acinar gas exchange. Respir Physiol 62:257–272
Perry SF (1986) Carbon dioxide excretion in fishes. Can J Zool 64:565–572
Perry SF, Davis PS, Daxboeck C, Randall DJ (1982) A comparison of CO_2 excretion in a spontaneously ventilating blood-perfused trout preparation and saline-perfused gill preparations: contribution of the branchial epithelium and red blood cell. J Exp Biol 101:47–60
Piiper J (1982) Respiratory gas exchange at lungs, gills, and tissues: mechanisms and adjustments. J Exp Biol 100:5–22
Piiper J (1986) Blood-gas equilibrium of carbon dioxide in lungs: a continuing controversy. J Appl Physiol 60:1–8
Piiper J, Scheid P (1975) Gas transport efficacy of gills, lungs and skin: theory and experimental data. Respir Physiol 23:209–221
Piiper J, Scheid P (1980) Blood-gas equilibration in lungs. In: West JB (ed) Pulmonary gas exchange, vol I. Academic Press, New York, pp 121–171
Piiper J, Scheid P (1981) Model for capillary-alveolar equilibration with special reference to O_2 uptake in hypoxia. Respir Physiol 46:193–208

Piiper J, Scheid P (1984) Model analysis of gas transfer in fish gills. In: Hoar WS, Randall DJ (eds) Fish physiology, vol 10, Part A. Academic Press, New York London, pp 230–262
Piiper J, Scheid P (1987) Diffusion and convection in intrapulmonary gas mixing. In: Fahri LE, Tenney SM (eds) Handbook of physiology: the respiratory system, vol IV. American Physiological Society, Bethesda, pp 51–69
Piiper J, Meyer M, Marconi C, Scheid P (1980) Alveolar-capillary equilibration kinetics of $^{13}CO_2$ in human lungs studied by rebreathing. Respir Physiol 42:29–41
Piiper J, Meyer M, Scheid P (1984) Dual role of diffusion in tissue gas exchange: blood tissue equilibration and diffusion shunt. Respir Physiol 56:131–144
Piiper J, Scheid P, Perry SF, Hughes GM (1986) Effective and morphometric oxygen-diffusing capacity of the gills of the elasmobranch *Scyliorhinus stellaris*. J Exp Biol 123:27–41
Ponte J, Purves MJ (1980) Changes in pH and PCO_2 with time in pulmonary post-capillary blood in cats. In: Bauer C, Gros G, Bartels H (eds) Biophysics and physiology of carbon dioxide. Springer, Berlin Heidelberg New York, pp 315–320
Powell FL, Mazzone RW (1983) Morphometrics of rapidly frozen goose lungs. Respir Physiol 51:319–332
Powell FL, Wagner PD (1982) Ventilation-perfusion inequality in avian lungs. Respir Physiol 48:233–241
Powell FL, Hastings RH, Mazzone RW (1985) Pulmonary vascular resistance during unilateral pulmonary arterial occlusion in ducks. Am J Physiol 18:R39–R43
Rahim SM, Delany JP, Larent P (1988) Identification and immunocytochemical localization of two different carbonic anhydrase isoenzymes in teleosestean fish erythrocytes and gill epithelium. Histochemistry 89:451–459
Rahn H, Paganelli C (1982) Role of diffusion in gas exchange of the avian egg. Fed Proc 41:2134–2136
Randall D, Daxboeck C (1984) Oxygen and carbon dioxide transfer across fish gills. In: Hoar WS, Randall DJ (eds) Fish physiology, vol 10, Part A. Academic Press, New York London, pp 263–314
Randall DJ, Farrell AP, Haswell MS (1978) Carbon dioxide excretion in the jeju, *Hoplerythrinus unitaeniatus*, a facultative air-breathing teleost. Can J Zool 56:970–973
Randall RF, Maren TH (1972) Absence of carbonic anhydrase in membranes. Biochim Biophys Acta 268:730–732
Reeves RB, Heidelberger E (1987) Rates of oxygen uptake and release of naked films of human whole blood one red cell in thickness. Physiologist 30:A231
Rice SA (1980) Hydrodynamic and diffusion considerations of rapid-mix experiments with red blood cells. Biophys J 29:65–78
Rispens P, Hessels J, Zwart A, Zijlstra WG (1985) Inhibition of carbonic anhydrase in dog plasma. Pflügers Arch 404:344–347
Rispens R, Oeseburg B, Zock JP, Zijlstra WG (1980) Intra-aortic decrease in blood plasma pH. Pflügers Arch 386:97–99
Robertson HT, Whitehead J, Hlastala MP (1986) Diffusion-related differences in the elimination of inert gases from the lung. J Appl Physiol 61:1162–1172
Romano L, Passow H (1984) Characterization of anion transport system in trout red blood cell. Am J Physiol 15:C330–C338
Rosen S, Friedley NJ (1973) Carbonic anhydrase activity in *Rana pipiens* skin: Biochemical and histochemical analysis. Histochemie 36:1–4
Rotman H, Ikeda I, Chiu CS, Kramer E, Aminoff D (1980) Resistance of red blood cell membrane to oxygen uptake. J Appl Physiol 49:306–310
Roughton FJW (1935) Recent work on carbon dioxide transport by the blood. Physiol Rev 15:241–296
Roughton FJW, Forster RE (1957) Relative importance of diffusion and chemical reaction rates in determining rate of O_2 exchange of pulmonary membrane and volume of blood in lung capillaries. J Appl Physiol 11:290–302
Ryan US, Whitney PL, Ryan JW (1982) Localization of carbonic anhydrase on pulmonary artery endothelial cells in culture. J Appl Physiol 53:914–919
Sanyal G, Swenson ER, Maren TH (1984) The isolation of carbonic anhydrase from the muscle of *Squalus acanthias* and *Scomber scombrus*. Bull MDIBL 24:66–68

Scheid P (1978) Analysis of gas exchange between air capillaries and blood capillaries in avian lungs. Respir Physiol 32:27–49

Scheid P, Siffert W (1985) Effects of inhibiting carbonic anhydrase on isometric contraction of frog skeletal muscle. J Physiol 361:91–101

Scheid P, Hlastala MP, Piiper J (1981) Inert gas elimination from lungs with stratified inhomogeneity: theory. Respir Physiol 44:199–309

Scheid P, Hook C, Piiper J (1986) Model for analysis of counter-current gas transfer in fish gills. Respir Physiol 64:365–374

Schuster KD (1985) Kinetics of pulmonary CO_2 transfer studied by using labeled carbon dioxide $C^{16}O^{18}O$. Respir Physiol 60:21–37

Schuster KD (1987) Diffusion limitation and limitation by chemical reactions during alveolar-capillary transfer of oxygen-labeled CO_2. Respir Physiol 67:13–22

Schuster KD, Meister R, Nanassy E, Heller H, Pusch C (1988) Diffusing capacity of 18-O labeled carbon dioxide between alveolar gas and blood during ergometer work. Pflügers Arch 411:(Suppl 1) R55

Sha'afi RI, Rich GI, Sidel VN, Bossent W, Solomon AK (1967) The effect of unstirred layers on human red cell water permeability. J Gen Physiol 50:1377–1399

Shields A, Jeffrey S, Wilson C, Carter N (1984) Radioimmunoassay of carbonic anhydrase III in rat tissue. Biochem J 281:281–284

Silverman DN, Tu C, Wynns GC (1976) Depletion of ^{18}O from $C^{18}O_2$ in erythrocyte suspensions. J Biol Chem 251:4428–4434

Silverman DN, Tu C, Wynns GC (1978) Proton transfer between hemoglobin and the carbonic anhydrase active site. J Biol Chem 253:2563–2567

Silverman DN, Tu CK, Roessler N (1981) Diffusion-limited exchange of ^{18}O between CO_2 and water in red cell suspensions. Respir Physiol 44:285–298

Sinha AK (1969) O_2 uptake and release by red cells through capillary wall and plasma layer. PhD Thesis University of California, San Francisco. (Xeron Univ Microfilms) Ann Arbor, Michigan

Sirs JA (1970) The interaction of carbon dioxide with the rate of exchange of oxygen by red blood cells. In: Hersey D (ed) Blood oxygenation. Plenum, New York, pp 116–136

Sly WS, Whyte V, Tashian RE, Hewett-Emmett D, Guibaud P, Vainsel M, Baluarte HJ, Gruskin A, Al-Mosawi M, Sakati N, Ohlsson A (1985) Carbonic anhydrase II deficiency in 12 families with the autosomal recessive syndrome of osteopetrosis with renal tubular acidosis and cerebral calcification. New Eng J Med 313:139–145

Smeda JS, Houston AH (1979) Carbonic anhydrase (Acetazolamide-sensitive esterase) activity in the red blood cells of thermally-acclimated rainbow trout, *Salmo gairdneri*. Comp Biochem Physiol 62A:719–723

Smith RG (1988) Inorganic carbon transport in biological systems. Comp Biochem Physiol 90B:639–654

Stein TR, Martin JC, Keller KH (1971) Steady-state oxygen transport through red blood cell suspensions. J Appl Physiol 31:397–402

Strome DR, Clancy RL, Gonazalez NC (1976) Myocardial CO_2 buffering: role of transmembrane transport of H^+ or HCO_3^- ions. Am J Physiol 230:1037–1041

Sugai N, Ninomiya Y, Oosaki T (1981) Localization of carbonic anhydrase in the rat lung. Histochemistry 72:415–424

Swenson ER (1984) The respiratory aspects of carbonic anhydrase. Ann NY Acad Sci 429:547–560

Swenson ER, Evans DH (1984) Inhibition of gill carbonic anhydrase blocks CO_2 excretion in the isolated perfused head of the dogfish "pup", *Squalus acanthias*. Bull MDIBL 24:45

Swenson ER, Maren TH (1978) A quantitative analysis of CO_2 transport at rest and during maximal exercise. Respir Physiol 35:129–159

Swenson ER, Maren TH (1984) Effects of acidosis and carbonic anhydrase inhibition in the elasmobranch rectal gland. Am J Physiol 247:F86–F92

Swenson ER, Maren TH (1987) Roles of gill and red cell carbonic anhydrase in elasmobranch HCO_3^- and CO_2 excretion. Am J Physiol 253:R450–R458

Swenson ER, Rhodes CG, Araujo L, Pike V, LeBars D, Effros RM, Hughes JMB (1989) In vivo localization of tissue carbonic anhydrase studied with positron emission tomography. Physiologist 32:210

Swenson ER, Robertson HT, Middaugh ME, Hlastala MP (1986) Carbonic anhydrase inhibition impairs ventilation-perfusion matching. Proc XXX Inter Cong Physiol Sci XVI:179

Tanashita K, Tanasawa I, Yamaguchi T, Sugawara M (1985) Facilitated diffusion of carbon dioxide in whole blood and hemoglobin solutions. Pflügers Arch 405:83–90

Tazawa H, Mochizuki M (1976) Rates of oxygenation and Bohr shift of capillary blood in chick embryos. Nature 261:509–511

Tazawa H, Ono T, Mochizuki M (1976) Oxygenation and deoxygenation velocity factors of chorioallantoic capillary blood. J Appl Physiol 40:399–403

Tenney SM (1974) A theoretical analysis of the relationship between venous blood and mean tissue oxygen pressures. Respir Physiol 30:283–296

Thews G (1959) Untersuchung der Sauerstoffaufnahme und -abgabe sehr dünner Blutlamellen. Pflügers Arch 268:308–317

Toews D, Boutilier R, Todd L, Fuller N (1978) Carbonic anhydrase membrane determined with [^{14}C] bicarbonate. J Physiol 294:521–539

Tokuda S, Iiboshi A, Suenaga M, Otsuji S (1984) Changes in erythrocyte carbonic anhydrase activity due to physical exercise. Eur J Appl Physiol 52:249–254

Troug WE, Hlastala MP, Standaert TA (1979) Oxygen-induced alteration of ventilation-perfusion relationships in rats. J Appl Physiol 47:1112–1117

Tu C, Wynns GC, McMurray RE, Silverman DN (1978) CO_2 kinetics in red cell suspensions measured by ^{18}O exchange. J Biol Chem 253:8178–8184

Tufts BL, Boutilier RG (1989) The absence of rapid chloride/bicarbonate exchange in lamprey erythrocytes: Implications for CO_2 transport and ion distributions between plasma and erythrocytes of Petromyzon marinus. J Exp Biol 144:565–576

Tyuma I (1984) The Bohr effect and the Haldane effect in human hemoglobin. Jpn J Physiol 34:205–216

Vandergriff KD, Olson JS (1984a) The kinetics of O_2 release by human red blood cells in the presence of external sodium dithionite. J Biol Chem 259:12609–12618

Vandergriff KD, Olson JS (1984b) Morphological and physiological factors affecting oxygen uptake and release by red blood cells. J Biol Chem 259:12619–12627

Vandergriff KD, Olson JS (1984c) A quantitative description in three dimensions of oxygen uptake by human red blood cells. Biophys J 45:825–835

Wagner PD (1977) Diffusion and chemical reaction of pulmonary gas exchange. Physiol Rev 57:257–312

Wagner PD, West JB (1972) Effects of diffusion impairment on O_2 and CO_2 time courses in pulmonary capillaries. J Appl Physiol 33:62–71

Wagner PD, Gale GE, Moon RE, Torre-Bueno JR, Stolp BW, Saltzman HA (1986) Pulmonary gas exchange in humans exercising at sea level and simulated altitude. J Appl Physiol 61:260–270

Wagner PD, Gillespie JR, Landgren GL, Fedde MR, Jones BW, DeBowes RM, Pieschl RL, Erickson HH (1988) Mechanism of exercise-induced hypoxemia in the horse. J Appl Physiol 66:1227–1233

Weibel ER (1984) The pathway for oxygen: structure and function in the mammalian respiratory system. Harvard University Press, Cambridge

West JB, Wagner PD (1980) Predicted gas exchange on the summit of Mt Everest. Respir Physiol 42:1–16

Whitney PL, Briggle TV (1982) Membrane-associated carbonic anhydrase purified from bovine lung. J Biol Chem 257:12056–12059

Wieth JO (1979) Bicarbonate exchange through the human red cell in the amphibia. Comp Biochem Physiol 59A:211–213

Wieth JO, Anderson OS, Brahm J, Bjerrum PJ, Borders CL (1982) Chloride-bicarbonate exchange in red blood cells: physiology of transport and chemical modification of binding sites. Philos Trans R Soc Lond B 299:383–399

Wildeboer-Venema F (1984) Influence of nitrogen, air and alveolar gas upon surface tension of lung surfactant. Respir Physiol 58:1–14

Williams JB, Kutchai H (1986) Use of a membrane-bound fluorophore to characterize diffusion boundary layers around human erythrocytes. Biophys J 49:453–458

Wistrand PJ (1981) The importance of carbonic anhydrase B and C for the unloading of CO_2 by the human erythrocyte. Acta Physiol Scand 113:417–426

Wittenberg JB, Wittenberg BA (1962) Active secretion of oxygen into the eye of fish. Nature 194:106–107

Wright PA, Heming T, Randall D (1986) Downstream pH changes in water flowing over the gills of rainbow trout. J Exp Biol 126:499–512

Wright PA, Randall DJ, Perry SF (1988) Fish gill water boundary layer: a site of linkage between carbon dioxide and ammonia excretion. J Comp Physiol B 158:627–635

Wyman J (1964) Linked functions and reciprocal effects in hemoglobin: a second look. Adv Protein Chem 19:223–286

Yamaguchi K, Nguyen-Phu D, Scheid P, Piiper J (1985) Kinetics of O_2 uptake and release by human erythrocytes studied by a stopped-flow technique. J Appl Physiol 58:1215–1224

Yamaguchi K, Glahn J, Scheid P, Piiper J (1987) Oxygen transfer conductance of human red blood cells at varied pH and temperature. Respir Physiol 65:209–223

Yamagucki K, Glass ML, Scheid P, Piiper J (1989) Oxygen transfer kinetics of red blood cells of the turtle Pseudemys scripta elegans. Respir Physiol 75:371–384

Zander R, Schmid-Schönbein H (1973) Intracellular mechanisms of oxygen transport in flowing blood. Respir Physiol 19:279–289

Zbrorowska-Sluis DT, L'Abbate A, Klassen GA (1974) Evidence of carbonic anhydrase activity in skeletal muscle: a role for facilitative carbon dioxide transport. Respir Physiol 21:341–350

Zborowska-Sluis DT, L'Abbate A, Milderberger RR, Klassen GA (1975) The effect of acetazolamide on myocardial carbon dioxide space. Respir Physiol 23:311–316

Zock JP, Rispens P, Zijlstra WG (1981) Calculated changes in pH and P_{CO_2} in arterial blood plasma assuming absence of ion and water exchange between plasma and erythrocytes during their equilibration with alveolar gas. Pflügers Arch 391:159–161

Chapter 7

Acid-Base Regulation: Interrelationships Between Gaseous and Ionic Exchange

N. Heisler[1]

Contents

1 Introduction

The metabolic activity required for energy production in vertebrate tissues results in a number of metabolic end products, which have to be eliminated in order to avoid the disturbing and toxic effects of accumulation. Most of those end pro-

[1] Max-Planck-Institut für experimentelle Medizin, Abteilung Physiologie, Hermann Rein Str. 3, D-3400 Göttingen, FRG

Advances in Comparative and Environmental Physiology
Vol. 6. Edited by R. G. Boutilier

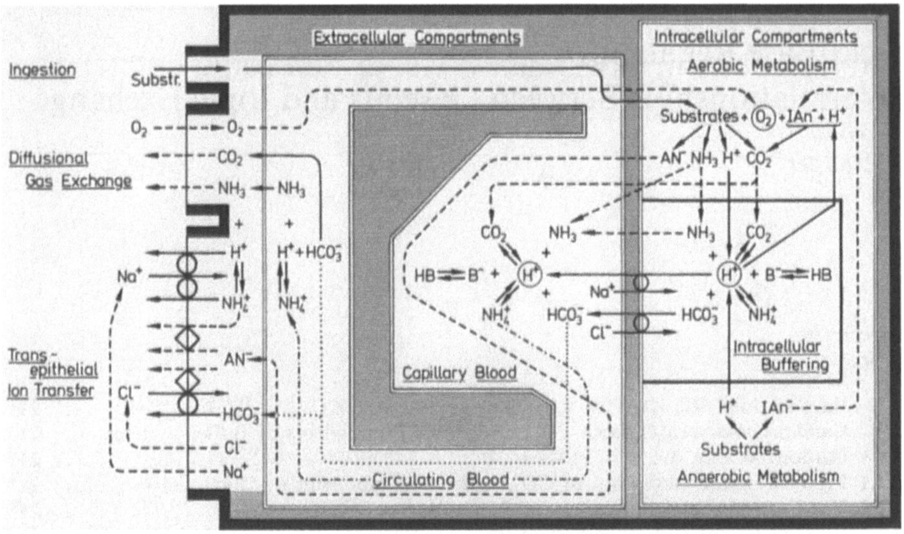

Fig. 1. Highly schematic general representation of the interaction of gaseous metabolic end products, CO_2 and NH_3, with the ionic regulation in intracellular and extracellular body fluid compartments, during transport in the circulating blood, with the diffusional gas exchange and with transmembrane and transepithelial ion transfer processes via their respective buffer equilibria. For details, see text of the respective sections

ducts are active in terms of acid-base regulation in one way or another, and are non-volatile ionic substances. The few volatile, non-dissociated end products, however, represent on a molar basis by far the largest fraction of the overall production. CO_2 as the quantitatively most important end product of aerobic glycolysis, and ammonia as the main nitrogenous metabolic end product of ammoniotelic organisms (particularly teleost fishes and aquatic amphibia), interact and affect the acid-base status via buffer reactions, resulting in dissociation to, or removal of H^+ ions from the body fluids (Fig. 1). These ionization processes take place at the site of production in tissue cells or in the capillary blood, but may be reversed for diffusive nonionic transfer at certain sites of the transport chain (e.g. cell and vascular membranes) to the environment. The elimination of such volatile, though ionizable substances, will therefore affect the ionic composition and the acid-base status at various sites of the organism, following the criteria of the respective buffer reactions (solubility coefficients, pK values, association/dissociation contants, etc.). Changes in the elimination of acid-base relevant volatile metabolic end products may cause vast disturbances of the acid-base status of body fluids. Even during normal steady-state conditions the unevenly distributed production of acid-base relevant volatile substances may locally induce limiting border conditions for mechanisms of acid-base-relevant ionic regulation. This chapter is intended to delineate the interrelationships and the interaction of acid-base relevant ionic regulatory mechanisms with production, ionization, intracorporeal transport and the elimination of gaseous metabolic end products from the body fluids.

2 Equilibria Between Gaseous and Nonvolatile Components of Buffer Systems

2.1 Carbon Dioxide

2.1.1 Production and Ionization of CO_2, and the Components of 'Total CO_2'

During aerobic steady-state energy production, CO_2 is produced in the organism at a fractional molar rate of 0.7 to 1 of the oxygen consumption (respiratory ratio), depending on the type of metabolic substrate. This amount is generally more than an order of magnitude larger than the net load of nonvolatile acidic equivalents (which is, with very few exceptions, less than 2% of the molar CO_2 production, cf. Heisler 1984, 1986c; 1988a), a fraction emphasizing the predominant role of CO_2 as an acid-base relevant metabolic end product.

CO_2 is in equilibrium with the common pool of H^+ ions and with bicarbonate via its hydrated counterpart, H_2CO_3. Bicarbonate in turn is in equilibrium with carbonate and sodium carbonate:

$$CO_2+H_2O \rightleftarrows H_2CO_3 \rightleftarrows H^+ + HCO_3^- \rightleftarrows 2H^+ + \underset{\displaystyle Na^+}{\underset{+}{CO_3^{2-}}} \rightleftarrows NaCO_3^- \qquad (1)$$

B (spanning $H_2CO_3 \rightleftarrows H^+ + HCO_3^-$ to CO_3^{2-})

A (spanning $CO_2+H_2O \rightleftarrows H_2CO_3 \rightleftarrows H^+ + HCO_3^-$) — C (spanning $CO_3^{2-} \rightleftarrows NaCO_3^-$)

Since the hydrated fraction (H_2CO_3) represents only about 0.3% of the physically dissolved CO_2 ($pK_h = -\log([H_2CO_3]/[CO_2]) \approx 2.6$; Edsall and Wyman 1958), both components and the pK values of their reactions are usually lumped together, resulting in an overall pK value of slightly higher than 6 for the combined hydration/dissociation reaction of CO_2 to H^+ ions and bicarbonate. The other two main dissociation/association products (CO_3^{2-} and $NaCO_3^-$) have similar effects for buffer reactions as bicarbonate, and, for reasons of simplicity, are treated as part of the bicarbonate component in this context. The condensed relationship between gaseous CO_2, pH and 'bicarbonate' concentration is expressed by the common form of the Henderson-Hasselbalch equation:

$$pH = pK_1''' + \log \frac{[HCO_3^-]'}{[CO_2]'} = pK_1''' + \log \frac{[HCO_3^-]'}{\alpha_{CO_2} \cdot P_{CO_2}}, \qquad (2)$$

where $[HCO_3^-]' = [HCO_3^-] + [CO_3^{2-}] + [NaCO_3^{2-}]$; $[CO_2]' = \alpha_{cCO_2} \cdot P_{CO_2}$, and represents the sum of hydrated and nonhydrated physically dissolved CO_2.

As long as referred to the free water phase, α is a function of only temperature and the molarity of the fluid system (Heisler 1984[2], 1986b):

[2] Note: the last line term of the α-formula is misprinted in Heisler (1984) and in Boutilier et al. (1984) and should read '+'.

$$\alpha_{CO_2} = 0.1008 - 29.80\times10^{-3}\,[M] + (1.218\times10^{-3}\,[M] - 3.639\times10^{-3})\,t - (19.57\times10^{-6}\,[M] - 69.59\times10^{-6})\,t^2 + (71.71\times10^{-9}\,[M] - 559.6\times10^{-9})\,t^3 \quad [\text{mmol/(L}\cdot\text{mmHg)}]\ , \qquad (3)$$

where: t = temperature (°C, range 0°–40 °C);
M = molarity of dissolved species (mol/L) referred to the volume of solute water.

pK_1' is affected, in addition to temperature, by a number of physicochemical parameters, particularly by ionic strength and sodium concentration. Other ions have a significant effect only if their concentration largely exceeds the typically low values in biological systems (e.g. Mg^{2+} and Ca^{2+} in long-term submerged turtles; Jackson and Heisler 1982, 1983). If the $[HCO_3^-]$ is determined as 'total CO_2 minus physically dissolved CO_2, thus including CO_3^{2-} and $NaCO_3^-$, and the CO_2 directly bound to terminal amino groups (carbamino-CO_2), then pK_1''' is also a function of pH and protein concentration (Heisler 1986b):

$$pK'''_{1\,app} = \underbrace{6.583 - 13.41\times10^{-3}\,t + 228.2\times10^{-6}\,t^2 - 1.516\times10^{-6}\,t^3 - 0.341\,I^{0.323}}_{A}$$
$$\underbrace{-\log\{1 + 0.00039\,[Pr] + (10^{pH - 10.64 + 0.011\,t + 0.737\,I^{0.323}})}_{B}$$
$$\underbrace{\times\,(1 + 10^{1.92 - 0.01\,t - 0.737\,I^{0.323} + \log[Na^+] + (-0.494\,I + 0.651)\,(1 + 0.0065\,[Pr])})\}}_{C}\ , \qquad (4)$$

where t = temperature (°C, range 0°–40 °C),
I = ionic strength of nonprotein ions ($I = 0.5\ \Sigma\ ([x]\,Z^2)$, with [x] = concentration in mol/L and Z = number of charges of the respective ion,
$[Na^+]$ = sodium concentration (mol/L),
[Pr] = protein concentration (g/L).

Term A represents the pK_1' for bicarbonate only, terms B and C account for the contributions of carbonate and $NaCO_3^-$ according to the correspondingly labelled reactions in Eq. (1).

According to these buffer equilibria and their respective constants, 'total CO_2' is represented at physiological mammalian plasma pH and $[Na^+]$ (7.4, 140 mM) mainly by bicarbonate (relative concentrations):

CO_2	H_2CO_3	HCO_3^-	CO_3^{2-}	$NaCO_3^-$
0.053	1.7×10^{-4}	1	0.004	0.034 .

At higher pH and $[Na^+]$ (e.g. pH 8.5 and 540 mM $[Na^+]$ of normal seawater) the concentrations of carbonate and $NaCO_3^-$ represent a sizable fraction of the total CO_2 content. Their concentration is then larger than that of physically dissolved CO_2 and the CO_2 output per unit respiratory water flow through fish gills, and may thus play a significant role in buffering of CO_2 eliminated from aquatic species:

CO_2	H_2CO_3	HCO_3^-	CO_3^{2-}	$NaCO_3^-$
0.002	7×10^{-6}	1	0.053	0.739

2.1.2 The CO_2/Bicarbonate System and Nonbicarbonate Buffers

As indicated by the equilibria of a pure CO_2/bicarbonate buffer system outlined above, changes in the concentration of physically dissolved CO_2 will result in dissociation of H^+ ions and bicarbonate into equimolar amounts. Therefore, changes due to dissociation of CO_2 in the concentration of free H^+, bicarbonate, and of physically dissolved CO_2, are limited to around 10^{-7} M in the physiological pH range, which is 4 to 5 orders of magnitude less than the in vivo [HCO_3^-]. Only with nonbicarbonate buffers (B_1^-, B_2^-) being in equilibrium with the CO_2/bicarbonate system will the transfer of free H^+ ions from the common pool into the nondissociated state, or vice versa, attain an appreciable extent by shifting the reaction to the right:

$$\begin{array}{rcl} & & B_1^- \rightleftarrows HB_1 \\ & & + \\ CO_2 + H_2O & \rightleftarrows & H^+ + HCO_3^- \\ & & + \\ & & B_2^- \rightleftarrows HB_2 \; . \end{array} \tag{5}$$

The changes in bicarbonate concentration are equimolar to the changes in buffer ions ($-\Delta B^-_{total}$ or ΔHB_{total}) and are utilized as an indicator for the nonbicarbonate buffer value (β_{NB}), determined by CO_2 titration (e.g. Heisler and Piiper 1971, 1972) on the basis of the appropriately extended (cf. Heisler 1986b) original definition by Van Slyke (1922):

$$\beta_{NB} = \frac{\Delta[H^+]_b}{-\Delta pH} = \frac{\Delta[HB]}{-\Delta pH} = \frac{-\Delta[B^-]}{-\Delta pH} = \frac{\Delta[HCO_3^-]}{-\Delta pH} \; . \tag{6}$$

In turn, with nonbicarbonate buffer values of biological fluids in the range of 35–70 mEq/(pH·L) in intracellular tissue compartments, and 5–30 mEq/(pH·L) in blood (for review: Heisler 1986b, e), considerable changes in [HCO_3^-] have to be expected during changes in P_{CO_2}, a factor of some importance in context with the mechanism of facilitated diffusion (see below). The interaction of the CO_2/bicarbonate buffer system with nonbicarbonate buffers is particularly advantageous when nonvolatile H^+ ions are to be buffered in a buffer system closed for molecular CO_2 (see below).

2.1.3 Open vs Closed CO_2/Bicarbonate Buffer System

The overall concentration (ionized plus nonionized components) of a typical nonbicarbonate buffer system is maintained constant during buffering, whereas the volatile component of the CO_2/bicarbonate system is adjusted by ventilation of the gas exchange system (or by changes in skin perfusion in skin-breathers). This mechanism allows the CO_2/bicarbonate system, which is principally inefficient at physiological pH values (pH − pK typically ≈ 1–1.5 units), to buffer nonvola-

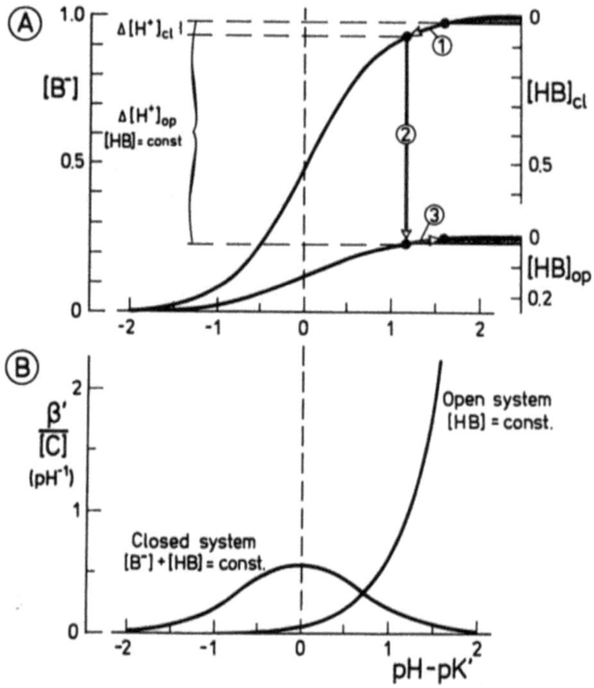

Fig. 2. Characteristics of the CO_2/bicarbonate buffer system. *A* Buffer curves for closed (index *cl*, $[CO_2]+[HCO_3^-] = \text{const.}$) and open-system buffering (index *op*, $[CO_2]$ variable). *B* Specific buffer value as a function of the difference between pH and pK′ for open- and closed-system buffering. For a given change in pH, buffering of nonvolatile H^+ ions in a closed system is rather poor *(1)* due to the large difference of pH from the pK′ value (cf. *B*). Keeping the concentration of the acid form ([HB] = $[CO_2]$, *shaded*) constant (by removal of molecular CO_2 produced from HCO_3^- and H^+ ions) reduces the total buffer concentration (cf. y-axis), and leads to a tremendous rise in buffering capability (ΔH^+_{op}, *(1)+(2)* opposed to ΔH^+_{cl}, *(1)*, cf. *B*). The original pH can be restored by reducing *(3)* the concentration of the acid form of the buffer system ([*HB*]) by the same factor as the reduction of the base form ($B^- = [HCO_3^-]$). For details, see text

tile H^+ ions to a great extent and still restore pH to values close to the original by reduction of the volatile component, CO_2, to appropriately lower values (Fig. 2, *(3)*). The effective specific buffer value (buffer value/buffer concentration, cf. Heisler 1986b) of the closed CO_2/bicarbonate system is $<0.1\ pH^{-1}$ at a pH 1.5 units higher than pK'_1, whereas the system open for CO_2 will provide about 1.8 pH^{-1}, when P_{CO_2} is kept constant (Fig. 2). Even more effective pH regulation is achieved when P_{CO_2} is reduced by compensatory hyperventilation (Fig. 2, *(3)*; cf. Woodbury 1965). This mechanism is particularly important and effective in the extracellular fluid compartments, because of the much higher initial bicarbonate concentration, which obviously is one of the limiting factors for buffering of H^+ ions by this system (Fig. 2; cf. Heisler 1986b). Under certain conditions, and in certain body fluid compartments, the bicarbonate buffer system reacts as a closed system, due to the impossibility of removing CO_2. This causes a severe reduction in the buffer value (cf. Fig. 2) and may lead to considerable limitations in the elimination of surplus nonvolatile H^+ ions (see below).

2.1.4 The Role of Carbonic Anhydrase

Conversion of volatile CO_2 into the various nonvolatile and acid-base-relevant buffer products outlined above takes place via hydration to H_2CO_3 and dissociation into HCO_3^- and H^+ ions. Hydration and dehydration of CO_2 is a relatively slow process with a half time in the order of minutes (cf. Kern 1960), depending on temperature and pH. Such kinetics are certainly too slow to allow any significant conversion of CO_2 into nonvolatile components of the buffer system during the very limited contact time between blood and tissues. In vivo, however, the process is considerably speeded up (e.g. in the order of 0.01 s in mammalian blood) by the action of carbonic anhydrase, an enzyme rather ubiquitously distributed in vertebrate erythrocytes, but also available, at least to a certain extent, in intracellular and extracellular spaces of certain tissues (for review: Gros and Dodgson 1988). The availability of carbonic anhydrase is indispensable not only for the convective transport of CO_2 by the blood to the elimination site, but because it has a key role for the mechanism of facilitated diffusion of CO_2 (see below).

2.2 Ammonia: Production Rate and Characteristics of Ionization

Ammonia is produced as main nitrogenous metabolic end product in ammoniotelic animals (e.g. teleost fishes) in a relative amount of about 12% of the oxygen consumption, (cf. Heisler 1984, 1986c, 1988a), which is about 60%–90% of their nitrogenous waste (Smith 1929; Wood 1958; Fromm 1963). The net production of the volatile ammonia is generally larger than the net production of nonvolatile acid-base-relevant end products in aquatic and semi-aquatic animals (Heisler 1988a; re: amphibia: Heisler et al. 1982; Boutilier and Heisler 1988; re: fishes: Heisler 1986c). This even holds for ureotelic aquatic organisms like the elasmobranchs (Heisler 1984, 1986c), whereas the reverse applies for terrestrial animals (cf. Heisler 1988a). This is likely related to the more favourable conditions for ammonia elimination via gills and skin by the rather unlimited volume of environmental water in contact with the elimination site (Heisler 1984), and the high aqueous solubility of NH_3 and its dissociated counterpart, NH_4^+ (Cameron and Heisler 1983, 1985). In contrast, voiding of ammonia is limited in terrestrial animals by factors like small urine flow rate, and the overall net renal H^+ ion excretion (Heisler 1984).

After production by deamination of α-amino acids (mainly glutamine) at various sites like liver, kidneys and gills (Goldstein et al. 1964; Pequin 1962; Pequin and Serfaty 1963; Aster et al. 1982; Cameron and Heisler 1983), ammonia is ionized by uptake of H^+ ions from the body fluids according to the Henderson-Hasselbach equation:

$$pH = pK' + \log \frac{[NH_3]}{[NH_4^+]} = pK' + \log \frac{\alpha_{NH_3} \cdot P_{NH_3}}{[NH_4^+]} \ . \qquad (7)$$

The pK′ value for this reaction is in the range of 9 to 9.8, depending on temperature, salinity and protein content (nomogram provided by Cameron and Heisler

1983). Due to this high pK′ value, about 99% of the ammonia is stored in biological fluids in the form of NH_4^+. The solubility of ammonia, α_{NH_3}, is larger than that for CO_2 by a factor of about 1000 in aqueous fluids (Cameron and Heisler 1983). According to this fact and to the high pK, the plasma P_{NH_3} at physiological pH values and ammonium concentrations in fishes is in the range of only about 50×10^{-6} mmHg (e.g. Cameron and Heisler 1983). In contrast to the conditions for the CO_2/bicarbonate system (see above), ionization of ammonia takes place instantaneously, and is not limited by any catalyzing enzyme.

3 Transfer from the Site of Production to the Site of Elimination

3.1 Diffusion of Volatile and Ionic Components

3.1.1 CO_2 and HCO_3^-

Molecular CO_2 is undoubtedly the form in which metabolically produced CO_2 is transferred across cell membranes. The passive permeability of cell membranes (except erythrocytes) for charged substances like HCO_3^- and H^+ ions is too low to allow any significant transfer of such dissociation products. Another factor to be considered in this context is the minute diffusion gradient for H^+ ions across the cell membrane, which renders simultaneous diffusion of HCO_3^- and H^+ ions (required for CO_2 elimination) impossible. Involvement of active, energy-consuming mechanisms cannot be taken into consideration because of quantitative and energetic aspects. Such mechanisms for H^+ and HCO_3^- certainly have an important role in the elimination of nonvolatile metabolic end products, but, being carrier-mediated, they are usually limited in their capacity to much lower transfer rates than required for ionic elimination of the aerobic metabolic end product, CO_2.

Diffusive movement of substances follows Fick's diffusion equation:

$$\dot{M} = D \cdot F \frac{\Delta C}{\Delta d} = D \cdot F \frac{\alpha_x \cdot \Delta P_x}{\Delta d}, \tag{8}$$

where $\dot{M}$ = the amount of substance transferred along the distance 'Δd',
D = the diffusion coefficient of the respective substance,
F = the surface area available for diffusion,
ΔC = the concentration difference along the diffusion path,
α_x and P_x represent solubility and partial pressure of gas 'x'.

The aqueous diffusion coefficients of CO_2 (e.g. 1.7×10^{-5} cm^2/s, Washburn 1929; $1.72-1.75 \times 10^{-5}$ cm^2/s, Gros and Moll 1974) and HCO_3^- (e.g. $3.8-4.0 \times 10^{-6}$, Gros and Moll 1974) suggest that bicarbonate could well be involved in intracompartmental (as opposed to transmembrane!) diffusive transfer, if the concentration difference of HCO_3^- were large enough. As outlined above, significant changes in $[HCO_3^-]$ can be achieved in the physiological range only by buffering of CO_2 by nonbicarbonate buffers. If a ratio of $\Delta[HCO_3^-]/\Delta(\alpha_{CO_2} \cdot P_{CO_2})$ of higher than about 4 to 5 could then be achieved along the diffusion distance, a considerable fraction of CO_2 could be transferred by diffusion of HCO_3^-, provided that rapid hydration/dehydration of CO_2 would be facilitated by the presence of a sufficient activity of carbonic anhydrase, and provided that the dissociated H^+ ions could be transferred at the same velocity as HCO_3^-. Although the diffusion coefficient for H^+ ions ($9.2 \cdot 10^{-5}$ cm^2/s, Kortüm 1972) is larger than that of HCO_3^-, appreciable diffusion of H^+ ions cannot be expected in the physiological pH range, because of the minute diffusion gradient ($\Delta[H^+] < 10^{-7}$). Furthermore, the diffusion gradient in $[HCO_3^-]$ established by nonbicarbonate buffering of hydrated and dissociated CO_2 is principally inversely related to the resulting diffusion gradient of H^+ ions.

3.1.2 Facilitated Diffusion of CO_2

Experiments of Longmuir et al. (1966) indicated that CO_2 could be transferred at considerably higher speed in haemolyzed blood than expected from Krogh's diffusion constant ($K = \alpha_{CO_2} \cdot D_{CO_2}$], suggesting that diffusion of HCO_3^- and simultaneous transfer of H^+ ions was also involved. These findings were corroborated by results obtained in layers of red cells (Enns et al. 1966; Moll and Gros 1966), haemoglobin solutions free of other nonbicarbonate buffers (Gros 1969), in inorganic phosphate buffer solution (Gros and Moll 1972; Gros et al. 1976), albumin solution (Gros and Moll 1974) and, finally, also in intact muscle tissue (Kawashiro and Scheid 1976). The enhancement of CO_2 elimination, which was more than that expected on the basis of simple aqueous diffusion, was attributed to a 'facilitated diffusion', by which the required simultaneous and equimolar transfer of HCO_3^- and H^+ ions was achieved by diffusion of HCO_3^-, and transfer of H^+ ions associated with H^+ ion-loaded nonbicarbonate buffer molecules (Fig. 3, upper panel). This implies that transfer of the H^+ ion/nonbicarbonate buffer complex can take place as fast as the diffusion of HCO_3^-, a condition possible for inorganic phosphate buffer ($D = 9.5 \times 10^{-6}$ for infinite dilution, Landolt and Börnstein 1960; $D = 7 \times 10^{-6}$ for 66 mM phosphate solution, Gros et al. 1976), but hardly conceivable for large molecule nonbicarbonate buffers like haemoglobin. This notion was verified by immobilization of phosphate buffer by fixation to large cellulose molecules, which eliminated the facilitated, enhanced diffusion of CO_2 (Gros et al. 1976). In order to explain the facilitated diffusion observed with large molecular size protein nonbicarbonate buffers, it was postulated that transfer of the associated H^+ ions was performed by rotational translocation (Fig. 3, lower panel) of the attached H^+ ions by spinning of the facilitating buffer molecules (Gros and Moll 1974; Gros et al. 1976), a hypothesis actually corroborated by recent data (Gros et al. 1984).

Such facilitated diffusion of HCO_3^- can certainly take place only if hydration and subsequent nonbicarbonate buffering, resulting in the relatively large elevation of $[HCO_3^-]$ required for this mechanism at the entrance of the aqueous diffusion pathway, and reversal of the process at the end of the path, is speeded up

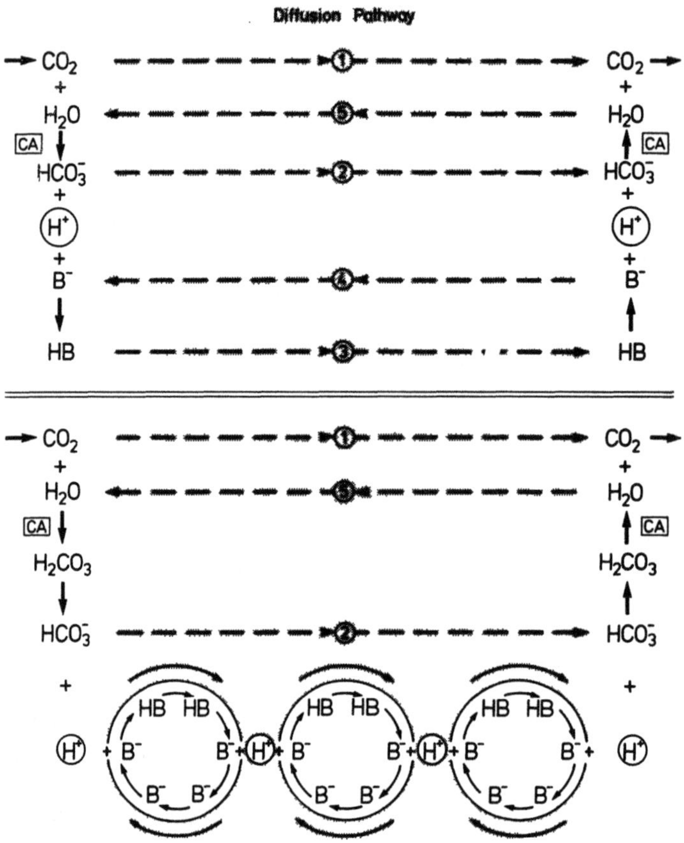

Fig. 3. Facilitated diffusion of CO_2. In addition to the diffusion of molecular CO_2 (*1*), diffusion of HCO_3^- (*2*) may contribute significantly to the transfer of CO_2, if nonbicarbonate buffering of CO_2 results in much larger diffusion gradients for HCO_3^- than for CO_2. Prerequisites for facilitated diffusion are rapid hydration and dehydration of CO_2 facilitated by carbonic anhydrase (*CA*), simultaneous transfer of the dissociated H^+ ions by diffusion of the H^+ ion-loaded nonbicarbonate buffers (*HB*, *3*) and equivalent back-diffusion of the buffer base (B^-, *4*) and of H_2O (*5*). Only nonbicarbonate buffers of small molecular weight can be expected to possess the required high diffusional mobility (*upper panel*), whereas large nonbicarbonate buffer molecules (such as haemoglobin) handle the transfer of H^+ ions by rotational translocation (*lower panel*). For details, see text

by the aid of carbonic anhydrase. In fact, blockade of CA with acetazolamide, or running in vitro experiments without carbonic anhydrase (see above references), reduces the diffusion of CO_2 to levels expected on the basis of pure molecular diffusion.

Applied to the conditions in vivo the influence of facilitated diffusion on the elimination kinetics of CO_2 will depend on three factors: a nonbicarbonate buffer value high enough for the establishment of a sufficient $[HCO_3^-]$ gradient along the diffusion distance, sufficient mobility, or apparent mobility as an expression of rotational H^+ ion translocation, of the involved nonbicarbonate buffers, and rapid hydration/dehydration of CO_2 at the beginning and the end of the facilitated diffusion pathway by carbonic anhydrase. The nonbicarbonate buffer values are high enough (cf. Heisler 1986b, e) and carbonic anhydrase is available to provide the required $[HCO_3^-]$ gradients and to support facilitated diffusion (e.g. mammalian muscle tissue: Gros and Dodgson 1988) to considerably speed up the intracompartmental CO_2 transfer in intracellular tissue compartments (e.g. by a factor of up to 5, found during very recent experiments on rabbit muscle; G. Gros, pers. comm.). But, although carbonic anhydrase is available in the extracellular space on the sarcolemmal surface (at least in muscle tissues of some species, cf. Gros and Dodgson 1988), and is 'considered by many researchers likely to be present on the luminal membrane of capillary endothelia' (Gros and Dodgson 1988), the extracellular, particularly the interstitial nonbicarbonate buffering (cf. Heisler 1986b, e) is insufficient to provide the basis for any significant contribution of facilitated diffusion in the CO_2 transfer across extracellular fluid compartments.

The conditions for facilitated diffusion are particularly favourable in animals with low levels of P_{CO_2} and $[HCO_3^-]$ in their body fluids, such as water-breathing species, especially fish. Although the nonbicarbonate buffer values in intracellular tissue compartments are generally lower in fish than in higher vertebrates by a factor of around 1.5–2 (cf. Heisler 1986b, e), the concentration difference of HCO_3^- produced for a given diffusion gradient of CO_2 will be much larger in relative terms on the background of values of P_{CO_2}, and $[HCO_3^-]$ lower by a factor of 10 or more as compared to mammals (cf. Heisler 1984, 1986a, c, d). This aspect of fish respiratory physiology has had little attention in the past, but will hopefully gain more during future analysis of CO_2 transfer in intracorporeal fluid compartments of fishes.

3.1.3 Diffusion of NH_3 and NH_4^+

In spite of the fact that the concentration of the ionized form is in the physiological pH range 100-fold higher than that of NH_3 transfer across cell membranes is generally considered to take place mainly by non-ionic diffusion, similar as for CO_2. The aqueous diffusion coefficient, D_{NH_3}, is in the same range (Radford 1964) as that for CO_2 (1.7×10^{-5} cm^2/s, Washburn 1929). However, the effective Krogh's diffusion constant ($K = D \cdot \alpha_{NH_3}$) is, in terms of P_{NH_3} larger by a factor of more than 1000 according to the much higher solubility of NH_3. The extremely high diffusivity of NH_3 in biological tissues is well documented for the mammalian kidney tubule system (e.g. Oelert et al. 1968), expressed by very small diffusion gradients of the non-ionized form throughout the tubular system (ΔP_{NH_3} about $20 \cdot 10^{-6}$ mmHg, variability by a factor of <1.2), whereas the concentrations of NH_4^+ vary by a factor of at least 2.5 (quite in accordance with the local distribution of pH). Also in rainbow trout P_{NH_3} gradients of only about $50 \cdot 10^{-6}$ mmHg across the branchial epithelium are sufficient for nonionic elimination of the endogenously produced ammonia (Cameron and Heisler 1983). The general conclusion regarding nonionic movement of ammonia across cell membranes has also been confirmed by experiments (1) on the kinetic effects of ammonium-chloride solutions in the extracellular space of isolated cell preparations on intracellular pH (initial very fast alkalinization due to diffusion and subsequent intracellular ionization of NH_3, and slow recovery due to the diffusion of NH_4^+ or simultaneous regulatory transfer of H^+ ion-equivalents; e.g. Boron and De Weer 1976), and (2) the acidifying effect of NH_4 Cl infusions on extracellular pH induced by nonvolatile H^+ ions left behind during preferential diffusion of NH_3 to other, intracellular and extracorporeal fluid compartments (Cameron and Heisler 1983; J.B. Claiborne and N. Heisler, unpublished data).

Recent estimates of the diffusion coefficients in turtle urinary bladder epithelium, however, suggest that the permeability of NH_4^+ is larger than that for K^+, and only about 50 times smaller than that for NH_3 (Schwartz and Tripolone 1983). Another study resulted in diffusion coefficients of NH_3 24 times and for NH_4^+ more than 310 times smaller than the equivalent aqueous values (Arruda et al. 1984). With such diffusivity ratios diffusion of NH_4^+ would actually be a good candidate for membrane transfer because of the ratio of $[NH_4^+]/[NH_3]$ of about 100 at physiological pH values, and equivalent concentration gradients along the diffusion distance. Diffusion of NH_4^+ through paracellular pathways in this specific preparation, however, may have provided a tendency towards elevated apparent membrane diffusion coefficients for NH_4^+ (Arruda et al. 1984). After analysis of the electrogenic and diffusive effects of ammonia transfer the authors concluded 'that NH_3 transport occurs via nonionic diffusion, whereas NH_4^+ transport is carrier-mediated' (Arruda et al. 1984). The biasing effect of the large fraction of paracellular pathways in the turtle bladder epithelium (Arruda et al. 1984) is evident by comparison with the ratios of $D_{NH_3}/D_{NH_4^+}$ determined in the gills of toadfish between ~100 (seawater-acclimated) and ~1200 (5% seawater-acclimated; Evans et al. 1989), in shark pup gills ~1000 (Evans and More 1988), and ~350 in rabbit proximal straight tubules (e.g. Garvin, Burg and Knepper 1987). These data indicate that a relatively high diffusivity of NH_4^+ (low D-ratios of 100 to 350) is always associated with 'leaky' paracellular pathways and corroborate the conclusion obtained from experiments in squid axons, where the effects of external NH_4Cl on the intracellular pH led to the conclusion that the diffusion coefficient of NH_4^+ is in the range of 10^{-3} of that for NH_3. The original notion that nonionic diffusion of NH_3 is the predominant form of ammonia transfer across cell membranes appears, therefore, to reflect the conditions in vivo.

If NH_3 is the main form in which ammonia is transferred across cell membranes, then the concentrations of intracellular and extracellular NH_4^+ should be in equilibrium with the pH difference, as long as the elimination of the end product is not diffusion- or perfusion-limited, and the elimination rate is not high enough to produce an appreciable gradient between intracellular space and venous plasma. Recent data on the distribution of NH_4^+ between intracellular and extracellular tissue compartments in erythrocytes and muscle tissue of two fish species, however, have indicated an ammonia distribution in equilibrium with pH in fish erythrocytes during control conditions, but deviations from the pH_{ery}/pH_{pl} relationship during hypercapnia (Wright et al. 1988a). This effect was even more pronounced in systemic tissues, the NH_4^+ concentrations being close to equilibrium with the estimated membrane potential (Wright et al. 1988b). These findings were interpreted by a diffusion coefficient of NH_4^+ by far larger than that for molecular NH_3, which is in contrast with reported values for the transmembrane mobility of NH_3 and NH_4^+, and the accumulated circumstantial evidence in this matter (see above). Also the energetic implications of largely different P_{NH_3} values between different fluid compartments and the load for the acid-base regulatory systems associated with the resulting ammonia shuttle across the cell membrane are not supporting preferential transmembrane diffusion of NH_4^+ rather than NH_3 (Heisler 1990). Determination of elevated ammonia concentrations in the tissue effected by ammonia-like substances (e.g. methylammonium) or by hydrolysis of terminal amino groups, active accumulation of NH_4^+ in the intracellular space, or larger than estimated gradients of ammonia between arterial plasma and the remote interstitial tissular fluid may be factors to explain these controversial data sets.

3.1.4 Facilitated Diffusion of Ammonia

With a ratio of 50 of the aqueous diffusion coefficients $D_{NH_3}/D_{NH_4^+}$ (Schwartz and Tripolone 1983), and the prevalent concentration ratio of about 100, diffusion of NH_4^+ could well contribute to intracompartmental transfer of ammonia. Although handling of the associated H^+ ions presents the same problems as for the CO_2/bicarbonate system, facilitated diffusion should be possible for ammonia with actually fewer restrictions than for the CO_2/HCO_3^- system (Fig. 4, upper panel). The production and thus transfer rate is, on average, lower by a factor of 10 (Heisler 1984, 1986c), and enzymatic catalysis is not required. Because of its flux rate, which is lower than that of CO_2, the transfer of H^+ ions back from the site of dehydration at the end of the facilitated diffusion pathway, to the site of hydration at the beginning of the NH_4^+ diffusion path (Fig. 4A) can be performed with a smaller nonbicarbonate buffer value than for CO_2. On the basis of the flux rates of ammonia which are eightfold lower than for CO_2, this mechanism may also contribute to the ammonia transfer in body compartments that are less well buffered, such as the extracellular space.

3.1.5 Simultaneous Diffusion of CO_2 and Ammonia

The production sites of CO_2 and ammonia are not necessarily the same, and ammonia is produced in appreciable quantities only in ammoniotelic organisms. During aerobic energy production, CO_2 is generated in ureotelic and ammoniotelic animals (although not homogeneously distributed) in all intracellular body fluid compartments, whereas during steady state, ammonia is produced in very few tissues, particularly liver and kidney (for ref. see above). A more homogeneous pattern may result from severe muscular activity, when ammonia as well as CO_2 is generated in large amounts in muscular tissues (Kutty 1972; Peer and Kutty 1981; Driedzic and Hochachka 1976; cf. Heisler 1984).

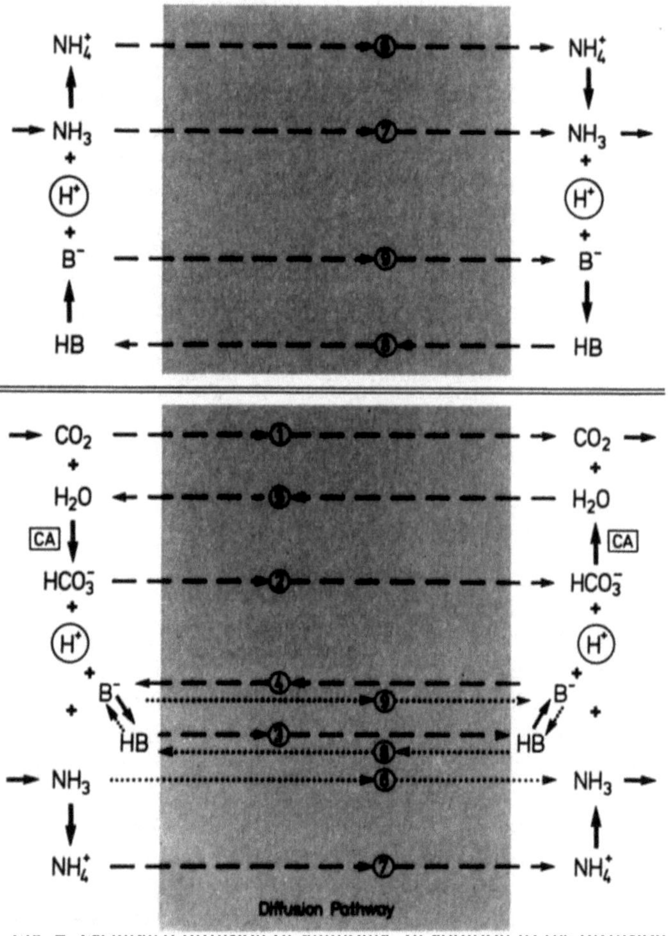

Fig. 4. Facilitated diffusion of ammonia. In addition to the diffusion of NH_3 *(7)*, diffusion of NH_4^+ *(6)* may contribute significantly to the overall ammonia transfer, if H^+ ions are provided by movement of acid *(8)* and base forms of nonbicarbonate buffers *(9)* via a shuttle mechanism similar to that for CO_2 (cf. Fig. 3). Prerequisites for this type of ammonia transfer are establishment of sufficient NH_4^+ gradients along the diffusion pathway by nonbicarbonate buffering, and an equivalent H^+ ion transfer rate by appropriate mobility of the nonbicarbonate buffers (or apparent mobility: rotational H^+ transfer, cf. Fig. 3) (*upper panel*).
Simultaneous facilitated diffusion of ammonia and CO_2 (*lower panel*) is highly advantageous. The transfer of the involved nonbicarbonate buffers is obsolete, because it would take place for ammonia and CO_2 in opposite direction (B^-: *9* for ammonia, *4* for CO_2; HB *8* for ammonia, *3* for CO_2). Note: this combined mechanism depends on the ammonia production, and may accordingly support less than about 10%–12% of the CO_2 flux. See text

If the elimination pathway is the same for CO_2 and ammonia, simultaneous transfer of both substances may support the elimination process considerably. Upon production, NH_3 will immediately buffer an equimolar number of H^+ ions generated from the dissociation of CO_2. It will, accordingly, alleviate the gradient of respiratory acidosis along the elimination path of CO_2.

Simultaneous elimination of CO_2 and ammonia will also provide support for facilitated diffusion of both substances (Fig. 4, lower panel). Facilitated diffusion includes translocation of H^+ ions by nonbicarbonate buffers, simultaneous with the diffusion of the ionized component of the respective substance. Transfer of H^+ ions and of the respective additional ionized and nonionized forms of nonbicarbonate buffers, however, has to occur in opposite directions for CO_2 and ammonia. Accordingly, diffusion of NH_4^+ and subsequent release of H^+ ions may substitute a fraction of the mobile or rotating nonbicarbonate buffer system responsible for H^+ ion translocation in accord with the diffusion of HCO_3^- (Fig. 3). In turn, diffusion of NH_4^+ will not be limited by the back-translocation of the associated H^+ ions, which will otherwise reduce local pH and thus ΔP_{NH_3} responsible for the transmembrane transfer of ammonia (Fig. 4).

3.2 Convective Transport in the Blood

3.2.1 Carbon Dioxide

After diffusive elimination from tissue cells, CO_2 will pass the interstial space, the plasma and the erythrocytic membrane by nonionic diffusion, before being hydrated and dissociated. This process is facilitated by the high activity of carbonic anhydrase, exclusively available in the erythrocytes, and by the very high intraerythrocytic nonbicarbonate buffer value (e.g. ≈ 95 mEq/(pH·L)) for mammalian blood (e.g. Nissen and Heisler 1973). As a result of the disturbed Donnan equilibrium, the major fraction of the HCO_3^- produced is then transferred in exchange with Cl^- to the plasma (Hamburger shift). The plasma $[HCO_3^-]$ is enhanced considerably by this mechanism, whereas the intraerythrocytic concentration falls accordingly (Fig. 5). The transfer of bicarbonate to the plasma induced by passive exchange against Cl^- enhances the transport capacity of the blood for CO_2 by the amount ($\Delta[HCO_3^-]_{ery \rightarrow pl} + \Delta[HCO_3^-]_1 - \Delta[HCO_3^-]_{NB\ cl}$) (Fig. 5), produced by further buffering of CO_2 induced by the transfer-associated reduction of the intraerythrocytic pH.

The carbonic anhydrase present in the interstitial space (cf. Gros and Dodgson 1988) will also facilitate hydration and dissociation of CO_2 in this fluid compartment. The low buffer value, however, will usually allow only a small fraction of the transported CO_2 to be converted into HCO_3^- and exchanged with the plasma space. During special conditions, however, like severe anaemia, or the complete lack of haemoglobin such as in the icefish, this may be a valuable mechanism in order to support hydration and dissociation of CO_2 to HCO_3^-, and thus enhance the limited transport capacity of the blood for molecular CO_2.

This additional amount of CO_2 cannot be transported in ionized form in some teleost fish species, where the intraerythrocytic pH is affected by the activity of circulating catecholamines, and is restored towards control values during conditions of acidified plasma pH (e.g. Primmett et al. 1986; Nikinmaa 1986; Boutilier et al. 1986b). This appears to be a protective mechanism, avoiding the displacement of the oxygen dissociation curve at lowered pH to the right and to generally lower oxygen concentration values at high P_{O_2} values (Bohr and Root

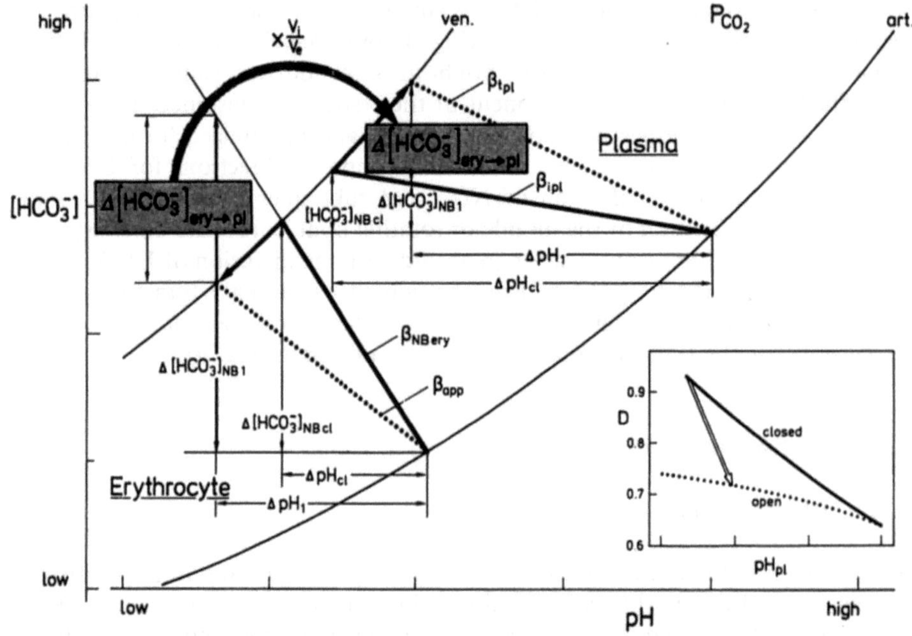

Fig. 5. Conversion of CO_2 into bicarbonate for convectional transport in the blood. CO_2 is hydrated and dissociated with the aid of CA mainly in the erythrocytes. Nonbicarbonate buffering of H^+ ions results in $\Delta[HCO_3^-]_{NB\ cl}$, of which a fraction is transferred to the plasma (Hamburger shift), enhancing the plasma $[HCO_3^-]$. The transfer of HCO_3^- to the plasma results in a further reduction of intraerythrocytic pH (from ΔpH_{cl} to ΔpH_1), which in turn facilitates ionization of an additional amount of CO_2 ($=\Delta[HCO_3^-]_{ery\rightarrow pl}+\Delta[HCO_3^-]_{NB_1}-\Delta[HCO_3^-]_{NBcl}$). See text

effects respectively), in order to allow complete oxygenation of the blood in the gas exchange organ. To date it remains unclear to what extent the associated transfer of bicarbonate-equivalent ions into the erythrocytes interferes with the elimination of CO_2 in the form of HCO_3^-. The effect of this mechanism on CO_2 transport will largely depend on the differential effects of the circulating catecholamines on arterial and venous intraerythrocytic pH values.

The fraction of CO_2 transported ionized as HCO_3^-, is primarily a function of the buffering capabilities, which are represented mainly by haemoglobin (80%–95%). Depending on the intraerythocytic buffer value (which is, with some exceptions, relatively constant among related species), the fraction of erythrocytes in the blood, the temperature-and species-specific plasma and intraerythrocytic pH values, and the above-delineated equilibria, about 60% to 90% of the CO_2 is transported in form of HCO_3^-. Up to 20% is bound as carbamino-CO_2 at terminal protein amino groups, and the remainder is transported in the form of physically dissolved CO_2 (with some notable exceptions, like the icefish). The distribution of HCO_3^- between plasma and erythrocytes ($R=[HCO_3^-]_{pl}/[HCO_3^-]_{tot}$) is about 0.7 in mammals (cf. Fig. 5, insert), but depends very much on species-specific pH_{ery}/pH_{pl} relationships and the overall charge state of nondiffusible erythrocytic buffer ions.

3.2.2 Ammonia

The mode of diffusion into the blood is not quite as clear for ammonia as for CO_2. Certainly, ammonia diffuses across cell membranes in the form of NH_3, but facilitated diffusion is a possible mechanism for passage of the interstitial space and the plasma. Since the ammonia buffer system has a high pK′ (see above), a larger fraction of total ammonia will be transported by the erythrocytes. This has recently been confirmed for blood of rainbow trout and lemon sole. Ammonia was found to be distributed according to the pH_{ery}/pH_{pl} relationship during control conditions, but during hypercapnia and after muscular activity, the intraerythrocytic ammonia concentrations were higher than predicted (Wright et al. 1988a, b). Generally, the intraerythrocytic $[NH_4^+]$ is higher than that of the plasma by factors between 1.5 (typical mammalian value) and 5 (lemon sole, Wright et al. 1988b). The higher values in fish are generally due to the larger pH_{pl}-pH_{ery} differences, but may also be related to additional factors leading to an accumulation in the cells (Wright et al. 1988a, b).

The concurrent transport of ammonia will certainly have a supportive effect on the transport of CO_2 in the blood. NH_3 that diffuses into the blood will quantitatively buffer H^+ ions dissociated from hydrated CO_2 and support the intraerythrocytic nonbicarbonate buffering. This factor is increasingly important in animals with low nonbicarbonate buffer values in their blood, the extreme of which is the haemoglobin-free blood of the icefish. The relatively small fraction of ammonia produced, as compared to CO_2 (in the range of up to 12%), may then considerably support the maintenance of relatively low tissue CO_2 levels.

4 Elimination of Ammonia and CO_2 from the Body Fluids

4.1 Air-Breathing Animals

Transfer of the main volatile metabolic end product, CO_2, from the blood to the site of gas exchange is generally considered to be performed by reversal of the same mechanisms as for the transfer from the cells into the blood. Carbonic anhydrase is available in the lung tissue, thus providing the basis for facilitated diffusion (e.g. Ryan et al. 1982; Whitney and Briggle 1982; Henry et al. 1986). In gas-exchange organs of air-breathers, further transfer to the environment, after diffusion into the gas phase, is performed by a combination of diffusion and convective aerial transport (e.g. by alveolar lung ventilation), as outlined in detail in other chapters of this volume (see chapters by Piiper, Boutilier and Milsom).

In principle, NH_3 could be eliminated by the same mechanisms, particularly facilitated by its high Krogh's diffusion constant (see above). However, since the capacitance of air is about the same for CO_2 and NH_3 (whereas the solubility in aqueous solutions is larger by a factor of 1000 for NH_3, see above), the partial pressure of NH_3 would have to be about 10% of that for CO_2 on the gas side of the body fluid/gas interface (according to the ratio of metabolic production). This type of elimination implies tremendous blood ammonia concentration values (see above description of equilibria), which are not feasible due to the toxicity of ammonia (e.g. Ball 1967; Hillaby and Randall 1979; Tomasso et al. 1980; Arillo et al. 1981; Thurston and Russo 1981). Circulating ammonia can, accordingly, only be voided to an aqueous medium through other organs, like kidneys, gills and skin. Because the amounts of ammonia produced by the nitrogen metabolism of ammoniotelic organisms can hardly be eliminated by exclusively air-breathing animals, other, less toxic and acid-base neutral (or much less active), nitrogenous end products are produced (urea and uric acid) and eliminated via renal pathways. This type of regulation observed in terms of phylogenetic development is also often found in facultative air-breathing fish, where ammoniogenesis is reduced during extended air-breathing periods to rates comparable with ureotelic animals. This likely occurs by switching to a different nitrogenous end product (e.g. Heisler 1982).

4.2 Water-Breathing Animals

4.2.1 Ammonia

Voiding of ammonia to the environmental water is facilitated by the high aqueous solubility of NH_3 and the relatively much higher ventilation rate of water breathing animals due to the lower oxygen content of the breathing medium, water (Rahn 1966; cf. Heisler 1986c). There is an ongoing discussion about the form, in which ammonia is actually transferred across the gill epithelium of fishes (for review: Evans 1977, 1979, 1980b, 1986; Heisler 1984, 1986c, Evans and Cameron 1986). Three possible ways may actually be involved: nonionic diffusion, ionic diffusion and ionic active exchange against Na^+, depending on the composition of the environmental water and further factors.

Nonionic Elimination of Ammonia

The diffusion coefficient certainly allows elimination by nonionic diffusion across the gill epithelium (Cameron and Heisler 1983), and this mechanism likely contributes predominantly to, or is even exclusively responsible for the elimination of NH_3 under normal conditions, i.e. low environmental ammonia concentrations and water pH values below 9 (Cameron and Heisler 1983; Holeton et al. 1983; Claiborne and Heisler 1984, 1986; Heisler 1990; for review: Heisler 1984, 1986c, e, 1988a). At the most usual environmental water pH values (7–9) the largest fraction of NH_3 combines immediately with CO_2 and water, producing HCO_3^- (cf. pK values above) and elevating the environmental concentrations of HCO_3^- and NH_4^+ (Fig. 6).

NH_4^+/Na^+ Ion Exchange

When the environmental ammonia concentration rises to values interfering with the diffusional elimination of NH_3, the plasma ammonia concentration is elevat-

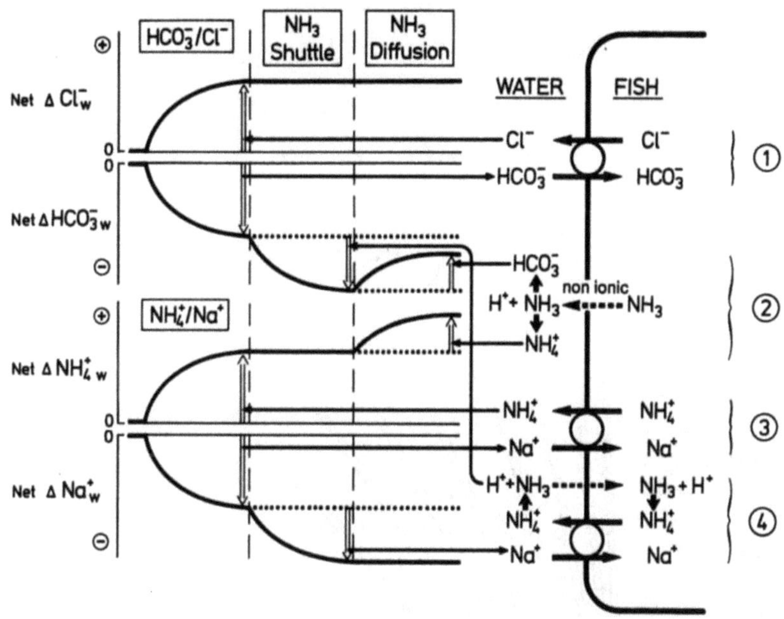

Fig. 6. Effect of various acid-base regulatory and nitrogenous waste mechanisms on the environmental water composition of aquatic animals. Nonionic diffusion of NH_3 *(2)* results in an elevation of both $[NH_4^+]$ and $[HCO_3^-]$ whereas ionic exchange of NH_4^+ against Na^+ *(3)* does not affect the $[HCO_3^-]$, but reduces the $[Na^+]$. Ammonia shuttling *(4)* reduces the water $[HCO_3^-]$ and $[Na^+]$, thus having the same effect as a H^+/Na^+ exchange (cf. Fig. 7). For details, see text

ed and active ion exchange mechanisms of NH_4^+/Na^+ (Fig. 6, *(3)*; well documented to exist in various fish species: e.g. Maetz 1973; Kerstetter et al. 1970; Evans 1977, 1980b; Payan and Maetz 1973; for review see also Maetz 1974; Evans 1979, 1980a, 1984, 1986; Evans and Cameron 1986) will be activated before toxic levels of ammonia are attained (Cameron and Heisler 1983, 1985). This mechanism is considered as a 1 : 1 ion exchange and affects the concentrations of both ammonia and Na^+ of the environmental water (Fig. 6, *(3)*). The accumulation of ammonia leads to an alkalinization of the body fluids (cf. Cameron and Heisler 1983) by ionization of NH_3 to H_4^+ and equimolar binding of NH^+ ions, which would normally be freed when NH_3 leaves the organism by nonionic diffusion (e.g. Heisler 1984, 1986c).

Ammonia Shuttle Mechanism

If the environmental ammonia concentrations are elevated to levels leading to a reversal of the P_{NH_3} gradient (water higher than plasma), then a NH_3/NH_4^+ shuttle mechanism may stimulate NH_4^+/Na^+ exchange to work at a rate higher than the metabolic production of ammonia, due to back-shuttle of NH_3 by nonionic diffusion, and subsequent intracorporeal ionization (Figs. 6, *(4)*; 7). The apparent net flux of ammonia would then be much less than the corresponding stimulation of the inward Na^+ flux, the discrepancy made up by the associated

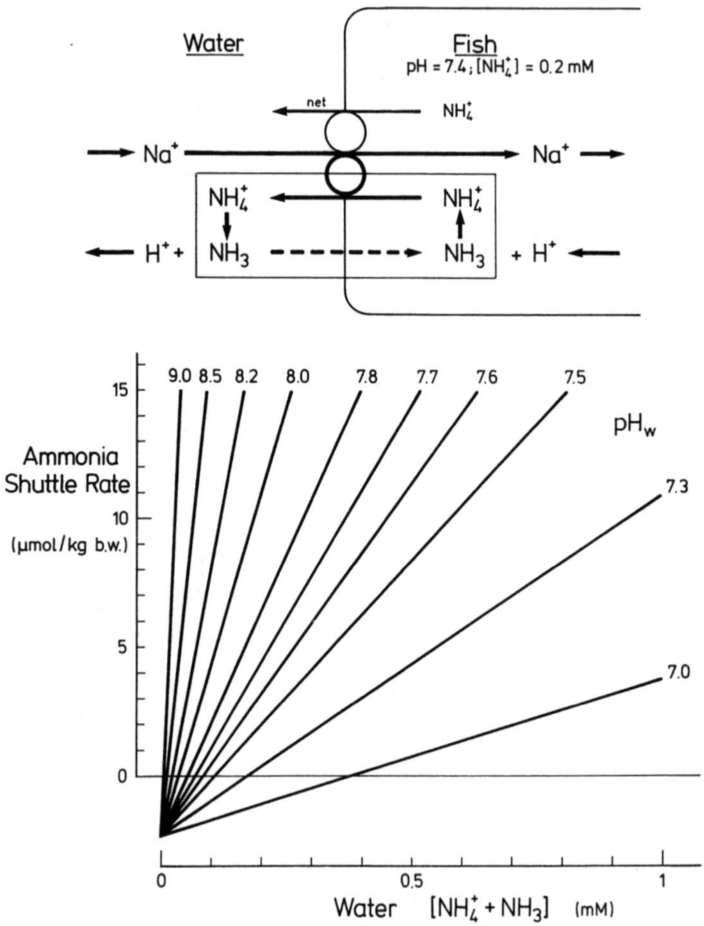

Fig. 7. Conditions for ammonia shuttling. The ammonia shuttle rate (circular flow of NH_3 and NH_4^+ with H^+ ion transporting characteristics, indicated by the *thin-line rectangle* in the *upper panel*) is largely dependent on the environmental ammonia concentration and pH, and is particularly pronounced during conditions of reduced body fluid pH values. The H^+ ion-transporting characteristics may potentially lead to a misestimate in the contribution of different acid-base relevant transepithelial ion transfer mechanisms. For details, see text. (Based on the ammonia diffusion coefficient of Cameron and Heisler, 1983)

apparent H^+ ion flux to the environment (H^+ ions dissociated from NH_4^+ in the water), affecting the environmental [HCO_3^-]. Since blood ammonia concentrations are normally quite low (100–300 µM; e.g. Cameron and Heisler 1983), special attention has to be paid to keep the environmental P_{NH_3} low during studies in closed-water recirculation systems required for advanced quantitative studies of ion transfer processes (for detailed discussion: Heisler 1984, 1986c, 1988a). The ammonia shuttle rate, induced by the high diffusivity of the volatile NH_3 (see above), may rise to considerable values (Fig. 7) even at relatively low environmental ammonia concentrations, when the water pH is relatively high, and the body

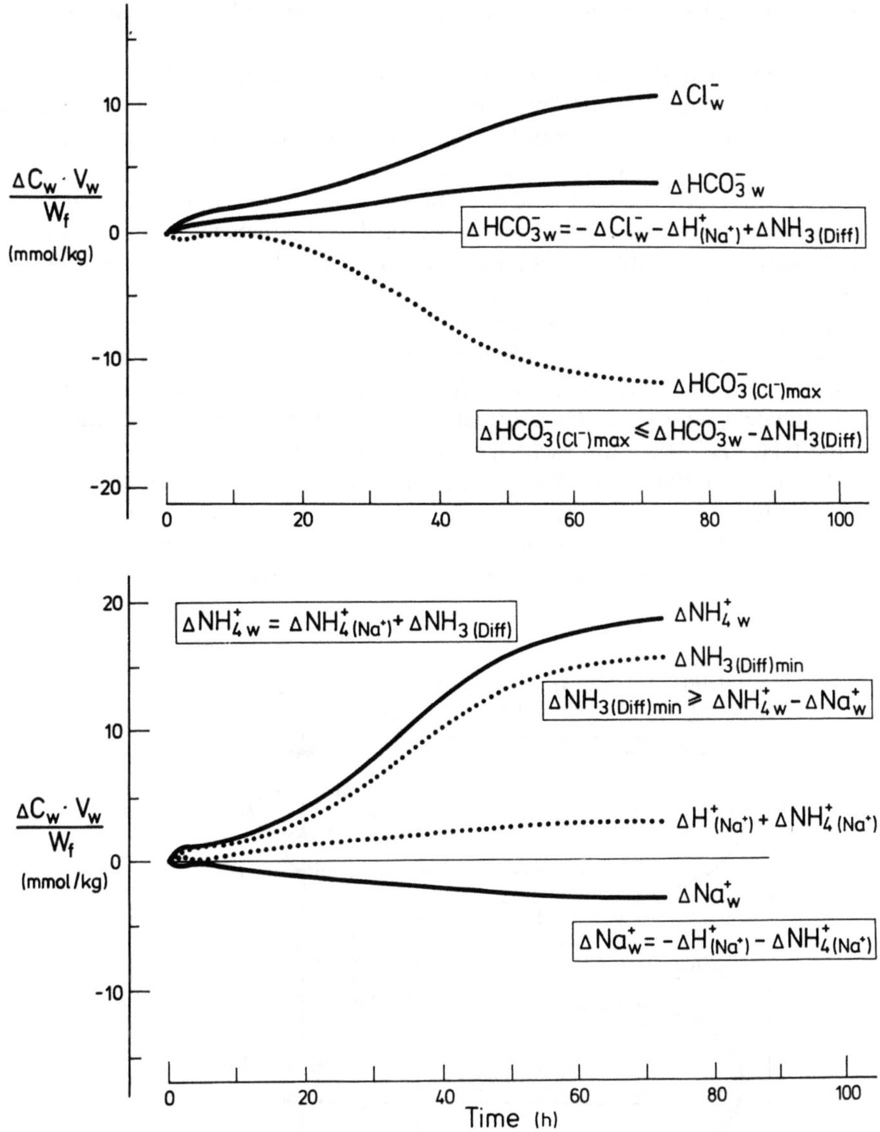

Fig. 8. Evaluation of net fluxes of acid-base relevant and counter-ions during 5% hypercapnia in carp. The changes in the environmental water concentration (ΔC_w) referred to the ratio of environmental water volume/animal weight (V_w/W_f) are balanced on the basis of a 1:1 ratio with their potential counter-ions. This approach allows an estimate of the minimal or maximal contributions, respectively, of either the HCO_3^-/Cl^-, the NH_4^+/Na^+ and/or the H^+/Na^+ ion-exchange mechanisms. It cannot indicate, however, any H^+/Na^+ ion exchange coupled with a Na^+-Cl^- codiffusion (cf. Fig. 9). For details, see text. (Based on data of Claiborne and Heisler (1986)

fluid pH in the fish is reduced due to a stress factor applied to stimulate ion transfer processes.

The ammonia shuttle mechanism results in an apparent H^+/Na^+ ion exchange, both for the internal fluid homeostasis of the fish, and for the environmental water (cf. Fig. 8). Since both H^+ ions and NH_4^+ are considered to be competing for the same carrier mechanism (for review: Maetz 1974 and Evans 1979, 1980a, 1984, 1986), the acid-base relevant ion transfer rate limitation, imposed by the carrier capacity, will a priori not be expected to be changed by ammonia shuttling. However, the affinity and K_m values of the carrier for transfer of H^+ ions and NH_4^+ may be considerably different. Also, the transfer process itself may require more energy and time for the larger NH_4^+ ion as compared to that for H^+. Accordingly, ammonia shuttling, although apparently conducting just the same transfer as the H^+/Na^+ exchange, may reduce the transfer capacity of the mechanism. Unfortunately, no data at all are available as to the above-mentioned characteristics of the $NH_4^+/H^+/Na^+$ ion exchange mechanism in vivo.

According to these considerations, ammonia shuttling may induce a contribution of apparent H^+/Na^+ exchange for acid-base regulatory purposes. It is unknown whether this contribution will reduce the role of the other main branchial acid-base relevant ion exchange mechanism, HCO_3^- against Cl^- (cf. Fig. 9), be-

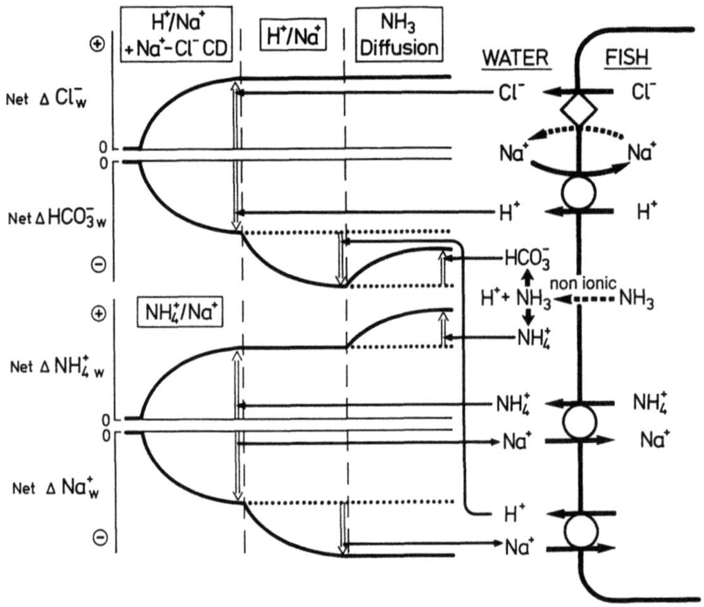

Fig. 9. Effect of various ionic and nonionic transfer mechanisms on the environmental water composition of aquatic organisms. The effect of a HCO_3^-/Cl^- ion exchange (cf. Fig. 6, *(1)*) may also be produced by a combination of a H^+/Na^+ ion exchange coupled with a Na^+-Cl^- codiffusion. Such a Na^+ shuttle mechanism could only be distinguished from a real HCO_3^-/Cl^- ion exchange by measurement of the unidirectional Na^+ fluxes. Unidirectional fluxes, however, are not suitable to distinguish between ionic and nonionic elimination of ammonia. See text

cause information about the extent to which the Cl^--related mechanism is linked to the Na^+-related mechanisms, apart from osmotic effects, is not available. A limitation in energy available to ion exchange processes in general, for instance, could reduce the contribution of HCO_3^-/Cl^- exchange.

Ionic vs Nonionic Elimination of Ammonia

When water P_{NH_3} is kept low during experimentation, by water pH being kept at the low end of the physiological range (≈ 7), the acid-base relevant transfer responsible for normalization of body fluid pH is mainly, if not completely, performed by stimulation of the HCO_3^-/Cl^- ion exchange. This has been indicated in studies with complete and simultaneous analysis of the patterns of acid-base- and potential counter-ions (Fig. 8) (e.g. Holeton et al. 1983; Claiborne and Heisler 1986).

Analysis of unidirectional fluxes of the potential counterions Na^+ and Cl^- may help to distinguish between utilization of Na^+- and Cl^--related ion exchange mechanisms. However, it does not provide any information about the differential contribution of H^+ and NH_4^+ transfer, and, in particular, it does not indicate the fraction of nonionic diffusion of NH_3. The analysis of net fluxes (Fig. 8), in turn, cannot eliminate the (although unlikely) possibility of a combination of H^+/Na^+ ion exchange with a Na^+-Cl^--co-diffusion (Fig. 9), re-establishing the osmotic equilibrium challenged by the H^+/Na^+ ion exchange mechanism (cf. Heisler 1989a). Such a Na^+ shuttle mechanism could not be picked up by analysis of net fluxes, indicating an apparent contribution of the HCO_3^-/Cl^- mechanism by appropriate changes of $[HCO_3^-]$ and $[Cl^-]$ in the water (Fig. 9). Recent simultaneous determinations of unidirectional Na^+ and Cl^- fluxes, however, have indicated (together with the analysis of net ionic transfer) that in carp (*Cyprinus carpio*) the Na^+-related transfer processes do not contribute at all, and acid-base regulation during hypercapnia is exclusively performed by HCO_3^-/Cl^- ion exchange (Fig. 10). The elimination of the tremendous amount of ammonia produced during hypercapnia (Claiborne and Heisler 1986) takes place accordingly, at least in this species, by nonionic diffusion (cf. Fig. 9). It is unknown to which extent these results can be extrapolated to other species, and how much of an apparent H^+/Na^+ exchange indicated by other experiments as the predominant contributor of acid-base regulation was actually effected by an artificially stimulated NH_4^+/H^+ exchange on the basis of ammonia shuttling (cf. Fig. 7).

Ionic Diffusion of NH_4^+

Elimination of ammonia from fish may also involve passive ionic diffusion of NH_4^+. As indicated by recent data obtained in marine teleost fish a significant fraction of the total ammonia release past the branchial epithelium may take place by passive movement of NH_4^+, likely through the relatively 'leaky' paracellular junctions (e.g. Goldstein et al. 1982; cf. Evans and Cameron 1986; Evans and More 1988; Evans et al. 1989). But even if a sizable portion of the ammonia is actually eliminated via ionic diffusion (estimated 30%, Evans and Cameron 1986), the major proportion of ammonia is still eliminated by nonionic

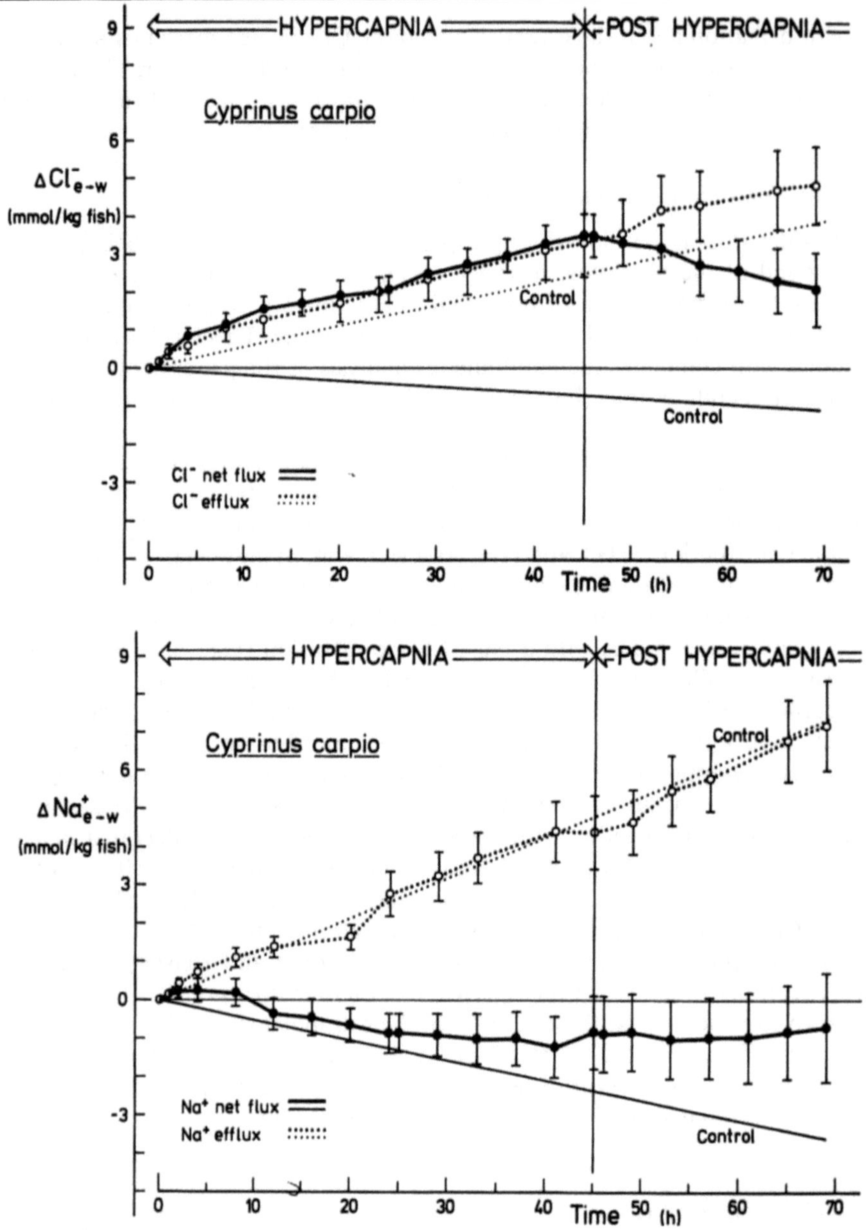

Fig. 10. Net fluxes and unidirectional efflux of Cl^- and Na^+ in carp (*Cyprinus carpia*) during and after exposure to environmental hypercapnia. The unidirectional efflux of Cl^- is essentially unaffected by hypercapnia, whereas the net flux is considerably enhanced as compared to the control conditions, indicating a stimulation of the HCO_3^-/Cl^- ion exchange mechanisms with practically complete inhibition of Cl^- influx during hypercapnia. The influx is resumed during the post-hypercapnia period. The unidirectional Na^+ efflux remains completely unchanged during all periods, the net flux being slightly enhanced during hypercapnia and post-hypercapnia, a maladaptive behavior in terms of acid-base regulation. These data confirm that none of the Na^+-related mechanisms was involved in accumulation of bicarbonate during hypercapnia and during restoration of the original status after termination of hypercapnia

diffusion (Evans and Cameron 1986; Evans and More 1988; Evans et al. 1989). Ionic diffusion is an unlikely contributor in freshwater fish with their very much smaller ion conductivity of their paracellular pathways, indicated by unidirectional flux rates that are orders of magnitude smaller than in marine teleost fish (cf. Evans 1979, 1980b, 1984, 1986; Fig. 10). The net effect of ionic diffusion of NH_4^+ is not quite clear, because the associated counter- or co-ions are still unknown. If NH_4^+ diffused together with HCO_3^- the process would be electroneutral as well as acid-base neutral, but have considerable effects for the osmoregulation of the fish. Any other combination of fluxes would have significant effects for acid-base regulation, except an exchange of NH_4^+ against H^+, which, however, cannot be taken into consideration on a passive basis because of the minute diffusion gradients for H^+ ions as compared to those for NH_4^+.

NH_4^+/H^+ Ion Exchange

Exchange of NH_4^+ against H^+ ions, although this time on an active basis, was recently proposed as an additional mechanism (Cameron 1986). The results of his study suggested that ammonia was extruded from a teleost (*Ictalurus punctatus*) against inward gradients of both NH_3 and NH_4^+, with the unidirectional Na^+ influx being unchanged, and H^+ equivalents removed from the external medium. This type of exchange mechanism will compensate for any passive movement of NH_3 into the fish. It will also balance the alkalinizing effect of ionization by addition of H^+ ions to the body fluid linked to the extrusion of NH_4^+, and thus be neutral for the acid-base status. Actually, during elevation of the external ammonia concentration, no change in plasma pH could be observed, supporting the proposed mechanism (Cameron 1986). The NH_4^+/H^+ exchange mechanism may provide a way around the alkalinizing effect of ammonia shuttling during normal acid-base status, which certainly has to be considered as maladaptive in face of control pH values. The data presented suggest that the ion exchange mechanisms may not be quite as selective in their respective counterions as considered to date, enhancing the regulatory band width of such transfer processes to a much larger variety of acid-base, metabolic and osmoregulatory disturbances.

4.2.2 Carbon Dioxide

The elimination of CO_2 from the blood to the environment of water-breathers is not quite as clear as in air-breathing species. Although the important role that fish erythrocytic carbonic anhydrase plays in the elimination of molecular CO_2 (similar to terrestrial animals) had been established by Maren in 1967, the role of intraerythrocytic carbonic anhydrase in this matter was discounted in favor of branchial CA (Haswell and Randall 1976, 1978; Haswell et al. 1980). This hypothesis was invalidated and the central role of red cell CA reestablished (Cameron 1978; Obaid et al. 1979; Perry et al. 1982). Even so, branchial carbonic anhydrase is still considered to play a significant role in CO_2 elimination (Swenson et al. 1984; Swenson and Evans 1984). Histochemically, CA can be detected at the respiratory epithelium and the chloride cells (Dimberg et al. 1981), but quantitative evidence as to the activity distribution of this enzyme has been provided only recently by Henry and co-workers (1988), concluding that very little, if any, 'membrane-associated' carbonic anhydrase is present in the gills. The overall branchial activity was less than one-fourth of the erythrocyte activity (Henry et al. 1988).

These findings suggest that facilitated diffusion (see above, cf. Fig. 3), if it takes place at all in the intracellular space of branchial cells, is probably less supported than in systemic intracellular fluid compartments. Nothing is known as to the buffer value and apparent mobility of branchial intracellular nonbicarbonate buffers, the critical prerequisites for facilitated diffusion. Most likely, the bulk of eliminated CO_2 is dehydrated in the erythrocytes, and transferred into the environmental water by nonionic diffusion.

Little is also known about the fraction of apical membrane-bound CA in the gills. If CA activity of sufficient amount were present at the branchial water interface, rapid hydration of CO_2 would certainly lead to a more complete equilibrium between blood and water with respect to P_{CO_2}, although this would also imply a larger inspired/expired difference in pH due to dissociation of H^+ ions.

4.3 Carbonic Anhydrase in the Water Phase and the Effect of Buffering

The enzymatic catalysis is a process facilitating hydration at a rate depending primarily on the available enzymatic activity, as long as no substrate or allosteric inhibition occurs. Disequilibrium with respect to pH in the expired water is, accordingly, a function of the amount per unit water volume of CO_2 to be hydrated in order to reach equilibrium. Therefore, the amount of CO_2 finally hydrated depends primarily on the nonbicarbonate buffer value of the respective water: no appreciable amount of CO_2 being hydrated in distilled water (Fig. 11). Although natural waters usually contain only a small amount of nonbicarbonate buffers, the formation of CO_3^{2-} and $NaCO_3^-$ in carbonated waters will increase the fraction of CO_2 hydrated at higher pH values to a significant extent (Fig. 11, see equilibria outlined above). An additional factor to be considered in this context is the nonionic elimination of NH_3, which certainly adds (although to a rather insignificant extent at physiological production and elimination rates) to the nonbicarbonate buffering, by ionization with associated uptake of H^+ ions (Fig. 11, *(1)*). The demand for carbonic anhydrase activity in order to achieve complete CO_2 hydration and pH equilibrium in the water leaving the secondary lamellae, may, accordingly, vary to a great extent due to differences in nonbicarbonate buffer value, carbonate content and P_{CO_2} of the inspired water, as well as with the amount of nonbicarbonate buffers added during gill passage, and their titration.

For a given CO_2 elimination rate, the change in P_{CO_2} of the water leaving the secondary lamellae will depend on the degree of hydration facilitated by water-phase oriented membrane-associated CA activity and the fraction of CO_2 finally hydrated (Fig. 11). As a function of the fraction of noncatalyzed hydration of CO_2 in the water after leaving the gills, P_{CO_2} will be reduced at constant total CO_2 content (e.g. during the process of measurement), and the apparent venous-expired P_{CO_2} difference will be enhanced. Relatively large venous-expired P_{CO_2} differences are actually observed in many fish species (e.g. Heisler et al. 1988;

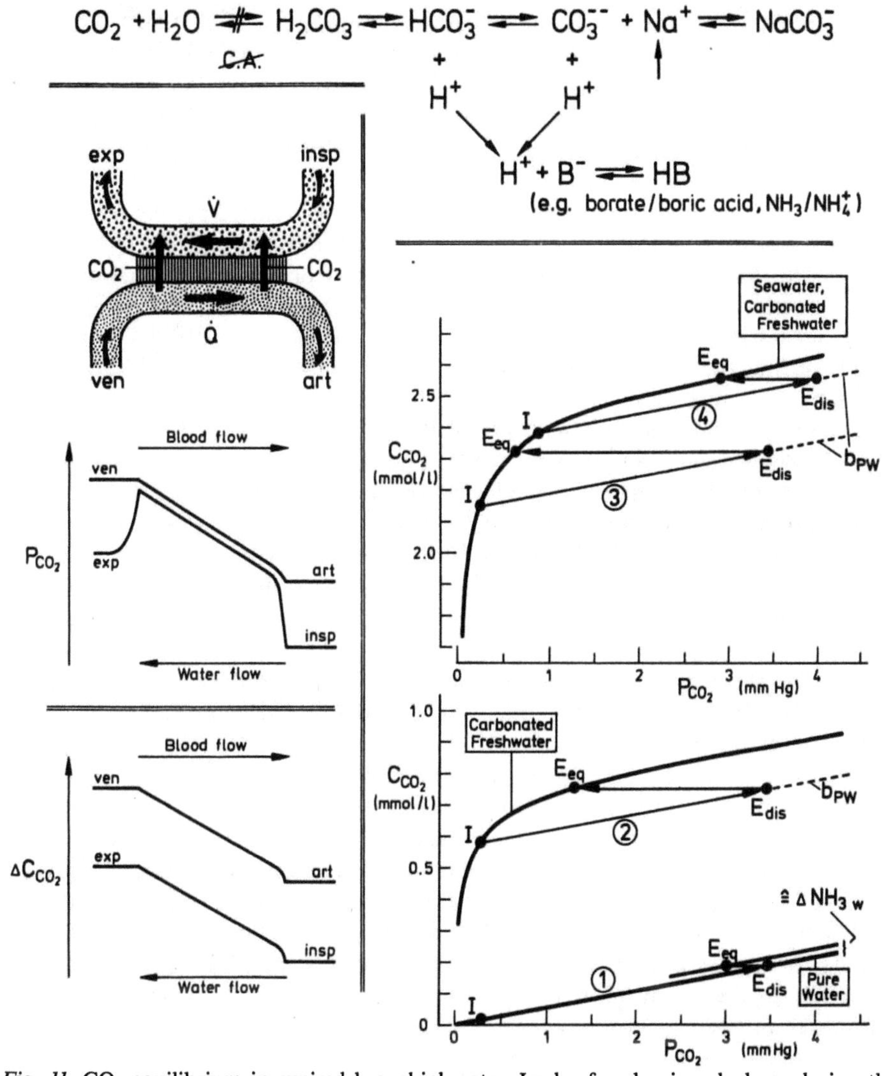

Fig. 11. CO_2 equilibrium in expired branchial water. Lack of carbonic anhydrase during the short lamellae-water contact time will result in complete equilibration of P_{CO_2} between gills and water. Hydration and dissociation, however, will take place in the water mainly after leaving the gills with constant total CO_2 content (C_{CO_2}), but falling P_{CO_2}. The extent of hydration, dissociation and disequilibrium depends primarily on the fraction of nonbicarbonate buffering of CO_2. For a given elimination of CO_2 P_{CO_2} will rise during gill passage in pure water (*1*) from the inspired value (*I*) along the capacitance line b_{PW} (equivalent to solubility of CO_2) to E_{dis}, independent of the amount of CA available. In ammoniotelic fish, the eliminated NH_3 adds a small amount of nonbicarbonate buffering. Without carbonic anhydrase, P_{CO_2} will be reduced to E_{eq} after gill passage, whereas with sufficient CA activity, P_{CO_2} equivalent to E_{eq} will be attained at the gill lamellae, and arterial P_{CO_2} will accordingly be lower. This effect, and the activity of CA required for complete equilibrium at the gill surface, is elevated with the slope of the CO_2 combination curve which is a function of C_{CO_2} at inspired P_{CO_2} (*(2)* medium carbonated freshwater; *(3)* seawater or highly carbonated freshwater), and of additional nonbicarbonate buffering. The effect of insufficient hydration and dissociation because of the lack of CA is diminished at elevated inspired P_{CO_2} *(4)* because of the slope of the CO_2 combining curve falling with rising P_{CO_2} (or falling pH). See text

Piiper and Schumann 1967; Baumgarten-Schumann and Piiper 1968), although the counter-current gas exchange system of fish is expected to result in almost complete equilibrium in terms of P_{CO_2} between venous blood and expired water. The observed differences could certainly also be due to inhomogeneities in blood/water flow distribution (cf. Piiper 1986), or to simple shunting of blood past the gas exchange organ (Heisler 1989a).

The post-branchial conversion of CO_2 with the associated pH changes is obviously a sensitive indicator for insufficient catalysis of the process by CA. Recently performed studies in the water downstream from the gill epithelium indicated that pH was not in equilibrium after leaving the gill chamber (Wright et al. 1986), suggesting that the eliminated CO_2 was incompletely hydrated (and dissociated) during the branchial epithelial contact time of 100 to 400 ms (Randall 1982a, b). The disequilibrium pH was enhanced by the addition of carbonic anhydrase inhibitors, and was eliminated by the addition of carbonic anhydrase to the inspired water (Wright et al. 1986). The latter was interpreted as an effect of CA in a thin unstirred mucus layer on the secondary lamellae (Wright et al. 1986). Although these data certainly indicate that some carbonic anhydrase activity is present at the water interface of the branchial epithelium, the experimental approach suggests that the observed activity may be rather small. For instance, during experimentation, the fish were kept in water having hardly any nonbicarbonate buffering ($\beta \approx 80\ \mu Eq/(pH \cdot L)$) (cf. Fig. 11) and bicarbonate/carbonate content. On the basis of the equilibria outlined above, and the experimental data (e.g. gill ventilation $\approx 0.1\ L/(kg \cdot min)$, $\dot{M}_{CO_2} \approx 40\ \mu mol/(kg \cdot min)$, $\Delta pH_W \approx 0.7$ (Wright et al. 1986); only about 14% of the eliminated CO_2 was hydrated. Assuming an additional buffering by NH_3 released at about $3\ \mu mol/(kg \cdot min)$ (determined at higher temperatures: Holeton et al. 1983; Wood et al. 1984), not even 20% of $\dot{M}_{CO_2}$ finally dissociated. Since disequilibrium could be observed in the expired water, only part of this roughly calculated fraction can have been hydrated by the aid of carbonic anhydrase. If the amount of carbonic anhydrase were constant with different water qualities (i.e. with higher fractions of CO_2 to be hydrated), then the available activity would apparently be insufficient for an appreciable CO_2 hydration in most natural waters, particularly seawater.

5 Accumulation of Volatile Metabolic End Products

The production of CO_2 per hour, at standard metabolic rate conditions, is in the same range as the total CO_2 content of the organism. Any changes in production rate and/or elimination characteristics will therefore result in considerable and often rather undelayed effects on the acid-base status. Such changes may be related to endogenous factors like elevated CO_2 production by muscular activity, bicarbonate buffering of H^+ ions and subsequent dehydration of HCO_3^- during anaerobic energy production, perfusion limitation of certain tissue areas and changes in ventilation as a result of regulatory processes, or may be imposed by exogenous factors like environmental hypercapnia. It shall be attempted in this section to delineate conditions and regulatory responses during some of the above-mentioned perturbations in the elimination of volatile metabolic end products. Since the problems involved in inhibition of ammonia elimination have been discussed in more detail in the preceding section, the following will focus mainly on disturbances by elevations of P_{CO_2} and the resulting regulatory measures. This discussion must necessarily be incomplete with respect to the range of covered species, and the limited selection of conditions (for more complete evaluation for all classes of animals: Heisler 1986a).

5.1 Tissue P_{CO_2} During Lactacidosis

Lactic acid is produced in vertebrates as the main end product of anaerobic glycolysis with intracellular concentrations in muscle tissue typically rising to values of 20 to 40 mM (in extreme cases up to 80 mM, Wardle 1978) during strenuous exercise (cf. Benadé and Heisler 1978; Heisler 1984, 1986c, d, 1988a, c). As a result of the low pK value (≈ 3.9) lactic acid is completely dissociated into lactate and H^+ ions. Although the majority of H^+ ions generated in the intracellular space are buffered by nonbicarbonate buffers, a large fraction of HCO_3^- is associated with H^+ ions and, with catalysis of carbonic anhydrase, dehydrated to molecular CO_2. This takes place with tissue HCO_3^-, but a large fraction of HCO_3^- transported into the musculature by blood perfusion is converted to molecular CO_2 as well.

The rate of CO_2 production by this mechanism may exceed by far the rate of aerobic CO_2 generation, and P_{CO_2} in mammalian muscle tissue may rise, according to its relatively low solubility (as compared to NH_3), to values of 120 mmHg or even higher (Fig. 12) depending on the ratio of CO_2 removal/production. Elimination of CO_2 from the musculature by diffusion into the perfusing blood, normally a very efficient mechanism supported and speeded up considerably by facilitated diffusion (see above), is impossible. In fact, CO_2 may even diffuse into the tissue during the beginning of the process, because the interstitial space and the perfusing blood are acidified by the very rapid transfer of H^+ ions to the extracellular space (Benadé and Heisler 1978; Holeton and Heisler 1983) and into the blood (cf. Heisler 1986d, 1988c).

During such conditions, the perfusing blood is also considerably acidified, by transfer of only a small fraction of the H^+ ions accumulated in the tissue, down to pH values in equilibrium with intracellular pH. This limits any further transfer to the buffer capacity (buffer value × volume) of the blood subsequently passing the muscle tissue (Fig. 12). Bicarbonate, the main buffer for nonvolatile H^+ ions in the blood ($\beta = 2.3 \times [HCO_3^-]$, cf. Heisler 1986b; open system, see above), is somewhat inefficient in the musculature, because the volatile component, CO_2, cannot be eliminated. Transport of surplus H^+ ions, therefore, takes place predominantly bound to blood nonbicarbonate buffers (Heisler 1988c). As a result, the release from muscle tissue of lactic acid-dissociated H^+ ions is to some extent delayed as compared to the membrane transfer capacity (Heisler 1986d, 1988c). The perfusion under such acidotic conditions is certainly enhanced, but the perfusion rates required to meet the membrane transfer capacity are manifold higher than can be achieved in vivo (cf. Heisler 1986d, 1988c).

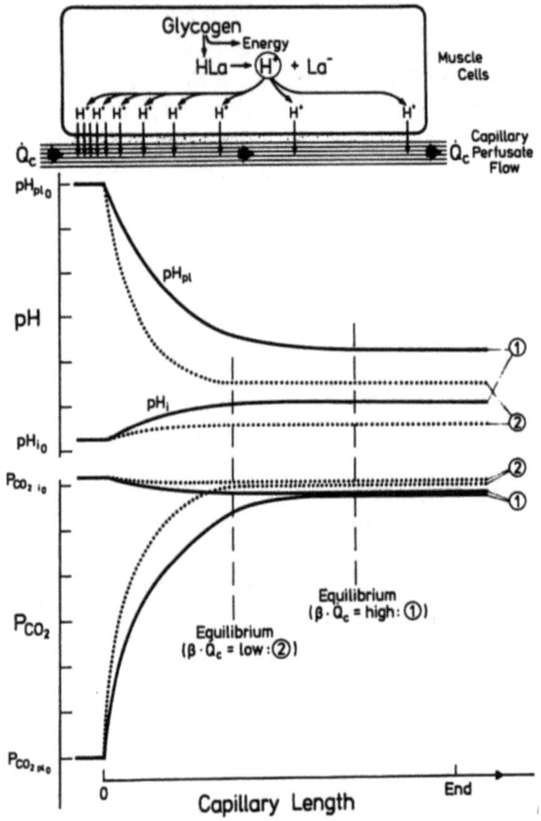

Fig. 12. Plasma pH and P_{CO_2} in capillary blood during passage of lactacidotic muscle tissue. pH in the blood is reduced by a relatively small amount of the accumulated H^+ ions to values in equilibrium with the interstitial and intracellular pH, while P_{CO_2} is largely increased due to conversion of HCO_3^- and H^+ ions to molecular CO_2 (closed buffer system). The fall of pH in the blood and the rise in P_{CO_2} is a function of the product of perfusion and blood buffer value, which is relatively low because of closed system buffering conditions for bicarbonate. Equilibrium between intracellular and plasma pH is attained earlier along the capillary length, with a lower product of $\beta \times \dot{Q}_c$, because the amount of H^+ ions to be transferred into the blood is less than with high $\beta \times \dot{Q}_c$ (cf. Heisler 1988e). See text

The accumulation of CO_2 in the muscle tissue is primarily the result of a misrelationship between production and elimination rate. Removal of molecular CO_2 is rather ineffective under such conditions. The high transfer of H^+ ions across the cell membrane acidifies the incoming blood considerably, shifting the CO_2 combination curve far down and to the right, resulting in a condition where most of the total CO_2 in the venous blood is present as physically dissolved species. Certainly, CO_2 is produced and eliminated from the muscle tissue at a much higher rate than normal, but, in contrast to aerobic conditions, it is mainly transported in the form of the physically dissolved molecule. This is due to the fact that nonbicarbonate buffers, normally responsible for conversion of hydrated and dissociated CO_2 into HCO_3^- (see above), are saturated with nonvolatile H^+ ions. The elevation of P_{CO_2} in the muscle tissue is then a simple function of the CO_2 solubility, the production rate and the muscle perfusion rate. Normalization of P_{CO_2} can, accordingly, not be expected before the fast release of nonvolatile H^+ ions from the tissues cells is complete, and is consequently perfusion- and pH equilibrium-limited by the same parameters responsible for the delay of intracellular pH normalization (Heisler 1986d, 1988c).

5.2 Environmental Hypercapnia

The elimination of CO_2 from the blood to the environment takes place via diffusional and convective mechanisms (see above). Changes in the composition of the environment, combined with an elevation of P_{CO_2}, can be compensated for by to a limited extent by increasing performance of the convective mechanisms (e.g. alveolar or branchial ventilation) via reduction of the partial pressure difference between the inspired breathing medium and the gas exchange site. If, however, the elevation of the inspired P_{CO_2} is too large for ventilatory compensation, the diffusion gradient at the gas exchange site (e.g. alveolar membrane or secondary lamellae) is diminished or reversed, resulting in a rise in P_{CO_2} and accumulation of CO_2 in the body fluids.

Due to the equilibria of the CO_2/HCO_3^- buffer system (see above) this generally results in a fall in plasma pH (Fig. 13). Most vertebrate species, however, possess mechanisms to alleviate or to completely compensate for such hypercapnia-induced disturbances of the acid-base status. Plasma pH recovers towards control values, partially due to conversion of molecular CO_2 to HCO_3^- by nonbicarbonate buffering (see above, cf. Fig. 15), but mainly by accumulation of bicarbonate gained by ion transfer processes from the environment. In terrestrial animals this is performed by renal elimination of H^+ ions, whereas aquatic species preferentially utilize branchial and skin epithelia for this purpose.

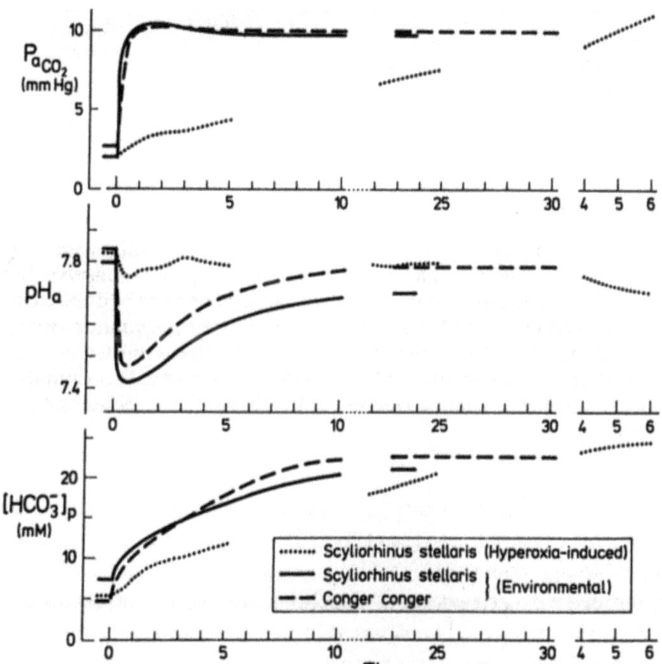

Fig. 13. Changes in arterial P_{CO_2}, pH and plasma $[HCO_3^-]$ in two fish species during environmental and hyperoxia-induced hypercapnia. For details, see text (Data from Heisler et al. 1976a, 1981, 1988; Heisler and Neumann 1977; Toews et al. 1983)

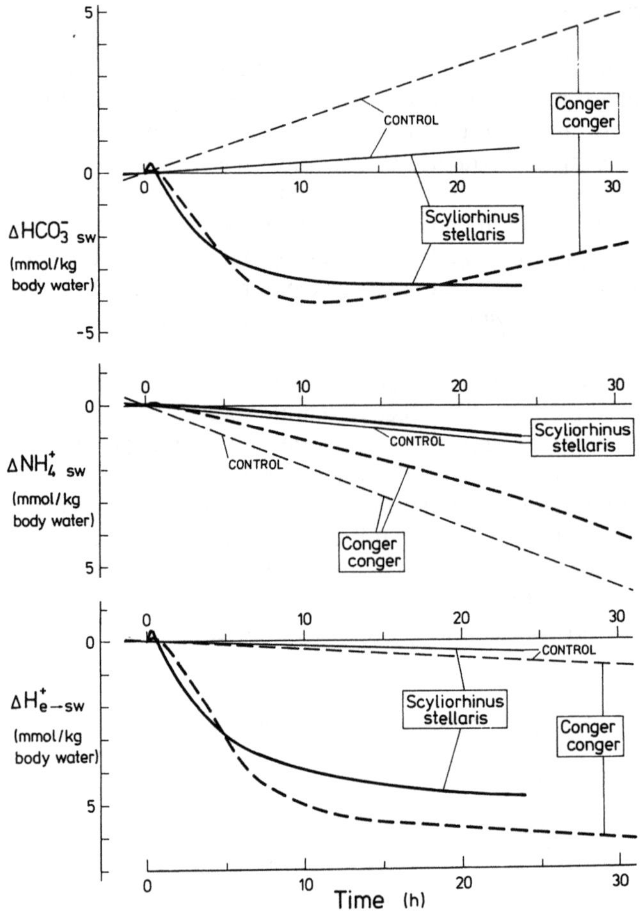

Fig. 14. Changes of HCO_3^- and NH_4^+ in the environmental water of an elasmobranch, *Scyliorhinus stellaris*, and a marine teleost fish, *Conger conger*, during environmental hypercapnia. The net elimination of H^+ ion equivalents as the difference between the net (actual minus control) release of NH_4^+ minus HCO_3^- to the environment takes place in both cases mainly via changes in HCO_3^- (the reduction in NH_4^+ release in *Conger* is maladaptive). Note the different control release rates of NH_4^+ between the ureotelic elasmobranch and the ammoniotelic teleost. See text. (Data of Heisler et al. 1976a, 1981; Heisler and Neumann 1977; Toews et al. 1983)

In the elasmobranch *Scyliorhinus stellaris* and the marine teleost fish *Conger*, typical examples for this type of CO_2 accumulation, restoration of plasma pH at continuing environmental hypercapnia to values close to normal (Fig. 13) is achieved by activation of acid-base-relevant ion exchange mechanisms (HCO_3^-/Cl^-, H^+/Na^+). The ammonia mechanism (NH_4^+/Na^+) certainly does not contribute in these species, because the ammonia elimination is small and constant in *Scyliorhinus*, and is actually reduced in *Conger* (Fig. 14). The water [HCO_3^-], however, is largely reduced in both species. This could have been effected by either the HCO_3^-/Cl^- or the H^+/Na^+ exchange mechanisms (cf. Figs. 6, 9), although the osmotic effects expected by the latter (cf. Heisler 1989a) could not be observed in any of the two species. Finally, about 4.3 (*Scyliorhinus*) and 5.2 mmol/kg (body water) (*Conger*) of H^+ equivalents are released to the environment in addition to the control rates (which are considerably different between the ureotelic elasmobranch and the ammoniotelic teleost, cf. Fig. 14), equivalent to an equimolar gain of HCO_3^- in the body fluids.

The distribution of the bicarbonate from the environment among body fluid compartments is very similar in both the elasmobranch and the teleost. During the initial phase of hypercapnia the rapid rise of plasma HCO_3^- is mainly due to transfer of bicarbonate-equivalents from intracellular to extracellular space (Fig. 15; Toews et al. 1983), before mechanisms for bicarbonate-equivalent ion exchange are activated. Later, the intracellular space is replenished, and bicarbonate originating from the environment is transferred to the cells in addition to the amount originally released. Although the quantity gained by the much larger intracellular space is considerably less than that supporting compensation in the extracellular compartments, the compensation of pH is even more complete in the intracellular space (Heisler 1988b), facilitated by the much lower initial bicarbonate concentration (Heisler et al. 1976b, 1980) and the more potent intracellular nonbicarbonate buffering (Heisler et al. 1980, for review: Heisler 1986b). This type of preferential intracellular pH regulation is a common phenomenon (cf. Toews and Heisler 1982; Heisler 1986d), and particularly pronounced in those exceptional animals, which show hardly any extracellular pH compensation in response to hypercapnia; like the facultative air-breathing teleost *Synbranchus marmoratus* (Heisler 1982) and the urodele amphibian *Siren lacertina* (Heisler et al. 1982).

5.3 Hyperoxia-Induced Hypercapnia

Branchial ventilation in fish is primarily regulated in order to cover the oxygen demand of the animal (cf. Dejours 1975, 1981), which results, due to the much lower capacitance coefficient in water for oxygen than for CO_2, in comparatively low P_{CO_2} values in fish as compared to exclusively air-breathing species (cf. Rahn 1966; Dejours 1975, 1981; Heisler 1989a). When the primary stimulus for gill ventilation is largely reduced by environmental hyperoxia, the resulting reduction in ventilation leads to a rise in P_{CO_2} at the gas exchange site, and accordingly to a retention of CO_2 associated with a rise of P_{CO_2} in the body fluids, similar to environmental hypercapnia (e.g. Dejours 1973; Heisler et al. 1981, 1989a; for review: Heisler 1986c).

The rise in P_{CO_2}, however, occurred in *Scyliorhinus stellaris* much slower than during environmental hypercapnia (Fig. 13) and was obviously quite well matched to the gain of bicarbonate in the arterial plasma, such that little change in pH occurred during regulatory adaptation to the new environmental conditions. During the initial stages of hyperoxia, gill ventilation was not exclusively regulated in order to maintain the oxygen supply of the animal, but maintained at a higher level, indicated by arterial P_{O_2} values exceeding 200 mmHg during the first day of exposure (Heisler et al. 1988). Levels of P_{O_2} below 100 mmHg and close to the normoxic controls (quite sufficient for complete blood oxygenation) were attained only 5 and 6 days after initiation of hyperoxia. Gill ventilation dropped initially to less than 25% of the normoxic controls concurrent with the largest fall in plasma pH during the first day, and rose again to about 35% of the control ventilation, restoring pH to almost normal. pH during the following four days of hyperoxia remained within −0.05 below the control value, obviously by adjustment of ventilation, tracking the compensatory gain of bicarbonate from the environment by slow ventilatory adjustment of P_{CO_2} to less than 20% of the controls. The plasma $[HCO_3^-]$ steadily rose to values of almost 25 mM, concurrent with an elevation in P_{CO_2} from 2 to 11 mmHg (Fig. 13, Heisler et al. 1988).

Although the final status of animals with environmental and hyperoxia-induced hypercapnia was pretty much the same at the end of the experiments (Fig. 13), there are remarkable differences in time course and utilization of regulatory mechanisms between these two different kinds of hypercapnia (Fig. 15). During environmental hypercapnia, pH in all body compartments was largely affected (cf. Fig. 13), which resulted in a much higher contribution of nonbicarbonate buffering for the initial compensatory response. With continuing compensation, and

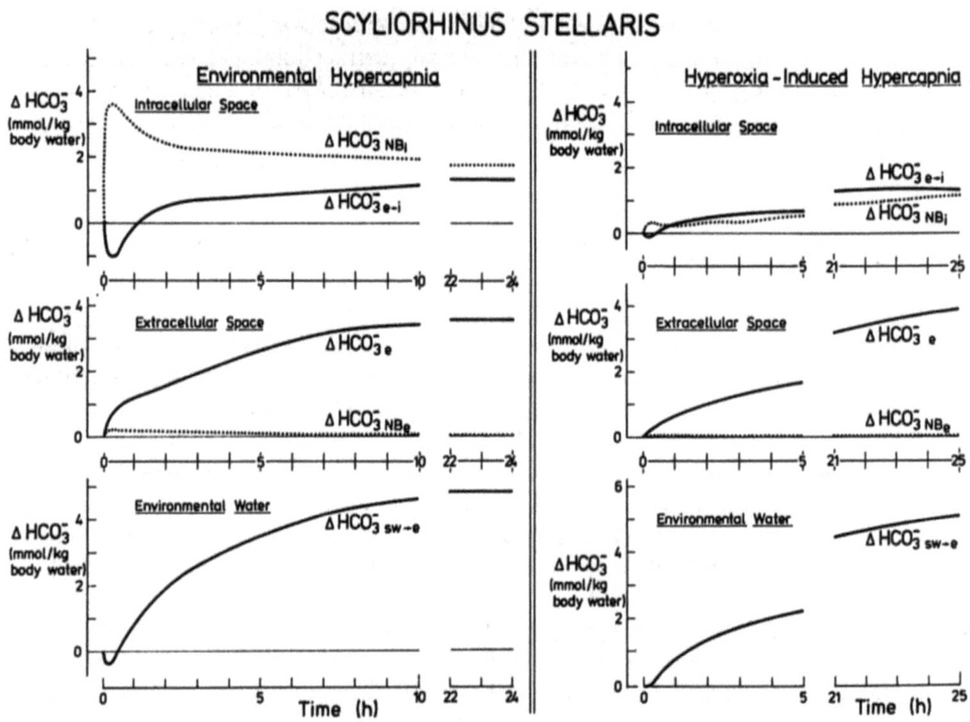

Fig. 15. Extent of intracellular and extracellular nonbicarbonate buffering (index *NB*, *dotted lines*), and transfer of bicarbonate-equivalent ions among intracellular and extracellular space, and the environmental water of the elasmobranch *Scyliorhinus stellaris* during environmental and hyperoxia-induced hypercapnia. See text (Data of Heisler et al. 1976, 1977, 1980, 1981, 1988)

concurrent normalization of pH, the contribution of nonbicarbonate buffering diminished, finally attaining low fractions similar to those during hyperoxia-induced hypercapnia (Fig. 15). The distribution of bicarbonate gained from the environment, among various body fluid compartments, was similar in both types of experiments, and again: intracellular pH was regulated even tighter than extracellular pH throughout the experiment (Heisler et al. 1988), although the changes in plasma pH were more than an order of magnitude smaller than those found during environmental hypercapnia.

Obviously the different type of hypercapnia considerably affected the pattern of regulation. Besides the minute utilization of buffering as a regulatory measure, the transepithelial transfer responsible for the final status also took place at a much slower rate. The maximal transfer rate of 15 μEq/(kg·min) of this species for bicarbonate equivalent resorption from the environment (Heisler et al. 1976a; Heisler and Neumann 1977; Holeton and Heisler 1983; cf. Heisler 1988b, was utilized only during the initial stages of hyperoxia with the associated, although slight, deflection in plasma pH (cf. Fig. 15). The time required to attain the same plasma bicarbonate concentration as attained during environmental hypercapnia was, accordingly, at least one order of magnitude longer during hyperoxia. Regulatory emphasis in order to keep deviations of pH within narrow limits was obviously more related to the adjustment of gill ventilation. These results suggest that with reduced oxygen-sensitive ventilatory drive, acid-base parameters may play an additional role as stimuli for gill ventilation in these animals.

This notion was actually confirmed by recent experiments. Specimens of *Scyliorhinus stellaris*, exposed to environmental hyperoxia, were additionally subjected to randomly chosen combinations of metabolic and respiratory acid-base disturbances by manipulating arterial pH, P_{CO_2}, and $[HCO_3^-]$ in wide ranges via changes in inspired P_{CO_2} and intra-arterial infusion of bicarbonate or, in some cases, HCl. It turned out that the hyperoxia-reduced gill ventilation was highly correlated with arterial plasma pH in an exponential fashion (Fig. 16), whereas

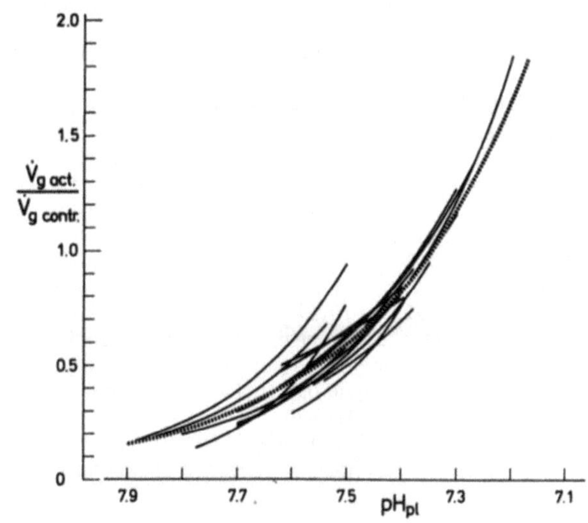

Fig. 16. Gill ventilation in the elasmobranch *Scyliorhinus stellaris* during hyperoxia ($\dot{V}_{g\,act.}$) relative to the controls during normoxic-normocapnia ($\dot{V}_{g\,contr.}$) as a function of plasma pH. See text (Data of N. Heisler, N.A. Andersen, R.G. Boutilier, J.B. Claiborne and G. Iwama, unpubl.)

gill ventilation was largely independent of plasma P_{CO_2} and plasma $[HCO_3^-]$ (N. Heisler, N.A. Andersen, R.G. Boutilier, J.B. Claiborne and G. Iwama, unpubl. data). It remains to be seen whether this type of regulatory stimulus, well known to exist in higher vertebrates, is an isolated phenomenon of elasmobranch fish (which are considered phylogenically less developed) or can also be found in teleost fish species.

6 Concluding Remarks

CO_2 and ammonia, as the quantitatively most important metabolic end products, are in equilibrium with acid-base and electrolyte status via buffer reactions of their hydrated counterparts. Although significant transfer is performed in the organism by nonionic diffusion, the main fraction of total CO_2 and ammonia is present in the organism in form of ionic species, interacting with ionic regulation. Except for convective blood transport, elimination from the organism takes place mainly by passive diffusion across cell membranes and through aqueous body fluids, sometimes considerably speeded up by facilitated diffusion. Changes in the production rates and/or elimination characteristics may result in considerable, and often immediate, disturbances of acid-base and electrolyte status. Limited accumulation in the body fluids, induced by interferences with the mechanisms of elimination, can be compensated for by active and energy-consuming compensatory mechanisms.

References

Arillo A, Margiocco C, Melodia F, Mensi P, Schenone G (1981) Ammonia toxicity mechanism in fish: studies on rainbow trout (*Salmo gairdneri* Rich). Ecotoxicol Environ Saf 5:316–328

Arruda JAL, Dytko G, Withers L (1984) Ammonia transport by the turtle urinary bladder. Am J Physiol 246:F635–F647

Aster PL, Casey CA, Vorhaben JE, Campell JW (1982) Hepatic ammoniogenesis in the catfish, *Ictalurus punctatus*. Am Zool 21:1029

Ball IR (1967) The relative susceptibilities of some species of fresh-water fish to poisons. I. Ammonia. Water Res 1:767–775

Baumgarten-Schumann D, Piiper J (1968) Gas exchange in the gills of resting unanaesthetized dogfish (*Scyliorhinus stellaris*). Respir Physiol 5:317–325

Benadé AJS, Heisler N (1978) Comparison of efflux rates of hydrogen and lactate ions from isolated muscles in vitro. Respir Physiol 32:369–380

Boron FW, De Weer P (1976) Intracellular pH transients in squid giant axons caused by CO_2, NH_3 and metabolic inhibitors. J Gen Physiol 67:91–112

Boutilier RG, Heisler N (1988) Acid-base regulation and blood gases in the anuran amphibian, *Bufo marinus*, during environmental hypercapnia. J Exp Biol 134:79–98

Boutilier RG, Heming TA, Iwama GK (1984) Appendix: physicochemical parameters for use in fish respiratory physiology. In: Hoar WS, Randall DJ (eds) Fish physiology, vol XA. Academic Press, NY, pp 403–430

Boutilier RG, Glass ML, Heisler N (1986a) The relative distribution of pulmocutaneous blood flow in *Rana catesbeiana*: effects of pulmonary or cutaneous hypoxia. J Exp Biol 126: 33–39

Boutilier RG, Iwama GK, Randall DJ (1986b) The promotion of catecholamine release in rainbow trout, *Salmo gairdneri*, by acute acidosis: interactions between red cell pH and haemoglobin oxygen-carrying capacity. J Exp Biol 123:145–157

Cameron JN (1978) Chloride shift in fish blood. J Exp Zool 206:289–295

Cameron JN (1986) Responses to reversed NH_3 and NH_4^+ gradients in a teleost (*Ictalurus punctatus*), and elasmobranch (*Raja erinacea*), and a crustacean (*Callinectes sapidus*): evidence for $NH_4{}^+/H^+$ exchange in the teleost and the elasmobranch. J Exp Zool 239:183–195

Cameron JN, Heisler N (1983) Studies of ammonia in the rainbow trout: physico-chemical parameters, acid-base behavior and respiratory clearance. J Exp Biol 105:107–125

Cameron JN, Heisler N (1985) Ammonia transfer across fish gills: a review. In: Gilles R (ed) Circulation, respiration and metabolism. Springer, Berlin Heidelberg New York Tokyo, pp 91–100

Claiborne JB, Heisler N (1984) Acid-base regulation and ion transfers in the carp (*Cyprinus carpio*) during and after exposure to environmental hypercapnia. J Exp Biol 108:25–43

Claiborne JB, Heisler N (1986) Acid-base regulation and ion transfers in the carp (*Cyprinus carpio*): pH compensation during graded long- and short-term environmental hypercapnia and the effect of bicarbonate infusion. J Exp Biol 126:41–61

Dejours P (1973) Problems of control of breathing in fishes. In: Bolis L, Schmidt-Nielsen K, Maddrell SHP (eds) Comparative physiology. Elsevier, North-Holland, Amsterdam, pp 117–133

Dejours P (1975/81) Principles of comparative respiratory physiology. Elsevier, North-Holland, Amsterdam

Dimberg K, Høglund LB, Knutson PG, Ridderstrale Y (1981) Histochemical localization of carbonic anhydrase in gill lamellae from young salmon (*Salmo salar* L.) adapted to fresh and salt water. Acta Physiol Scand 112:218–220

Driedzic WR, Hochachka PW (1976) Control of energy metabolism in fish white muscle. Am J Physiol 230:579–582

Edsall JT, Wyman J (1958) Biophysical chemistry. Academic Press, New York

Enns T, Sham GB, Andersen S (1966) Carbonic anhydrase diffusion of CO_2 in whole blood. Physiologist 9:176

Evans DH (1977) Further evidence for NH_4^+/Na^+ exchange in marine teleost fish. J Exp Biol 70:213–220

Evans DH (1979) Fish. In: Maloiy GMO (ed) Comparative physiology of osmoregulation in animals, vol 1. Academic Press, New York, pp 305–390

Evans DH (1980a) Kinetic studies of ion transport by fish gill epithelium. Am J Physiol 238:R224–R230

Evans DH (1980b) Na^+/NH_4^+ exchange in the marine teleost, *Opsanus beta*: Stoichiometry and role in Na^+ balance. In: Lahlou B (ed) Epithelial transport in the lower vertebrates. Cambridge University Press, London New York, pp 197–205

Evans DH (1984) The roles of gill permeability and transport mechanisms in euryhalinity. In: Hoar WS, Randall DJ (eds) Fish physiology, vol. XB. Academic Press, New York London, pp 239–283

Evans DH (1986) The role of branchial and dermal epithelia in acid-base regulation in aquatic animals. In: Heisler N (ed) Acid-base regulation in animals. Elsevier Biomedical Press, Amsterdam, pp 139–172

Evans DH, Cameron JN (1986) Gill ammonia transport. J Exp Zool 239:17–23

Evans DH, More KJ (1988) Modes of ammonia transport across the gill epithelium of the dogfish pub (*Squalus acanthias*). J Exp Biol 138:375–397

Evans DH, More KJ, Robbins SL (1989) Modes of ammonia transport across the gill epithelium of the marine teleost fish *Opsanus beta*. J Exp Biol 144:339–356

Fromm PO (1963) Studies on renal and extrarenal excretion in a freshwater teleost, *Salmo gairdneri*. Comp Biochem Physiol 10:121–128

Goldstein L, Forster G, Fanelli GM Jr (1964) Gill blood flow and ammonia excretion in the marine teleost, *Myoxocephalus scorpius*. Comp Biochem Physiol 12:489–499

Goldstein L, Claiborne JB, Evans DH (1982) Ammonia excretion by two marine teleost fish: the importance of NH_4^+ permeance. J Exp Zool 219:395–397

Gros G (1969) Die freie und erleichterte Diffusion von Kohlendioxid in Erythrozyten und Haemoglobinlösungen. MD Thesis, Universität Tübingen, Tübingen
Gros G, Dodgson SJ (1988) Velocity of CO_2 exchange in muscle and liver. Annu Rev Physiol 50:669–694
Gros G, Moll W (1972) The facilitated diffusion of CO_2 in hemoglobin and phosphate solutions. In: Rørth M, Astrup P (eds) Oxygen affinity of hemoglobin and red cell acid-base status. Munksgaard, Copenhagen
Gros G, Moll W (1974) Facilitated diffusion of CO_2 across albumin solutions. J Gen Physiol 64:356–371
Gros G, Moll W, Hoppe H, Gros H (1976) Proton transport by phosphate diffusion – a mechanism of facilitated CO_2 transfer. J Gen Physiol 67:773–790
Gros G, Lavalette D, Moll W, Gros H, Amand B, Pochon F (1984) Evidence for rotational contribution to protein-facilitated proton transport. Proc Natl Acad Sci USA 81:1710–1714
Haswell MS, Randall DJ (1976) Carbonic anhydrase inhibitor in trout plasma. Respir Physiol 28:17–27
Haswell MS, Randall DJ (1978) The pattern of carbon dioxide excretion in the rainbow trout, *Salmo gairdneri*. J Exp Biol 72:17–24
Haswell MS, Randall DJ, Perry SF (1980) Fish gill carbonic anhydrase: acid-base regulation or salt transport? Am J Physiol 238:R240–R245
Heisler N (1982) Intracellular and extracellular acid-base regulation in the tropical fresh-water teleost fish *Synbranchus marmoratus* in response to the transition from water breathing to air breathing. J Exp Biol 99:9–28
Heisler N (1984) Acid-base regulation in fishes. In: Hoar WS, Randall DJ (eds) Fish physiology, vol XA. Academic Press, New York, pp 315–401
Heisler N (ed) (1986a) Acid-base regulation in animals. Elsevier Science, Amsterdam, 492 pp
Heisler N (1986b) Buffering and transmembrane ion transfer processes. In: Heisler N (ed) Acid-base regulation in animals. Elsevier Science, Amsterdam, pp 3–47
Heisler N (1986c) Acid-base regulation in fishes. In: Heisler N (ed) Acid-base regulation in animals. Elsevier Science, Amsterdam, pp 309–356
Heisler N (1986d) Comparative aspects of acid-base regulation. In: Heisler N (ed) Acid-base regulation in animals. Elsevier Science, Amsterdam, pp 397–450
Heisler N (1986e) Mechanisms and limitations of fish acid-base regulation. In: Nilson S, Holmgren S (eds) Fish physiology: recent advances. Croom-Helm, London, pp 24–49
Heisler N (1988a) Acid-base regulation in elasmobranch fishes. In: Shuttleworth TJ (ed) Physiology of elasmobranch fishes. Springer, Berlin Heidelberg New York, pp 215–252
Heisler N (1988b) Ion transfer processes as mechanisms for acid-base regulation. In: Wood SC (ed) (Lenfent C, exec ed) Comparative pulmonary physiology: current concepts. Lung biology in heart and disease, vol 39. Marcel Dekker, New York, pp 539–583
Heisler N (1988c) Lactic acid elimination from muscle cells. In: Grothe J (ed) Aktuelle Probleme der Atmungsphysiologie Funktionsanalyse biologischer Systeme, Vol. 18. Gustav Fischer Verlag, Stuttgart New York, pp 241–251
Heisler N (1989a) Acid-base regulation in fishes. I. Mechanisms. In: Moris R, Taylor EW, Brown DJA, and Brown JA (eds) Acid toxicity and aquatic animals. Society of Experimental Biology Seminar Series, Cambridge University Press, Cambridge, pp 85–97
Heisler N (1989b) Parameters and methods in acid-base physiology. In: Bridges CR, Butler PJ (eds) Techniques in comparative respiratory physiology – an experimental approach. Society of Experimental Biology Seminar Series, Cambridge University Press, Cambridge, pp 305–332
Heisler N (1990) Mechanisms of ammonia elimination in fishes. In: Mellinger J, Truchot JP, Lahlou B (eds) Comparative Physiologie: Animal nutrition and transport processes: II. Transport, respiration and excretion. Basel, Karger
Heisler N, Piiper J (1971) The buffer value of rat diaphragm muscle tissue determined by P_{CO_2} equilibration of homogenates. Respir Physiol 12:169–178
Heisler N, Piiper J (1972) Determination of intracellular buffering properties in rat diaphragm muscle. Am J Physiol 222:747–753

Heisler N, Neumann P (1977) Influence of sea-water pH upon bicarbonate uptake induced by hypercapnia in an elasmobranch fish (*Scyliorhinus stellaris*). Pflügers Arch Suppl 368:R 19
Heisler N, Neumann P (1980) The role of physico-chemical buffering and of bicarbonate transfer processes in intracellular pH regulation in response to changes of temperature in the larger spotted dogfish (*Scyliorhinus stellaris*). J Exp Biol 85:99–110
Heisler N, Weitz H, Weitz AM (1976a) Hypercapnia and resultant bicarbonate transfer processes in an elasmobranch fish. Bull Eur Physiopath Respir 12:77–85
Heisler N, Weitz H, Weitz AM (1976b) Extracellular and intracellular pH with changes of temperature in the dogfish *Scyliorhinus stellaris*. Respir Physiol 26:249–263
Heisler N, Neumann P, Holeton GF (1980) Mechanisms of acid-base adjustment in dogfish (*Scyliorhinus stellaris*) subjected to long-term temperature acclimation. J Exp Biol 85:89–98
Heisler N, Holeton GF, Toews DP (1981) Regulation of gill ventilation and acid-base status in hyperoxia-induced hypercapnia in the larger spotted dogfish (*Scyliorhinus stellaris*). Physiologist 24:58, 305
Heisler N, Forcht G, Ultsch GR, Anderson JF (1982) Acid-base regulation in response to environmental hypercapnia in two aquatic salamanders, *Siren lacertina* and *Amphiuma means*. Respir Physiol 49:141–158
Heisler N, Toews DP, Holeton GF (1988) Regulation of ventilation and acid-base status in the elasmobranch *Scyliorhinus stellaris* during hyperoxia-induced hypercapnia. Respir Physiol 71:227–246
Henry RP, Dodgson SJ, Forster RE, Storey BT (1986) Rat lung carbonic anhydrase: activity, localization and isozymes. J Appl Physiol 60:638–645
Henry RP, Smatresk NJ, Cameron JN (1988) The distribution of branchial carbonic anhydrase and the effects of gill and erythrocyte carbonic anhydrase inhibition in the channel catfish, *Ictalurus punctatus*. J Exp Biol 134:201–218
Hillaby BA, Randall DJ (1979) Acute ammonia toxicity and ammonia excretion in rainbow trout (*Salmo gairdneri*). J Fish Res Board Can 36:621–629
Holeton GF, Heisler N (1983) Contribution of net ion transfer mechanisms to the acid-base regulation after exhausting activity in the Larger Spotted Dogfish (*Scyliorhinus stellaris*). J Exp Biol 103:31–46
Holeton GF, Neumann P, Heisler P (1983) Branchial ion exchange and acid-base regulation after strenuous exercise in rainbow trout (*Salmo gairdneri*). Respir Physiol 51:303–318
Jackson DC, Heisler N (1982) Plasma ion balance of submerged anoxic turtles at 3 °C: the role of calcium lactate formation. Respir Physiol 49:159–174
Jackson DC, Heisler N (1983) Intracellular and extracellular acid-base and electrolyte status of submerged anoxic turtles at 3 °C. Respir Physiol 53:187–201
Kawashiro T, Scheid P (1976) Measurement of Krogh's diffusion constant of CO_2 in respiring muscle at various CO_2 levels: evidence for facilitated diffusion. Pflügers Arch 362:127–133
Kern DM (1960) The hydration of carbon dioxide. J Chem Educ 37:14–23
Kerstetter FH, Kirschner LB (1972) Active chloride transport by the gills of rainbow trout (*Salmo gairdneri*). J Exp Biol 56:263–272
Kerstetter FH, Kirschner LB, Rafuse DD (1970) On the mechanism of sodium ion transport by the irrigated gills of rainbow trout (*Salmo gairdneri*). J Gen Physiol 56:342–359
Kutty MN (1972) Respiratory quotient and ammonia excretion in *Tilapia mossambica*. Mar Biol 16:126–133
Kortüm G (1972) Lehrbuch der Elektrochemie, 5th edn Verlag Chemie, Weinheim, p 263
Landolt H, Börnstein R (1960) Zahlenwerte und Funktionen, vol II (7). In: Hellwege KH, Hellwege AM, Schafer K, Lax E (eds) Elektrische Eigenschaften II (Elektrochemische Systeme). Springer, Berlin Göttingen Heidelberg 259:843–844
Longmuir IS, Forster RE, Woo CY (1966) Diffusion of carbon dioxide through thin layers of solution. Nature Lond 209:393–394
Maetz J (1973) Na^+/NH_4^+, Na^+/H^+ exchanges and NH_3 movement across the gill of *Carassius auratus*. J Exp Biol 58:255–275
Maetz J (1974) Adaptation to hyper-osmotic environments. Biochem Biophys Perspect Mar Biol 1:91–149
Maren TH (1967) Carbonic anhydrase: chemistry, physiology and inhibition. Physiol Rev 47:595–781

Moll W, Gros G (1966) Untersuchungen über den Gasaustausch der Erythrozyten. Pflügers Arch 291:54
Nikinmaa M (1983) Adrenergic regulation of haemoglobin-oxygen affinity in rainbow trout red cells. J Comp Physiol B152:67–72
Nikinmaa M (1986) Control of red cell pH in teleost fish. Ann Zool Fenn 23:223–235
Nissen P, Heisler N (1973) Pufferung und Verteilung von Bikarbonat in Erythrocytensuspensionen mit verschiedenen Zellkonzentrationen. Pflügers Arch 343:221–234
Obaid AL, Critz AM, Crandall ED (1979) Kinetics of bicarbonate/chloride exchange in dogfish erythrocytes. Am J Physiol 237:132–138
Oelert H, Uhlich E, Hills AG (1968) Messungen des Ammoniakdruckes in den cortikalen Tubuli der Rattenniere. Pflügers Arch 300:35–48
Payan P, Maetz J (1973) Branchial sodium transport mechanisms in *Scyliorhinus canicula*: evidence for Na^+/NH_4^+ and Na^+/H^+ exchanges and for a role of carbonic anhydrase. J Exp Biol 58:487–502
Peer MM, Kutty MN (1981) Respiratory quotient and ammonia quotient in *Tilapia mossambica* (Peters) with special reference to hypoxia and recovery. Hydrobiologia 76:3–9
Pequin L (1962) Les teneurs en azote ammoniacal du sang chez la carpe (*Cyprinus carpio* L.). CR Hebd Seances Acad Sci 255:1795–1797
Pequin L, Serfaty A (1963) L'excretion ammoniacale chez un teleosteen dulcicole: *Cyprinus carpio* L. Comp Biochem Physiol 10:315–324
Perry SF, Davie PS, Daxboeck C, Randall DJ (1982) A comparison of CO_2 excretion in a spontaneously ventilating blood-perfused trout preparation and saline-perfused gill preparations: contribution of the branchial epithelium and the red blood cell. J Exp Biol 101:47–60
Piiper J (1986) Gas exchange and acid-base status. In: Heisler N (ed) Acid-base regulation in animals. Elsevier Science, Amsterdam, pp 49–81
Piiper J, Schumann D (1967) Efficiency of O_2 exchange in the gills of the dogfish, *Scyliorhinus stellaris*. Respir Physiol 2:135–148
Primmett DRN, Randall DJ, Mazeaud M, Boutilier RG (1986) The role of catecholamines in erythrocytic pH regulation and oxygen transport in rainbow trout (*Salmo gairdneri*) during exercise. J Exp Biol 122:139–142
Radford EP (1964) The physics of gases. In: Fenn WO, Rahn H (eds) Handbook of physiology, Sect 3, vol I: Respiration. American Physiological Society, Washington, pp 125–152
Rahn H (1966) Aquatic gas exchange: theory. Respir Physiol 1:1–12
Randall DJ (1982a) The control of respiration and circulation in fish during exercise and hypoxia. J Exp Biol 100:275–288
Randall DJ (1982b) Blood flow through fish gills. Soc Exp Biol Semin Ser 16:173–191
Ryan US, Whitney PL, Ryan JW (1982) Localization of carbonic anhydrase in pulmonary artery endothelial cells in culture. J Appl Physiol 53:914–919
Schwartz JH, Tripolone M (1983) Characteristics of NH_4^+ and NH_3 transport across the isolated turtle urinary bladder. Am J Physiol 245:F210–F216
Smith HW (1929) The excretion of ammonia and urea by the gills of fish. J Biol Chem 81:727–742
Swenson ER, Evans DH (1984) Inhibition of gill carbonic anhydrase blocks CO_2 excretion in the isolated perfused head of the dogfish 'pup', *Squalus acanthias*. Bull Mt Desert Isl Biol Lab 24:45–49
Swenson ER, Maren TH, Hildesley MA (1984) Further studies on elasmobranch CO_2 and HCO_3^- excretion: roles of branchial and erythrocytic carbonic anhydrase in *Squalus acanthias*. Bull Mt Desert Isl Biol Lab 24:72–75
Thurston RV, Russo RC (1981) Ammonia toxicity to fishes. Effect of pH on the toxicity of the un-ionized ammonia species. Environ Sci Technol 15:837–840
Toews DP, Heisler N (1982) The effects of hypercapnia on intracellular and extracellular acid-base status in the toad *Bufo marinus*. J Exp Biol 97:79–86
Toews DP, Holeton GF, Heisler N (1983) Regulation of the acid-base status during environmental hypercapnia in the marine teleost fish *Conger conger*. J Exp Biol 107:9–20
Tomasso JR, Boudie CA, Simco BA, Davis KB (1980) Effects of environmental pH and calcium on ammonia toxicity in channel catfish. Trans Am Fish Soc 109:229–234

Van Slyke DD (1922) On the measurement of buffer values and on the relationship of buffer value to the dissociation constant of the buffer and the concentration and the reaction of the buffer system. J Biol Chem 52:525–570
Wardle CS (1978) Non-release of lactic acid from anaerobic swimming muscle of plaice *Pleuronectes platessa* L.: A stress reaction. J Exp Biol 77:141–155
Washburn EW (1929) International critical tables of numerical data. Physics, chemistry and technology, vol 3, 1st edn. McGraw-Hill, New York, pp 259
Whitney PL, Briggle TV (1982) Membrane-associated carbonic anhydrase purified from bovine lung. J Biol Chem 257:12056–12059
Wood JD (1958) Nitrogen excretion in some marine teleosts. Can J Biochem Physiol 36:1237–1242
Wood CM, Wheatly MG, Hōbe H (1984) The mechanisms of acid-base and ionoregulation in the freshwater rainbow trout during environmental hyperoxia and subsequent normoxia. III. Branchial exchanges. Respir Physiol 55:175–192
Woodbury JW (1965) Regulation of pH. In: Ruch TC, Patton HD (eds) Physiology and biophysics. Saunders, Philadelphia, pp 899–934
Wright PA, Heming T, Randall DJ (1986) Downstream pH changes in water flowing over the gills of rainbow trout. J Exp Biol 126:499–512
Wright PA, Randall DJ, Wood CM (1988a) An in vitro and in vivo study of the distribution of ammonia between plasma and red cells of rainbow trout (*Salmo gairdneri*). J Exp Biol 134:423–428
Wright PA, Randall DJ, Wood CM (1988b) The distribution of ammonia and H^+ between tissue compartments in lemon sole (*Parophrys vetulus*) at rest, during hypercapnia and following exercise. J Exp Biol 136:149–175

Chapter 8

Control and Co-Ordination of Gas Exchange in Water Breathers

D. Randall[1]

Contents

1 Introduction

Approximately equal amounts of oxygen and carbon dioxide[2] are transferred across the gills of fish, but in opposite directions, whereas ammonia[1] excretion is only 10 to 30% of oxygen uptake. This is, of course, a reflection of the metabolic utilization and production of these compounds by fish. The carbon dioxide/oxygen exchange ratio is usually between 0.7 and 1, whereas the ammonia/carbon dioxide exchange ratio is around 0.1 to 0.3. Variations in exchange ratios outside these levels result from changes in body stores of carbon dioxide and/or ammonia, but not oxygen, because these are too small to have a significant effect on oxygen uptake. Body stores of carbon dioxide and ammonia are about three times the excretion rate per hour, so changes in stores can have a marked, but neverthe-

[1] Zoology Department, University of British Columbia, 6270 University Boulevard, Vancouver, BC, V6T 2A9, Canada
[2] The term carbon dioxide refers to $(CO_2+H_2CO_3+HCO_3^-+CO_3^{2-})$, and ammonia to $(NH_3+NH_4^+)$

Advances in Comparative and Environmental Physiology
Vol. 6. Edited by R. G. Boutilier

less short-term effect on exchange ratios. Oxygen stores in fish, ignoring that in the swim bladder, are low and would supply the requirements of the fish for only a few minutes, so oxygen uptake from the environment is approximately equal to oxygen utilization. The swim bladder, however, can act as an oxygen store (as well as a float, sound guide and sound generating organ), the oxygen in the bladder being utilized during hypoxia (see Randall and Daxboeck 1984). Assuming that the swim bladder contains only oxygen and that the fish can utilize all the oxygen present, then the swim bladder oxygen could maintain the tissue oxygen requirements for several hours.

CO_2 is between 20 and 25 times more soluble than O_2 in water and tissues, and NH_3 is an order of magnitude more soluble than CO_2. Thus, the ratio of the permeation coefficients (diffusion coefficient × solubility) in water is around 1:23:32, 600 for O_2:CO_2:NH_3 respectively. Given flux rate ratios of 1:1:0.1 for O_2:CO_2:NH_3, then the ratio of the partial pressure gradients will be 1:1/23:1/32, 6000 or 1:0.0435:0.0000031. That is, assuming similar ratios for permeation coefficients for these compounds in gill tissue, if the O_2 gradient across the gills is 50 mmHg, then the CO_2 gradient will be 2.2 mmHg and the NH_3 gradient will be 16 μTorr. Thus, given a countercurrent exchange of blood and water at the gills, the arterial blood PCO_2 level will only be a few mmHg above that in the water, i.e. about 3 mmHg, as water levels are often close to zero. Blood NH_3 levels in fish are around 100 μTorr, higher than predicted from considerations of aqueous permeation coefficients and ammonia excretion rates. This difference is probably a reflection of the fact that lipid solubility for NH_3 in artificial membranes and probably, therefore, biological membranes, is an order of magnitude less than that for CO_2 (see review by Knepper et al. 1989). Thus, Cameron and Heisler (1983) found that the carbon dioxide and ammonia diffusing capacities of trout gills were similar.

Changes in gill ventilation will have a marked effect on blood oxygen levels, a minor effect on blood CO_2 levels, especially as HCO_3^-/Cl^- exchange across the red blood cell membrane is the rate-limiting step in CO_2 excretion (Perry et al. 1982), and a negligible effect on blood ammonia levels. Not surprisingly, the control of ventilation is strongly coupled to the maintenance of oxygen transfer requirements of the fish (Randall and Cameron 1973).

2 Oxygen Transfer

Not all the oxygen taken up by the fish enters the blood (Randall 1985). The skin, for example, exchanges gases directly with the environment but is not involved in gas transfer between the blood and water. Kirsch and Nonnotte (1977) estimated that trout skin consumes 15% and Johansen and Pettersson (1981) reported that cod gills consumed about 7% of the resting oxygen uptake of the fish. The skin is very active and derives nutrients from the blood but exchanges gases with the environment; in fact the blood vessels under the skin are part of the secondary circulation and contain plasma with a very low or zero hematocrit (Vogel 1985). Thus, only 80% of the oxygen consumed by the resting fish is transported in the blood, the remainder is consumed by surface structures which obtain oxygen directly from the environment.

Oxygen uptake across the gills is both perfusion and diffusion limited (see Piiper, this Vol., and review by Randall and Daxboeck 1984). Increases in oxygen uptake are achieved by increases in both blood and water flow and the diffusing capacity of the gills. Trout increase cardiac output with increases in swimming speed and oxygen uptake. In addition, erythrocytes are released from the spleen to raise hematocrit and increase blood oxygen capacity. Oxygen utilization by the tissues is ensured by adjusting conditions for oxygen transport in the blood and the distribution of blood flow.

Oxygen is carried in the blood between the gills and tissues bound to hemoglobin. The packed red cell volume (hematocrit) in fish is usually between 20% and 40% of blood volume. Oxygen delivery to the tissues is determined by cardiac output and arterial blood oxygen content. Many teleost fish have blood with a marked Root shift; that is, acid conditions reduce blood-oxygen capacity even at high PO_2 levels, a response important in unloading oxygen into the swim bladder. Acidotic conditions, therefore, could lead to a reduction in oxygen delivery to the tissues. This is ameliorated following burst swimming, hypercapnia and hypoxia, when acidotic conditions prevail, by the release of catecholamines.

Red blood cell pH decreases linearly with plasma pH in fish (Heming et al. 1986). Thus, if the hemoglobin has a Root shift, plasma acidosis will reduce red blood cell pH and cause a marked reduction in blood-oxygen content. Catecholamines, however, stimulate Na^+/H^+ exchange via β-adrenergic receptors in the red blood cell membrane of many teleosts (Nikinmaa 1982; Heming et al. 1987), resulting in an increase in red blood cell pH. This response is generally correlated with the presence of a Root shift (Tufts and Randall 1989). Circulating catecholamines are elevated following burst swimming, with a time course similar to that seen for the subsequent acidosis (Primmett et al. 1986). This release of catecholamines maintains red blood cell pH and ameliorates the effect of the acidosis on oxygen transfer, such that the fish can maintain oxygen delivery. The end result is that aerobic exercise is unimpaired following burst activity by the fish (Randall et al. 1987). The release of catecholamines has a large number of effects, many of which are directed towards enhancing oxygen transfer. These actions include gill hyperventilation (Peyraud-Waitzenegger 1979), an increase in gill epithelial permeability (Isaia et al. 1978), a more complete perfusion of the gills with blood (see review by Randall and Daxboeck 1984), increases in cardiac output (Farrell 1984), blood redistribution (Wood 1976), an elevation of red blood cell pH (Nikinmaa 1982; Primmett et al. 1986) and the release of erythrocytes from the spleen (Perry and Kinkead 1989). Thus, it appears that the effect of the increase in circulating catecholamines is to enhance oxygen transport from the water to the tissues.

Many fish have multiple hemoglobins within their erythrocytes (Vanstone et al. 1964). In coho salmon, however, only 3 of 18 hemoglobin components are present throughout the entire life cycle (Giles and Vanstone 1976). The three major components found in all life stages comprise 95% of the hemoglobin in fry blood, but only 55% of hemoglobin in adult blood (Giles and Vanstone 1976). The three ubiquitous forms are anodal starch-gel electrophoretic hemoglobin components, whereas the additional forms found in adults are predominantly cathodal components (Giles and Randall 1980; Sauer and Harrington 1988). The anodal forms have a high oxygen affinity, a large Bohr shift, a marked Root shift, and oxygen binding is sensitive to changes in temperature, organic phosphate levels and ionic strength (Giles and Randall 1980; Sauer and Harrington 1988). Conversely, oxygen binding by the cathodal hemoglobin forms is independent of pH, organic phosphate levels and temperature. Thus, salmonid freshwater fry have blood-oxygen dissociation curves that are very sensitive to changes in pH, organic phosphate levels and temperature, whereas adult blood curves are less sensitive to these changes.

Clearly, by adjusting the relative concentration of different hemoglobin components, the system can respond in a more complex way than would be allowed if the animal possessed only a single component. The probable functional significance of having hemoglobins in adult fish that are insensitive to pH and organic phosphate levels is that oxygen transfer can be maintained in the face of a marked acidosis resulting, for example, from burst activity (Giles and Randall 1980). High-affinity hemoglobins that are sensitive to changes in pH and organic phosphate levels allow fry to closely couple O_2 and CO_2 transfer and maintain an elevated flux rate of oxygen, especially in small fish where there is a more rapid transit of blood through the gills and tissues. In addition, fish can modulate hemoglobin : oxygen affinity and therefore maintain oxygen transfer during hypoxia (Boutilier et al. 1988).

Red blood cell size does not appear to play a critical role in oxygen transfer, because triploid salmon swim just as well as their diploid siblings (Small and Randall 1989). These triploid fish had red blood cells that were 1.5 times the size of normal erythrocytes, but had a lower total blood hemoglobin concentration than their diploid siblings, although they had the same hematocrit.

3 Carbon Dioxide Transfer

CO_2 represents only 1%–5% of total carbon dioxide, and there are large stores of HCO_3^- in the body. Factors determining HCO_3^- distribution in tissues, other than blood, of aquatic vertebrates have not been investigated.

Carbon dioxide is excreted largely as molecular CO_2 by fish but stored as bicarbonate within the body. Thus, bicarbonate dehydration is involved in carbon dioxide excretion. In trout, and probably other fish, erythrocytes are required for carbon dioxide excretion across the gills and, under conditions of controlled blood flow and blood carbon dioxide content, excretion is proportional to hematocrit (Perry et al. 1982; Perry 1986). There appears to be no functional carbonic anhydrase activity in the plasma or on the respiratory endothelium in fish. Carbonic anhydrase activity in fish red blood cells is high and is required to catalyze carbon dioxide hydration/dehydration reactions in the plasma. Blood transit time through the gills is around 100 ms (Randall 1982), whereas the uncatalyzed rate of bicarbonate dehydration is of the order of seconds. Thus, plasma bicarbonate must enter the red blood cell to be excreted across the gills. The trout

Table 1. Characterization of the anion transport system in trout red blood cells (Romano and Passow 1984)

	Trout	Human
Band 3 molecules ($cell^{-1}$)	8×10^6	1×10^6
Cell surface (cm^2)	2.67×10^{-6}	1.42×10^{-6}
Band 3 molecules (cm^{-2})	30×10^{11}	7×10^{11}
Half time for Cl^- ion exchange (s)		
0 °C	3.42	17.2
10 °C	1.29	2.32
15 °C	0.81	0.89
38 °C		0.05

erythrocyte contains high levels of band 3 protein (Romano and Passow 1984), which transfers chloride for bicarbonate across the red blood cell membrane (Table 1). Trout rbc membranes have approximately three times the density of this protein than human rbc membranes, and transfer chloride at a faster rate at temperatures below 15 °C. Lampreys, on the other hand, appear to lack band 3 protein in their red blood cells and plasma bicarbonate dehydration is not facilitated by carbonic anhydrase in the red blood cell (Nikinmaa and Railo 1987). Patterns of carbon dioxide excretion have not been investigated in lampreys, nor in antarctic fish which lack red blood cells. The latter are most interesting because they exist at temperatures around 0 °C and the uncatalyzed carbon dioxide hydration/dehydration reaction is very slow.

Tufts et al. (1989) investigated carbon dioxide flux through trout red blood cells in vitro, and found that blood taken from exercised fish had a more rapid bicarbonate/carbon dioxide transfer through the erythrocyte than blood taken from resting fish. This observation indicated stimulation of Cl^-/HCO_3^- exchange and/or bicarbonate dehydration in erythrocytes of exercising fish. Tufts et al. (1989) could find no indication of catecholamine inhibition of erythrocytic bicarbonate dehydration, suggested by Perry and Wood (1989) as an explanation for the elevated $PaCO_2$ observed following exercise in fish. Steffensen et al. (1987) were also unable to detect any reduction in carbon dioxide excretion following burst swimming in trout, when circulating levels of catecholamines were increased. In addition, hematocrit is elevated during exercise, due to release of red blood cells from the spleen, and this should augment carbon dioxide excretion. Blood transit time through the gills, however, is reduced during exercise and carbon dioxide content of venous blood is probably increased, in addition, the bicarbonate gradient between plasma and erythrocytes may be reduced (due to a catecholamine induced rise in erythrocytic pH), resulting in a reduced bicarbonate flux through each RBC. All of these factors may contribute to the elevation of $PaCO_2$ seen after exercise in fish.

Carbamino formation appears to play a minor role in carbon dioxide transfer in fish (Heming et al. 1986); the terminal NH_2 group of the α-chain of the hemoglobin molecule in many fish is unavailable for carbon dioxide binding, and carbamino formation is restricted to the terminal NH_2 group of the β-chain, in competition with organic phosphates (Riggs 1979). Carbamino formation increased, as expected, with pH and carbon dioxide levels in the blood (Heming et al. 1986), but no studies have been carried out on the interaction between carbamino formation and organic phosphate binding to hemoglobin in fish blood.

4 Ammonia Transfer

At a blood pH of 7.9, 98% of ammonia exists as NH_4^+, with the small remainder as NH_3. Because NH_3 is a nonpolar gas it diffuses rapidly across biological membranes down its partial pressure gradient at about the same rate as CO_2 (Thomas 1974; Cameron and Heisler 1983). Membranes are generally less permeable to NH_4^+, with its large hydrated diameter and net charge (Jacobs 1940). Thus, ammonia could be partitioned between tissue compartments, with NH_3 in equilibrium across the membrane and $[NH_4^+]$ determined by pH of the compartment. In tissue compartments of lemon sole (Wright et al. 1988) and rainbow trout (Wright and Wood 1988), however, intracellular ammonia levels are considerably greater than predicted from transmembrane pH gradients. In fact, these studies indicate that fish muscle membranes have a considerable permeability to NH_4^+, and the distribution of ammonia is not determinded by pH gradients, but rather, the cell membrane potential. During severe exercise, muscle adenylates are broken down (Driedzic and Hochachka 1976), resulting in large increases in muscle ammonia levels. A portion of the ammonia produced during exhaustive exercise is re-utilized in situ (Mommsen and Hochachka 1989), but the remainder is excreted. The fact that muscle membranes are permeable to NH_4^+ leads to enhanced ammonia removal following violent exercise when muscle ammonia levels are high, but leads to considerable NH_3 and NH_4^+ cycling in the resting animal and an increased load on the proton extrusion system (Wright et al. 1988; Fig. 1).

Biological membranes are readily permeable to NH_3, but NH_3 gradients across the gills are often very small, so fish probably modulate NH_4^+ transfer (via changes in either Na^+/NH_4^+ exchange, NH_4^+ membrane permeability, or NH_4^+ substitution for K^+ in Na^+/K^+ ATPase or Na^+-K^+-$2Cl^-$ cotransporter, see Knepper et al. 1989 for review) to ensure adequate rates of ammonia excretion. Little is known about the modulation of NH_4^+ transport in fish except that the gill Na^+/NH_4^+ exchange mechanism is inhibited at low and high water pH (Wright and Wood 1985).

At high pH, NH_3 levels in the blood rise with increases in the NH_3/NH_4^+ ratio in the water next to the gills (Lin and Randall 1990). A small increase in $[NH_3]$ results in a large increase in total ammonia in the blood. Ammonia is toxic to all vertebrates; however, fish appear to be less sensitive than most (Wright 1987). Ammonia toxicity can be ameliorated by converting ammonia to less toxic forms like urea and glutamine. Teleosts produced some urea via uricolytic pathways, the ornithine-urea cycle being repressed. An exception has been reported for a single aquatic species of Tilapia, existing in very alkaline conditions (Randall et al. 1989). This fish excretes only urea and has a very high capacity to produce

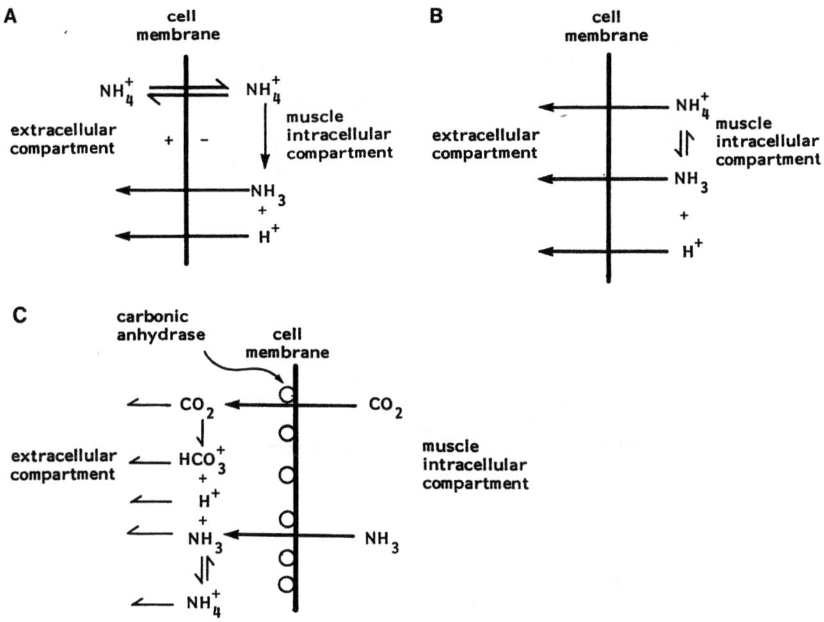

Fig. 1. A model of ammonia (NH_3 and NH_4^+) movements between the extracellular and intracellular compartments of muscle tissue where NH_4^+ ions are passively distributed across the membrane according to the membrane potential (Wright et al. 1988). *A* In the intracellular compartment where there is no ammonia production, the direction of the intracellular $NH_4^+ \rightleftharpoons NH_3$ reaction will always be towards the formation of NH_3 and H^+, because intracellular H^+ ions are lower than expected for a passive equilibrium. The net result is an increase in intracellular H^+ ions levels and NH_3 gradients directed from cell to plasma. *B* In the active state, there may be both NH_3 and NH_4^+ gradients from cell to plasma. Intracellular NH_3 production will trap H^+ ions and ammonia leaving as NH_4^+ will act to export H^+ ions from the cell, while NH_3 diffusion out of the cell will not affect H^+ ions levels. *C* Carbonic anhydrase, bound to the sarcolemma, catalyzes the hydration of excreted CO_2 from the muscle tissue to form HCO_3^- and H^+ ions. This CO_2 hydration reaction facilitates ammonia efflux from the muscle by forming an acid boundary layer near the sarcolemma. Capillary NH_3 levels will be low, as excreted NH_3 combines with H^+ ions to form NH_4^+ ions, thereby enhancing NH_3 diffusion gradients from the muscle (Randall and Wright 1990)

urea via the ornithine-urea cycle. Elasmobranchs also produce urea via the ornithine-urea cycle, but utilize it as an osmotic filler. Some fish can ameliorate oscillations in body ammonia levels by glutamine formation (see review by Randall and Wright 1987).

5 Interactions Between Carbon Dioxide and Ammonia Excretion

Carbon dioxide and ammonia exist in the body in both ionized and unionized forms. The ratio of carbon dioxide (CO_2) to bicarbonate (HCO_3^-) and ammonia (NH_3) to ammonium ion (NH_4^+) varies with pH (Fig. 2), but the pK of these re-

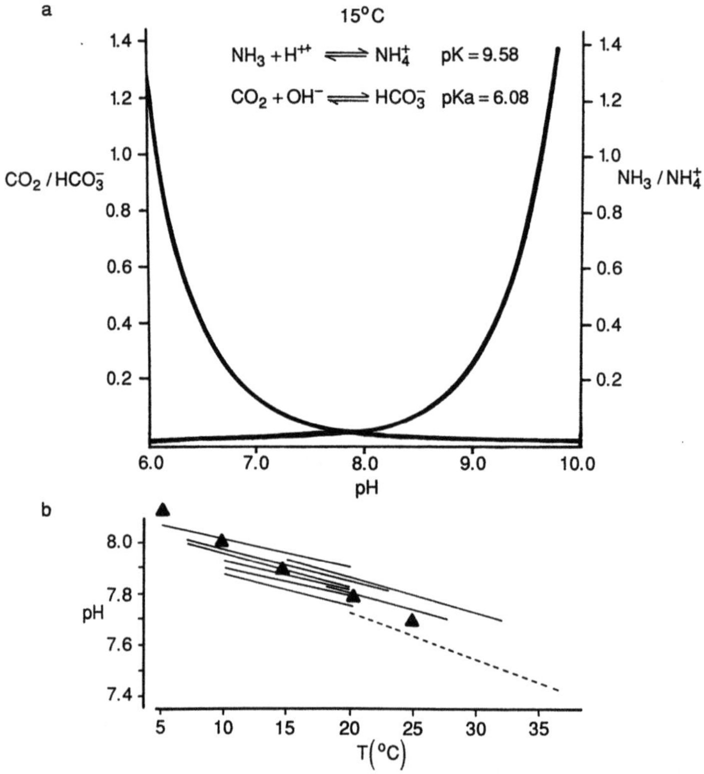

Fig. 2. A The effect of varying pH on the CO_2/HCO_3^- and NH_3/NH_4^+ ratio in trout plasma at 15 °C. *B* The pH where the ratios are equal have been calculated at different temperatures and added to a graph of variations in plasma pH with temperature for several fish (Randall and Wright 1989)

actions is very different. The pK of the NH_3/NH_4^+ reaction is about 9.5, whereas the apparent pK of the CO_2/HCO_3^- reactions is around 6.1 (Boutilier et al. 1985). The CO_2/HCO_3^- and NH_3/NH_4^+ ratios are equal at the pH where the two lines cross in Fig. 2, at the midpoint between the pK of the NH_3/NH_4^+ reaction and the apparent pK of the CO_2/HCO_3^- reaction. The pH at this point is similar to the pH of fish blood at the same temperature. The midpoint between these two pKs varies with temperature in much the same way as blood pH changes in fish. The temperature coefficients are not identical, but in general there is a similarity between blood pH and the midpoint between the pK of these two reactions over a range of temperatures (Fig. 2). Epithelia are usually not very permeable to either HCO_3^- or NH_4^+ but are very permeable to CO_2 and NH_3. Thus, CO_2 and NH_3 will be the predominant forms excreted, as long as adequate blood-to-water CO_2 and NH_3 gradients exist. This will occur if blood pH is maintained at the midpoint between the pK of the two reactions and environmental levels of NH_3 and CO_2 remain low.

Excretion of CO_2 and NH_3 will be influenced by the composition of water flowing over the gill. CO_2 and NH_3, along with some HCO_3^- (Perry et al. 1982)

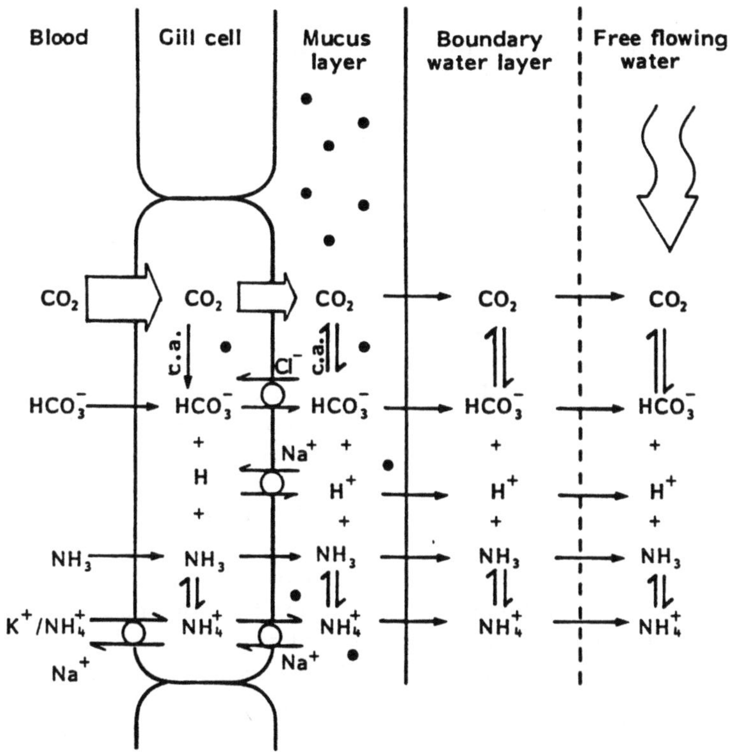

Fig. 3. Schematic cross-section through the gill epithelium, mucus and water boundary layer, containing carbonic anhydrase (●). The *thickness of the arrows* denotes the approximate magnitude of the particular process illustrated (Wright 1987)

and NH_4^+ (Wright and Wood 1985), are excreted into the mucous and water boundary layer next to the gill surface (Fig. 3). Bicarbonate ions represent only a small portion of the carbon dioxide excreted, probably less than 10% (see review by Randall and Wright 1989), and the amount of ammonia excreted as NH_3 varies between 45% and 100% (Cameron and Heisler 1983; Wright and Wood 1985). As long as the pH is above 6.1, then most of the CO_2 excreted into the mucous and boundary layer will form HCO_3^- and acidify the water. Water is in contact with the gill for no longer than a few hundred milliseconds, whereas the uncatalyzed CO_2 hydration reaction velocity is of the order of minutes. Wright et al. (1986), however, showed that this reaction was catalyzed by carbonic anhydrase in the mucous layer. The mucous layer contains a large number of damaged gill epithelial cells, which are known to contain high levels of carbonic anhydrase (Haswell et al. 1980; Lacy 1983), and the external apical surface of the gill also contains carbonic anhydrase activity (Rahim et al. 1988). Thus, CO_2 entering the boundary layer will acidify the mucous and boundary layer next to the gill surface (Wright et al. 1986). The extent of acidification will be determined by the rate of carbon dioxide excretion in relation to water flow and the buffering capacity and pH of the water. NH_3 excretion will have the opposite effect, but because more

CO_2 than NH_3 is excreted across the gills, the overall effect will be an acidification of the boundary layer, when water pH is above 6.

NH_3 entering this acidified layer next to the gill surface will be converted to NH_4^+ and will diffuse out of the boundary water layer into the bulk medium. The conversion of NH_3 to NH_4^+ will reduce NH_3 levels in the water and, therefore, NH_3 levels in the blood. Starving trout have elevated blood NH_3 levels (Hillaby and Randall 1979) and this can be correlated with a decreased CO_2 excretion and, therefore, a reduced acidification of the water boundary layer. Wright et al. (1989) were able to show a direct interaction between CO_2 and NH_3 in the boundary layer of the trout gill, using a perfused trout head preparation. They showed that a reduction in carbon dioxide excretion reduced the acidification of the mucus and water boundary layer next to the gill and this decreased ammonium ion formation and, therefore, NH_3 excretion.

Ammonia transfer from muscle tissue also may be facilitated by an acid boundary layer next to the sarcolemma (Fig. 1 c). It has been shown in mammals that carbonic anhydrase facilitates CO_2 excretion from muscle tissue by catalyzing the CO_2 hydration reaction (Zborowska-Sluis et al. 1974). It is possible that NH_3 diffuses out of the muscle into an acid boundary layer and forms NH_4^+, which is then carried away in the blood. The maintenance of low extracellular NH_3 levels would enhance NH_3 diffusion from the muscle, especially during exhaustive exercise when there are marked increases in muscle ammonia levels. The significance of this, however, may be less in fish than in mammals, because fish appear to have a much higher membrane permeability to NH_4^+. Nevertheless, trout pre-injected with carbonic anhydrase do have lower muscle ammonia levels following exercise than those fish pre-injected with saline (C. M. Wood, pers. commun.).

6 Sites of Control of Gas Transfer

In order to increase the transfer of oxygen from the water to the tissues, the animal must increase gill water flow, gill diffusing capacity, blood flow, blood oxygen transport and conditions for transfer in tissues. These are kept at a low level when oxygen requirements are low, in order to save energy and reduce the exchange of ions and water with the environment. For example, when trout exercise, the turnover of water increases and this is reflected in a linear increase in urine production with exercise (Wood and Randall 1973).

If conditions are set for oxygen transfer then, because carbon dioxide and ammonia are more soluble in aqueous solutions and have larger permeation coefficients than oxygen, conditions will be more than adequate for carbon dioxide and ammonia transfer. In fact, increases in gill water flow have only a minor effect on carbon dioxide flux (Iwama et al. 1987); the rate-limiting step in carbon dioxide excretion appears to be HCO_3^-/Cl^- exchange across red blood cell membrane (Perry et al. 1982). Thus, one would expect any control of carbon dioxide excretion to be centred on regulation of band III protein in the red blood cell membrane. Cl^-/HCO_3^- exchange across the apical surface of gill epithelium is only a minor component of carbon dioxide excretion and is probably modulated to maintain blood pH (Cameron 1976). Little is known about the regulation of ammonia excretion. Na^+/NH_4^+ exchange is inhibited by high and low water pH and amiloride (Wright and Wood 1985), but there is no information on modulation of either passive permeability or transport of NH_4^+ across gill membranes to enhance ammonia excretion. Some fish convert toxic ammonia and excrete the less toxic compound, urea (Randall et al. 1989), or buffer oscillations in ammonia levels by converting ammonia to glutamine (Randall and Wright 1987).

The control sites determining the rates of oxygen, carbon dioxide and ammonia transfer in the body are clearly different. For example, gill water and blood flow and gill diffusing capacity, as well as blood oxygen transport characteristics, must be adjusted to meet the requirements for oxygen transfer. Thus, it seems probable that there is direct or indirect monitoring of aspects of oxygen transfer, in order to adjust the system to meet the oxygen transfer requirements of the animal. On the other hand, regulation of carbon dioxide transfer is likely to be related to modulation of chloride/bicarbonate exchange across the red blood cell membrane, whereas ammonia transfer may well be adjusted by changes in NH_4^+ permeability and/or transport across membranes.

7 Control Systems

7.1 Gill Ventilation

In general, aquatic vertebrates breathe continuously, maintaining a unidirectional flow of water over their gills to ensure adequate rates of oxygen, carbon dioxide and ammonia transfer (Figs. 4, 5). There are exceptions: Goldfish and other cyprinids, for example, stop breathing for prolonged periods during anoxia and supply their energy requirements anaerobically (Shoubridge and Hochachka 1979), with ethanol as an end product and presumably no production of CO_2 or NH_3. Many fish show periodic breathing when at rest and undisturbed, especially in hyperoxic water. Lamprey exhibit tidal ventilation of their gill pouches (Randall 1972), an adaptation associated with the fact that their mouth is often buried in the flesh of other animals. Similarly, sturgeon ventilate their gills tidally through the opercular flaps, when their mouth is buried in mud (Burggren 1978). The general rule, however, is a continuous, unidirectional flow of water over the gills.

The general assumption is that ventilation is adjusted to meet the metabolic requirements of the animal. In some cases, however, metabolism is adjusted to the availability of oxygen. For example, sturgeon decrease metabolism during hypoxia, such that energy expenditures are determined by oxygen supply. In other instances, energy expenditures, in the face of a reduction in aerobic metabolism due to environmental hypoxia, are maintained by anaerobic metabolism. Also, in some fish, increases in energy expenditure are achieved by anaerobic as well as aerobic means. Thus, there is not always a simple relationship between gill ventilation and the energy requirements of the fish.

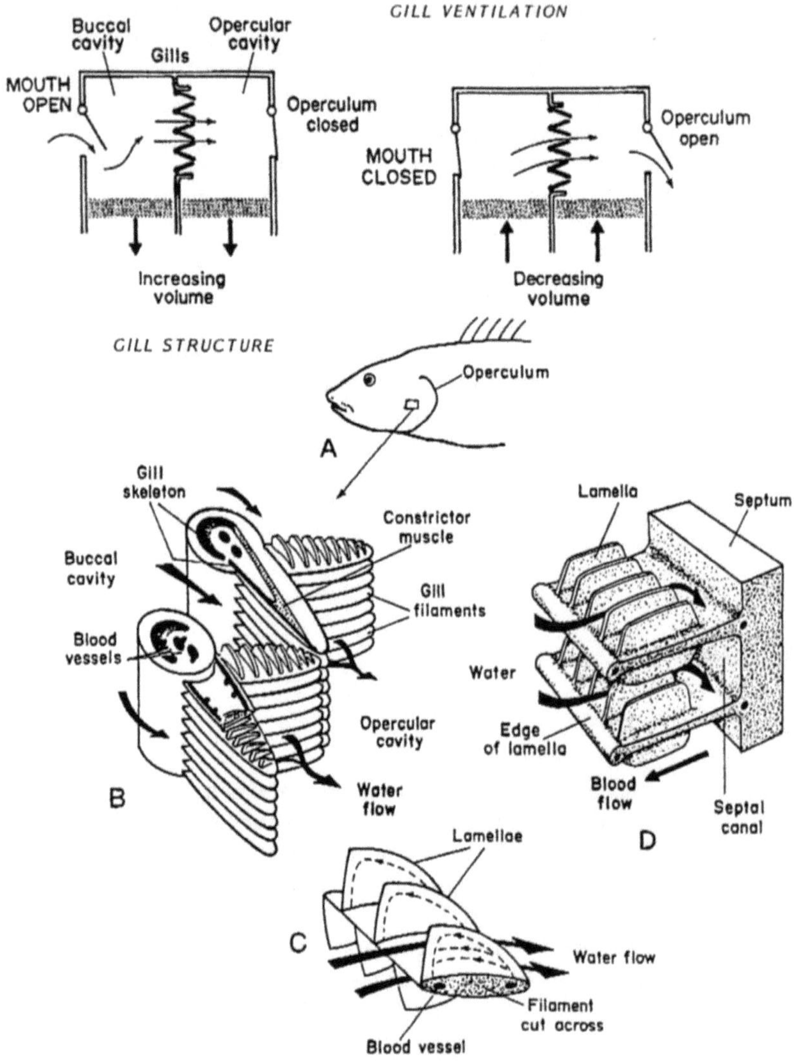

Fig. 4. The structure and mode of ventilation of fish gills. *A*, *B* and *C* are increasing magnifications of portions of a teleost gill; *D* is a diagram of a portion of an elasmobranch gill (After Eckert et al. 1988)

Water flow over the gills of fish is maintained by the combined action of buccal and opercular pumps. These pumps depend on the contractions of skeletal muscles, controlled by motoneurons with cell bodies in the medulla, as in other vertebrates. Forward motion of fish, if the mouth is suitably placed facing forward, either enhances or maintains gill water flow; that is, gill water flow is aided or maintained by ram ventilation.

It has been estimated (see the review by Shelton et al. 1986) that there are at least 2000 neurons in the medulla of tench that show rhythmic activity in phase

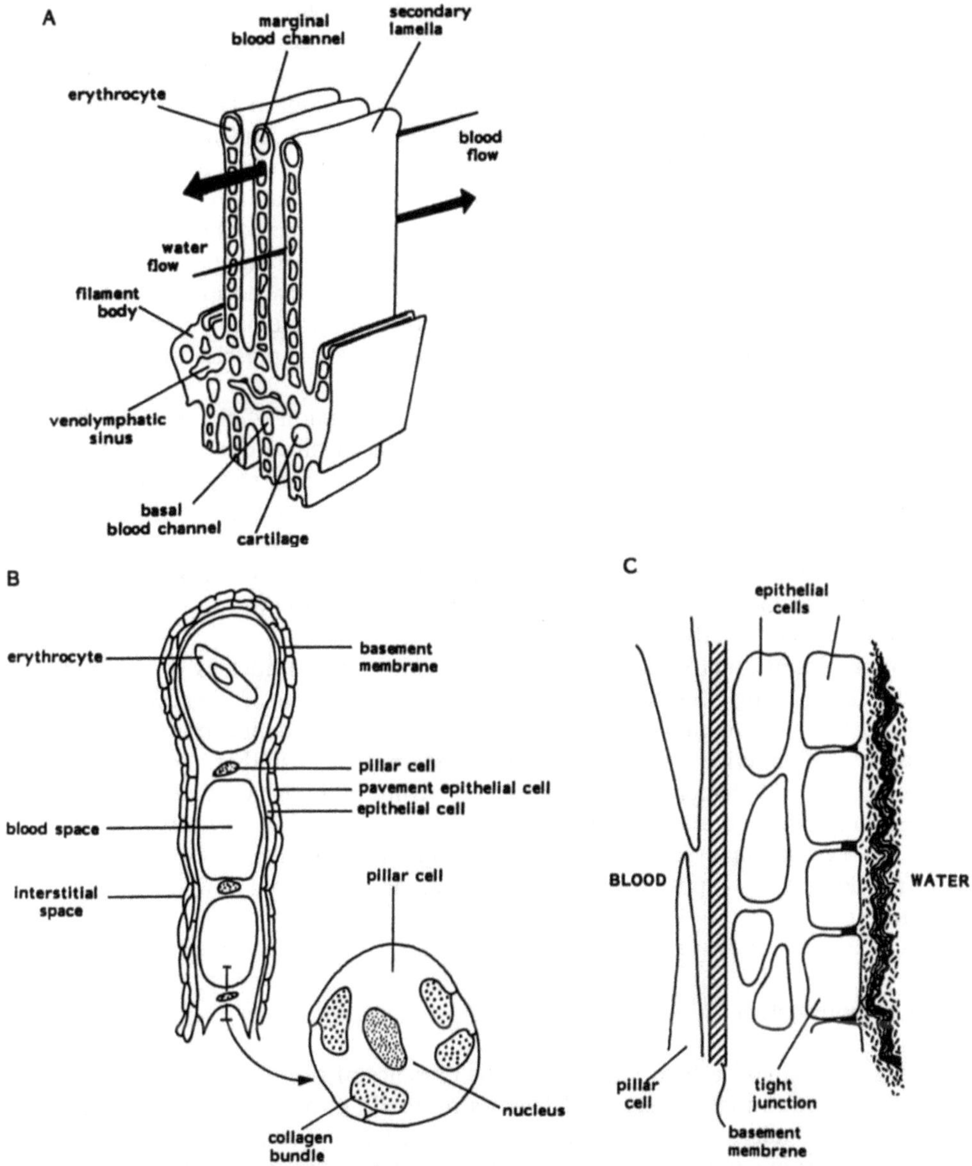

Fig. 5. Schematic of sections through a trout gill drawn to scale. *A* Section through a portion of a filament and several lamellae; *B* section through a lamella and a pillar cell; *C* section through the gill lamellar epithelium (After Randall 1982 and Randall 1989)

with breathing and that between 5% and 40% are branchial motoneurons (Ballintijn and Alink 1977; Waldron 1972). Activity in these neurons maintains the sequential contraction of a series of muscles controlling the buccal and opercular pumps. In some instances neuronal activity is confined to a discrete phase of the breathing cycle, but, in other cases, a neuron may be active for most of the cycle.

This, to some extent, reflects the timing of muscular activity; some muscles like the adductor mandibulae show strong phasic activity, whereas the sternohyoideus can be active throughout most of the breathing cycle. The motoneurons extend dorso-laterally throughout the medulla (Shelton 1970; Waldron 1972) and comprise the trigeminal Vth, facial VIIth, glossopharyngeal IXth, and vagal Xth motor nuclei. These motonuclei and the descending trigeminal nucleus and the reticular formation constitute the medullary respiratory centre (Ballintijn 1982). Barrett and Taylor (1985) showed that the respiratory motoneurons of the dogfish fire in order of the sequential rostro-caudal distribution of their motonuclei in the brainstem. In fact, there are extensive interconnections between all respiratory motonuclei and the reticular formation (Ballintijn 1988), presumably to maintain the complex sequencing of activation of the various skeletal muscles involved in gill ventilation.

The rhythmic neuronal activity continues after isolation of the medulla by brain and nerve transection, and following paralysis of all respiratory muscles. Ballintijn (1988) has suggested that there is a central pattern generator in the reticular formation that remains active during anesthesia and in the absence of peripheral input. Paralysis of respiratory muscles with curare slows the respiratory rhythm in teleosts (Ballintijn 1972) but increases the rhythm in elasmobranchs (Barrett and Taylor 1985), indicating that peripheral inputs have an overall excitatory and inhibitory effect respectively.

The respiratory moto-nuclei receive inputs via the facial nucleus from vagal afferents innervating gill mechanoreceptors. In addition, areas of the mesencephalic tegmentum in the midbrain, with afferent and efferent connections to the reticular formation, initiate periodic breathing seen in some fish that are either inactive or exposed to hyperoxic conditions (Ballintijn 1988). Fish change breathing pattern in response to exercise, hypoxia, feeding and probably many other conditions, so there are many other inputs to the respiratory neurons, presumably including descending inputs from the forebrain and midbrain and peripheral chemoreceptors.

7.2 Chemoreceptors

There is considerable evidence indicating the presence of two groups of chemoreceptors located in the gill region of teleosts responding to hypoxia (see Shelton et al. 1986 for references). The external oxygen-sensitive chemoreceptors on the gill surface respond to aquatic hypoxia and, when stimulated, cause a bradycardia (Smith and Jones 1978). There is some evidence to indicate that they may be more concentrated on the first gill arch (Daxboeck and Holeton 1978). The internal oxygen-sensitive chemoreceptors appear to be blood-oxygen content receptors, mediating increases in gill ventilation in response to hypoxemia (Fig. 6; see Randall 1982 for references). The major reasons for assuming two groups of receptors is based on observations that aquatic hypoxia generates a bradycardia, whereas hypoxemia in the presence of aquatic normoxia does not, but causes an increase in gill ventilation. In addition, externally applied NaCN results in a bradycardia in catfish, whereas internally applied NaCN causes an increase in breathing rate

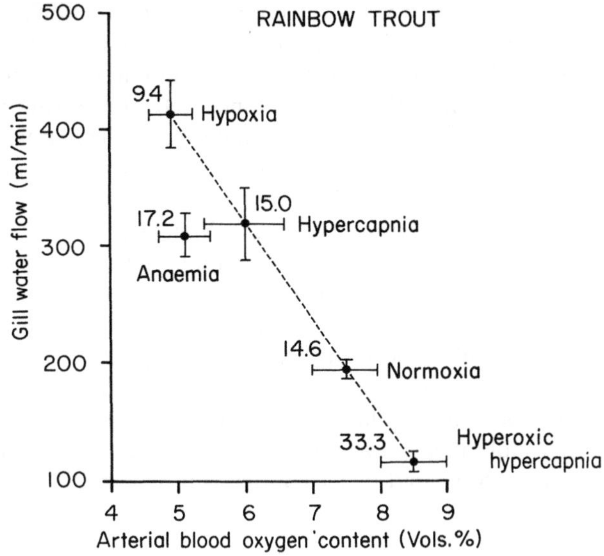

Fig. 6. The relationship between water flow over the gills and arterial oxygen content in trout. The *lines* indicate ± SE. The *number* by each point is the corresponding oxygen tension in KPa (Randall 1982, based on data from Smith and Jones 1982)

(Burleson and Smatresk 1989). There is a possibility that the external receptors have a small hyperventilatory effect (Saunders and Sutterlin 1971; Eclancher 1972); however, under the conditions of their experiments, internal as well as external receptors could have been stimulated. The effects of carbon dioxide on these chemoreceptors, either directly or in modifying the response to hypoxia, has not been investigated.

Both internal and external oxygen-sensitive chemoreceptors are located in the gills, the external receptors facing the water flow, the internal receptors monitoring blood oxygen content and flow. Dunel-Erb et al. (1982) described a system of neuroepithelial cells resembling the oxygen-sensitive glomus cells of mammals; these could be the internal oxygen-sensitive chemoreceptors in fish. Milsom and Sadig (1983) found that the β-adrenergic receptor blocker propranolol depressed the rabbit chemoreceptor response to hypoxia. Ventilatory responses of trout to hypoxia are also blocked by pretreatment with propranolol (Aota et al. 1990). Thus, it is possible that fish peripheral chemoreceptors, like those of mammals, are blocked by propranolol, and that similarities exist between chemoreceptors in these two groups. The pseudobranch has chemosensitive elements (Laurent and Rouzeau 1972) but these do no appear to be involved in the responses to hypoxia in trout (Randall and Jones 1973).

These oxygen chemoreceptors cause changes in breathing and/or heart rate via neuronal pathways because the responses are rapid, occurring within seconds of the appearance of hypoxia (Holeton and Randall 1967; Butler and Taylor 1971; Eclancher and Dejours 1975). The ventilatory response of the intact animal, however, is likely to involve more than a simple chemoreceptor reflex. Catecholamines are released by fish during hypoxia (Boutilier et al. 1988) and are known to stimulate breathing (Peyraud-Waitzenegger 1979). Thus, inhibition of the ventilatory response of trout to hypoxia by propranolol is likely to be more complex than a simple block of chemoreceptor input (Aota et al. 1990). As in mammals (Folgering 1980), propranolol has a central respiratory depressive effect in fish (Taylor and Randall 1989) and blocks the central stimulation of breathing caused by increased circulating levels of catecholamines, released during hypoxia (see below).

Chemoreceptors may not be involved in the hypoxic stimulation of ventilation in elasmobranchs. There is little change in ventilation in stressed elasmobranchs in response to hypoxia, but when care is taken not to disturb the fish, there is an increase in ventilation (Butler and Metcalfe 1983). The increase in ventilation in elasmobranchs during hypoxia could be due entirely to an increase in circulating levels of catecholamines, which increase in hypoxia in undisturbed fish but are already high in stressed fish, when no ventilatory response to hypoxia is observed. Thus, the gill blood oxygen content receptors, if present in elasmobranchs, may be involved in catecholamine release rather than having any direct action on the respiratory centre. The external group of gill oxygen receptors presumably exists in elasmobranchs, however, because hypoxia is associated with a marked bradycardia with a rapid onset (Butler and Taylor 1971).

7.3 Catecholamines

Catecholamines stimulate ventilation in fish (Peyraud-Waitzenegger 1979; Aota et al. 1990; Taylor and Randall 1989) but the response is seasonal. Peyraud-Waitzenegger et al. (1980) showed that winter eels responded to catecholamine infusion by hypoventilation, modulated via α-adrenergic receptors, whereas summer eels showed hyperventilation dependent on stimulation of β-adrenergic recep-

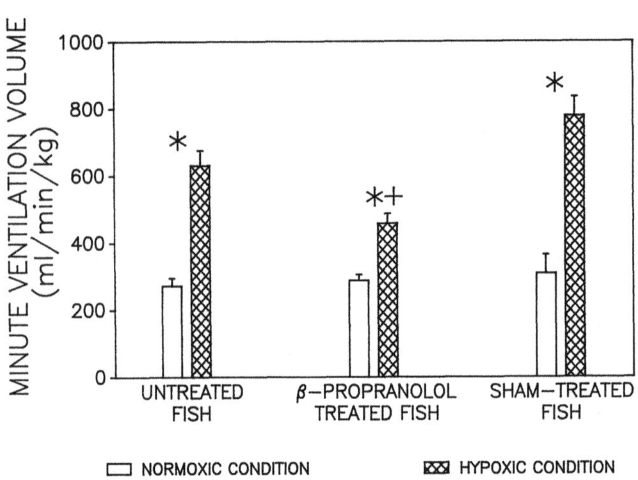

Fig. 7. The effect of the β-adrenoceptor antagonist, propranolol, on the minute ventilation volume of rainbow trout, *Oncorhynchus mykiss*. Hypoxia measurements were taken 30 min after the onset of hypoxia, 1 h after propranolol or sham treatment. * Indicates a significant increase of minute ventilation volume during hypoxia as Compared to control. + indicates a significant difference between treatments in the hypoxic condition. A significance level of α = 5% was used (Aota et al. 1990)

tors. Catecholamines are released into the circulation of fish in response to a wide variety of stresses, including hypoxia, hypercapnia, anaemia, blood acidosis, air exposure, and both exhausting and violent exercise (see Perry et al. 1989 for references). However, the proximate cause for the release appears to be hypoxemia (Perry et al. 1989), perhaps modulated by the acid-base status of the animal.

Acid infusion and hypoxia (Aota et al. 1990) cause an increase in both circulating catecholamines and gill ventilation, the hyperventilation being blocked by pretreating the fish with propranolol (Fig. 7). The simplest explanation of these observations is that hypoxia and acid infusion result in the release of catechol-

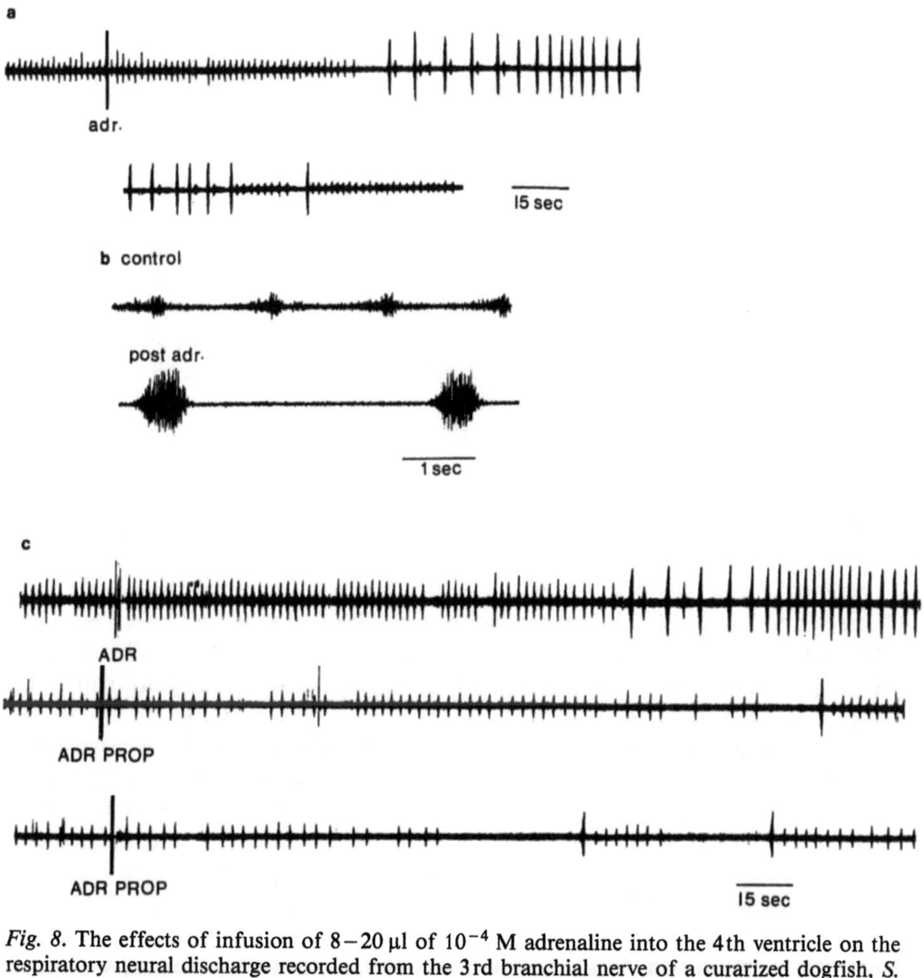

Fig. 8. The effects of infusion of 8–20 µl of 10^{-4} M adrenaline into the 4th ventricle on the respiratory neural discharge recorded from the 3rd branchial nerve of a curarized dogfish, *S. acanthius. a* Burst activity associated with breathing recorded from the 3rd branchial nerve, before and after injection of adrenaline (adr.) into the 4th ventricle. After several seconds the rate slows but the amplitude of each burst increases. *b* Comparison of bursts of activity in the 3rd branchial nerve recorded before and after injection of adrenaline into the 4th ventricle. *c* The effects of adrenaline (*ADR*) on branchial nerve discharge are blocked by the simultaneous injection of propranolol (*PROP*) (After Taylor and Randall 1989)

amines which in turn increase ventilation; the pathway is blocked by propranolol and so involves β-adrenergic receptors. Taylor and Randall (1989) reported that intravenous injection of noradrenaline into paralyzed, hyperoxic dogfish caused a marked but transient increase in respiratory drive. There was also a spread of this drive into the hypobranchial nerve. Injections of small volumes of catecholamines into the 4th ventricle of dogfish also stimulated respiratory drive, initially increasing rate but then slowing rate, accompanied by a huge increase in the activity within each burst (Fig. 8). The response was inhibited by a simultaneous injection of propranolol, which, as in mammals, depresses ventilation. Noradrenaline is known to cross the blood brain barrier of fish (Peyraud-Waitzenegger et al. 1979; Nekvasil and Olson 1986), so elevations in blood catecholamine levels will lead to a hyperventilation by direct stimulation of respiratory drive in the medulla. As outlined above, many stresses cause the release of catecholamines into the blood, and this release could be an important factor in mediating any associated hyperventilation. In the case of hypoxia, the increase in ventilation in the trout is far too rapid to be solely due to the central effect of circulating catecholamines. Stimulation of gill oxygen chemoreceptors presumably drive the initial increase in ventilation. The subsequent rise in catecholamines during severe hypoxia (Boutilier et al. 1988), however, could play an important role in the stimulation of ventilation during maintained hypoxia. Propranolol blocks the ventilatory increase presumably by blocking both peripheral chemoreceptors and central sites of action. In elasmobranchs the onset of hyperventilation in response to hypoxia is slower than in teleosts, and in stressed fish there is no hypoxic hyperventilation. It is possible that, in elasmobranchs, the increased release of catecholamines into the circulation may account for all of the increase in breathing.

7.4 CO_2/H^+ Receptors

Hypercapnic and acidotic states cause an increase in ventilation in teleosts (Janssen and Randall 1975) and elasmobranchs (Randall et al. 1976; see Perry and Wood 1989 for other references). Hypercapnic hyperventilation in teleosts is, at least in part, a response to hypoxemia associated with a Root and Bohr shifts and the developing blood acidosis. Smith and Jones (1982) were able to ameliorate the ventilatory response to hypercapnia in trout by exposing the fish to hyperoxia and so raising blood-oxygen content. Not all of the ventilatory response to hypercapnia, however, can be accounted for by changes in blood-oxygen content; there is no Root shift in elasmobranchs, and there is not a clear relationship between blood content and ventilation under all conditions of hypercapnia (see Perry and Wood 1989 for further discussion and references). As a result, it is possible that some sort of central H^+ receptor, along the lines of that postulated to exist in mammals, exists in fish. Gill ventilation, however, is not well correlated with cerebral spinal fluid pH (Perry and Wood 1989). In addition, hyperoxic hypercapnia, although causing an acidosis, can result in hypoventilation (Smith and Jones 1982) or hyperventilation (Thomas et al. 1983). Acid injections into trout cause a hyperventilation and increased blood catecholamine levels, the breathing response being blocked by propranolol (Aota et al. 1990). Acid injections into fish exposed to hyperoxic conditions generated an acidosis but no liberation of catecholamines and no increase in ventilation. The simplest explanation is that the increase in ventilation during hypercapnia or an acidotic state is due to the increase in circulating catecholamines rather than any central H^+ receptor stimulation. In fact, there is no direct evidence for the existence of any such receptor in fish. It may be that in aquatic vertebrates ventilation is controlled via peripheral oxygen chemoreceptors, central (and possibly peripheral) catecholamine responses and mechanoreceptor reflexes (see below) acting on a central medullary respiratory control system, modulated by higher order neuronal activity originating in the mid- and forebrain. Acid-base regulation via changes in ventilation are only a reality in air-breathing vertebrates (see, however, Randall and Cameron 1973; Iwama et al. 1987), and perhaps the existence of central H^+ receptors is coupled to the movement of vertebrates onto land and the evolution of air breathing.

7.5 Gill Mechanoreceptors

The gill region of both elasmobranchs and teleosts contains large numbers of mechanoreceptors, innervated by the IXth and Xth cranial nerves and stimulated by movements of the branchial apparatus. These receptors modulate the breathing pattern and adjust the geometry of the gill apparatus to the water flow (see reviews by Jones and Milsom 1982; Shelton et al. 1986; Ballintijn 1988). The mechanoreceptors show a wide variety of activity patterns, both tonic and phasic. The intermediate facial nucleus, which receives afferent information from gill mechanoreceptors, projects to the respiratory motonuclei (Ballintijn 1988). Paralysis of the gills with curare, which presumably removes much of the mechanoreceptor input seen during normal breathing, results in a decrease in breathing rate in teleosts (Ballintijn 1972) but an increase in elasmobranchs (Barrett and Taylor 1985). Sectioning the IXth and Xth cranial nerves has a similar effect as paralysis in elasmobranchs (Satchell 1959) and increases the amplitude of breathing in teleosts (Shelton 1959). Thus, the probable overall effect of mechanoreceptor input is to adjust the relationship between breathing rate and amplitude, inhibiting breathing frequency in elasmobranchs but increasing rate while reducing amplitude in teleosts. If the jaw or operculum are mechanically interfered with, or respiratory muscles caused to contract out of phase, then rapid reflex responses in breathing occur which are presumably of mechanoreceptor origin. Some of these reflexes operate as load-matching, stretch reflexes (Luiten 1979), while others seem to involve higher levels of integration and can have effects lasting for several breaths (Ballintijn and Bamford 1975). Other reflexes control the angle of the filament to the gill arch to adjust the geometry of the gill apparatus to the water flow and prevent water spillage between the arches (Pasztor and Kleerekoper 1962).

Many fish ram ventilate their gills when swimming at high speed or when placed in a rapid water flow. That is, they stop rhythmic breathing and open their mouth, and flow over the gills is dependent on the forward motion of the fish or the water flow past the fish. A *Remora* attached to a shark will stop rhythmic breathing once its host achieves speeds of 50–80 cm/s (Muir and Buckley 1967).

The switch from rhythmic to ram ventilation occurs at a specific water velocity, depending on the fish, temperature and oxygen content of the water. If the fish swims at speeds around the transition point, the fish will switch from rhythmic to ram ventilation very rapidly. This observation has been a major factor used to support the idea that mechanoreceptive input, located in the gills, acousticolateralis system, or skin and detecting water velocity, may trigger the switch (see review by Shelton et al. 1986). These inputs may operate through the mesencephalic tegmentum in the midbrain, which has been shown to initiate bursts of breathing activity in otherwise apneic carp (Ballintijn 1988). Oxygen levels modify the water velocity at which the switch occurs. In continuously ram-ventilating tuna, aquatic hypoxia results in a larger mouth gape (Brown and Muir 1970). Thus, chemoreceptors appear to play a modulating, rather than initiating, role in ram ventilation.

Nociceptors have been located in dogfish gills (Poole and Satchell 1979) that show similarities to type J receptors found in mammalian lungs. The receptors are located between blood vessels and the epithelium in the gill filaments. They respond to many of the same stimuli as J receptors, have fairly large receptive fields, and cause apnea, inhibition of swimming, bradycardia and hypotension (Satchell 1974). It has been suggested that this response is protective, reducing the possibility of gill edema (Satchell 1974).

8 Gill Diffusing Capacity

Gill diffusing capacity probably increases in response to hypoxia and exercise (see review by Randall and Daxboeck 1984). Increases in blood pressure increase the width of the gill blood sheet and result in a more even distribution of blood throughout lamellae, but have little effect on lamellar height and length (Farrell et al. 1980). The gills are innervated by adrenergic and cholinergic fibres, and respond with β-adrenergic vasodilation and α-adrenergic and cholinergic vasoconstriction (see review by Nilsson 1984). Adenosine and hypoxia may act as local vasodilatory agents (Colin and Leray 1981). At rest, only about 60% of the lamellae in a trout are perfused with blood. This increases to 100% during exercise, hypoxia and, perhaps, hypercapnia (Randall and Daxboeck 1984). Gill blood volume is somewhat larger than heart stroke volume, so each stroke volume remains in the gills for one heartbeat. Blood residence time in the gills is about 3 s at rest, decreasing to 1 s during exercise. At low blood flows, oxygen uptake is largely perfusion limited, but at higher rates of oxygen uptake there are both diffusion and perfusion limitations (Randall 1982). Increases in diffusing capacity probably enhance oxygen uptake during exercise and hypoxia. Many factors, including changes in both water and blood flow, increases in gill epithelial permeability and a thinning of the epithelium, probably contribute to this increase in diffusing capacity.

9 Blood Flow

9.1 Cardiac Output

Fish alter cardiac output through changes in both stroke volume and heart rate (Randall 1970). Rate is regulated via excitatory adrenergic and inhibitory cholinergic nerves innervating the sinus node (Laurent et al. 1983). Stroke volume is augmented by increased venous filling, modulated by changes in the level of circulating catecholamines and/or activity in adrenergic fibres innervating the ventricle (Farrell et al. 1988). Acidotic states can reduce contractility of the heart, but this effect can be compensated for by either increased intracellular Ca^{2+} and/or elevated circulating catecholamine levels (Farrell 1985). The atrium has about the same volume as the ventricle, so ventricular filling is achieved through atrial contraction rather than direct venous inflow, as in mammals. Thus, changes in venous pressure have effects on cardiac output via atrial filling, and these effects are then amplified through filling and the subsequent contraction of the ventricle. There is both a vis a tergo and a vis a fronte effect of cardiac contraction on blood flow in fish. The vis a fronte effect is mediated through the presence of somewhat inflexible pericardium and/or surrounding tissue (Farrell et al. 1988).

Cardiac output increases with exercise, due to an increase in heart rate and stroke volume (Randall 1982), the exact contribution of each varying with the species and conditions. This is achieved by a reduction in cholinergic and an increase in adrenergic activity in nerves innervating the heart (Jones and Randall 1978). There is an increase in venous return and an elevation in circulating catecholamines (Primmett et al. 1986) that probably also contribute to the increase in cardiac output, as well as ameliorating any adverse effects of an acidotic state on cardiac contraction.

Hypoxia results in a marked bradycardia and sinus arrhythmia, both due to increased cholinergic inhibitory activity in the vagus innervating the heart (Randall 1970). The afferent arm of this reflex is from external oxygen chemoreceptors in the gills, detecting a reduction in oxygen in the water flowing over the gills (Daxboeck and Holeton 1978). The bradycardia is offset by an increase in stroke volume in trout, such that there is little change in cardiac output until oxygen levels are low, when cardiac output is reduced. The changes in cardiac function with hypoxia alter the pattern of filling of the gills and the relationship between water and blood flow. These changes presumably enhance gill diffusing capacity and, therefore, oxygen transfer (Randall 1982).

Stimulation of nociceptors on the gills (Satchell 1974) causes a bradycardia and a reduction in blood pressure in fish.

9.2 Blood Distribution

There is an increase in blood flow to the working muscles, but a decrease in gut blood flow, during exercise in fish (Randall and Daxboeck 1982). There is a contraction of the spleen during both exercise (Jones and Randall 1978) and hypoxia (Perry and Kinkead 1989), increasing the number of circulating erythrocytes.

There is some evidence for a reduction in blood volume, due to both water loss and transfer to other tissues, during exercise which also raises hematocrit (Wood and Randall 1973). The peripheral circulation has an α-adrenergic vasoconstrictor and a β-adrenergic dilator response. In general, however, little is known of the control of the distribution of blood flow in the peripheral circulation of fish, especially in terms of the regulation of gas transfer to the tissues.

References

Aota S, Holmgren K, Gallaugher P, Randall DJ (1990) A possible role for catecholamines in the ventilatory responses associated with internal acidosis or external hypoxia in rainbow trout, *Oncorhynchus mykiss*. J Exp Biol (in press)

Ballintijn CM (1972) Efficiency, mechanics and motor control of fish respiration. Respir Physiol 14:125–141

Ballintijn CM (1982) Neural control of respiration in fishes and mammals. In: Addink ADF, Spronk N (eds) Exogenous and endogenous influences on metabolic and neural control. Pergamon, Oxford, pp 127–140

Ballintijn CM (1988) Evolution of central nervous control of ventilation in vertebrates. In: Taylor EW (ed) The neurobiology of the cardiorespiratory system, Chap 1. Manchester University Press, Manchester

Ballintijn CM, Alink GM (1977) Identification of respiratory motor neurons in the carp and determination of their firing characteristics and interconnections. Brain Res 136:261–276

Ballintijn CM, Bamford OS (1975) Proprioceptive motor control in fish respiration. J Exp Biol 62:99–114

Barrett DJ, Taylor EW (1985) Spontaneous efferent activity in branches of the vagus nerve controlling heart rate and ventilation in the dogfish. J Exp Biol 117:433–448

Boutilier RG, Iwama GK, Heming TA, Randall DJ (1985) The apparent pK of carbonic acid in rainbow trout plasma between 5 and 15 °C. Respir Physiol 61:237–254

Boutilier RG, Iwama GK, Randall DJ (1986) Acute extracellular acidosis promote catecholamine release in rainbow trout (*Salmo gairdneri*): Interactions between red cell pH and O_2-Hb carrying capacity. J Exp Biol 123:145–157

Boutilier RG, Dobson G, Hoeger U, Randall DJ (1988) Acute exposure to graded levels of hypoxia in rainbow trout (*Salmo gairdneri*): metabolic and respiratory adaptations. Respir Physiol 71:69–82

Brown CE, Muir BS (1970) Analysis of ram ventilation of fish gills with application to skipjack tuna (*Katsuwonis pelamis*). J Fish Res Board Can 27:1637–1652

Burggren WW (1978) Gill ventilation in the sturgeon, *Acipenser transmontanus*: unusual adaptations for bottom dwelling. Respir Physiol 34:153–170

Burleson M, Smatresk NJ (1989) Evidence for two oxygen sensitive chemoreceptor Poci in channel catfish, *Ictalurus punctatus*. Physiol Zool 63:208–221

Butler PJ, Metcalfe JD (1983) Control of respiration and circulation. In: Rankin JC, Pitcher TJ, Duggan R (eds) Control processes in fish physiology. Wiley-Interscience, New York, pp 41–65

Butler PJ, Taylor EW (1971) Response of the dogfish (*Scyliorhinus canicula* L.) to slowly induced and rapidly induced hypoxia. Comp Biochem Physiol 39A:307–323

Cameron JN (1976) Branchial ion uptake in Arctic grayling: resting values and effects of acid-base disturbance. J Exp Biol 64:711–725

Cameron JN, Heisler N (1983) Studies of ammonia in the rainbow trout: physico-chemical parameters, acid-base behavior, and respiratory clearance. J Exp Biol 105:107–125

Colin DA, Leray C (1981) Vasoactivities of adenosine analogues in trout gill (*Salmo gairdneri* R.). Biochem Pharmacol 30:2971–2978

Daxboeck C, Holeton GF (1978) Oxygen receptors in the rainbow trout, *Salmo gairdneri*. Can J Zool 56:1254–1259

Driedzic WR, Hochachka PW (1976) Control of energy metabolism in fish white muscle. Am J Physiol 230:579–582
Dunel-Erb S, Bailly Y, Laurent P (1982) Neuroepithelial cells in fish gill primary lamellae. J Appl Physiol Respir Environ Exercise Physiol 53:1342–1353
Eckert R, Randall DJ, Augustine G (1988) Animal physiology: mechanisms and adaptations, 3rd edn. Freeman, San Francisco, 683 pp
Eclancher B (1972) Action des changements rapides de P_{O_2} de l'eau sur la ventilation de la truite et de la tanche. J Physiol Paris 65:397 A
Eclancher B, Dejours P (1975) Control de la respiration chez les poissons teleosteens: existence de chemorecepteurs physiologiquement analogues aux chemorecepteurs des vertebres superieur. CR Acad Sci Paris Ser D 280:451–453
Farrell AP (1984) A review of cardiac performance in the teleost heart: intrinsic and humoral regulation. Can J Zool 62:523–536
Farrell AP (1985) A protective effect of adrenaline on the acidotic teleost heart. J Exp Biol 116:503–508
Farrell AP, Sobin SS, Randall DJ, Crosby S (1980) Intralamellar blood flow patterns in fish gills. Am J Physiol 239:R428–R436
Farrell AP, Johansen JA, Graham MS (1988) The role of the pericardium in cardiac performance of trout (*Salmo gairdneri*). Physiol Zool 61:213–221
Folgering H (1980) Central β-adrenergic effects on the control of ventilation in cats. Respiration 39:131–138
Giles MA, Randall DJ (1980) Oxygenation characteristics of the polymorphic hemoglobins of coho salmon (*Oncorhynchus kisutch*) at different developmental stages. Comp Biochem Physiol 65A:265–271
Giles MA, Vanstone WE (1976) Ontogenetic variation in the multiple hemoglobins of coho salmon (*Oncorhynchus kisutch*) and effect of environmental factors on their expression. J Fish Res Board Can 33:1144–1149
Haswell MS, Randall DJ, Perry SF (1980) Fish gill carbonic anhydrase: acid-base regulation or salt transport? Am J Physiol 238:R240–R245
Heming TA, Randall DJ, Boutilier RG, Iwama GK, Primmett D (1986) Ionic equilibria in red blood cells of rainbow trout (*Salmo gairdneri*): Cl^-, HCO_3^- and H^+. Respir Physiol 65:223–234
Heming TA, Randall DJ, Madzeaud MM (1987) Effects of adrenaline on ionic equilibria in the red blood cells of rainbow trout (*Salmo gairdneri*). Fish Physiol Biochem 3:83–90
Hillaby BA, Randall DJ (1979) Acute ammonia toxicity and ammonia excretion in rainbow trout (*Salmo gairdneri*). J Fish Res Board Can 36:621–629
Holeton GF, Randall DJ (1967) The effect of hypoxia upon the partial pressure of gases in blood and water afferent and efferent to the gills of rainbow trout. J Exp Biol 46:317–327
Isaia J, Maetz J, Haywood GP (1978) Effects of epinephrine on branchial nonelectrolyte permeability in trout. J Exp Biol 74:227–237
Iwama GK, Boutilier RG, Heming TA, Randall DJ, Mazeaud M (1987) The effects of altering gill water flow on gas transfer in rainbow trout. Can J Zool 65:2466–2470
Jacobs MH (1940) Some aspects of cell permeability to weak electrolytes. Cold Spring Harbour Symp Quant Biol 8:30–39
Janssen RG, Randall DJ (1975) The effect of changes in pH and PCO_2 in blood and water on breathing in rainbow trout (*Salmo gairdneri*). Respir Physiol 25:235–245
Johansen K, Pettersson K (1981) Gill O_2 consumption in a teleost fish, *Gadus morhua*. Respir Physiol 44:277–284
Jones DR, Milsom WK (1982) Peripheral receptors affecting breathing and cardiovascular function in non-mammalian vertebrates. J Exp Biol 100:59–91
Jones DR, Randall DJ (1978) The respiratory and circulatory systems during exercise. In: Hoar WS, Randall DJ (eds) Fish physiology, vol 7. Academic Press, New York, pp 425–501
Kirsch R, Nonnotte G (1977) Cutaneous respiration in three freshwater teleosts. Respir Physiol 29:339–354
Knepper MA, Packer R, Good DW (1989) Ammonium transport in the kidney. Physiol Rev 69:179–249

Lacy ER (1983) Histochemical and biochemical studies of carbonic anhydrase activity in the opercular epithelium of the euryhaline teleost, *Fundulus heteroclitus*. Am J Anat 166:19–39

Laurent P, Rouzeau JD (1972) Afferent neural activity from the pseudobranch of teleosts. Respir Physiol 14:307–331

Laurent P, Holmgren S, Nilsson S (1983) Nervous and humoral control of the heart: structure and function. Comp Biochem Physiol 76A:525–542

Lin H, Randall DJ (1990) The effect of varying water pH on the acidification of water as it passes over the gills of rainbow trout. J Exp Biol (in press)

Luiten PGM (1979) Proprioceptive reflex connections of head musculature and the mesencephalic trigeminal nucleus in the carp. J Comp Neurol 183:903–912

Milsom WK, Sadig T (1983) Interaction between norepinephrine and hypoxia on carotid body chemoreception in rabbits. J Appl Physiol Respir Environ Exercise Physiol 55:1893–1898

Mommsen TP, Hochachka PW (1989) The purine nucleotide cycle as a metabolic nightmare – a study on trout muscle. Metabolism (in press)

Muir BS, Buckley RM (1967) Gill ventilation in *Remora remora*. Copeia 3:581–586

Nekvasil NP, Olson KR (1986) Extraction and metabolism of circulating catecholamines by the trout gill. Am J Physiol 250:R526–R531

Nikinmaa M (1982) Effects of adrenaline on red cell volume and concentration gradient of protons across the red cell membrane in the rainbow trout, *Salmo gairdneri*. Mol Physiol 2:287–297

Nikinmaa M, Railo E (1987) Anion movement across lamprey (*Lampetra fluviatilis*) red cell membrane. Biochim Biophys Acta 899:134–136

Nilsson S (1984) Innervation and pharmacology of the gills. In: Hoar WS, Randall DJ (eds) Fish physiology, vol 10A. Academic Press, New York, pp 185–227

Pasztor VM, Kleerekoper H (1962) The role of the gill filament musculature in teleosts. Can J Zool 40:785–802

Perry SF (1986) Carbon dioxide excretion in fishes. Can J Zool 64:565–572

Perry SF, Kinkead R (1989) The role of catecholamines in regulating arterial oxygen content during acute hypercapnic acidosis in rainbow trout (*Salmo gairdneri*). Respir Physiol 77:365–377

Perry SF, Wood CM (1989) Control and co-ordination of gas transfer in fishes. Can J Zool (in press)

Perry SF, Davie PS, Daxboeck C, Randall DJ (1982) A comparison of CO_2 excretion in a spontaneously ventilating blood-perfused trout preparation and saline-perfused gill preparations: contribution of the branchial epithelium and red blood cell. J Exp Biol 101:47–60

Perry SF, Kinkead R, Gallaugher P, Randall DJ (1989) Evidence that hypoxemia promotes catecholamine release during hypercapnic acidosis in rainbow trout (*Salmo gairdneri*). Respir Physiol 77:351–363

Peyraud-Waitzenegger M (1979) Simultaneous modifications of ventilation and arterial P_{O_2} by catecholamines in the eel, *Anguilla anguilla* L.: participation of A and B effects. J Comp Physiol 129:343–354

Peyraud-Waitzenegger M, Savina A, Laparra J, Morfin R (1979) Blood brain barrier for epinephrine in the eel *Anguilla anguilla* L. Comp Biochem Physiol 63:35–38

Peyraud-Waitzenegger M, Barthelemy L, Peyraud C (1980) Cardiovascular and ventilatory effects of catecholamines in unrestrained eels (*Anguilla anguilla*): a study of seasonal changes in reactivity. J Comp Physiol 138B:365–375

Poole CA, Satchell GH (1979) Nociceptors in the gills of the dogfish (*Squalus acanthias*). J Comp Physiol 130:1–7

Primmett DRN, Randall DJ, Mazeaud M, Boutilier RG (1986) The role of catecholamines in erythrocyte pH regulation and oxygen transport in rainbow trout (*Salmo gairdneri*) during exercise. J Exp Biol 122:139–148

Rahim SM, Delaunoy J, Laurent P (1988) Identification and immunocytochemical localization of two different carbonic anhydrase isoenzymes in teleostean fish erythrocytes and gill epithelia. Histochemistry 89:451–459

Randall DJ (1970) Gas exchange in fish. In: Hoar WS, Randall DJ (eds) Fish Physiology, vol 4. Academic Press, New York, pp 253–292

Randall DJ (1972) Respiration. In: Hardisty MW, Potter IC (eds) The biology of lampreys, vol 2. Academic Press, London, pp 287–306
Randall DJ (1982) The control of respiration and circulation in fish during exercise and hypoxia. J Exp Biol 100:275–288
Randall DJ (1985) Shunts in fish gills. In: Johansen K, Burggren W (eds) Cardiovascular shunts: phylogenetic, ontogenetic and clinical aspects. Alfred Benzon Symposium No 21. Munksgaard, Copenhagen, pp 71–87
Randall DJ (1989) The impact of variations in water pH on fish. In: Sandifer P (ed) Advances in aquaculture (in press)
Randall DJ, Cameron JN (1973) Respiratory control of arterial pH as temperature changes in rainbow trout (*Salmo gairdneri*). Am J Physiol 225:997–1001
Randall DJ, Daxboeck C (1982) Circulatory changes during exercise in fish. Can J Zool 60:1135–1140
Randall DJ, Daxboeck C (1984) Oxygen and carbon dioxide transfer across fish gills. In: Hoar WS, Randall DJ (eds) Fish physiology, vol 10A. Academic Press, New York, pp 263–314
Randall DJ, Jones DR (1973) The effect of deafferentation of the pseudobranch on the respiratory response to hypoxia and hyperoxia in the trout (*Salmo gairdneri*). Respir Physiol 17:291–301
Randall DJ, Wright PA (1987) Ammonia distribution and excretion in fish. Fish Physiol Biochem 3:107–120
Randall DJ, Wright PA (1989) The interaction between carbon dioxide and ammonia excretion and water pH in fish. Can J Zool (in press)
Randall DJ, Wright PA (1990) Circulation and gas transfer in Pacific Salmon. In: Brett JR, Clarke WC (eds) Physiological ecology of Pacific salmon. Scientific Information and Publications Branch. Fisheries and Oceans, Canada (in press)
Randall DJ, Heisler N, Drees F (1976) Ventilatory response to hypercapnia in the larger spotted dogfish, *Scyliorhinus stellaris*. Am J Physiol 230:590–594
Randall DJ, Mense D, Boutilier RG (1987) The effects of burst swimming on aerobic swimming in chinook salmon (*Oncorhynchus tshawytscha*). Mar Behav Physiol 13:77–88
Randall DJ, Wood CM, Perry SF, Bergman H, Maloiy GMO, Mommsen TP, Wright PA (1989) Urea excretion as a strategy for survival in a fish living in a very alkaline environment. Nature 337:165–166
Riggs A (1979) Studies of the hemoglobins of Amazonian fishes: an overview. Comp Biochem Physiol 62A:257–271
Romano L, Passow H (1984) Characterization of anion transport system in trout red blood cell. Am J Physiol 246:C330–C338
Satchell GH (1959) Respiratory reflexes in the dogfish. J Exp Biol 36:531–543
Satchell GH (1974) The J reflex in the fish. Proc Int Union Physiol Sci 11:26th Int Congr Physiol Sci, New Dehli
Sauer J, Harrington JP (1988) Hemoglobins of the sockeye salmon, *Oncorhynchus nerka*. Comp Biochem Physiol 91A:109–114
Saunders RL, Sutterlin AM (1971) Cardiac and respiratory responses to hypoxia in the sea raven, *Hemitripterus americanus*, and an investigation of possible control mechanisms. J Fish Res Board Can 28:491–503
Shelton G (1959) The respiratory centre in the tench (*Tinca tinca* L.). I. The effects of brain transection on respiration. J Exp Biol 36:191–202
Shelton G (1970) The regulation of breathing. In: Hoar WS, Randall DJ (eds) Fish physiology, vol 4. Academic Press, New York, pp 293–359
Shelton G, Jones DR, Milsom WK (1986) Control of breathing in ectothermic vertebrates. In: Cherniak NS, Widdicombe JG (eds) Handbook of physiology, Section 3. The respiratory system, vol 2. Control of Breathing. Am Physiol Soc, Bethesda, pp 857–909
Shoubridge E, Hochachka PW (1979) Lactate oxidation in the anoxic goldfish. Int Congr Biochem 13:1–R123
Small SA, Randall DJ (1989) The effects of triploidy on the swimming performance of coho salmon (*Oncorhynchus kisutch*). Can J Fish Aquat Sci 46:243–245
Smith FM, Jones DR (1978) Localization of receptors causing hypoxic bradycardia in trout (*Salmo gairdneri*). Can J Zool 56:1260–1265

Smith FM, Jones DR (1982) The effect of changes in blood oxygen-carrying capacity on ventilation volume in the rainbow trout (*Salmo gairdneri*). J Exp Biol 97:325–334
Steffensen JF, Tufts BL, Randall DJ (1987) Effect of burst swimming and adrenaline infusion on O_2 consumption and CO_2 excretion in rainbow trout, *Salmo gairdneri*. J Exp Biol 131:427–434
Taylor EW, Randall DJ (1989) Control of ventilation in fish. In: Fish physiology, fish toxicology and fisheries management. Int Symp Guangzhou, PRC Sept 13–16, 1988. EPA (in press)
Thomas RC (1974) Intracellular pH of snail neurons measured with a new pH-sensitive glass microelectrode. J Physiol 238:159–180
Thomas S, Fievet B, Barthelemy L, Peyraud C (1983) Comparison of the effects of exogenous and endogenous hypercapnia on ventilation and oxygen uptake in the rainbow trout (*Salmo gairdneri*). J Comp Physiol 151:185–190
Tufts BL, Randall DJ (1989) The functional significance of adrenergic pH regulation in fish erythrocytes. Can J Zool 67:235–238
Tufts BL, Ferguson RA, Boutilier RG (1989) In vivo and in vitro effects of adrenergic stimulation on chloride/bicarbonate exchange in rainbow trout erythrocytes. J Exp Biol 140: 301–312
Vanstone WE, Roberts E, Tsuyuki H (1964) Changes in the multiple hemoglobin patterns of some Pacific salmon, Genus *Oncorhynchus*, during the parr-smolt transformation. Can J Physiol Pharmacol 42:697–703
Vogel WOP (1985) Systematic vascular anastomoses, primary and secondary vessels in fish, and the phylogeny of lymphatics. In: Johansen K, Burggren W (eds) Cardiovascular shunts: phylogenetic, ontogenetic and clinical aspects. Alfred Benzon Symposium No 21. Munskgaard, Copenhagen, pp 143–151
Waldron I (1972) Spatial organisation of respiratory neurones in the medulla of tench and goldfish. J Exp Biol 57:449–459
Wood CM (1976) Pharmacological properties of the adrenergic receptors regulating systemic vascular resistance in the rainbow trout. J Comp Physiol B 107:211–228
Wood CM, Randall DJ (1973) The influence of swimming activity on water balance in the rainbow trout (*Salmo gairdneri*). J Comp Physiol 82:257–276
Wright PA (1987) Ammonia stores and excretion in fish: relationship to pH. PhD Thesis, University of British Columbia, Vancouver, Canada
Wright PA, Wood CM (1985) An analysis of branchial ammonia excretion in the freshwater rainbow trout: effects of environmental pH change and sodium uptake blockade. J Exp Biol 114:329–353
Wright PA, Wood CM (1988) Muscle ammonia stores are not determined by pH gradients. Fish Physiol Biochem 5:159–162
Wright PA, Heming TA, Randall DJ (1986) Downstream pH changes in water flowing over the gills of rainbow trout. J Exp Biol 126:499–512
Wright PA, Randall DJ, Wood CM (1988) The distribution of ammonia and H^+ between tissue compartments in lemon sole (*Parophyrs vetulus*) at rest, during hypercapnia and following exercise. J Exp Biol 136:149–175
Wright PA, Randall DJ, Perry SF (1989) Fish gill water boundary layer: site of linkage between carbon dioxide and ammonia excretion. J Comp Physiol 158:627–635
Zborowska-Sluis DT, L'Abbate A, Klassen GA (1974) Evidence of carbonic anhydrase activity in skeletal muscle: a role for facilitative carbon dioxide transport. Respir Physiol 21:341–350

Chapter 9

Control and Co-Ordination of Gas Exchange in Bimodal Breathers

R. G. Boutilier[1]

Contents

[1] Department of Biology, Dalhousie University, Halifax, Nova Scotia, Canada B3H 4J1

Advances in Comparative and Environmental Physiology
Vol. 6. Edited by R. G. Boutilier

1 Introduction to Bimodal Systems

The ability to exchange respiratory gases effectively in either air or water has been exploited by a wide variety of amphibious vertebrates. The partitioning of O_2 and CO_2 transfer between aerial and aquatic exchange sites is a function of the exchange organs' surface area, blood-to-medium diffusion distances and ventilation-perfusion ratios (see Piiper, this Vol.). In addition, the physical properties of water and air place different demands on the respiratory organs (Dejours 1981; Piiper 1982), and this is thought to have had important consequences, during evolution, on the design and performance of gas exchange organs in bimodal breathers (Johansen 1970; Randall et al. 1981; Shelton and Boutilier 1982; Shelton et al. 1986). It is generally accepted that a major selective force in the evolution of the air-breathing habit was aquatic hypoxia (Packard 1974), enabling those animals with air-breathing organs to remain in the warm, O_2-deficient waters that are thought to have existed in the Upper Devonian (Inger 1957). Indeed, the intermittent use of air-breathing organs by extant vertebrates is indicative of this, where constraints on aquatic gas exchange lead to a periodic need for supplemental forms of gas exchange.

1.1 Water vs Air as a Respiratory Medium

In general, bimodal breathers utilize their air-breathing organs for supplementing O_2 uptake, particularly during periods of aquatic hypoxia. CO_2 excretion, on the other hand, is primarily accounted for by the aquatic exchange organs, where the physico-chemical conditions for its release are more favourable. Indeed, differences in the respiratory properties of water and air are of particular importance in understanding the aerial/aquatic partitioning of respiratory gases in these animals. In comparing air and water, it is necessary to consider the capacitance coefficient (β) of each medium, defined as the increment of concentration per increment of partial pressure ($\beta = \Delta C/\Delta P$; Piiper et al. 1971, Fig. 1). In the gas phase, β is equal for all (ideal) gases (i.e. $\beta CO_2 : \beta O_2 = 1$), however, in water, the capacitance coefficients for O_2 and CO_2 are markedly different (e.g. $\beta CO_2/\beta O_2 \sim 30$ depending on salinity, temperature and buffering capacity; c.f. Dejours 1981). Moreover, whereas the ratio of $\beta_{(water)}/\beta_{(gas)}$ is near unity for CO_2, it is only about 0.033 for O_2. As a consequence, this means that water-breathing animals need to convect relatively large quantities of water across their respiratory epithelium in order to achieve an O_2 uptake equivalent to that of an air breather. Because βCO_2 exceeds βO_2 in water, the comparatively high ventilatory conductance for O_2 in the water-breathing animal means an equally increased conductance for CO_2, leading to very low partial pressures of CO_2 (PCO_2) in expired water and arterial blood, even when O_2 extraction rates are high (Rahn 1966; Piiper 1982).

The consequence of these differences in β between the respiratory media, air and water, can be seen in the CO_2/O_2 diagram (Fig. 1), in which the relationships between inspired (i) and expired (e) PO_2 and PCO_2 are plotted for aquatic

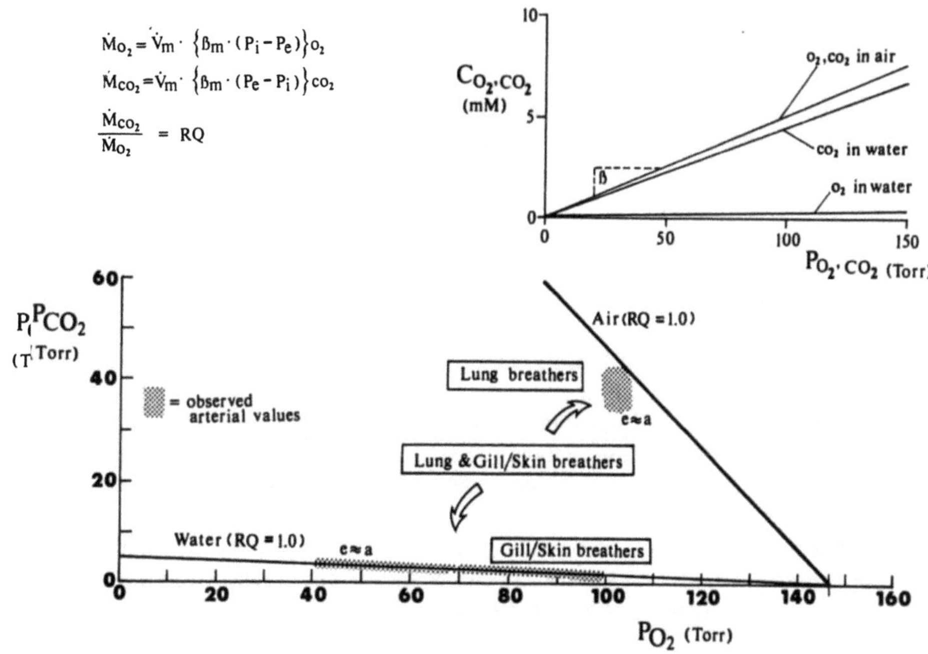

Fig. 1. The *upper diagram* shows the relationship between respiratory gas content (*C*) and partial pressure (*P*) for air and water. The slopes of the lines ($\Delta C/\Delta P$) define the capacitance coefficient, *β*. The PO_2-PCO_2 plot in the *lower diagram* shows all possible combinatitons of PCO_2 and PO_2 in unimodal lung breathers and unimodal water breathers operating in a steady state at an RQ of 1. *Shaded areas* are arterial values (*a*), which are close to expired values (*e*). It is not possible to derive theoretically the PCO_2-PO_2 relationships in bimodal breathers. Arterial values for these animals will fall within an intermediate range. See text for further details (Scheid 1982)

and aerial convective systems. These relationships are described for each medium (m) by the equation:

$$\frac{(P_e-P_i)_{CO_2}}{(P_i-P_e)_{O_2}}=\frac{(\beta_m)_{O_2}}{(\beta_m)_{CO_2}}\cdot RQ\ , \qquad (1)$$

where the respiratory quotient, $RQ = \dot{M}CO_2/\dot{M}O_2$, is normally between 0.65 and 1 for animals at rest. The equation above can be used to predict the theoretical combination of O_2 and CO_2 partial pressures that can exist at the lungs of a mammal or the gills of a unimodally breathing fish exchanging gases continuously and having a respiratory quotient of 1 (Fig. 1). Note that when water is the breathing medium, the slope is approximately 30 times smaller than for air (Fig. 1), as one would predict according to the differences in the capacitance coefficients of the media concerned. Since expired PO_2 and PCO_2 values are similar to arterial (a) blood values, these lines also predict, and set limits for, the $PaCO_2$ levels in water breathers and air breathers.

1.2 Steady vs Unsteady States

The relationships between O_2 and CO_2 as plotted in Fig. 1 represent conditions of gas exchange in the steady state (Scheid 1982; Piiper, this Vol.), and therefore, this type of analysis cannot be usefully applied to unsteady-state conditions such as occur in periodic bimodal breathers. For example, even in unimodal systems of gas exchange, oscillations in the respiratory gases, e.g. due to breath-holding (Fig. 1, bottom panel), do not occur as simple movements along the respiratory exchange lines. The reason for this has to do with differences in the rates at which O_2 and CO_2 move into and out of gas stores in the body, the tissue CO_2 stores being larger, and the time for filling and emptying being longer, than the respective O_2 stores. The intermittent patterns of ventilation of many air-breathing fish and amphibians lead to even greater deviations from the ideal steady-state model as depicted in Fig. 1, owing to unevenly balanced exchanges of O_2 and CO_2 with the environment. As a consequence, the acquisition of O_2 exceeds elimination of CO_2 at the aerial sites of gas exchange, whereas the opposite is true of the aquatic exchangers.

Aquatic exchange sites also participate in transepithelial ion transfers with environmental water, and such transfers can be important in regulating the acid-base balance of the animals (see Heisler, this Vol.).

1.3 Acid-Base Balance

Because CO_2 is an acid, the differences in $PaCO_2$ between water- and air-breathing animals have important consequences on the acid-base balance of these two groups of animals. According to the Henderson-Hasselbalch equation:

$$pH = pK'_{app} + \log \frac{[HCO_3^-]}{\alpha_{CO_2} \cdot PCO_2}, \tag{2}$$

where pK'_{app} is the apparent first dissociation constant of carbonic acid and α_{CO_2} is the solubility of CO_2 in plasma, a much higher plasma pH is predicted in fish, compared to mammals, for the same plasma bicarbonate concentrations $[HCO_3^-]$. However, low plasma HCO_3^- concentrations effectively offset the low PCO_2 levels of fish, such that the pH of water- and air-breathing animals are comparable at any given body temperature (Rahn 1967).

2 Structural Adaptations for Bimodal Gas Exchange

Amphibious vertebrates exhibit a remarkable diversity in the form and function of their gas exchange organs. Modifications of the gills, skin, buccopharyngeal chamber, digestive tract and swim bladder are known to contribute significantly to aerial gas exchange in various air-breathing fish. This increased specialization and reliance on the air-breathing organ (ABO) is often associated with a reduction in the efficacy of the aquatic exchanger (e.g. the gills become more rudimentary and the musculoskeletal system associated with the gills becomes reduced or modified for aerial respiration). Amphibians too utilize a broad spectrum of gas exchange structures, including lungs, gills, skin and buccopharyngeal chamber, the latter two of which can operate effectively in both air and water. Depending on the species or the life history stage, various combinations can apply, ranging from almost exclusive use of the skin in Plethodontid salamanders, to the utilization of all four sites in certain neotenic urodeles. Structures for non-pulmonary gas exchange are also evident in various aquatic species of turtles and snakes, where bimodal breathing can be of considerable importance (see reviews by Seymour 1982, 1989).

2.1 Gills

The numerous and closely spaced lamellae aligned on the gill filaments of strictly aquatic-breathing fish tend to stick together and/or collapse under the force of gravity in air. However, a few quite remarkable species of fish (e.g. *Clarius batrachus*, *Symbranchus*) are able to use their gills for breathing either water or air. Such fish have modified gills featuring thickened and widely spaced lamellae that reside in suprabranchial chambers, and whose morphology ensures adequate support in air (Johansen 1968; Jordan 1976). Air-breathing marine species are characterized by a lack of the additional air-breathing structures characteristic of freshwater forms (Randall et al. 1981). The absence of such structures is thought to reflect an incompatibility between exposure of large surface areas to air and the conservation of water in these hyporegulating fishes (Graham 1976). Adaptations for aerial gas exchange by the gills of marine intertidal species include reductions in the number of lamellae per filament, ensuring that the gills will not collapse in air, and guarding against large evaporative water losses (Graham 1976). In fact, among air-breathing intertidal species, the total gill area appears to decrease with increasing terrestrial capability (Schottle 1932; Graham 1973, 1976). In some species, there is evidence that this decrease in total gill area is paralleled by a compensatory increase in the amount of vascularized buccopharyngeal epithelium (e.g. Schottle 1932; Todd and Ebeling 1966; Singh 1976).

The trend towards decreased gill surface area with increased reliance on aerial gas exchange is a common feature for both marine (Graham 1976) and freshwater (Johansen 1970) fish species. This is clearly seen in the dipnoan lungfish, where comparatively fewer lamellae are found on the gill arches than in other fish. Indeed, the trend is seen among the lungfish themselves, i.e. the facultative air-breather, *Neoceratodus*, has lamellae on all of its gill arches (Sewertzoff 1924), whereas several arches are devoid of lamellae in the obligate air-breathers, *Protopterus* and *Lepidosiren* (Johansen 1970; Laurent et al. 1978; Fishman et al. 1989).

Unlike fish, amphibian gills are not specialized for gas exchange in air, nor do they ever constitute the sole organ of gas exchange. When they are present (e.g. in neotenic urodeles and larvae), the gills of amphibians function in association with the skin, lungs and buccopharyngeal cavity. The complexity of the amphibian gill as well as its contribution to overall gas exchange varies enormously, both within and between species. Most studies on the respiratory physiology of amphibian gills have been carried out on neotenic salamanders (Guimond and Hutchison 1972, 1973, 1976; Heath 1975; Shield and Bentley 1973; Malvin 1989) or anuran tadpoles (Burggren et al. 1983; Burggren and West 1982; West and Burggren 1982). A major paper addressing the morphology and cardiorespiratory physiology of amphibian gills has recently appeared (Malvin 1989).

2.2 Skin

In marine intertidal fish, cutaneous respiration plays a major role in gas exchange, as it does in amphibians (see review by Feder and Burggren 1985). Rich capillarization is observed in the skin of several intertidal species (Rao and Hora 1938; Zander 1972; Graham 1976). Unlike amphibians, little is known about the ability of the skin to recruit cutaneous vessels during air exposure, though it seems a promising avenue to explore, given that the efficacy of the gills for gas exchange is thought to decline upon air-exposure in some species (Randall et al. 1981). Indeed, several air-breathing fish exhibit consistent increases in cutaneous O_2 uptake following exposure to air (Berg and Steen 1965; Tamura et al. 1976; Sacca and Burggren 1982). Responses such as these have often been taken to reflect uptake and carriage of oxygen to internal sites, rather than for skin metabolism alone (see Feder and Burggren 1985). In the case of the eel (*Anguilla rostrata*), however, O_2 uptake following transfer to 10 °C air is not sufficient to meet with the aerobic demands of the resting animal (Hyde et al. 1987). The marked decreases in arterial PO_2 and concommitant increases in PCO_2 within the first hours of air-exposure reflect the eels' low capacitance for aerial gas transfer (Hyde et al. 1987), contrary to earlier claims (Berg and Steen 1965). Thus, the ability of eels to survive prolonged periods of air-exposure is related to their tolerance to hypoxaemia and anaerobiosis, and not to any unusual modifications of their external gas exchanger.

Morphological adaptations for cutaneous gas exchange in amphibians have long been recognized (see review by Feder and Burggren 1985). For example, capillaries within the skin of various amphibian species can account for as little as 20% to as much as 95% of the total respiratory vasculature (Czopek 1965). However, despite this rich and varied level of vascularization, there appears to be little correlation between capillary density itself and dependence upon cutaneous respiration (Feder and Burggren 1985). Recently, there has been considerable interest in various 'active' components of cutaneous respiration that can lead to changes in the 'functional surface area' of the skin of amphibians (Burggren and Moalli 1984; Feder and Pinder 1988; Malvin 1988). This has also prompted renewed interest in cutaneous respiration in certain reptiles, deep diving sea turtles and sea snakes in particular. For example, the high levels of cutaneous gas exchange seen in sea snakes can be attributed to adaptations of their skin morphology, such as superficial capillarization (Rosenberg and Voris 1980) and spacing of vessels underneath the scale hinges or on the skin between the scales (Rauch 1978), as well as to active redistribution of blood flow to the skin (Seymour 1982, 1989).

2.3 Buccal Cavity

Extensive interfolding of the epithelia of the mouth cavity is seen in the electric eel (*Electrophorus electricus*), which uses its buccal cavity for a major portion of its overall oxygen uptake. Indeed, the electric eel is an obligate air breather and will succumb if access to air is cut off. The delicate papillations and rich vascularization of the mouth cavity must be avoided in most fish that also use their mouth for biting and swallowing prey. Because the electric eel stuns its prey with an electrical shock, and then swallows it whole, it manages to avoid damage to the delicate respiratory epithelium that lines the mouth cavity (Johansen 1968, 1970). Morphometric measurements indicate that the surface area of the buccal cavity of the electric eel can be 15% of the total surface area of the animal. It is also important to note that the gills are essentially vestigial, indicating again that as the aerial gas exchanger becomes more important in overall gas exchange, the branchial apparatus becomes reduced.

It has been difficult to accurately quantify the importance of the buccopharyngeal cavity in overall gas exchange in amphibians. However, Plethodontid salamanders actively ventilate the cavity with air, and increase the frequency of ventilation when O_2 levels decrease (D. P. Toews, per. commun.). Moreover, studies on the cutaneous/buccopharyngeal distribution of respiratory capillaries in certain Plethodontidae (Czopek 1961; Foxon 1964) have shown that up to 10% of these vessels can be found in the epithelia lining the mouth. Although it would be difficult to argue a major role for the buccal cavity in the gas exchange of anuran amphibians, there is good reason to suspect that buccopharyngeal respiration may play an important role in the overall gas exchange of lungless and gill-less forms. The comparatively large, predominantly skin-breathing urodele, *Cryptobranchus*, is known to ventilate its buccal cavity with water while submerged (Boutilier and Toews 1981 b). It also has a rich supply of blood vessels to the buccal cavity area. Certain reptiles ventilate their pharyngeal (and cloacal) cavities with water, and morphological adaptations for gas exchange include increases in vascularization and surface area attributed to thin filamentous projections on the walls lining the cavities (Winokur 1973; Legler and Cann 1980).

2.4 Alimentary Canal

Some fish have portions of their alimentary canal specialized for the uptake of oxygen. Tropical catfish such as *Hoplosternum*, *Ancistrus* and *Plecostomus* periodically rise to the surface and swallow air into their digestive tract. The gas bubbles are subsequently exhaled either through the mouth or anus (Gee and Graham 1978; Kramer and McLure 1981; Graham and Baird 1982). In some species, distinct areas of the alimentary canal are thin-walled and richly vascularized, while in others, the distinction between respiratory and digestive/absorptive areas are less clear cut. The extent of vascularization can be as much as twofold higher in specific 'respiratory' regions of the alimentary canal, and diffusion paths for respiratory gas exchange can be as little as 2 μm (McMahon and Burggren 1987). The apparent incompatibility between feeding and respiration has often been alluded to (e.g. Johansen 1968, 1970) but little studied.

Recently, McMahon and Burggren (1987) examined the respiratory efficacy of intestinal gas exchange in the Japanese weatherloach, *Misgurnus anguillicaudatus*. They showed that the alimentary canal was divided into three discrete regions: a small glandular anterior portion used for digestion, a middle transition zone and a large posterior region specialized for gas exchange. Moreover, they found that the degree of spiral rotation of the gut can be modulated by contractions of smooth muscle in the middle transition zone, such that food is compressed into a narrow string before entering the posterior 'respiratory' section of the gut. While packaging the foodstuff in this way may reduce the efficiency of food assimilation, it undoubtedly serves to maximize the surface area of the intestinal epithelium that is in contact with the gas bubble (McMahon and Burggren 1987). It was also noted that a single air breath replaces at least 50% – 80% of the total volume of the gut and that this is approximately equal to the total capacity of the respiratory portion of the intestine. Since air breathing occurs by unidirectional ventilation of the alimentary canal, McMahon and Burggren (1987) concluded that each breath should lead to a considerable turnover of the gas tensions so that favourable diffusion gradients for O_2 uptake and CO_2 removal are maintained.

2.5 Diverticula of Opercular and Pharyngeal Cavities in Air-Breathing Fish

Several fish species utilize regions of their buccopharyngeal cavities for air-breathing. These include members of the orders Symbranchiformes, such as *Monopterus albus* (Liem 1967), *Symbranchus marmoratus* (Johansen 1966), *Amphipnous cuchia* (Lomholt and Johansen 1976) and Channiformes (e.g. *Channa argus*: Ishimatzu and Itazawa 1981; Yu and Wo 1985; Glass et al. 1986). Other lineages such as the groups Claridae (e.g. *Clarius batrachus*) and Anabantoidae (e.g. *Trichogaster trichopterus*; Burggren 1979) also develop suprabranchial chambers that are effective in gas exchange. *Amphipnous*, for example, is a well adapted bimodal breather that depends on a pair of specialized air sacs that extend postero-laterally from the pharynx (Munshi and Singh 1968; Lomholt and Johansen 1976). *Channa* also has a pair of specialized air breathing organs that extend as pouches from the dorsal pharynx (Liem 1984). The accessory air-breathing organs of the Anabantidae and certain Claridae, on the other hand, are labyrinth-like or coral-like organs that are derived from the gill arches, and which extend along the dorsal regions of the opercular cavity (Munshi 1968; Burggren 1979).

2.6 Air Bladders and Lungs

The swimbladder of many fish is used as an accessory respiratory organ, and the lungs of dipnoan fish (i.e. *Protopterus*, *Lepidosiren*, and *Neoceratodus*) are thought to be the precursors of the amphibian lung (Johansen 1970). For example, the fine structure of the lung of *Protopterus* is not unlike that of higher vertebrates (DeGroodt et al. 1960), and is reported to have a blood-to-air diffusion distance of approximately 0.5 μm (Johansen 1970). Whereas the facultative air breather, *Neoceratodus*, has a single lung, both *Protopterus* and *Lepidosiren* have paired structures that meet anteriorly where the pneumatic duct and the pneumatic sphincter connect the lungs with the esophagus. The degree of internal septation, vascularization and innervation is apparently greater in *Protopterus* and *Lepidosiren* than in *Neoceratodus* (Johansen 1970; Fishman et al. 1989).

While not being as architecturally developed as the dipnoan 'lungs', certain species within the Chondrostei (*Polypterus*), Holostei (*Amia*, *Lepisosteus*) and Teleostei (*Arapaima*, *Hoplerythrinus* and *Erythrinus*) make effective use of their swim bladders for aerial gas exchange (Johansen 1970; Randall et al. 1981). Indeed, the swim bladder of the obligate air-breathing teleost, *Arapaima gigas*, is a large, highly vascularized and subdivided chamber that enables the animal to obtain as much as 80% of its O_2 requirements from air (Stevens and Holeton 1978; Randall et al. 1981).

Within the Amphibia, certain aquatic urodeles (e.g. *Amphiuma*) and apodans (e.g. *Typhlonectes*) possess lungs that are extremely large, in many cases extending for two-thirds of the body length (Toews 1971; Guimond and Hutchison 1976; Toews and Macintyre 1978; Martin and Hutchison 1979). The lungs of these species are also highly septate, and play a major role in gas exchange. Other urodeles such as *Necturus* and *Cryptobranchus* have smaller, far simpler lungs, with little or no septation of the inner lung surface. Anuran lungs, on the other hand, are more globular in shape and, for the most part, are sac-like structures with various degrees of internal septation, depending on the species and their habits. As pointed out by Smith and Rapson (1977) and Burggren (1989), the greatest levels of internal septation of lungs of anurans are often found in those animals that live in poorly oxygenated habitats (e.g. *Xenopus*) or whose skin is more keratinized and therefore less permeable to gases (e.g. *Bufo*). By far, the least complex lungs of any anuran must be those of the predominantly skin-breathing (Lake Titicaca) frog *Telmatobius culeus*. They are not reputed to ventilate their lungs, even though they are chronically exposed to high altitude hypoxia (Hutchison et al. 1976).

The lung volumes of various bimodal breathing reptiles, such as sea snakes and river snakes, are evidently much larger than in many terrestrial species (Wood and Lenfant 1976), meaning that there will be larger stores of oxygen for periods of submergence. In fact, the ratio of ABO storage to oxygen utilization is very high in most bimodal breathing fish, amphibians and reptiles. Because most of these animals can also regulate the flow of blood to the ABO (Sect. 6), they also share an ability to control the rate at which O_2 is removed from the store during periods of underwater activity.

3 Respiratory Gas Transfer

3.1 Bimodal Gas Exchange

In bimodal breathers, most of the metabolically produced CO_2 is almost always eliminated via the aquatic site for gas exchange, whereas the aerial exchanger usually accounts for the bulk of the O_2 taken up. In a compilation of data for more

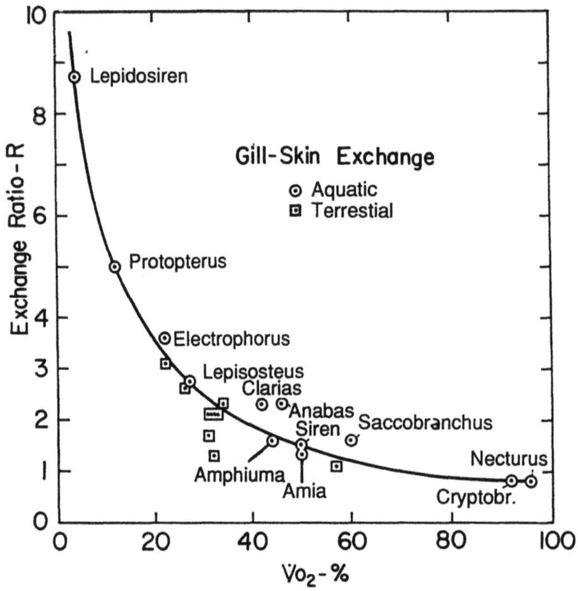

Fig. 2. The respiratory exchange ratio (R) of the gill-skin system as a function of the % O_2 uptake of the gill-skin system in a number of amphibious vertebrates (Rahn and Howell 1976)

than 20 bimodal breathing fish and amphibians, the gill-skin system accounted on average for 76% of the CO_2 eliminated (Rahn and Howell 1976). Indeed, the CO_2 excreted by the gill-skin system is rather constant in the vast majority of the species considered in the survey. What differs largely between the species, however, is the amount of oxygen taken up by the nonpulmonary exchanger. Thus, as the gill-skin site becomes less important as an O_2 exchanging system, its respiratory exchange ratio ($R = \%\dot{V}co_2/\%\dot{V}o_2$) increases to exceedingly high values (e.g. greater than 8 for *Lepidosiren*; Fig. 2). This shows that despite there being decreased use of the exchanger for O_2 uptake, CO_2 excretion by this route remains unaltered. By far, the greatest discrepancies between O_2 uptake and CO_2 excretion are found across the gills of the lungfish ($\sim 5-8$), with the gill-skin R values of other air-breathing fish and amphibians ranging between 0.9 and 3.6 (Rahn and Howell 1976). It is of interest to note that when the gill and/or skin system predominates in both O_2 uptake and CO_2 removal, such as in *Necturus* and *Cryptobranchus* (Guimond and Hutchison 1976), the R value of the gill-skin site is essentially the same as the metabolic RQ (~ 0.9).

Temperature plays an important role in determining the mode of gas exchange in a great many air-breathing fish (Johansen et al. 1970a; Rahn et al. 1971; Smatresk 1988), aquatic salamanders (Lenfant et al. 1970; Guimond and Hutchison 1976) and adult anurans (Hutchison et al. 1968; Hutchison and Miller 1979). In general, these animals are thought to be facultative air breathers at species-specific low temperatures and obligate air breathers at high temperatures. However, this is not invariably the case. For example, when larvae of either *Rana berlanderi* or *Rana catesbeiana* were exposed to acute temperature change (Burggren et al. 1983), the partitioning of overall O_2 consumption between skin (70%), gills (20%) and lungs (10%) remained essentially the same at experimental temperatures ranging from 15° to 33 °C. Burggren et al. (1983) suggested that because most of the O_2 uptake is cutaneous, energetically costly movements associated with branchial irrigation or with travelling to and from the surface for pulmonary gas exchange can be avoided.

Bimodal gas exchange is also important in many reptiles. The capacity for extrapulmonary gas exchange in sea snakes appears to be related to the degree of aquatic or terrestrial adaptation of individual species. Snakes that occasionally emerge onto land as a part of their normal behaviour (*Cerberus rhynchops*, *Laticauda* sp.) have relatively low rates of O_2 uptake in comparison to strictly aquatic forms such as Hydrophidae (Heatwole and Seymour 1978). The more terrestrial forms spend much of their life in shallow waters, near the surface, whereas exclusively aquatic species are often submerged at great depths, in O_2-rich water, for prolonged time periods. In the latter, the ability to respire cutaneously may be of considerable advantage, as in amphibians, in providing an effective store of O_2 and release point for CO_2, enabling the animal to remain submerged.

Non-pulmonary O_2 uptake in turtles is generally considered to be low, whereas non-pulmonary CO_2 losses appear again to be related to habitat, with the more aquatic species exhibiting greater CO_2 losses via this route (Jackson 1976; Jackson et al. 1976). For example, non-pulmonary CO_2 loss in the aquatic *Trionyx* accounts for 64% of the total CO_2 production, whereas CO_2 losses amount to less than 3% in the terrestrial *Geochelone denticulata*. Non-pulmonary CO_2 losses are also related to temperature. In the box turtle (*Terrapene ornata*), for example, these losses can increase from 35% of the $\dot{V}CO_2$ at 25 °C to over 75% at 5 °C (Glass et al. 1979). In this case, it was apparent that aquatic CO_2 losses were important in preventing any large buildup of blood PCO_2 levels at low temperatures, when the frequency of air breathing is less than 4 breaths/h (Glass et al. 1979). As with most air-breathing fish and amphibians, the frequency of air breathing in many reptiles becomes reduced as temperatures and metabolic rates decrease (Glass et al. 1983; Milsom, this Vol.). Though many air-breathing fish and amphibians are capable of fulfilling all of their metabolic requirements aerobically through non-pulmonary gas exchange at low temperatures (Johansen 1970; Pinder 1987), this has yet to be substantiated unequivocally for any reptile.

3.2 Controlling CO_2 Losses

Although the gill-skin system of bimodal breathers often accounts for the bulk of the metabolically produced CO_2 that is removed, there is little evidence to indicate that it plays any significant role in controlling the CO_2 losses. In bimodal breathing amphibians such as *Rana catesbeiana* and *Xenopus laevis*, there is good evidence to show that CO_2 removal is controlled through pulmonary ventilation (Jackson 1978; Boutilier and Shelton 1986c). For example, measurements of cutaneous and pulmonary CO_2 losses in *Rana catesbeiana* showed that as temperatures increased from 10° to 30 °C, over 90% of the metabolically-induced increase in $\dot{V}CO_2$ could be accounted for by increased pulmonary ventilation, with cutaneous CO_2 removal remaining rather constant (Mackenzie and Jackson 1978). In a subsequent experimental series, Jackson and Braun (1979) examined the CO_2

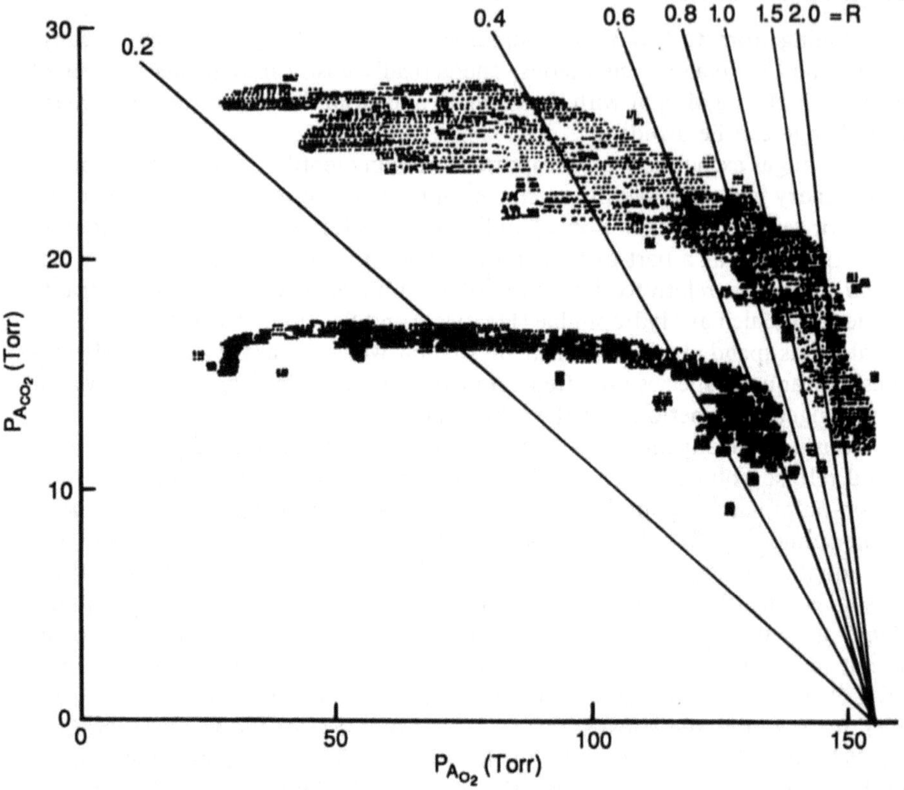

Fig. 3. The relationships between alveolar PO_2 ($PACO_2$) and PCO_2 ($PAco_2$) in *Xenopus laevis* during voluntary diving and surfacing in air-equilibrated conditions (*lower plot*) and subjected on the next day to aquatic hypercapnia (*upper plot*). Data were collected continuously by an extracorporeal loop that incorporated O_2 and CO_2 electrodes to measure lung gases. Hypercapnia was administered by equilibrating the water with a 1% CO_2/99% air mixture, but the animal was free to breathe fresh air at a surface blowhole. Lung R lines are shown from 0.2 to 2 for reference. Temperature, 25 °C (Boutilier and Shelton 1986c)

conductances of the aquatic exchanger in bullfrogs exposed to elevated levels of CO_2 delivered either to the lungs or skin. These experiments showed that when the CO_2 was delivered to the skin, the animals were able to maintain their arterial PCO_2 at a near constant level, by increasing pulmonary ventilation. This demonstrated that the lungs could effectively control the respiratory acid-base balance of the animal by eliminating both the metabolically produced CO_2 plus the additional CO_2 that was entering through the skin. Similar experiments, in which CO_2 was delivered to the skin of *Xenopus laevis*, have shown that the marked rise in pulmonary ventilation leads to increases in the pulmonary respiratory exchange quotient (R), to values in excess of 1.5 (Fig. 3). The elevated R values can be attributed to the different rates at which the O_2 and CO_2 stores are adjusted during ventilation (Sect. 5.2.2). When, on the other hand, the CO_2 was delivered via the lungs, arterial PCO_2 levels increased, resulting in the development of a respiratory acidosis (Mackenzie and Jackson 1978; Boutilier 1988). Both studies concluded that the skin was unable to respond in a manner that approximated the capacity of the lungs, and that CO_2 losses in these animals were primarily controlled by pulmonary ventilation. The notion that cutaneous CO_2 removal is a relatively ineffective regulator of respiratory acid-base balance was first suggested by Piiper et al. (1976) in their studies of diffusion-limited gas exchange in the lungless and gill-less Plethodontid salamander *Desmognathus fuscus*.

Arterial PCO_2 levels in air-breathing fish are generally higher than those of exclusively water-breathing species (Johansen 1970). Presumably, this reflects the comparatively smaller surface area of the gills of air-breathing fish (Laurent et al. 1978), and their reduced requirements for O_2 uptake via gill convection. When the aquatic CO_2 excretion of such animals is blocked either by elevated ambient CO_2 levels or exposure to air, arterial PCO_2 levels increase, indicating that the aerial exchange organs cannot offset the increased demands for CO_2 elimination (Hughes and Singh 1971; Singh and Hughes 1971; Randall et al. 1978a, b; 1981; Wright and Raymond 1978; Heisler 1982). For example, when the bowfin, *Amia calva*, is exposed to air, the rise in arterial PCO_2 leads to a large respiratory acidosis (Daxboeck et al. 1981). Similar effects have been noted during both air-exposure and aquatic hypercapnia in the lungfish *Neoceratodus forsteri* (Johansen et al. 1967; Johansen 1970) and in air-exposed *Symbranchus* (Johansen 1966; Heisler 1982), *Protopterus* (Lenfant and Johansen 1968) and *Hoplerythrinus* (Randall et al. 1978a). The rise in arterial PCO_2 that occurs when the animals are deprived of their gills as a means for eliminating CO_2 results in an increased PCO_2 gradient between blood and air. The establishment of this gradient means that movement of CO_2 from blood to air-breathing organ will be enhanced, resulting in a rise in the gas exchange ratio of the aerial exchanger. Such increases in R for several air-exposed teleosts have been reported (see review by Singh 1976).

Little is known of the effects that changes in ventilation and perfusion of the gas exchange organs might have on overall CO_2 excretion in bimodal breathing fish. In the air-exposed snakehead fish, *Channa argus*, reduced air convection is thought to play a major role in modulating PCO_2 and therefore extracellular pH (Ishimatzu and Itazawa 1983). In the same species, Glass et al. (1986) found that the respiratory gas exchange ratio rose from 0.16 at 15 °C to 0.29 at 25 °C, indicating greater importance of aerial elimination of CO_2 as temperatures increase, as reported above for amphibians (see Jackson 1978). There is also clear evidence of a marked air-breathing response to aquatic hypercapnia in *Channa* (Glass et al. 1986), as well as in the Cuchia eel *Amphipnous cuchia* (Lomholt and Johansen 1974), the Australian lungfish *Neoceratodus* (Johansen et al. 1967) and the blue gourami *Trichogaster* (Burggren 1979). The ventilatory response in *Trichogaster* was not unlike that found when the toad, *Xenopus laevis*, was subjected to aquatic hypercapnia (Fig. 3). In both, the aerial respiratory exchange quotient increased to values in excess of 1.

Very little is known of whether any control processes operate to regulate CO_2 losses by non-pulmonary routes in reptiles. Although the skin is considered to be the sole route for extrapulmonary CO_2 losses in sea snakes (Graham 1974), certain turtles respire across their buccal cavity and/or cloaca. Pharyngeal respiration in certain species (*Trionyx triunguis*, *Sternothernus odoratus*) can account for as much as 30%–35% of their extrapulmonary uptake of oxygen (Root 1949;

Girgis 1961). However, it is not known whether any of these animals can effectively modulate the rates of CO_2 loss or O_2 uptake by changing the rates of water convection into and out of either the pharyngeal or cloacal cavities. As with amphibians, the skin of reptiles is considered to be primarily diffusion limited with little inherent capacity for effective homeostatic control. This is not to say that cutaneous respiration cannot be modulated and that gas exchange cannot be influenced through changes in perfusion and disruption of boundary layers (Feder and Pinder 1988; Malvin 1988; Piiper 1988). Indeed, the impressive survival capacity that many amphibians and reptiles have for prolonged submergence at low temperatures must underline an important, perhaps controlling, role for cutaneous gas exchange during such times. As pointed out by Feder and Burggren (1985), cutaneous gas exchange in many animals is looked upon rather more as a curiosity than an effective and essential part of the animals' overall gas exchange capacity. Amphibians have generally been looked upon as the champions of non-pulmonary gas exchange, even though it is obvious that certain species of reptiles, such as aquatic snakes, have impressive capabilities in this regard.

3.3 The Role of Carbonic Anhydrase in CO_2 Elimination

Respiratory epithelia from a number of vertebrates contain appreciable quantities of carbonic anhydrase. The enzyme is found in the gills and aerial exchange organs of air-breathing fish (Burggren and Haswell 1979; Randall et al. 1981), in the gills, skin and lungs of amphibians (Toews et al. 1978) and in the lungs of many higher vertebrates (e.g. humans; Bidani and Crandall 1988). The presence or absence of this enzyme in the respiratory epithelia of certain species is often considered indicative of the relative role that the epithelium plays in overall CO_2 removal. However, many such arguments are conjecture, based on the assumption that epithelial carbonic anhydrase activity is available to extracellular CO_2-H_2CO_3 reactions. So far, this issue has only been adequately described and quantified for mammalian lung (Crandall and Bidani 1981; Heming et al. 1986).

Because the aquatic exchange organs of bimodal breathers are often responsible for the majority of the CO_2 that is excreted, there have been several attempts to see whether the aerial exchange organs are capable of compensating for this CO_2 removal when the aquatic route is effectively blocked (Sect. 3.2). Some workers have suggested that the absence of carbonic anhydrase (CA) in the epithelia of gas exchange organs of fish might limit the removal of CO_2 by this route (Randall et al. 1978a, b; Burggren and Haswell 1979; Randall et al. 1981). Indirect evidence that CO_2 excretion is limited by the absence of carbonic anhydrase comes from experiments on *Hoplerythrinus* (Randall et al. 1978a), in which the respiratory acidosis that developed upon air exposure could be ameliorated after adding carbonic anhydrase to the blood. In this case, the reduction of arterial PCO_2 following carbonic anhydrase infusion occurred in concert with a doubling of the gas exchange ratio of the swim bladder, indicative of enhanced aerial CO_2 excretion.

Burggren and Haswell (1979) proposed that the extraordinarily high aerial CO_2 excretion of the air-exposed anabantid, *Trichogaster*, could be accounted for by the presence of carbonic anhydrase in the labyrinth epithelium. Randall et al. (1981) suggested further that carbonic anhydrase would only be found in the epithelia of animals whose air-breathing organs were embryonically derived from gill tissue, as in *Trichogaster*. Heming and Watson (1986) recently showed that the bowfin, *Amia*, has a carbonic anhydrase (CA) activity ratio of swim bladder: red cells of 1 : 3, similar to the ratios observed in the bimodal breathing teleosts, *Trichogaster trichopterus* and *Clarius batrachus* (Burggren and Haswell 1979). Moreover, when compared to the ~ 1 : 80 ratio of lung CA: red cells CA in the rat (Maren 1967), the air-breathing organs of fish would appear to have relatively higher enzyme activities than mammalian lung (Heming and Watson 1986). The occurrence of extracellular CA activity in aerial exchange organs of air-breathing fish, its availability to plasma HCO_3^- in vivo, and the relative magnitude of its importance to overall CO_2 excretion are important gaps in our knowledge of gas exchange in these animals.

Carbonic anhydrase has been measured manometrically in the red blood cells, lungs and skin of several species of bimodal-breathing amphibians (Toews et al. 1978). It has also been detected histochemically in cutaneous tissues from the frog, *Rana pipiens* (Rosen and Friedley 1973). Erythrocytic carbonic anhydrase facilitates CO_2 excretion in amphibians such as *Bufo marinus* and *Cryptobranchus alleganiensis* in much the same manner as in mammals. In these animals, most of the carbon dioxide in the blood is transported in the plasma as bicarbonate. As venous blood enters the lungs and/or skin, it begins to give up CO_2 to the environment; i.e. the PCO_2 of the blood becomes reduced, erythrocyte bicarbonate concentrations fall, and plasma bicarbonate moves into the red cell in exchange for chloride (Boutilier and Toews 1981 a, b; Boutilier et al. 1979 a; McDonald et al. 1980). As far as is known, erythrocytic Cl^-/HCO_3^- exchange facilitates CO_2 removal in bimodal breathing vertebrates, though it should be emphasized that vertebrates lacking rapid anion exchange in their erythrocytes have been documented (Nikinmaa and Railo 1987; Tufts and Boutilier 1989; Nikinmaa and Tufts 1989). An apparent anomaly in the Amphibia is the mudpuppy, *Necturus maculosus*, whose erythrocytes evidently lack any carbonic anhydrase (Toews et al. 1978; T. H. Maren, pers. commun.). This is a surprising result, but one which has been independently identified by two laboratories. Though absent in the red blood cells of *Necturus*, carbonic anhydrase was detected in the lungs and skin (Toews et al. 1978), where one might predict that it facilitates CO_2 removal. If carbonic anhydrase is indeed absent from the erythrocytes of *Necturus*, and not just undetectably low, this would mean that the dehydration of plasma bicarbonate should proceed at the uncatalyzed rate. Thus, if not for extracellular catalysis, one would predict there to be extensive disequilibria in extracellular CO_2-H_2CO_3 reactions in *Necturus*.

In their analysis of external gas exchange organs (e.g. skin), Piiper and Scheid (1977) coined the phrase 'infinite pool' to mean that pool of air or water with which cutaneous exchange of gases may occur. Even though the external pool may be well mixed, there is always an unstirred diffusion boundary layer close to the skin, which can limit the exchange of gases (Feder and Pinder 1988; Piiper, this Vol.). For a quiescent animal in still water, it is important to consider the fate of the dissolved CO_2 after it crosses the skin and/or gills and enters the water. It will be important to learn whether CA is present in the unstirred layer about the skin of amphibians and, if so, how fast it establishes chemical equilibrium between CO_2 and HCO_3^- within the layer. Since the uncatalyzed hydration of CO_2 is in the order of minutes at the temperatures and pH of waters in which these animals often reside (Kern 1960), the absence of CA in the unstirred layer could lead to an increase in the PCO_2 about the animal, and thereby to a decrease in the PCO_2 gradient across the skin. Carbonic anhydrase on either the apical surface of the epithelial cells or contained in the skin mucus could, alternatively, lead to an increase in the transcutaneous PCO_2 gradient, as the PCO_2 immediately adjacent to the skin would decrease upon HCO_3^- and H^+ formation. The effects that either of these conditions might have on overall CO_2 excretion in bimodal-breathing animals is not presently known. However, Wright et al. (1986) have recent evidence which indicates that CA activity in the gill mucus of rainbow trout effects almost complete CO_2 hydration of water exiting the gills. In the case of the water-breathing fish, whose interlamellar transit time of gill water is 100–400 ms (Randall 1982a, b), the effects that an uncatalyzed CO_2 hydration reaction would have on CO_2 excretion would presumably be less than in cases where the exchange surfaces lack any specialized form of convection (e.g. skin, certain gills of air-breathing fish and urodele amphibians). A great deal remains to be resolved about the effects that catalyzed or uncatalyzed boundary layers will have on aquatic CO_2 excretion in bimodal breathers.

4 Co-Ordination of Aquatic and Aerial Breathing

4.1 Ventilatory Mechanisms

Air-breathing fish utilize the buccal force pump to ventilate the air-breathing organ (ABO) as well as irrigate the gills. The same basic mechanism also operates to ventilate the lungs of amphibians, and comparisons of the buccopharyngeal pump of representative fish and amphibians reveal strikingly similar mechanisms of operation (Randall et al. 1981).

4.1.1 The Buccal Force Pump: Transition from Water- to Air-Breathing

The activities of muscles concerned with branchial ventilation are very similar in both lungfish (i.e. *Protopterus*: McMahon 1969) and exclusively water-breathing teleosts (Ballintijn and Hughges 1965; McMahon 1969). Modifications of the pump for breathing air constitute small variations on a common theme (Randall et al. 1981). In water breathing, the object is for the buccal-opercular pump to provide a continuous unidirectional stream of water over the gills, whereas for air-breathing, the buccal pump is used simply to force air into the ABO.

Electromyograms of muscle activities during breathing have been carried out in exclusively water-breathing fish (Ballintijn and Hughes 1965) and lungfish (McMahon 1969). Whereas the muscles of the buccal floor and lower jaw are activated in synchrony during an air-breath, the water-breathing mode is characterized by some degree of temporal separation between the contraction of the same muscle groups. The opercular pump of air breathing fish must also be uncoupled from the buccal pump during air breathing, so that the buccal cavity can work to force air into the lung without loss of pressure and air to the opercular cleft. The lungfish solves this problem by encouraging ventral rather than lateral expansion of the buccal cavity during an air-breath, the opposite being true during the water-breathing phase. This effectively reduces the interaction between buccal and opercular pumps.

Modifications to prevent air loss during force ventilation of the ABO include comparatively reduced opercular slits in air-breathing fish (Carter and Beadle 1931), active closure of the slit, e.g. hyoideus muscle contraction in lungfish (McMahon 1969), or sealing of the operculum to the cleithrum with mucus during air-exposure (Johansen 1970). This, in addition to the synchronous contraction of buccal and jaw muscles during air breathing, evidently contributes to the comparatively higher pressures in the air- vs water-breathing mode. High pressures are thought to be required in order to overcome resistances associated with filling the ABO (Randall et al. 1981). Factors such as air bladder compliance, stiffness of body walls, and diameter and complexity of airways may all contribute to this resistance.

The turnover of gases will depend to a large extent on the amount of residual gas left in the air bladder following the exhalation phase of an air breath. This emptying process may be aided in some air-breathing fish and amphibians by contraction of musculature in proximity to the ABO, in addition to elastic recoil (Randall et al. 1981; Shelton et al. 1986). In the lungfish *Protopterus* (McMahon 1969) and *Neoceratodus* (Johansen et al. 1967), exhalation is thought to be aided by contraction of smooth muscle of the ABO itself. Indeed, many fish species have smooth or striated muscle associated with their ABO (e.g. *Saccobranchus*, *Amia* and *Amphipnous*, Johansen 1970; Randall et al. 1981), and this musculature is also presumed to contribute to an 'active' form of exhalation. Similarly, contractures of the flank musculature in frogs (e.g. *Rana catesbeiana*) and toads (*Bufo marinus*, *Xenopus laevis*) are known to produce more forceful expirations (DeJongh 1972; DeJongh and Gans 1969; Macintyre and Toews 1976; Brett 1980). Boutilier (1984) noted that the first breath following a prolonged dive in *Xenopus* often exhibited higher flow rates and lower expiratory durations, indicative of a more forceful expulsion of lung gas. Measurements by Brett (1980) have shown that *Xenopus* can indeed contract its flank muscles in order to perform greater expiratory volumes. It was suggested that the more forceful expiration following a prolonged dive in *Xenopus* would serve to empty the lung more completely, thereby ensuring a more rapid and efficient turnover of lung gases during fleeting visits to the surface (Fig. 4C, Boutilier 1984). Similar events presumably aid in the nearly complete collapse of the lung that precedes the lung-filling phase of both *Amphiuma tridactylum* and *Siren lacertina* (Martin and Hutchison 1979, Sect. 5.2.2).

In addition to the lungfish (*Protopterus aethiopicus*), whose mechanical basis of air ventilation is well understood (McMahon 1969), the air-breathing mechanisms of a number of bimodal-breathing teleosts have also been examined (e.g.

Lepisosteus osseus, Rahn et al. 1971; *Anabas & Helostoma*, Liem 1980; *Channa*, Liem 1984; *Amia calva*, Deyst and Liem 1985; among others, see Liem 1987). Pressure and air-flow characteristics of the air-breathing cycle of *Amia* are very similar to those seen in the lungfish, *Protopterus* (McMahon 1969), suggesting either an early common evolutionary origin or convergent evolution of a system that is inherently restricted to a single functional design (Deyst and Liem 1985).

4.1.2 Patterns of Air Flow During Ventilation of the ABO

Recent studies using high-speed cinematography and/or cineradiography (Rahn et al. 1971; Ishimatzu and Itazawa 1981; Peters 1978; Liem 1980, 1987) have focused on the structural and functional designs of the air-breathing apparatus in a variety of air-breathing teleosts. Such studies, particularly when accompanied by electromyographical and pressure recordings (Liem 1984, 1987), have clearly demonstrated that several air-breathing fish use water to expel the air bubble from the mouth cavity prior to the inhalation of fresh air. This form of air breathing in teleosts is referred to as quadraphasic breathing, owing to there being four distinct phases in the air ventilation cycle (see Liem 1987). Other forms show three discrete phases in their breathing cycles, and are termed 'triphasic' breathers. These fish ventilate their air cavity by creating a flow of fresh air that proceeds from the buccal cavity into the suprabranchial chamber, replacing the stale air which then exits via the opercular cavity. This 'draft' of fresh air is created without the use of water, and this is considered by some to be a preadaptation to emergence onto land (Liem 1987). The use of water in the quadraphasic breathing cycle appears to be restricted to members of the Anabantoidei and Clariidae and is thought to have chained many of them to an aquatic existence (Ishimatzu and Itazawa 1981).

In some air-breathing fish (*Hoplerythrinus*, Farrell and Randall 1978; *Piabucina*, Graham et al. 1977) as well as in certain anuran amphibians (*Rana catesbeiana*, DeJongh and Gans 1969; *Rana pipiens*, West and Jones 1975; *Bufo marinus*, Macintyre and Toews 1976) the ventilatory sequence is one in which inspiration precedes expiration. This inevitably leads to some degree of mixing between inhalent and exhalant gases. However, in both *Hoplerythrinus* and *Rana*, such mixing is thought to be minimized by selective 'jet-streaming' of either inhalent or exhalent air. In *Hoplerythrinus*, inhaled gas appears to be selectively channeled into a highly compliant anterior section of the gas bladder, thereby reducing the possibility for mixing with residual gas contained in the posterior-most 'respiratory' chamber (Kramer 1978). Immediately following this inspiratory phase, exhalation from the posterior chamber then proceeds, again with very little anterior-posterior mixing. This is followed by active filling of the posterior chamber with the nearly fresh air that was originally inhaled. The *non-respiratory anterior chamber* is separated from the *posterior respiratory chamber* by a distinct muscular sphincter which evidently regulates gas flow. A similar sequence of events is believed to occur in Ranid and Bufonid amphibians (DeJongh and Gans 1969; Macintyre and Toews 1976). The anterior chamber in this case is the buccal cavity and the posterior chamber is the lung, with air flow between the two being co-ordinated by the glottis. Air is drawn through the nares and into the buccal cavity as the latter becomes distended ventrally by muscular contraction. The glottis then opens and the air contained within the lung is thought to be channeled in a 'jet-stream' along the dorsal roof of the buccal cavity and out through the nares, causing little mixing with previously inspired air (DeJongh and Gans 1969). The nares then close and the lungs are inflated to positive pressures by the buccal force pump.

Unlike the semi-terrestrial anurans, all movements of the buccal floor of *Xenopus* are concerned with ventilating the lung. Moreover, the buccal phase of inspiration only occurs after expired air has passed from the lung and buccal cavity (Brett and Shelton 1979). Following a single expiration, the buccal pump then fills the lungs with either a single, double or triple stroke of the pump. When more than one inspiration occurs, they are always separated in time by short inspiratory pauses (Boutilier 1984). The sequence of expiration preceding inspiration ensures that there will be little reinspiration of exhalent air. Brett and Shelton (1979) postulated that this sequence of events may confer important advantages to an animal whose acquisition of fresh air may be limited to brief episodes at the air-water interface (Fig. 4). On the other hand, the opposite has been argued, in the sense that the inspiratory-expiratory sequence of certain air-breathing fish may enable

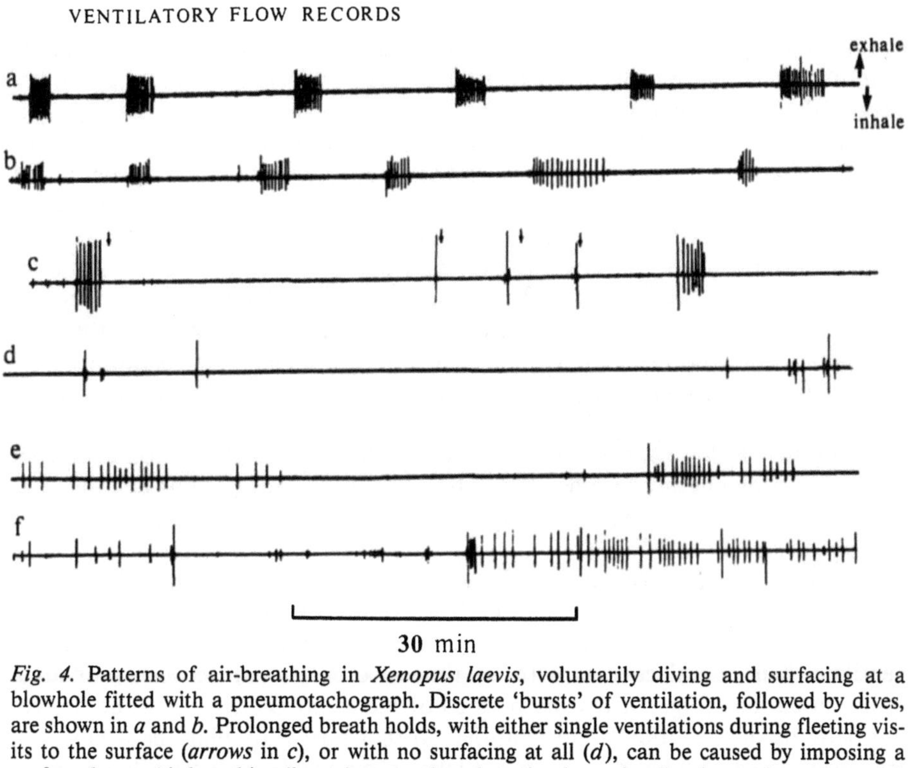

Fig. 4. Patterns of air-breathing in *Xenopus laevis*, voluntarily diving and surfacing at a blowhole fitted with a pneumotachograph. Discrete 'bursts' of ventilation, followed by dives, are shown in *a* and *b*. Prolonged breath holds, with either single ventilations during fleeting visits to the surface (*arrows* in *c*), or with no surfacing at all (*d*), can be caused by imposing a surface threat. Air-breathing 'bouts' occur when the animal remains floating at the surface with its nostrils in air, ventilating its lungs intermittently (*e* and *f*). All animals were in air-saturated water and breathing air at 25 °C. (Boutilier 1984).

such animals to acquire fresh air, even if the breathing cycle becomes terminated before exhalation (Randall et al. 1981).

4.2 Patterns of Air Breathing in Amphibious Vertebrates

Bimodal gas exchange is well developed in all air-breathing fish and amphibians, as well as in certain aquatic reptiles (e.g. aquatic snakes). Ventilation of the air-breathing apparatus in all such animals is a highly intermittent process, with individual breaths, or a series of breaths, being separated in time by periods of breath-holding of variable duration. The intervals between individual breaths may be very long, representing periods of submergence, or constitute shorter periods of time between the individual breaths of a ventilatory episode at the air-water interface. Some animals rise to the surface only to take a single breath (e.g. *Electrophorus*, *Amphiuma*, certain sea snakes: Johansen et al. 1968 b; Farber and Rahn 1970; Toews 1971; Martin and Hutchison 1979; Seymour 1982, 1989), whereas others ventilate their lungs several times before submerging (e.g. *Amphipnous*, *Xenopus*, *Acrochordus*: Lomholt and Johansen 1974, 1976; Glass and Johansen 1976; Boutilier 1984). To a large extent, the timing and frequency of air breaths is correlated with the degree of aquatic or terrestrial adaptation of the animals themselves or of the prevailing environmental conditions. Thus, breath-holding intervals in bimodal breathers can range from seconds to hours, depending on a whole host of interactions, including whether the animals are primarily aquatic or semi-terrestrial, whether they are highly active or sedentary, whether their environment is cool and well oxygenated or warm and hypoxic. The common feature of the air-breathing habit in these animals is its arrhythmicity, and the fact that ventilation of the ABO is turned on by some variable need for additional gas exchange.

Very few studies have carried out detailed analyses of the temporal patterns of air breathing in amphibious vertebrates (Ackerman and White 1979; Boutilier 1984; Milsom, this Vol.). Most studies on air-breathing fish have concentrated upon the periods of breathing, with little emphasis on the periodicity of the non-ventilatory time intervals. Unlike the single air gulp of many air-breathing fish, the Cuchia eel, *Amphipnous*, ventilates its ABO in a series of breaths while at the surface (Lomholt and Johansen 1974). Upon breaking the surface, a small inspiration is held for approximately one min followed by a succession of deeper breaths of progressively increasing duration. These surface periods can be separated in time by dives of 30 min or more, depending on the temperature (Lomholt and Johansen 1974, 1976).

The breathing patterns of amphibians, particularly anurans, have been more extensively analyzed than air-breathing fish. Anurans such as *Rana* and *Bufo* exhibit the most continuous patterns of breathing, whereby the animals continue to ventilate their buccal cavities with air (buccal oscillations) between individual lung ventilations (Fig. 5). During periods of increased O_2 demands, the breathing patterns of the animals are modified so that buccal oscillations decrease in frequency and are replaced by lung ventilations, many of which are often grouped into discrete episodes of lung inflation (DeJongh and Gans 1969; West and Jones 1975; Macintyre and Toews 1976; Boutilier and Toews 1977; Jones 1982; Pinder

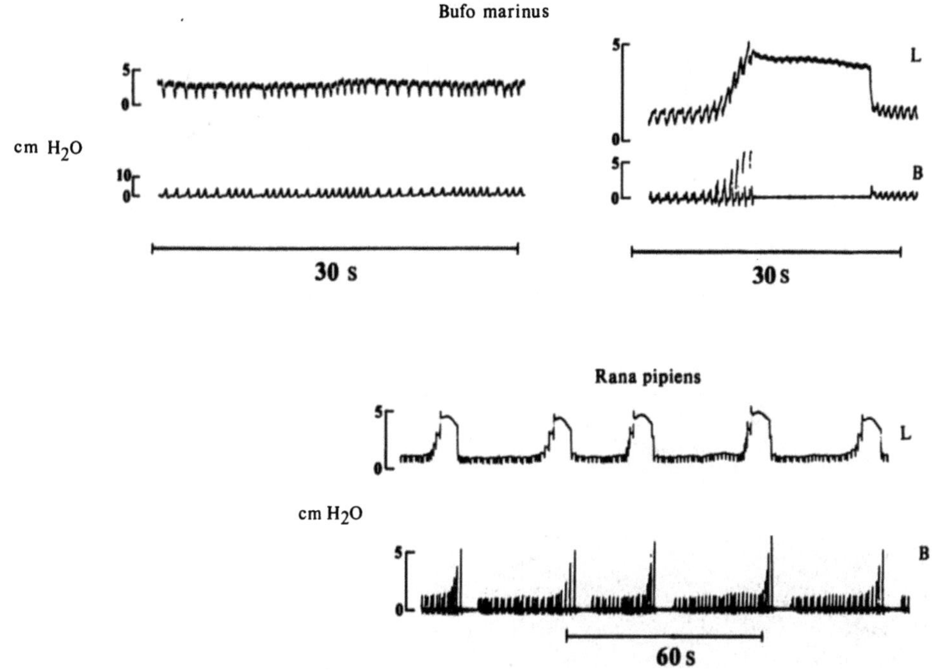

Fig. 5. Top: Simultaneous pressure recordings from the lung (*L*) and buccal cavity (*B*) of the toad, *Bufo marinus*, ventilating its lungs with room air at a slightly elevated rate (*left*) due to disturbance by investigators. *Right*, the buccal and lung pressures that occur during a single inflation cycle recorded when the animal was breathing a 5% CO_2/95% air mixture (Macintyre and Toews 1976). *Bottom*: Air-breathing records of buccal (*B*) and lung (*L*) pressures in the leopard frog, *Rana pipiens*, showing sequences of lung inflation and deflation. It should be noted that the frogs were restrained during these measurements and, with less disturbance, the rates of breathing would probably be less than are shown here (e.g. cf. Pinder and Burggren 1986). Temperatures, 23°–24 °C (West and Jones 1975)

and Burggren 1986; Kruhøffer et al. 1987). The patterns of ventilation change abruptly when well-adapted divers such as *Rana* are in water (Sect. 7.1.1.2). Unlike the Bufonids, which rarely dive, many species of *Rana* will alternate series of breathing movements at the air-water interface, with prolonged sessions of diving. This is certainly true for the exclusively aquatic anuran *Xenopus laevis*. Experiments on *Xenopus* have revealed two discrete breathing patterns (Boutilier 1984). In one (burst breathing), long periods of diving are periodically interrupted with brief visits to the surface, whereupon a discrete episode of lung ventilations takes place (Fig. 4A, B). At other times, the same animal might remain at the surface for more prolonged intervals, ventilating its lungs in a more arrhythmic fashion (Fig. 4E, F).

Using hypercapnia as a stimulus for air breathing (Sect. 7.1.4), the amounts of gases exchanged in the Cuchia eel (*Amphipnous*) and the clawed toad (*Xenopus*) were analyzed over extended time periods, in order to see whether changes in minute ventilation ($\dot{V}_I$) occur due to the frequency (f), or the depth (V_I), of

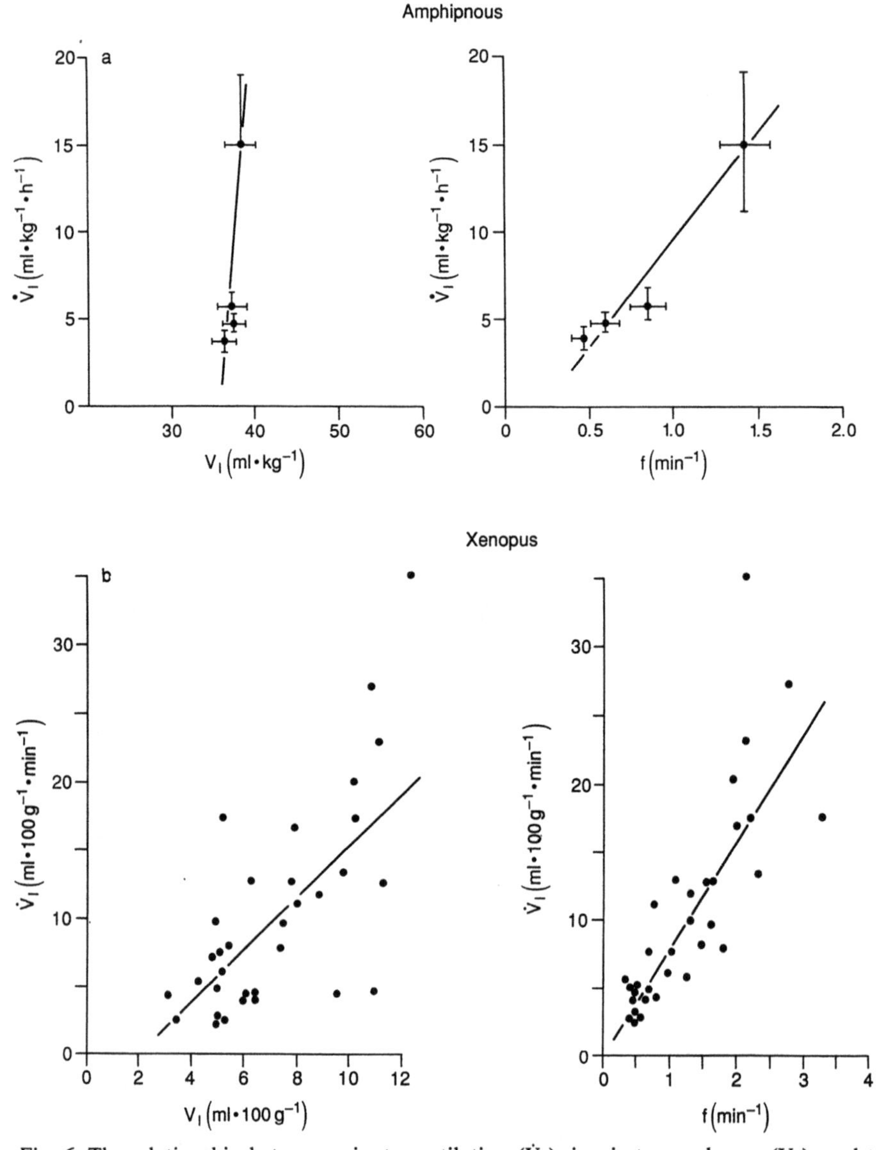

Fig. 6. The relationship between minute ventilation ($\dot{V}_I$), inspiratory volumes (V_I), and total respiratory frequency (*f*), in the amphibious fish, *Amphipnous cuchia* (*A*) and in the African clawed toad, *Xenopus laevis* (*B*). In both instances, the animals were diving voluntarily and free to surface at a blowhole. The range of breathing frequencies and inspired volumes were achieved by exposing the animals to hypercapnic environments at 30 °C (*A*) or 25 °C (*B*). Mean values for *Amphipnous* were computed from data presented for individual fish in Fig. 8 of Lomholt and Johansen (1974) (*Xenopus* data from Boutilier 1988)

ventilation (Fig. 6; Lomholt and Johansen 1974; Boutilier 1988). The depth of breathing in both animals is effectively limited by the (essentially) fixed-volume buccal pump, though *Xenopus* often increases its inspired volume through multiple inspirations (Boutilier 1984). Regulation of the amounts of gases exchanged in *Xenopus* and *Amphipnous* result primarily from manipulations of the temporal patterns of lung ventilation (i.e. the breath-hold durations), rather than through changes in the depth of breathing (Fig. 6). Similar results have been described for *Rana pipiens* (Pinder and Burggren 1986) and for *Bufo paracnemius* (Kruhøffer et al. 1987). In all of these animals, it appears that the non-respiratory time of the animal, and not the composition of its individual breaths, is the main 'ventilatory reserve', a conclusion also reached by Glass and Johansen (1976) in the bimodal-breathing snake, *Acrochordus javanicus*.

4.3 Patterns of Aquatic Convection

4.3.1 Air-Breathing Fish

Many air-breathing fish convect both water and air, with the frequency of branchial pumping generally exceeding that of air breathing. Rates of branchial pumping of approximately 20–40 cycles/min are observed in species such as *Erythrinus* (Stevens and Holeton 1978), *Neoceratodus* (Johansen et al. 1967), *Channa maculata* (Yu and Woo 1985) and *Lepisosteus oculatus* (Smatresk 1988), where irrigation of the gills is rather continuous. More arrhythmic patterns of gill irrigation and generally lower rates of branchial pumping are found in the lungfish *Protopterus* (Johansen and Lenfant 1968) and *Lepidosiren* (Johansen and Lenfant 1967). Part of the arrhythmicity of gill irrigation evidently arises as a consequence of air breathing, at which time gill ventilation is temporarily suspended while the buccal pump is used to ventilate the ABO with air. Following an air breath, the rate of branchial pumping is often slow to return to the pre-breath rate, suggesting that air breathing effectively reduces the drive for branchial breathing (Johansen et al. 1970a; Johansen 1970; Sect. 7.1.1.1).

Some species of fish (e.g. *Electrophorus*: Johansen 1970) do not actively ventilate their gills with water. In such cases, the gills are comparatively reduced in size, and participate little towards the overall gas exchange requirement. In *Amphipnous*, active branchial ventilation is only observed if the animal is deprived of air breathing (Lomholt and Johansen 1974). During these times, its seemingly vestigial gills, having only a few coarse filaments on the second branchial arch (Munshi and Singh 1968; Hughes and Munshi 1973), are capable of extracting up to 25% oxygen from the water (Lomholt and Johansen 1976). Branchial breathing in *Amphipnous* is thought to be of greater importance to overall gas exchange in larval and juvenile stages (Johansen 1970; Lomholt and Johansen 1974), which is also the case in amphibians (reviewed by Malvin 1989).

4.3.2 Amphibians

Branchial irrigation in the neotenic salamander, *Necturus maculosus*, occurs as the result of rhythmic beating of its externalized gills back and forth through the water. Although many other urodeles possess external gills, they are rarely 'irrigated' in this or any other fashion, and simply hang in the water. Such gills may become passively irrigated when animals position themselves in water currents or when they move about. Disruption of boundary layers through spontaneous movements (including external gill movements) facilitates increased diffusive conductance for oxygen, by increasing the PO_2 of the water immediately adjacent to the site of gas exchange (Feder and Pinder 1988). This is perhaps exemplified by the exclusively aquatic and predominantly skin-breathing amphibians, *Cryptobranchus alleganiensis* and *Telmatobius culeus*, both of whom rock or sway their entire bodies back and forth in water when it becomes either low in O_2 or high in CO_2 (Hutchison et al. 1976; Boutilier and Toews 1981b).

Active irrigation of the gills of *Necturus* can account for up to 60% of the animals' total O_2 uptake at 25 °C (Guimond and Hutchison 1973). In contrast, the inactive gills of *Siren lacertina* account for less than 3% of the O_2 uptake at comparable temperatures. The frequency of branchial movements in *Necturus* is known to vary with the metabolic rate, increasing from approximately 10 beats/min to more than 50 beats/min as temperature increases from 5 ° to 25 °C (Guimond and Hutchison 1973). Increased activity, as well as environmental hypoxia or hypercapnia, leads to increased branchial beating in *Necturus* (Guimond and Hutchison 1972, pers. observ.). Though external gills are also found in anuran tadpoles, nothing is known about their role in overall gas exchange.

The internalized gills of lungless tadpoles of *Rana catesbeiana* and *R. berlandieri* do, however, play a major role in gas exchange, accounting for up to 40% of these animals' total O_2 and CO_2 exchange at 20 °C (Burggren et al. 1983; Burggren and West 1982; West and Burggren 1982). As the animals develop lungs, however, the total irrigation of the gills decreases by some 50% (Burggren and West 1982), and cutaneous respiration begins to predominate. The internal gills of anuran tadpoles are irrigated by a unidirectional water flow generated by muscles of the buccopharyngeal cavity. Ventilatory rates of 90 cycles/min are not uncommon at 20 °C (Burggren and West 1982), and studies by Wasserzug et al. (1981) show that the frequency of gill irrigation is synchronous with heart beat in a number of species. In the later stages of development, both the branchial irrigation rate and the stroke volume of the pump become reduced following an air breath (West and Burggren 1982, 1983), reminiscent of the situation in air-breathing fish (Johansen 1970). Thus, in both air-breathing fish and anuran tadpoles, ventilation of the ABO appears to temporarily inhibit branchial pumping (Sect. 7.1.1). Thereafter, the rate of branchial breathing progressively increases with time elapsed, until the frequency and irrigation volume appears continuous and rhythmic once again (West and Burggren 1982, 1983).

5 Diving: Alternate Storage and Utilization of Oxygen

5.1 Behavioural Regulation

The relative balance between aquatic and aerial gas exchange is very often determined by the behaviour of the animal, particularly in exclusively aquatic forms where underwater activities may take precedence over the drive to surface and breathe. As pointed out by Kramer (1988) and Boutilier (1989), among others, the terms 'diver' or 'diving response', in describing the physiology of such animals, are anthropocentric and might be more appropriately altered to 'surfacer' or 'emergence response'.

5.1.1 The Relative Cost of Managing the O_2 Resource

Recently, Kramer (1988) has used models of behavioural ecology to investigate patterns of air breathing by aquatic animals. In this context, the resource (oxygen) is available at a single location (at the surface) to which the animal must periodically travel at some cost. An optimal foraging model of resource gain maximization can therefore be used to predict patterns of energy acquisition (Kramer 1988) and resource partitioning (Boutilier et al. 1986; Boutilier and Shelton 1986c; Shelton 1985; Boutilier 1988, 1989; Burggren 1988).

In general, the energy acquisition model predicts that as the cost of travelling to the surface (i.e. resource) increases, the length of time at the surface (for replenishing O_2 stores) and the interval between surface periods will both increase (Kramer 1988). These principles appear to apply equally to unimodal as well as bimodal breathers, however, in the latter, the capacity for aquatic gas exchange can clearly affect the pattern of resource gain; e.g. in the most extreme case, the aquatic mode of resource gain may be sufficient to meet the total energy demand of the animal (e.g. at low temperatures). In less extreme cases, it is important to understand the capacity for storing the resource while at the surface (Sect. 5.2.2) and for managing the resource during submergence (Sect. 6.4). Bimodal breathers might therefore be viewed as organisms that can optimize the way in which they exploit the resource by adjusting the relative proportions of aquatic and aerial gas exchange. Theoretical approaches (e.g. Andersson 1978) indicate that the use of a particular mode will decrease as the cost of the mode increases. In the case of a submerged animal, the prediction is that the proportion of O_2 obtained by aerial respiration will decrease as the cost of travelling to the site increases (Kramer 1983). This model holds true for several air-breathing fish (e.g. Bevan and Kramer 1986, 1987) as well as for the clawed toad *Xenopus* (Shannon and Kramer 1988). In both groups, as the distance to the surface was increased (by increasing the depth of the water column), air-breathing frequency declined. Unfortunately, none of these studies have directly assessed the ventilation volumes to see if the decrease in frequency is correlated with a decrease in minute ventilation. Nor do we know whether the increased periods of submergence can be accounted for by increased storage of oxygen at the surface or through increased reliance on O_2 supplies in the water. It is clear, nevertheless, that depth plays an important modulating role in the patterns of cyclic gas exchange and should be considered as a major component of future studies on bimodal gas exchange in vertebrates.

5.1.2 Competitive Interactions Between Behaviour and Air Breathing

Competition for food resources, predator avoidance and mating are but a few examples of behaviours which have the potential to interfere with air breathing. A clear example of the competitive interaction between behaviour and air breathing is that of sexual display in the aquatic newt, *Triturus vulgaris* (Halliday and Sweatman 1976). Males of these animals will forego aerial respiration when presented with the opportunity to display to a female. Removal of the stimulus leads to an immediate resumption of their normal periodic air-breathing habit (Halliday and Sweatman 1976). It has been suggested too, that the enlarged surface area of the tail fin of male newts during the breeding season may be important in facilitating uptake of dissolved oxygen (Czopek 1959, 1965; Bennett and Houck 1983), thereby enabling the animal to remain aerobic while submerged. Apart from seasonal changes in cutaneous surfaces, the functional surface area of the skin can be modulated in a more immediate fashion by altering the numbers of open capillaries (Burggren and Moalli 1984; Feder and Burggren 1985; Feder and Pinder 1988; Malvin 1988).

While extrapulmonary gas exchange is undoubtedly important, part of the capacity for extending certain underwater activities must also come from the relative ease with which anaerobiosis can be recruited into the energy budgets of these animals (Bennett 1978; McDonald et al. 1980; Boutilier et al. 1980; Boutilier and Shelton 1986a, b, c). In voluntarily diving *Xenopus*, for example, threatening conditions at the surface lead to longer periods of submergence and elevated levels of blood lactate, indicative of behaviourally-induced anaerobic energy production (Boutilier 1984, 1989). An analogous situation occurs when *Bufo marinus* assumes its defensive posture upon being threatened; the lungs are immediately over-inflated so as to increase body size, and the animal maintains the breath-hold for the duration of the disturbance. At such times, O_2 levels in the lung rapidly decline and CO_2 levels rise (Macintyre 1975), making the animal both hypoxic and hypercapnic. The alveolar PO_2 and PCO_2 levels observed during threat-induced breath-holds in both *Xenopus* and *Bufo* are far in excess of those that would normally stimulate air breathing (West et al. 1987; Boutilier and Shelton 1986c; Boutilier 1988, 1989; Sect. 7.1.5).

5.2 Aquatic and Aerial Gas Exchange Efficacy: Impact on Diving

The ability of certain bimodal-breathing animals to remain submerged for prolonged time periods is largely attributable to their capacity for aquatic gas exchange. Thus, many air-breathing fish with well-developed gills can remain submerged indefinitely, and only surface to breath air when oxygen in the aquatic environment is reduced to some species-specific threshold. At low temperatures, some aquatic amphibians (e.g. *Xenopus*, *Rana*) and reptiles (e.g. *Chrysemys*) can remain submerged for several months, during which a significant fraction of their metabolic energy requirement is evidently accrued through non-pulmonary gas exchange (Jackson 1986; Pinder 1987). The ability to uptake O_2 from the water is clearly advantageous, since it constitutes an additional O_2 store that the animal can draw on during submergence. Moreover, the greater the efficacy with which CO_2 can be eliminated by the aquatic exchanger, the less likely is the possibility of a significant respiratory acidosis. Such is the case in *Xenopus laevis*, when, during the progressive hypoxaemia of a voluntary dive, cutaneous CO_2 removal effectively offsets any significant buildup of arterial PCO_2 (Boutilier and Shelton 1986b).

In comparison to amphibians, the thicker and more cornified skin of reptiles is generally considered to have made their cutaneous surfaces less permeable to water and therefore to respiratory gases. However, several aquatic reptiles show marked non-pulmonary gas exchange. Numerous studies report significant losses of carbon dioxide and/or uptake of oxygen through non-pulmonary routes (Belkin 1968; Jackson et al. 1976; Crawford and Schultetus 1970; Graham 1974; Standaert and Johansen 1974; Heatwole and Seymour 1976, 1978; Glass et al. 1979; reviews by Seymour 1982, 1989; and Feder and Burggren 1985). Thus, the lungs of reptiles often do not operate at R values that are equivalent to the metabolic RQ, as would be expected of a fully terrestrial animal with a water-tight epithelium (cf. Fig. 1). A particularly good example of this is the large capacity that certain aquatic snakes have for CO_2 elimination (Standaert and Johansen 1974). In fact, the release of CO_2 via cutaneous routes in *Acrochordus* is so prodigious that even when breathing hypercapnic air the animal can maintain arterial PCO_2 at levels substantially lower than those inspired, provided of course that the water is normocapnic (Glass and Johansen 1976).

As shown in Fig. 7A, the breathing pattern of *Acrochordus*, voluntarily diving and surfacing at 25 °C, is characterized by relatively short periods of intermittent ventilation separated by prolonged breath-holds (up to 50 min). During the initial stages of such breath-holds, arterial PCO_2 levels rise but then are maintained relatively constant throughout the period of submergence (Glass and Johansen 1976). Similar patterns are found in certain air-breathing fish and amphibians (Johansen 1970; Randall et al. 1981; Boutilier and Shelton 1986a, b, c). The capacity for non-pulmonary CO_2 release in *Acrochordus* is clearly evident in Fig. 7B, where the volume of the animal is shown to change during the non-ventilatory period (cf. Fig. 7A) as CO_2 gas leaves the animal. Note also that the change in volume during the breath-hold increases even more as the CO_2 load imposed by hypercapnic breathing is effectively eliminated by the skin.

5.2.1 Factors Influencing the Respiratory Exchange Ratio of the ABO During Submergence

Changes in the respiratory exchange quotient of the ABO during a dive serve as a useful index of the ability of the aquatic exchanger to remove CO_2. Upon submergence, the PCO_2 in the ABO of several bimodal breathers (*Electrophorus*, *Arapaima*, *Amphipnous*, *Xenopus*) initially increases at a much greater rate than

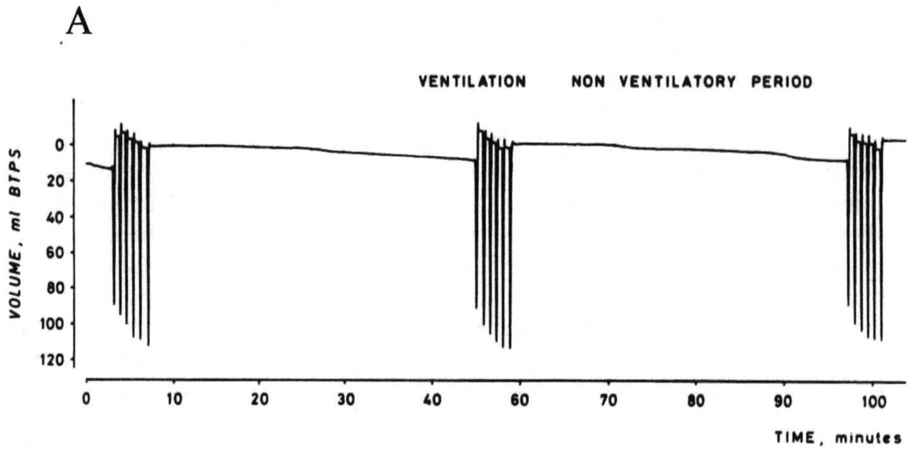

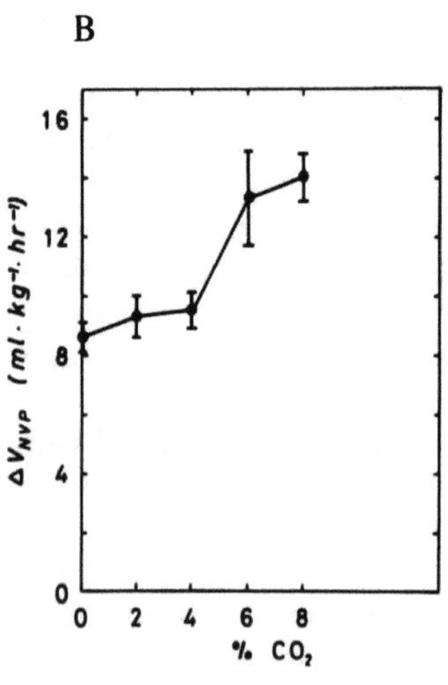

Fig. 7. A Continuous recording of ventilation volumes in the elephant trunk snake (*Acrochordus javanicus*) voluntarily diving and surfacing at a blowhole to breathe fresh air. *B* Changes recorded in the volume of the animals during non-ventilatory periods (*NVP*), as a function of the %CO_2 of inspired air. Note that the higher the CO_2 level inspired, the greater the change in the total volume of the animal during the non-ventilatory period that followed. Both experiments at 25 °C (Glass and Johansen 1976)

it does in the latter stages of the dive (Johansen et al. 1968b; Johansen 1970; Farber and Rahn 1970; Randall et al. 1978b; Lomholt and Johansen 1976; Fig. 8). However, because oxygen continues to be taken from the ABO as the dive progresses, the result is a marked decline in the respiratory quotient (i.e. the instantaneous exchange ratio, R_i, as defined in this volume by Piiper; cf. Fig. 3). Because the aquatic exchanger of all of these animals takes up less O_2 than the CO_2 it removes, the relative magnitude of the decline in R_i thus serves as a general measure of the relative efficacy of CO_2 removal by the aquatic mode.

The changes in the slopes of the lines in the CO_2-O_2 diagram in Fig. 9 show the way that R_i varies in the air-breathing organs of a number of bimodal breathers. Despite there being large oscillations in the PO_2 of the air-breathing organs of all four species (e.g. see Fig. 8 for *Xenopus*), the variations in PCO_2 are much smaller, moreso in some animals (*Xenopus* & *Amphipnous*) than in

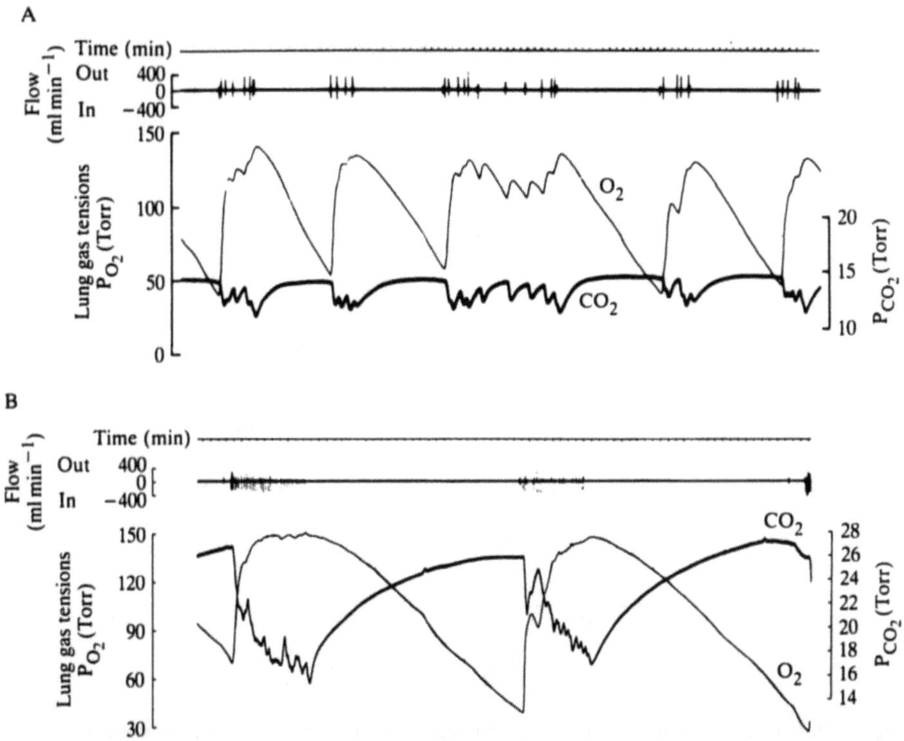

Fig. 8. Continuous recordings of lung ventilation rates (shown as vertical deflections recorded when the animal surfaced to a blowhole pneumotachograph), and partial pressures of alveolar gases (PAo_2, $PAco_2$; measured by an extracorporeal loop), in voluntarily diving and surfacing *Xenopus laevis* (25 °C). The water in the tank was equilibrated with air (*A*), or with 1% CO_2 in air (*B*). In both *A* and *B*, fresh air was flushing through the blowhole at 25 °C (Boutilier and Shelton 1986c)

others (*Electophorus* & *Arapaima*). These dissimilar responses between species could reflect individual differences in the relative efficacy of their aquatic exchangers for CO_2 removal, but could also represent differences in their tissue storage capacities for CO_2. For example, the high capacitance of the tissues for CO_2, as compared to O_2 (Piiper 1982), has been used as the sole explanation for the reductions in R_i observed during breath-holding in certain Chelonian reptiles (Ackerman and White 1979; Burggren and Shelton 1979; White 1989).

5.2.2 Ventilatory Readjustment of O_2 and CO_2 Stores upon Air Breathing

As animals periodically surface to breathe, there is a need to recharge the O_2 stores (contained primarily in the blood and ABO), as well as to rid the body of any built-up levels of CO_2 (contained primarily in blood and tissues). In some air-breathing fish and amphibians, aquatic hypercapnia (or activity-related anaerobiosis) leads to increased aerial ventilation when the R_i of the ABO can become

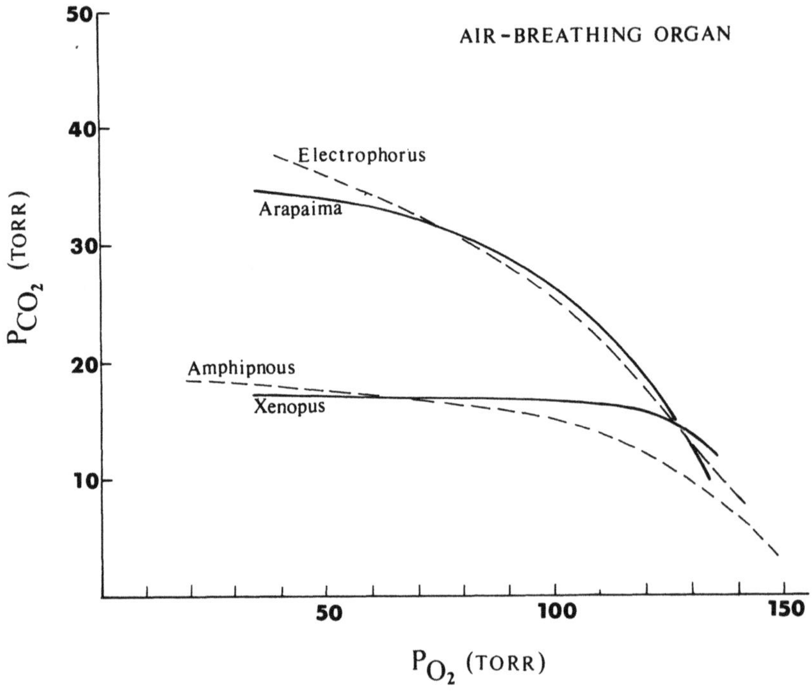

Fig. 9. Plots of PCO_2 and corresponding PO_2 values measured in gas samples taken from the ABO of 4 bimodal breathing species during routine voluntary behaviour in normoxic and normocapnic conditions (i.e. periods of submergence interrupted periodically by periods of air-breathing). Lines were fitted by eye from data presented in Johansen (1970) and Farber and Rahn (1970) for *Electrophorus*, in Randall et al. (1978b) for *Arapaima*, in Lomholt and Johansen (1976) for *Amphipnous*; and in Boutilier and Shelton (1986c) for *Xenopus*

higher than that of the metabolic RQ of the animal (Burggren 1979; Boutilier and Shelton 1986c; Figs. 3, 8). One explanation is that the elevated R_i values reflect differences in the rates at which the O_2 and CO_2 stores can be adjusted during breathing. Thus, because most of the diving O_2 store is contained in the lungs and blood, these stores can be readjusted much faster than the CO_2 stores; i.e. the latter must be mobilized from stores contained in the blood, tissues and cells.

The extent to which the gas stores in the body are readjusted during air breathing depends on a number of factors: i.e. on the time spent ventilating at the surface, on the efficiency of ventilation-perfusion matching (Sect. 6) and on the gas-carrying properties of the blood. Aerial ventilation in many air-breathing fish and urodele amphibians is a very short-lived event, often taking less than a second or two to complete. It is obvious that under these conditions the opportunities for effective turnover of gas stores in the blood or tissues (i.e. bringing these compartments into equilibrium with lung gases) will be severely restricted. The adaptations in these animals appear to be focused on effective turnover of the gas stores of the ABO, rather than on the attempt to establish a new steady state in the blood and tissues prior to diving. Because of the low dead space of the aerial exchanger of most air-breathing fish, they can charge their ABO with (essentially) fresh air, even with a single breath. Though the urodele amphibians *Amphiuma tridactylum* and *Siren lacertina* have a larger dead space than most air-breathing fish, they share the fishes' effectiveness for gas turnover of the ABO as a result of a complete collapse of their high volume lungs (Sect. 2.6) during the expiration that precedes the lung-filling phase (Martin and Hutchison 1979). Indeed, the entire ventilatory cycle in these urodeles lasts only 4–5 s, leaving little time for equilibration of blood and tissue compartments with lung gases (Toews et al. 1971). Other bimodal breathers, such as the clawed toad *Xenopus* (Boutilier 1984), the Cuchia eel *Amphipnous* (Lomholt and Johansen 1974) and the water-snake *Acrochordus* (Glass and Johansen 1976), ventilate their lungs over several cycles upon surfacing to breathe (e.g. Figs. 4, 7). Presumably, this series of lung ventilations allows time for gas storage compartments in both the blood and tissues to come into equilibrium with the lung gas pool. In *Xenopus*, the duration of the breathing episode following a voluntary dive is often prolonged beyond the point at which alveolar and arterial PO_2 should be fully readjusted (Boutilier and Shelton 1986a, b, c; Fig. 8A). This could mean that the ventilatory control system is one that monitors the levels of oxygen in all parts of the body, and that breathing only becomes terminated after they are fully recharged. However, the length of time spent breathing at the surface can be increased even further by exposing the animals to aquatic hypercapnia (Boutilier 1988, 1989; Fig. 8B), suggesting that breathing continues until the excess CO_2 has been removed (Boutilier and Shelton 1986b, c).

5.3 Anaerobiosis During Prolonged Periods of Submergence

As outlined earlier, certain underwater behaviours may take precedence over surfacing to breathe and, under these conditions, anaerobiosis may begin to be recruited as O_2 supplies run down. Anaerobiosis can also be called upon to supplement aerobic metabolism when sudden bursts of mechanical power are necessary in order to escape predators or persue prey, much like the condition for terrestrial animals (Bennett 1978). Whatever the reason, the immediate advantage is that the animal can utilize the energy gained from anaerobic metabolism to extend its underwater activity range. The extent to which anaerobiosis need be recruited in bimodal breathers will depend in large part on whether O_2 uptake by the aquatic exchanger can satisfy the total aerobic metabolic demands. This in turn can be influenced by environmental temperature, dissolved O_2 levels and metabolic rate. Regardless, it is this facility for aquatic O_2 uptake that separates the bimodal breathers from other diving vertebrates, whose sole O_2 stores are contained in the body.

There are many instances where strenuous activity leads to marked increases in the ventilatory rate and/or minute ventilation of certain air-breathing fish (Johansen 1970; Smatresk 1988) and amphibians (McDonald et al. 1980; Boutilier et al. 1980; Boutilier and Shelton 1986a, b, c). The oxygen debt and acid-base disturbance associated with such anaerobic activity have a major impact on the frequency and duration of subsequent dives in *Xenopus* (Fig. 10), effectively reduc-

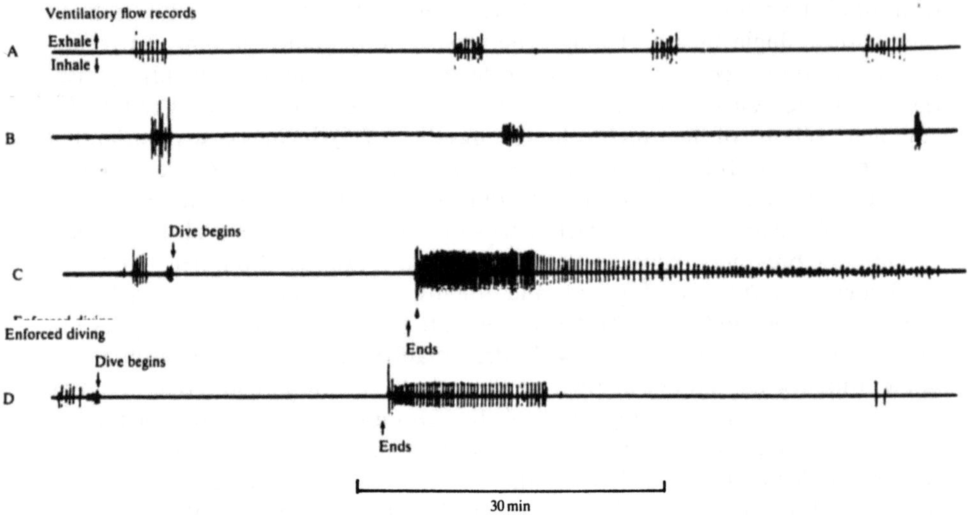

Fig. 10. Lung ventilations and periods of breath holding in *Xenopus laevis*, showing breathing patterns when animals are diving voluntarily (*A*, *B*), and before and after two enforced dives of approximately 30 min duration (*C*, *D*). Breathing movements were detected when animals surfaced to breathe at a blowhole fitted with a pneumotachograph. The large increase in ventilation following enforced dives is due entirely to the extra amount of time spent at the surface since breathing frequency within the surface period as well as inspired volume per breath, were not significantly different between the pre- and post-dive breathing bursts in e.g. *C* and *D* (Boutilier and Shelton 1986b)

ing the time that can be devoted to underwater persuits for some considerable duration. Restoration of acid-base balance, for example, can take upwards of 8 – 16 h (depending on the species) following strenuous activity (Boutilier et al. 1980; Boutilier and Shelton 1986b; Toews and Boutilier 1986), and up to 4 h following a 30-min enforced dive in *Xenopus* (Boutilier and Shelton 1986b). During the period of recovery in *Xenopus*, minute ventilation increases primarily as a consequence of the animal spending more time at the surface (Fig. 10), as also occurs when the animals are exposed to hypercapnia (Fig. 8B).

5.4 Effects of Temperature

Few studies have examined the influence of temperature on the control of periodic ventilation in voluntarily diving bimodal breathers. Of those that have, there is a general trend for diving durations to increase as temperatures fall, and for aerial ventilation to be correspondingly reduced (Johansen 1970; Smatresk 1988; Glass et al. 1986; Kruhøffer et al. 1987). At very low temperatures (i.e. 5 °C and below), it seems that aquatic gas exchange can meet all of the aerobic metabolic requirements in various air-breathing fish (Johansen 1970), as well as in certain amphibians (Pinder 1987). Control systems, particularly those involved in regulating pH and PCO_2, are temperature dependent. While there is something known of the effects of temperature on certain receptor mechanisms (Jones and Milsom 1982), little is known about how such receptors interact to control the rate of air breathing when temperatures change (Sect. 7.1.3).

6 Ventilation-Perfusion Relationships in Bimodal Systems

Virtually all bimodal-breathing vertebrates practise some form of intermittent ventilation of the aerial exchange organ, in which there is alternate storage and utilization of oxygen. Changes in perfusion accompany the periods of ventilation and most commonly involve increases in cardiac output and/or selective distribution of blood flow to the air-breathing organ. This ventilation-perfusion matching has been likened to the conditions observed for avian and mammalian lungs, where regional differences in ventilation are matched by perfusion changes that optimize gas exchange (Johansen and Burggren 1980; Johansen 1985). In higher vertebrates with completely divided circulations, lung ventilation and perfusion occur continuously (see Milsom, this Vol.), with the ratio between them showing little variability in the steady state. Moreover, because the left and right ventricles of endotherms are placed in series, rather than in parallel as in many lower vertebrates, the opportunities for controlling pulmonary blood flow are much more restricted. Thus, the undivided ventricle of animals such as air-breathing fish and amphibians makes it possible to adjust the relative amounts of blood flowing to the ABO and body over an infinitely variable range. Optimizing gas exchange in this context means evolving a system that can match increases in pulmonary blood flow with periods of air breathing.

Presumably, an important design feature in the evolution of cardiorespiratory systems of the kind seen in bimodal breathers is towards minimizing the energetic costs of the gas transfer process. Performance of the ABO can be assessed as the rate of gas transfer achieved per unit of energy expended in ventilating and perfusing the ABO. Clearly, it would be inefficient to perfuse an unventilated organ unless it was being used as an O_2 store. Thus, the progressive decrease in blood flow to the lungs of bimodal breathers as alveolar O_2 tensions fall (Johansen et al. 1968a; Johansen et al. 1970b; Shelton 1970; West and Burggren 1984) is consistent with the notion that these systems have evolved to maximize gas exchange per unit of energy expended.

6.1 Evolution of the Cardiovascular Apparatus: Structure and Function

Associated with the broad diversity of structures for aerial gas exchange in fish (Sect. 2) is an equally diverse assortment of circulatory connections to and from the air-breathing organs (Fig. 11). All such organs have evidently evolved to facilitate periodic O_2 uptake from the relatively O_2-rich aerial environment. Compared to the gas exchange systems in tetrapods, the air-breathing organs of non-dipnoan fish are rather less effective, owing to the fact that there are no structural divisions in the central circulation to allow for selective distribution of oxygenated

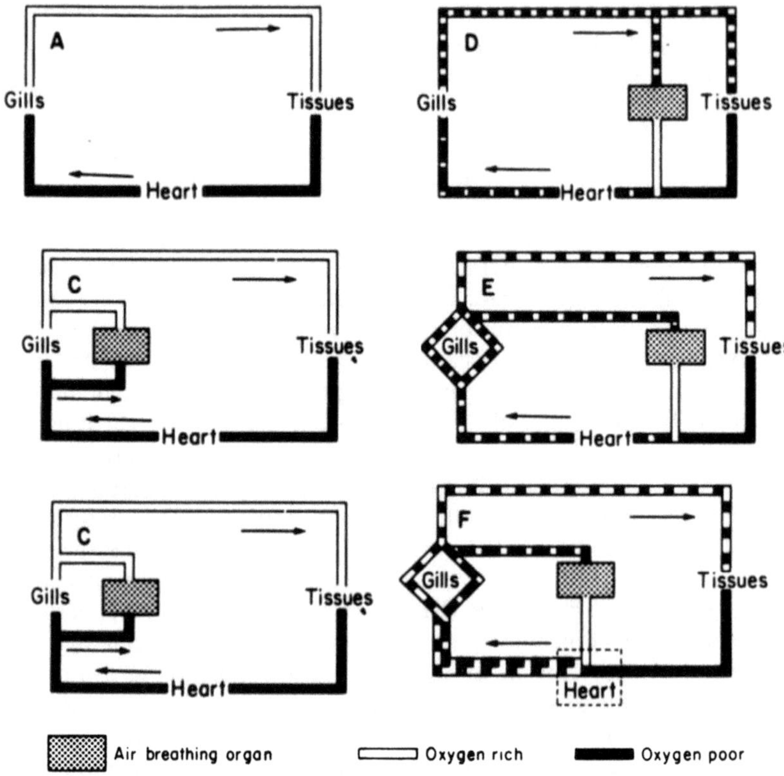

Fig. 11. Schematic of the flow of oxygenated (*white*) and deoxygenated (*black*) blood through the circulation of various fishes. *A* Typical arrangement in an exclusive water breather. *B* Air-breathing fish whose air-breathing organ (ABO) is derived from pharyngeal or opercular mucosa, e.g. *Amphipnous, Electrophorus, Monopterus, Periophthalmus. C* Gills, buccal mucosa, or chambers extending from the opercular cavity serve as ABO, e.g. *Clarius, Saccobranchus. D* Intestinal tract used as ABO, e.g. *Hoplosternum, Plecostomus, Misgurnus. E* Air bladder serving as ABO, e.g. *Polypterus, Amia, Lepisosteus. F* Lung homologous to those of tetrapods serving as ABO, e.g. *Neoceratodus, Lepidosiren, Protopterus* (Johansen 1970)

and deoxygenated blood (see, however, Ishimatzu and Itazawa 1983). The composition of blood flowing to the air-breathing organ often consists of a deoxygenated systemic venous portion mixed with a bolus of blood that has been recently arterialized by passage through a gas exchanger. This condition exists in animals whose air-breathing organs are downstream and either in-series (Fig. 11 D, E) or in-parallel (Fig. 11 B, C) with the gills (Johansen 1968, 1970, 1982; Burggren and Johansen 1986).

In virtually all air-breathing fish, except for the true lungfish, oxygenated blood emanating from the air-breathing organ drains directly into the systemic veins, where it is mixed with deoxygenated systemic venous blood, before being returned to the heart (Fig. 11 B–E). For this reason, the air-breathing organ often receives blood that is already partially oxygenated, sometimes so much that it is questionable whether any further benefits to gas exchange are realized (Burggren and Johansen 1986). As pointed out by Johansen (1982), the reason that direct passage of blood between the efferent circulation of air-breathing organs and the systemic arterial network have never arisen is probably related to the hemodynamics of the associated vasculatures. Because there is a high resistance to blood flow across the aerial exchanger, a considerable pressure drop must occur as more and more blood is delivered to the organ. As a consequence, the low-pressure efferent circulation from the aerial exchanger is unable to reestablish itself with the comparatively high-pressure systemic arterial network. Thus, blood flowing from the aerial exchanger is resigned to the venous return.

On the surface, it appears that the aerial gas exchanger in the nondipnoan air breathers serves only to elevate the overall oxygen content of systemic venous blood (Johansen 1968). However, other cardiorespiratory mechanisms that serve to increase the efficacy of aerial gas exchange in these animals are in operation. For instance, selective perfusion of the air-breathing organ is possible in both the electric eel, *Electrophorus electricus* (Johansen et al. 1968b), and the Cuchia eel, *Amphipnous cuchia* (Lomholt and Johansen 1976), based on temporal separation of cardiac output between aerial gas exchanger and systemic arterial vasculature. More recently, it has been shown that selective distribution of blood is also possible based on streamlined flow patterns, in which oxygenated blood from the air-breathing organ is delivered, more or less intact, through the completely undivided heart of the snakehead fish *Channa argus* (Ishimatzu et al. 1979; Ishimatzu and Itazawa 1983; Andresen et al. 1987). Such flow patterns are also thought to be responsible for the separation of oxygenated and deoxygenated blood streams in certain amphibians (Shelton and Boutilier 1982).

6.2 Perfusion Adjustments During Air Breathing

Ventilation-perfusion matching occurs during periodic renewal of air in the mouth cavity of the electric eel (Fig. 12). During such renewal, the ratio of mouth cavity perfusion ($\dot{Q}_M$) to total flow ($\dot{Q}_T$) increases to as much as 0.8 (Johansen et al. 1968b). In the interim between breaths, $\dot{Q}_M/\dot{Q}_T$ progressively decreases as the availability of O_2 in the mouth cavity declines (Fig. 12). It is noteworthy that blood flow to the mouth cavity in this animal is always such a high fraction of the total. This may be related to the way in which the aerial O_2 store is rapidly attained. *Electrophorus* is an obligate air breather that periodically rises to the surface, where it expires almost completely before inhaling fresh air. The ventilatory exchange is over within four-tenths of a second (Farber and Rahn 1970), and each single breath is usually followed by a period of quiet submergence that lasts upwards of 2–3 min (Garey and Rahn 1970). Inhalation of hyperoxic gas mixtures leads not only to greater increases in $\dot{Q}_M/\dot{Q}_T$ (Fig. 12; Johansen et al. 1968b), but also to much longer periods of submergence (6–7 min according to the observations of Garey and Rahn 1970). Blood flow in these animals is obviously matched to O_2 availability in a way that maximizes use of the mouth cavity store. Clearly, this breathing pattern does not allow enough time for recharging the blood O_2 stores while the animal is at the surface and, as a result, the 'diving O_2 store' of the electric eel is not at maximum capacity (Sect. 5.2.2). In compari-

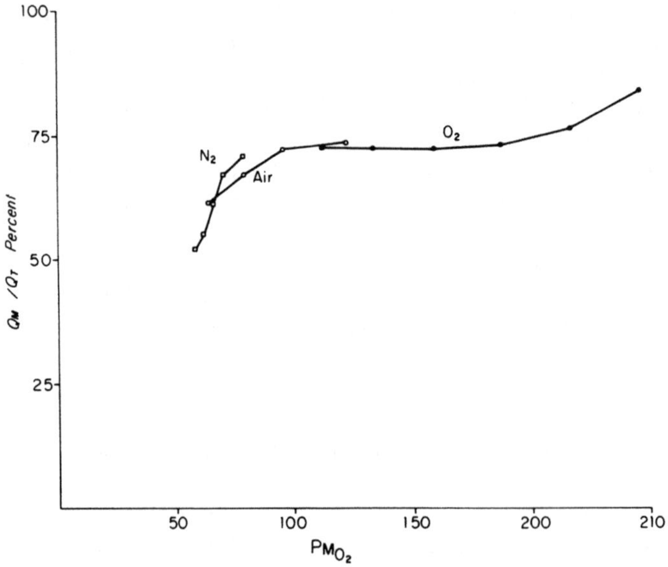

Fig. 12. Ratio between the blood flow going to the mouth ($\dot{Q}_m$) and the total cardiac output ($\dot{Q}_T$), as a function of the partial pressure of oxygen in the mouth gas (PMo_2). The relationships were determined under three experimental conditions, while the animals were breathing ambient air, hyperoxic gases (O_2) or hypoxic gases (N_2) (Johansen et al. 1968b)

son to other diving vertebrates that take multiple breaths upon surfacing (e.g. Figs. 4, 6, 7), the single-breath strategy of the electric eel effectively reduces the aerobic diving time of the animal.

The structural basis for the temporal separation of blood flow to the air sac and systemic circulation in *Amphipnous* is shown in Fig. 13. Shown too, are estimates of blood perfusion rates during breathing and breath-holding periods, based on data in Lomholt and Johansen (1976). As with the electric eel, aerial ventilation leads to a greater proportion of the total blood flow being directed to the ABO, whereas during breath-holding, blood flow progressively shifts back to the systemic circuit. Though the comparatively large blood O_2 capacities (~20 vol%) of *Amphipnous* and *Electrophorus* (Johansen et al. 1968b; Lomholt and Johansen 1976) are clearly advantageous to O_2 storage, repeated ventilations by the Cuchia eel presumably allow for a greater buildup of diving O_2 stores (Sect. 5.2.2).

Apart from time-related separation of O_2-rich from O_2-poor blood, as occurs in the latter two examples, some air-breathing fish of the genus *Channa* evidently possess the capacity to selectively channel the oxygenated efferent blood supply from the ABO through the heart and into the systemic arterial circulation. The structural basis for this is a unique teleostean outflow system from the heart, in which two (rather than one) ventral aortas emanate from the bulbous arteriosis (Fig. 14). One vessel, the anterior ventral aorta, feeds the well-developed gills of arches 1 and 2, whose efferent blood then perfuses bilaterally situated suprabranchial ABOs. The posterior ventral aorta, on the other hand, perfuses less-

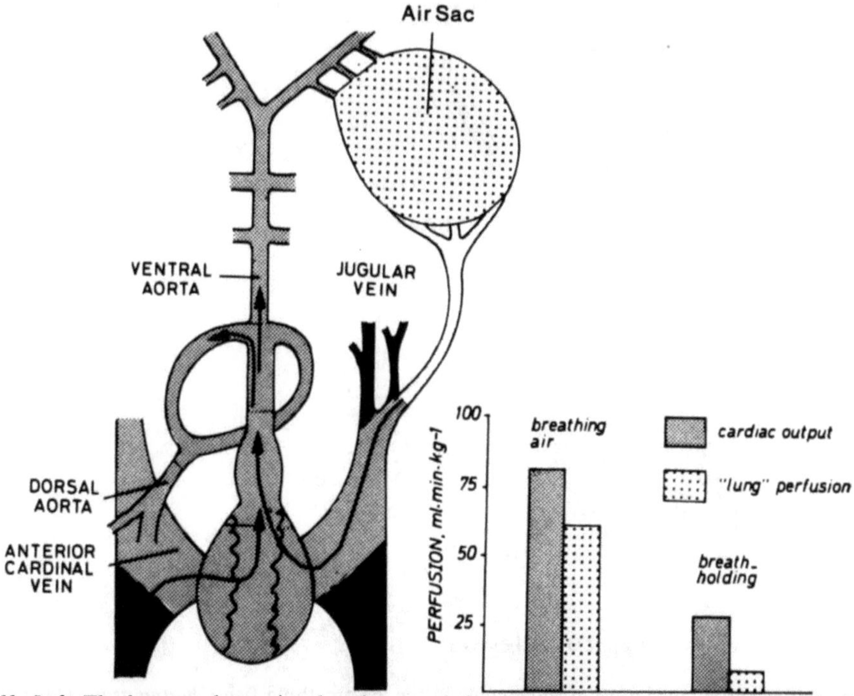

Fig. 13. Left: The heart and associated major vessels in *Amphipnous cuchia*. Note the paired shunt vessels that form the dorsal aorta. Blood from the air breathing organ converges with venous blood in the jugular vein. *Right*: During air breathing, cardiac output increases, as does the fraction of it that goes to the air sac. During breath holding, the major fraction of the cardiac output is shunted to the dorsal aorta (Johansen 1982)

developed gill arches, which thereafter converge to form the dorsal aorta. Ishimatzu and Itazawa (1983) have shown that the O_2 content of blood from the ventral aorta supplying the ABO is consistently lower than that supplying the dorsal aorta. How this separation is achieved in the presence of a total anatomical shunt was the basis for a recent angiocardiographic analysis of the central circulation of *Channa* (Andresen et al. 1987). By tracing the fate of a contrast medium injected 1–2 min following a spontaneous air breath, these authors searched for evidence of selective passage of blood from the ABO to systemic arterial circulation. Their results indicate that the basis for the separation may have to do with a comparatively greater vasodilatation of posterior than anterior ventral aorta, thereby resulting in preferential (greater) blood flow to the posterior ventral aorta. Whether factors such as ridge formations along the inner wall of the bulbous arteriosus (noted by Johansen 1985) are responsible for streamlining blood through the heart must await further examination. Experiments by Ishimatzu et al. (1986) have shown that the vascular resistance of the ABO increases following administration of adrenaline or acetylcholine, and that this leads to a redistribution of blood flow from the ABO to systemic circuit. Stimulation of the vagal branch that enters the ABO causes the same effect as the addition of adrenergic and cholinergic agonists.

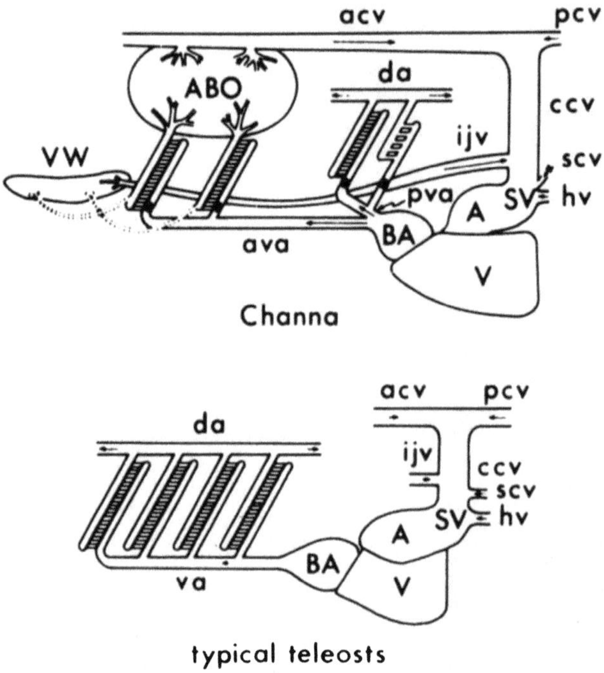

Fig. 14. Central circulatory arrangements in the air-breathing fish *Channa argus* (*top*) and in a typical teleost (*bottom*). The anterior cardinal vein (*acv*) of *Channa* drains both the head and the air-breathing organ (*ABO*). Unlike the typical teleost, *Channa* has two aortas, the anterior ventral aorta (*ava*) supplying gill arches 1 and 2 which drain blood to the *ABO*, and the posterior ventral aorta (*pva*) which supplies the less developed gills that drain blood to the dorsal aorta (*da*). Other symbols: *pcv* posterior cardinal vein; *CCV* common cardinal vein; *SCV* subclavian vein; *ijv* internal jugular vein; *hv* hepatic vein; *A* atrium; *Sv* sinus venosus; *V* ventricle; *BA* bulbus arteriosus; *VW* buccal vascular wall (Ishimatzu and Itazawa 1983)

In a phylogenetic context, the dipnoan lungfish are the first among the vertebrates to exhibit a direct return of blood from lung to heart (i.e. the first pulmonary vein; Fig. 11 F). They are also the first vertebrate group to show septation of atrial and ventricular spaces (Bugge 1960; Johansen and Burggren 1980), forecasting the complete double circulation of higher vertebrates. The morphological separations within the lungfish heart provide the basis for the partial separation of pulmonary venous and systemic venous returns (Johansen 1985). Within the three genera of lungfish, the degree of septation is greatest in *Lepidosiren* and least in *Neoceratodus*, the latter of which is largely dependent on the gills, rather than lung, for its gas exchange requirements (Johansen et al. 1967). Most of what is known about ventilation-perfusion characteristics in lungfish has, however, been determined on *Protopterus*. Thus, as shown in Fig. 15 A, cardiac output and pulmonary blood flow of *Protopterus* increase following an air breath and then progressively decline over the breath-holding period that follows; i.e. almost 100% of the dorsal aorta blood is derived from the pulmonary venous return. In concert with the large redistribution of blood flow to the lung is a considerable decrease in branchial vascular resistance (Johansen et al. 1968a).

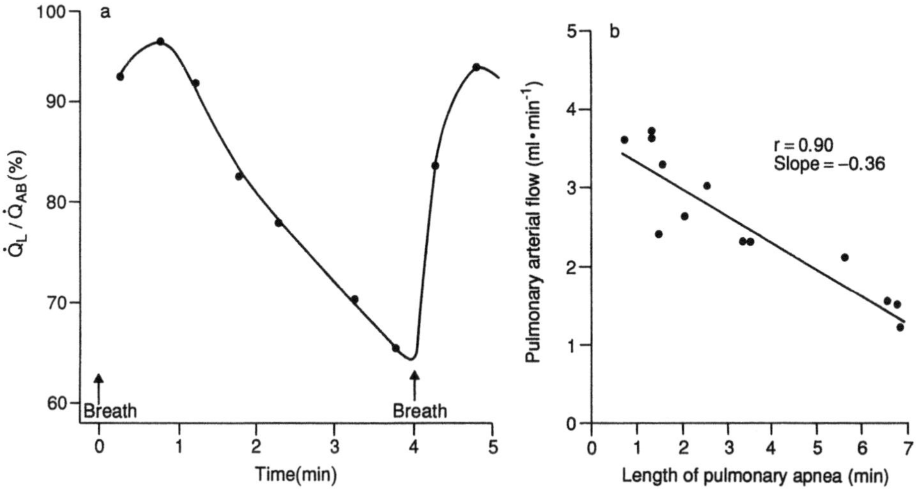

Fig. 15. A Moment-to-moment changes in the ratio of blood flow derived from the lung ($\dot{Q}_L$) relative to total blood flow perfusing the afferent branchial arteries forming the dorsal aorta ($\dot{Q}_{AB}$) during intermittent air breathing in *Protopterus*. The contribution from pulmonary venous blood to systemic arterial flow increases considerably with each breath (Johansen et al. 1968a). *B* Relationship between mean pulmocutaneous arterial blood flow at the end of a period of pulmonary apnea (i.e. breath holding), as a function of the duration of the apnea. Results are from a single specimen of the toad, *Bufo marinus*, recorded over a 1-h period at 22 °C (West and Burggren 1984)

As with air-breathing fish, similar matchings of intermittent ventilation with perfusion can be found in most amphibians and reptiles (Shelton 1970; Shelton and Boutilier 1982; Shelton et al. 1986). Like the situation seen in *Protopterus* (Fig. 15A), a matching of high pulmonary blood flow with ventilation occurs in amphibians such as *Xenopus laevis* (Shelton 1970) and *Bufo marinus* (Fig. 15B). The increased blood flow immediately following a breath is due, in both animals, to vasoactive changes in their pulmonary vasculature (Emilio and Shelton 1972; West and Burggren 1984). During periods of breath-holding, however, the lung vasculature is thought to progressively constrict so as to gradually reduce blood flow in accordance with the ever diminishing supplies of O_2 in the lungs.

6.3 Conserving the O_2 Stores via Adjustments in Perfusion

Unlike the aerial environment, aquatic gas exchange is often limited by low ambient PO_2 levels caused by a variety of biotic factors (Boutilier, this Vol.). Because the circulatory arrangement of several bimodal breathers delivers oxygenated blood to the gills and/or skin (where the majority of CO_2 is exchanged), the possibility arises that oxygen gained by air breathing could be lost to hypoxic water. Many of these animals evidently avoid such situations by redistributing blood flow, away from the O_2 sink. The control mechanisms involved in such redistribution processes are known to be responsive to both internal and external changes

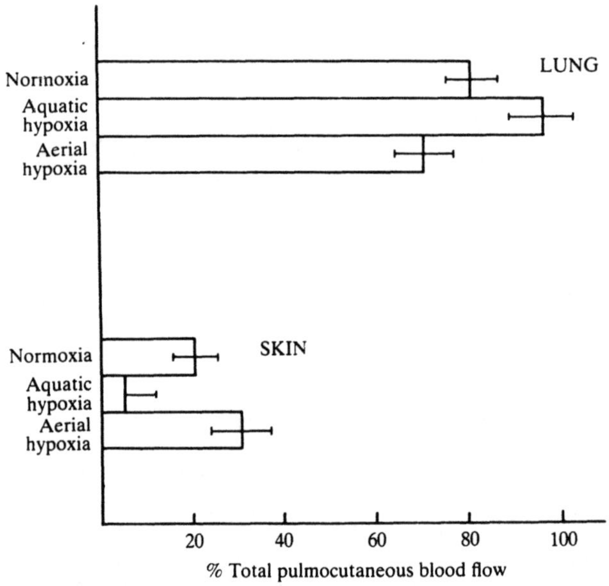

Fig. 16. Histogram showing the relative flow of pulmocutaneous blood in voluntarily diving bullfrogs (*Rana catesbeiana*), subjected to normoxia, aquatic hypoxia (delivered only at the skin), or aerial hypoxia (delivered only at the lungs). Normoxic $PO_2 \sim 155$ Torr; Hypoxic PO_2 levels were ~ 40 Torr. Relative flows were calculated from distributions of radioactive microspheres injected via a cannula, 10 min into a voluntary dive. Data are means from 7 experiments at 20 °C (Boutilier et al. 1986)

in O_2 tensions (Sect. 7.1), enabling these animals to exploit the medium with the greatest potential for O_2 uptake.

The incompletely divided circulation of amphibians allows for variable perfusion and selective distribution of blood to various sites of gas exchange (Shelton and Boutilier 1982; Malvin and Boutilier 1985). In experiments on voluntarily diving *Rana catesbeiana*, the relative perfusion of the lung increased during aquatic hypoxia and decreased during periods of lowered aerial PO_2 (Fig. 16). This can be interpreted as being (1) a readjustment of blood flow to the gas exchange site having the greatest potential for O_2 uptake, or (2) a mechanism for preventing O_2 losses from body O_2 stores at exchange sites of low O_2 tension. Thus, diverting blood flow from the skin to the lungs during aquatic hypoxia reduces the possibility that oxygen gained by air breathing will be lost transcutaneously (Johansen 1968; Boutilier et al. 1986). Control over the vascular resistance of the two circulatory networks may well reside with muscular sphincters located in both pulmonary and cutaneous vessels. In this regard, acetylcholine or vagal stimulation causes a strong pressor effect on the sphincter located in the distal segment of the pulmonary artery of *Rana temporaria* (de Saint-Aubain and Wingstrand 1979), whereas in *Bufo marinus*, the proximal segment of the cutaneous branch of the pulmocutaneous vessel receives a rich supply of adrenergic constrictory fibres (Smith 1976). Reciprocal innervation of the two circuits is thus a distinct possibility for controlling blood flow to the two vascular networks. Control

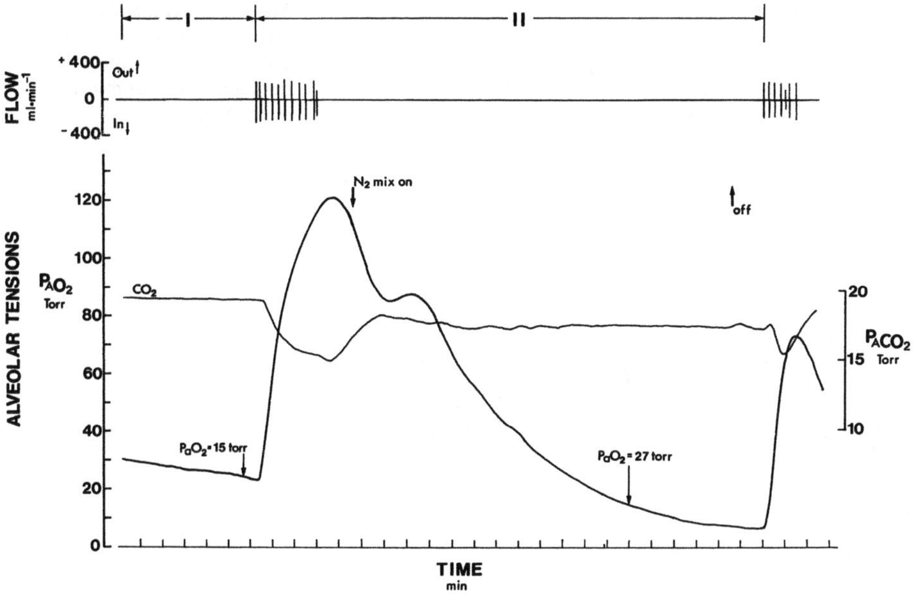

Fig. 17. Ventilatory air flow and alveolar gas tensions of a single *Xenopus laevis*, voluntarily diving and surfacing at a blowhole fitted with a pneumotachograph. Lung-gas tensions were continuously measured in an extracorporeal loop containing electrodes through which alveolar gases were circulated using a peristaltic pump. A second loop connected in series with the first allowed various mixtures to be introduced into the circulation. In the experiment shown, a N_2/CO_2 mixture was introduced soon after a voluntary dive, to purge the lungs of oxygen supplies while maintaining PCO_2 at physiological levels. Blood samples taken towards the end of a normal dive (*I*) and after purging the lungs to low O_2 tensions (*II*) were analyzed for arterial PO_2 (PaO_2). Temperature, 25 °C (R. G. Boutilier, P. J. Butler, and B. K. Evans, unpublished data)

over the pulmonary and cutaneous circuitry such as this may also be important for metering out lung oxygen during routine voluntary dives (e.g. Fig. 8).

The ability to selectively perfuse the lung circuit in response to hypoxia has been shown indirectly in experiments on the clawed toad, *Xenopus laevis* (R. G. Boutilier, P. J. Butler and B. K. Evans), in which the alveolar gas composition was continuously measured while experimentally manipulating the composition of lung gas (Fig. 17). The lung gas O_2 tensions in the latter part of a voluntary dive in *Xenopus* are known to be approximately 5 to 10 Torr higher than those of arterial blood (Boutilier and Shelton 1986c). A breathing episode immediately following one such dive is shown in Fig. 17 (panel II). Following the series of air breaths, and upon voluntary submergence, the lungs were rendered progressively less useful as an O_2 store by perfusing the lungs with a CO_2/N_2 mixture such that alveolar PO_2 decreased at a constant PCO_2. Soon after this procedure began, however, there was always a brief period when alveolar O_2 tensions would increase, indicative of O_2 loss from blood to lung. Immediately following this reversal, O_2 levels in the lung rapidly declined and the normal gradient between lung and arterial blood PO_2 was effectively reversed. Because arterial PO_2 was

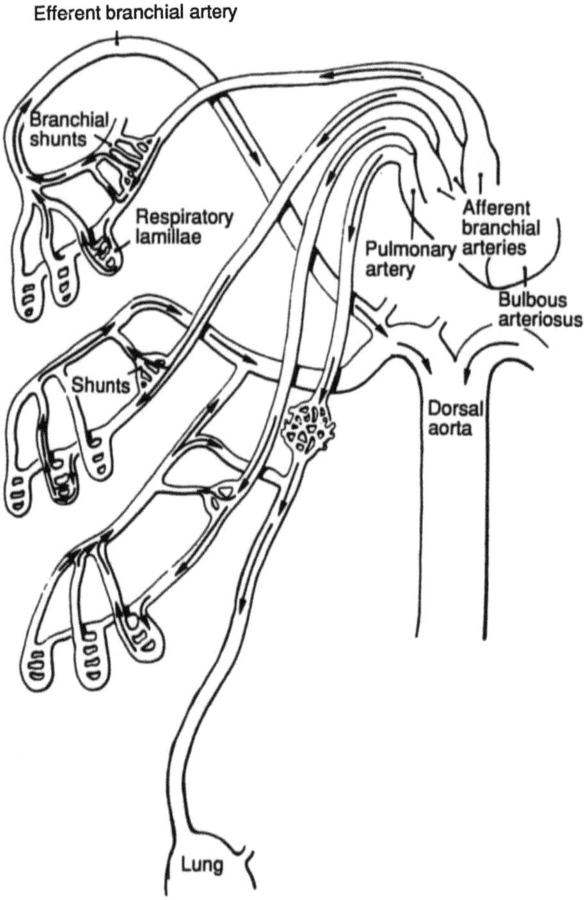

Fig. 18. Vascular arrangement of the aortic arches, gills and lung in the salamander, *Ambystoma tigrinum* (Malvin 1989)

now much higher than alveolar (Fig. 17), this indicated that pulmonary perfusion must have been markedly reduced as the lung became purged of its O_2 store, and exemplifies the ability such animals have for selectively perfusing their lungs in accordance with their potential as an O_2 store. It also demonstrates a general ability to conserve body O_2 stores during periods of submergence, a trait shared by many bimodal breathers.

Conservation of blood O_2 stores can also be predicted from recent experiments on the urodele amphibian, *Ambystoma tigrinum*, in which microspheres were used to characterize blood flow patterns during periods of severe aquatic hypoxia (Malvin 1985c). Under conditions such as these, when all of the O_2 uptake occurs via pulmonary ventilation, loss of oxygen across the gills is predicted. However, Malvin (1985c) has shown that during aquatic hypoxia, branchial blood flow is redistributed such that blood is diverted away from the first gill arch, to arches 2 and 3 (see Fig. 18 for vascular arrangement). The adaptive advantage of this is that more blood will be shunted to the lung (in series with the third gill arch), presumably so as to match the increased ventilatory effort with increased perfusion. In addition, because the first gill arch would normally receive blood that is high in oxygen, diversion of blood away from this circuit will probably reduce O_2 losses to the water (Malvin 1989). Though the mechanisms responsible for mediating changes in gill perfusion patterns are unknown, several associated studies on *Ambystoma* have shown that the gills are responsive to adrenergic, cholinergic and renin-angiotensin systems (Malvin 1985a, b, 1989). More work is needed to assess not only the integration of these control mechanisms but also to uncover the role that blood flow changes have on gas exchange in amphibian gills.

7 Regulation of Dual Gas Transfer Systems

During the evolutionary transition from unimodal to bimodal systems of gas exchange, the overall patterns of breathing changed from the continuously rhythmic branchial movements of water breathing fish, to the decidedly arrhythmic breathing patterns characteristic of many air breathing fish, amphibians and reptiles. In addition to the new cardiorespiratory structures that evolved to convect air and extract O_2, some reorganization of central and peripheral nervous systems must also have evolved in order to co-ordinate the activities and interactions of the dual exchangers.

The control of 'ventilation' in unimodal water-breathers is thought to be mediated centrally, as it is in birds and mammals (see Milsom, this Vol.). In these instances, peripheral inputs are considered to have mainly a modulating effect on a centrally determined pattern. Intermittent air-breathing, on the other hand, appears to be driven largely by a periodic need for additional gas exchange; e.g. on a model in which ventilation is turned on and off by signals originating from chemical and/or mechanical information during the periods of breathing or breath-holding. Milsom (1988, this Vol.) has recently advanced the notion that all vertebrates might possess a central rhythm generator from which both continuous or arrhythmic breathing patterns can arise, depending on the level of respiratory drive.

Most of what is known about the onset and termination of air breathing in bimodal breathers, or of shifts in emphasis between respiratory modes, comes from experiments that have focused on responses to (1) changes in the chemical composition of the aquatic or aerial environments, (2) changes in activity metabolism and/or temperature, and (3) acute changes in the volume of the air breathing organ.

7.1 Chemosensitive Control of the Respiratory Mode

7.1.1 Aquatic O_2 Levels: Impact on Breathing

Air-Breathing Fish

Animals found to be relatively insensitive to O_2 levels in the water are often 'obligate air breathers' such as *Electrophorus, Protopterus* and *Lepidosiren* (Johansen and Lenfant 1968; Johansen et al. 1968 a, b; Johansen 1970). In all such animals, aquatic hypoxia is without effect on either gill irrigation or air breathing. This insensitivity is thought to be related to the fact that their reduced gills participate little in O_2 uptake. In *Protopterus*, for example, deoxygenation of the water had no significant effect on systemic PO_2 (Johansen and Lenfant 1968), indicating an effective independence from aquatic PO_2. The fact that these animals also do not exhibit any avoidance activity to aquatic hypoxia indicates that they do not monitor the environmental PO_2, as has been suggested for other bimodal breathers (Johansen et al. 1967; Johansen 1970; Graham and Baird 1982; Smatresk et al. 1986; Smatresk 1986, 1988).

Some air-breathing fish (e.g. *Neoceratodus*) respond to hypoxic water by increasing gill irrigation (Johansen et al. 1967), a response similar to that observed in exclusively water-breathing species (Davis and Cameron 1971; Dejours 1973; Randall 1982 a). In *Neoceratodus*, aquatic hypoxia induced a three- to four fold increase in irrigation, which was the result of an increase in the volume of individual breaths, rather than in the rate of breathing. Though aquatic hypoxia also stimulated air breathing, the onset of this activity was slow to appear in comparison to the branchial responses. The prompt increase in gill irrigation was indicative of a response to changes in water PO_2 levels, whereas lung ventilation was evidently not stimulated until there were marked changes in the blood gas tensions (Johansen et al. 1967).

The immediate branchial responses of air-breathing fish to aquatic hypoxia (e.g. Johansen et al. 1967) have often been used to imply that there are 'water-facing' chemoreceptors which monitor the external environment of the animal. Johansen and Lenfant (1968) examined this directly by introducing the chemosensitive stimulant nicotine (nicetamide) under the opercula of *Protopterus*, where it promptly led to an increase in gill irrigation and air breathing. Recent studies by Smatresk (1986) have shown that bolus injections of the O_2-chemoreceptor probe, cyanide, into the water flowing over the gills of conscious gar (*Lepisosteus*), also causes an immediate increase in both gill irrigation and air breathing. Compelling evidence for externally located chemoreceptors in gar comes from further experiments (Smatresk et al. 1986) in which dynamic increases in the PO_2 of water flowing over the gills of artificially ventilated animals led to an immediate increase in opercular pressure amplitude.

Since O_2 tensions in the aquatic environment are subject to considerable fluctuation (cf. Boutilier, this Vol.), externally located chemoreceptors would provide a means for effective matching of the mode of gas exchange to the PO_2 of the respective medium. Many other air-breathing fish respond to a decrease in water PO_2 by initially increasing branchial irrigation (Johansen et al. 1970 a; Hughes and Singh 1971; Smatresk and Cameron 1982 a; Pettit and Beitinger 1985). As aquatic hypoxia ensues, however, and the animals make increased use of their ABO, branchial irrigation often becomes reduced, presumably so that the O_2 gained by air breathing is not lost across the gills. This suggests that there must be some adaptation of the receptor input over time, shifting emphasis between aerial and aquatic respiration so as to match the breathing mode to the relative availability of oxygen in either medium. Associated changes in perfusion are known to accompany this shift in emphasis from water to air breathing. For example, Johansen et al. (1967) calculated that blood flow to the lung of *Neoceratodus* could increase by as much as 25% after 25 min of exposure to aquatic hypoxia, and in gar, the increase in pulmonary perfusion during aquatic hypoxia leads to enhanced pulmonary O_2 uptake (Smatresk and Cameron 1982a).

Temperature also plays an important role in determining the relative sensitivity of the gas exchangers to aquatic hypoxia (Sect. 7.1.3). At 10 °C, the bowfin (*Amia calva*) is virtually an exclusive water breather, and while mild aquatic hypoxia (water $PO_2 \sim 70-80$ Torr) elicits an increase in gill irrigation, it does not cause any change in air breathing frequency (Johansen et al. 1970a). In fact, some animals utilized their gills as the sole organ of gas exchange at O_2 tensions as low as 40 Torr at 10 °C (Johansen et al. 1970a). At higher temperatures, however, when air breathing dominates overall gas exchange, mild levels of aquatic hypoxia cause marked increases in air breathing frequency, with no change in gill irrigation (Johansen et al. 1970a).

Amphibians

In comparison to air-breathing fish, only a few studies have examined the effects of aquatic hypoxia on lung ventilation in amphibians. Aquatic hypoxia stimulates air-breathing in bullfrogs (*Rana catesbeiana*) submerged at 5 °C (A. Pinder, pers. comm.) as well as when they are voluntarily diving at 20 °C (Fig. 19). In the latter

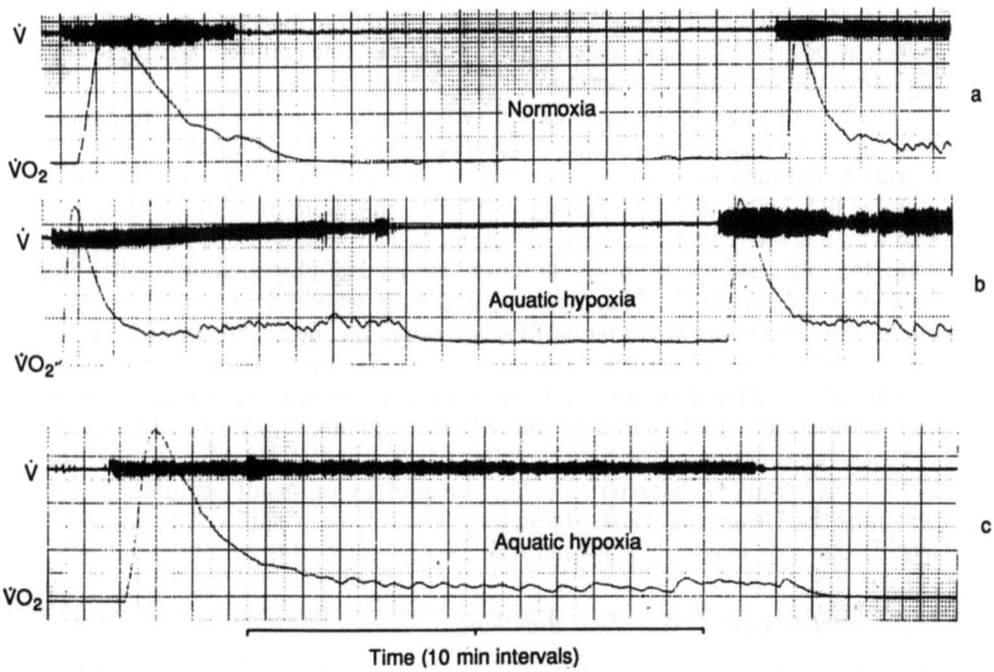

Fig. 19. Pneumotachograph recordings of lung ventilation ($\dot{V}$) in voluntarily diving *Rana catesbeiana* in normoxic conditions (*a*), and after 1 h (*b*) and 3 h (*c*) of aquatic hypoxia (water $PO_2 \sim 40$ Torr) at 20 °C. The uncalibrated output from an O_2 analyzer, measuring O_2 consumption ($\dot{V}O_2$) at the blowhole, is also shown (unpublished data)

case, the overall frequency of lung ventilations increased as a consequence of the animal spending more time at the surface (i.e. the frequency of breathing within the surface period remained the same). Moreover, the duration of the diving breath-holds decreased from ~25–30 min at a water PO_2 of 155 Torr and to ~10–15 min when the PO_2 of the water was lowered to 40 Torr (Fig. 19b, c). As shown earlier in Fig. 16, this increased lung ventilation during aquatic hypoxia is accompanied by an increased proportion of pulmocutaneous blood being directed toward the lungs. This also correlates well with the gas washout experiments of Malvin and Hlastala (1986) which show that cutaneous gas exchange decreases during voluntary air-breathing in *Rana pipiens*. This response appears to be O_2 mediated, since inflation of the lung with N_2 had no immediate effect on cutaneous gas exchange, whereas inflation with air led to a 30% decrease in inert gas washout, and, with pure O_2, to even greater decreases in cutaneous gas exchange (Malvin and Hlastala 1986).

Aquatic hypoxia has also been shown to increase the frequency of branchial movements in *Necturus*, and to increase air breathing (Guimond and Hutchison 1972). On the other hand, when *Necturus* was transferred to a moist air environment for 25 min, arterial PO_2 decreased rapidly in spite of increased lung ventilation (Lenfant and Johansen 1967). Upon recovery in well-aerated water, the animals made strong fanning motions with their external gills, and arterial PO_2 was soon raised to values in excess of those seen prior to air exposure. Although this would seem to suggest that the lungs are relatively ineffective in overall gas exchange, Miller and Hutchison (1979) have shown that the increased metabolic demands associated with spontaneous activity are supported entirely aerobically by the stimulation of lung ventilation.

The PO_2 of inspired water is certainly an important variable in the respiratory control of anuran tadpoles (*Rana catesbeiana*), effecting changes in both gill irrigation and lung ventilation (West and Burggren 1982). Progressive decreases in water PO_2 led to increases in lung ventilation that were most often accompanied with marked reductions in gill irrigation. Hyperoxic conditions, on the other hand, led to a cessation of lung ventilation *and* an overall reduction in gill water flow. West and Burggren (1982) proposed a receptor whose input leads to an inhibition of gill irrigation when blood PO_2 is elevated (i.e. by air breathing or hyperoxic water). Under this scheme, the elevated PO_2 following an air breath would initially inhibit gill irrigation until the subsequent progressive hypoxemia between air breaths elicits a gradual resumption of branchial pumping (West and Burggren 1982). This pattern of progressive elevation of gill irrigation frequency with time after an air breath, is characteristic of the breathing patterns of a number of air-breathing fish (Jesse et al. 1967; McMahon 1969; Johansen et al. 1970a; Johansen 1970; Singh and Hughes 1973; Smatresk 1988; Sect. 7.1.1).

7.1.2 Aerial Hypoxia: Impact on Breathing

Virtually all bimodal breathers increase ventilation of the ABO when arterial or mixed venous PO_2 levels fall as a result of breathing hypoxic gas mixtures. Elevated air breathing is also observed when O_2 delivery becomes threatened due to increases in temperature or levels of activity.

Air-Breathing Fish

Despite being emancipated from the effects of aquatic hypoxia (Sect. 7.1.1), inhalation of hypoxic gas mixtures by *Protopterus* and *Electrophorus* leads to abrupt increases in the frequency with which the ABO is ventilated (Jesse et al. 1967; Johansen and Lenfant 1968; Johansen et al. 1968b). When hypoxic mixtures were introduced into the lung of *Protopterus*, air breathing became stimulated at levels of arterial PO_2 that were higher than normal (Johansen and Lenfant 1968). From this, Johansen (1970) suggested that chemosensitive areas associated with triggering this response might be reacting to the rate of change rather than the absolute levels of PO_2. Other experiments on *Electrophorus* indicated a similar breath-to-breath responsiveness (i.e. a combined hyperventilation and tachycardia) to changes in aerial PO_2 (Johansen et al. 1968b). As shown in Fig. 12, there is a considerable redistribution of the cardiac output to the mouth and away from the body during these periods of breathing. Owing to this inbuilt delay between blood flowing to the ABO and body, it was considered unlikely that the immediate response to aerial hypoxia could be the result of a feedback mechanism in the systemic arterial circuit. Instead, the receptor system was thought more probably to reside in either the buccal mucosa or in its underlying vasculature (Johansen et al. 1968b). Although this acute ventilatory responsiveness to hypoxia exists, little is known about the mechanisms underlying its control. There is, however, some evidence that a chemoreceptive area may exist in association with the branchial arches of the lungfish *Protopterus*, since sectioning the nerves to the arches reduces the responsiveness to hypoxia (Lahiri et al. 1970). Even so, Fishman et al. (1989) report that there is no histological evidence of an organized chemoreceptive body in the vasculature of the branchial arches in this animal, thus leaving the whole question of mechanisms of action unsolved.

Evidence for intravascular O_2-chemosensitive elements in air breathing fish comes from experiments in which NaCN (Lahiri et al. 1970) or nicotine (Johansen 1970) were introduced into the bloodstream of *Protopterus*. Both of these substances are used as chemical probes for O_2-sensitive elements, and both led to increases in air breathing and gill irrigation. More recently, Smatresk et al. (1986) carried out experiments on the gar, *Lepisosteus*, in which injections of NaCN were performed in three distinct locations (dorsal aorta, ventral aorta or conus arteriosus). In all cases, the injections resulted in an increase in both gill irrigation and air breathing frequency. However, the response latency was shortest when NaCN was administered via the conus or ventral aorta, suggesting that intravascular chemoreceptor loci were located in the branchial and/or head region of the gar. As noted earlier, Smatresk et al. (1986) also provided compelling evidence for externally located (i.e. water facing) O_2 chemoreceptors. The role that the two putative chemoreceptors play in controlling aerial and aquatic breathing was examined in an elegant experimental series shown in Fig. 20. In these experiments, the possibility of interactive effects between internal and external loci was tested by independently altering the aquatic PO_2 and ventral aorta PO_2, while monitoring both air breathing and gill irrigation. Figure 20A shows that air breathing always increased as ventral O_2 tensions declined and that, for any given blood O_2 level, the rates of air breathing were significantly higher in hypoxic water. These results suggest that the internal and external O_2 loci can combine to drive venti-

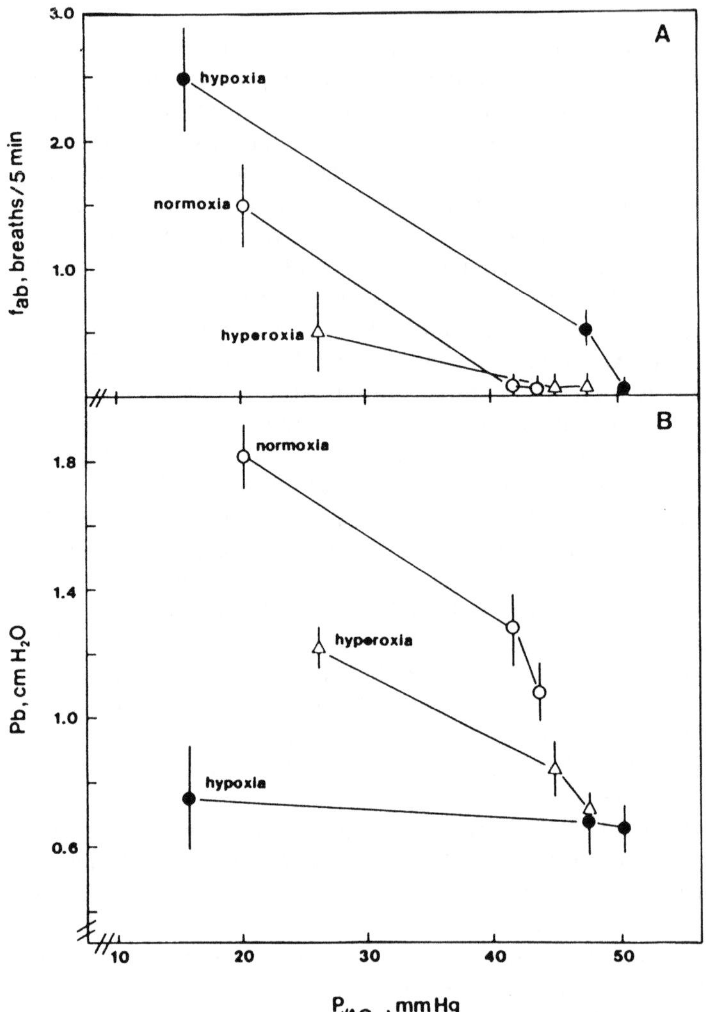

Fig. 20. Interactive effects of aquatic (external) and ventral aorta (internal) O_2 tensions on (*A*) the air-breathing frequency (f_{ab}) and (*B*) buccal pressure amplitude (Pb) in the longnose gar *Lepisosteus osseus*. Changes in water O_2 tensions are shown for normoxic ($PO_2 \sim 150$ Torr), hypoxic ($PO_2 \sim 25$ Torr) and hyperoxic ($PO_2 \sim 630$ Torr) conditions (20 °C). Changes in ventral aorta PO_2 ($PvaO_2$) were produced by unidirectional ventilation of the air-breathing organ with humidified nitrogen, air, or oxygen (Smatresk et al. 1986)

lation of the ABO in gar. Gill irrigation was also modulated by internal and external O_2 stimulus levels as the PO_2 of the water fell from hyperoxic to normoxic levels (Fig. 20B). Even so, aquatic hypoxia inhibited the response, suggesting that the external chemoreceptors may exert an overriding inhibitory effect on gill irrigation, independent of O_2 levels in the blood. Smatresk (1988) speculated that this sort of arrangement might be common among air-breathing fish, and that the interspecific variability in the branchial responses to hypoxia may reflect differences in the central integration of such peripheral chemosensory input.

Amphibians

Almost all amphibians respond to decreases in environmental O_2 levels by increasing their frequency of lung ventilation. In most cases, the increases in breathing occur as a consequence of a decrease in the duration of the breath-holding intervals (Boutilier and Toews 1977; West and Burggren 1982; Kruhøffer et al. 1987; Jones and Chu 1988). Hypoxia can also lead to changes in the way that the lungs are ventilated. For example, when exposed to severe hypoxia, *Bufo marinus* depart from their normal pattern of lung ventilation (i.e. buccal oscillations punctuated with single lung ventilations) and begin to group the lung ventilations in 'inflation cycles', as shown in Fig. 5. Similar hypoxic-induced changes in the co-ordination of the lung filling and emptying sequence were also observed in *Bufo paracnemius* (Kruhøffer et al. 1987) and *Rana pipiens* (Pinder and Burggren 1986).

Measurements of the amounts of gas exchanged during exposure to environmental hypoxia have only been made in a few instances. Relative to normoxic conditions, minute ventilation increased twofold when *Xenopus* were exposed to 10% O_2 mixtures (Brett 1980). The increased ventilation occurred almost entirely from a decrease in the duration of the breath-holding periods (i.e. an increase in lung ventilation frequency rather than in the depth of breathing; Jones and Chu 1988), as was also the case in *Rana pipiens* (Pinder and Burggren 1986). The same basic strategy was used by the toad, *Bufo paracnemius*, which increased its minute ventilation in response to both increases in temperature as well as decreases in oxygen (Kruhøffer et al. 1987). In *Rana pipiens*, environmental hypoxia caused a decrease in total oxygen consumption ($\dot{M}O_2$), caused mainly by a reduction in cutaneous $\dot{M}O_2$ (Fig. 21 A), and an increase in the frequency of lung ventilations (Fig. 21 B) such that there was a shift in the proportion of oxygen uptake towards the lung. In comparison to normoxic conditions, the lung ventilation frequency of *Rana* more than doubled, as environmental PO_2 was lowered to 52 Torr (Fig. 21 B). Recent studies on *Rana catesbeiana* (Boutilier et al. 1986) indicate that shifts in emphasis between aquatic and aerial O_2 uptake may depend on the relative supplies of O_2 at either exchange site (e.g. Fig. 16).

Although several chemoreceptive elements have been identified in amphibians (Ishii and Oosaki 1969; Rogers 1966; Ishii and Ishii 1976; Poullet 1977; Ishii and Kusakabe 1982; Jones and Milsom 1982), their role in ventilatory control is largely unkonwn. The carotid labyrinth in anurans, for example, has often been looked upon as an important chemosensitive element in the control of ventilation, largely because of its responsiveness to decreases in O_2 or NaCN in vitro (Ishii et al. 1966, 1985). Lillo (1980) also showed that injections of NaCN into the carotid artery of lightly anaesthetized bullfrogs caused an increase in ventilation after a latency of 2–5 s, well in line with the proximity of the injection site to the carotid labyrinth. The early observations of Smyth (1939) showed that denervation of the labyrinth inhibited the hyperpnea response to hypoxia in the frog, Rana temporaria. Recent studies on both voluntarily diving *Xenopus laevis* (Evans and Shelton 1984; Jones and Chu 1988) and unidirectionally ventilated *Bufo marinus* (West et al. 1987) have shown that denervation of the carotid labyrinths causes a reduc-

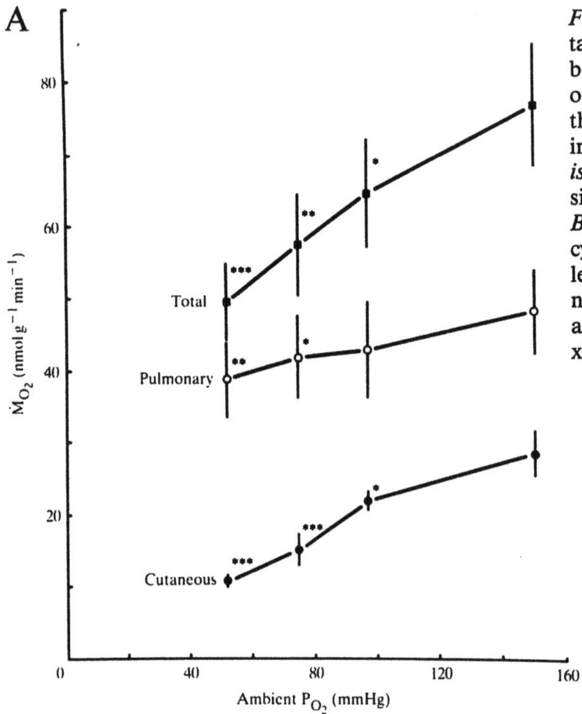

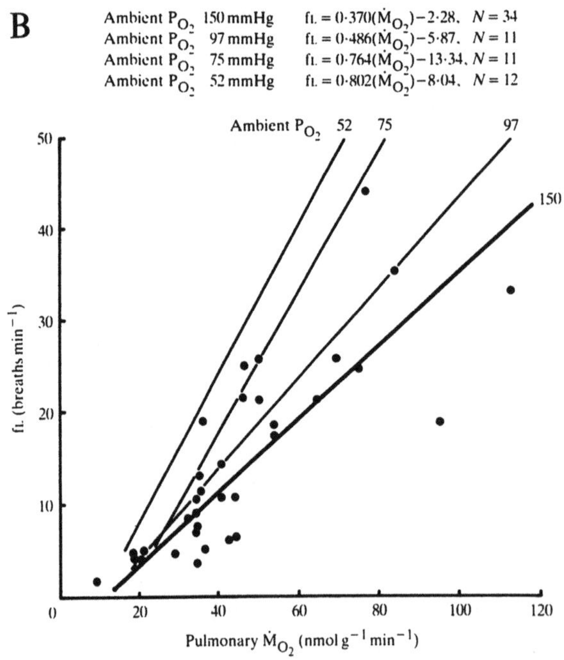

Fig. 21. A Partitioning of the total oxygen consumption ($\dot{M}O_2$) between pulmonary and cutaneous exchangers as a function of the ambient PO_2, in freely moving *Rana pipiens* at 25 °C. *Asterisks* denote when values differ significantly from normoxic level. *B* Changes in ventilation frequency (f_L), at various ambient PO_2 levels, as a function of pulmonary $\dot{M}O_2$. Regression data (*top*) are given for each level of hypoxia (Pinder and Burggren 1986)

tion in overall ventilation but does not abolish the sensitivity of the animals to hypoxia, and, at least in the case of *Xenopus*, to hypercapnia as well (Jones and Chu 1988). In addition, Van Vliet and West (1986) have shown that the ventilatory increases associated with NaCN or lobeline sulfate injections in *Bufo marinus* were uninfluenced by denervation of the glossopharyngeal nerve (a branch of which innervates the labyrinth). Nevertheless, the decrease in minute ventilation during normoxia, observed after denervation of the labyrinth in both *Bufo* (West et al. 1987) and *Xenopus* (Jones and Chu 1988), indicates that the structure does play a significant role in regulating gas exchange in these animals. In *Xenopus*, the decrease in ventilation occurred as a result of a significant reduction in the volume of each individual buccal pumping movement, whereas the frequency of ventilation remained unchanged (Jones and Chu 1988).

Because the main O_2 stores of amphibians are located in the blood and lungs, it is difficult to imagine how they could be effectively regulated with reference to only a single receptor site (e.g. the carotid labyrinth) unless information was also available about heart output and the levels of blood flow (and/or peripheral resistance) in the major divisions of the animals' incompletely divided circulations. Stimulation of ventilation in the bullfrog occurs 5–8 s after the start of injection of cyanide into the pulmocutaneous artery, indicating that this arch or its derivatives supplies a chemoreceptive area (Lillo 1980). In addition, chemosensitive elements within the aortic trunk of the toad, *Bufo vulgaris*, are apparently linked to respiratory control (Ishii et al. 1985). And, as pointed out by West and Burggren (1982), bullfrog tadpoles increase their ventilation in response to hypoxia at stages when the carotid labyrinth has yet to be developed. Little progress has been made in uncovering the way in which these 'additional' O_2-sensitive chemoreceptors might act to control the ventilation in amphibians.

7.1.3 Temperature-Induced Changes in the Hypoxic Threshold of Air-Breathing Fish and Amphibians

The magnitude of the air-breathing response to inhalation of hypoxic gas mixtures is highly dependent on temperature in a variety of animals including air-breathing fish (Glass et al. 1986, Fig. 22A), amphibians (Kruhøffer et al. 1987; Fig. 22B) and reptiles (Jackson 1973; Glass et al. 1983; reviewed in Milsom, this Vol.). In all of these groups of animals, the threshold of the hypoxic-induced hyperventilation (i.e. the arterial PO_2 or % Hb-O_2 saturation at which increased ventilation is initiated) is increased as body temperatures rise. Increases in arterial PO_2 with temperature occur at relatively constant arterial blood O_2 saturations in both *Xenopus laevis* and *Bufo marinus* (Boutilier et al. 1987), as they do in many reptiles (Glass et al. 1983; Wood and Hicks 1985; Milsom, this Vol.). Although these blood data have yet to be collected for an air-breathing fish, the ventilatory responses of *Channa* (Fig. 22A) are indicative of the same overall response. The increases in arterial PO_2 with temperature can be explained on the basis of existing models of O_2 transport in animals with cardiovascular shunts (Wood 1982, 1984), detailed by Milsom (this Vol.). In brief, the degree of arterial desaturation during air breathing depends on the extent of right-to-left intraven-

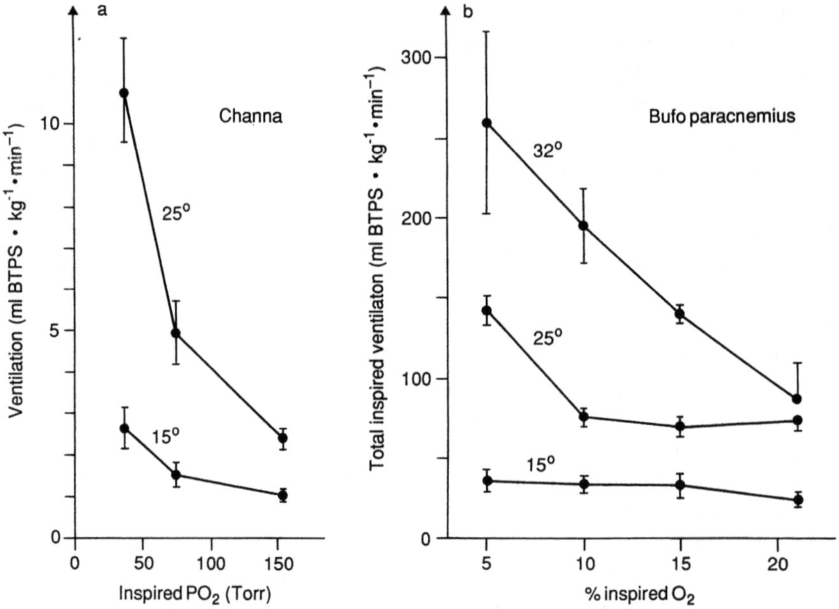

Fig. 22. The effect of changes in inspired oxygen levels on the total inspired ventilation in the air-breathing fish *Channa argus* (*A*), and in the toad *Bufo paracnemius* (*B*) at various temperatures (*Channa*, Glass et al. 1986; *Bufo*, Kruhøffer et al. 1987)

tricular shunting, with the arterial PO_2 becoming a function of the resulting O_2 saturation. This has been taken to suggest that O_2 saturation or content, rather than PO_2 may be the common denominator of oxygen homeostasis in these animals (Wood and Hicks 1985). Certainly, the mechanism(s) underlying this widespread phenomenon must be complex, possibly involving measurement of the volumes and O_2 contents of partially saturated and desaturated blood, as well as their mixing ratio within the heart.

7.1.4 Carbon Dioxide: Impact on Breathing

Air-Breathing Fish

Ventilatory responses to aquatic hypercapnia in air-breathing fish might be caused indirectly by changes in blood oxygenation (i.e. through pH-mediated Bohr and Root effects), or by a CO_2-induced depression of branchial respiration. Decreased branchial irrigation in response to aquatic hypercapnia has been observed in several air-breathing species (*Neoceratodus*, *Symbranchus*, *Amia*; Johansen 1970), where it is thought to be a response towards limiting CO_2 entry via the gills, thereby minimizing any respiratory acidosis. However, in species such as *Neoceratodus*, whose branchial respiration normally accounts for a large proportion of the animals O_2 uptake, the CO_2-induced reduction in gill irrigation might effectively constitute an O_2-driven stimulus for increased aerial ventilation (Johansen et al. 1967). Indeed, these authors suggested that this secondary stimulation of air breathing could be brought about by hypoxic stimulation of chemoreceptive sites located in the efferent branchial vessels.

Depending on the efficacy of the aerial exchanger for CO_2 removal, a hypercapnic acidosis could, through Bohr and Root effects, lead to a marked hypoxaemia. Certainly in the gourami, *Trichogaster*, the increased aerial ventilation associated with aquatic hypercapnia was suppressed when the water PO_2 was kept at hyperoxic levels (Burggren 1979), indicating that the ventilatory response was primarily O_2-mediated. Moreover, aerial hypercapnia alone, at moderate levels, had little or no effect on air breathing in either *Trichogaster* (Burggren 1979) or *Neoceratodus* (Johansen et al. 1967), whereas when aquatic hypercapnia was combined with hypoxia, marked increases in air breathing occurred.

The fact that aerial hypercapnia did not lead to increased air breathing in *Trichogaster* or *Neoceratodus* must mean that the aquatic exchanger was able to eliminate all of the CO_2 added via this route. Otherwise, it should have caused a hypercapnic acidosis and concomitant hypoxaemia-stimulated ventilatory response, as before. The efficacy of the gills in eliminating CO_2 was clearly demonstrated in experiments on *Neoceratodus*, in which pure CO_2 was injected into the lung. Following the injection, the PCO_2 of the lung gas and pulmonary venous blood increased markedly, whereas that of the pulmonary arterial blood showed only a modest rise (Johansen et al. 1967). Thus, despite there being a relatively large Bohr effect (Lenfant et al. 1966), the efficient removal of CO_2 via the gills effectively maintained arterial oxygenation at a constant level (Johansen et al. 1967). Other air-breathing fish such as *Channa* (Glass et al. 1986) and the Cuchia eel (Lomholt and Johansen 1974) increase ventilation of their air-breathing organs only when inspired PCO_2 exceeds 30 Torr. Presumably, aerial ventilation is stimulated at such times through pH-mediated effects on arterial oxygenation. On the other hand, the Japanese weatherloach (*Misgurnus*), while exhibiting a strong hypoxic drive for aerial ventilation of the intestine, is essentially unresponsive to elevated levels of CO_2 ($\sim$14 Torr) in both the aerial and aquatic environment (McMahon and Burggren 1987). Since the Bohr and Root effects of many of these fish are unknown, it is impossible to assess whether the variety of ventilatory responses observed reflect differences in CO_2-induced hypoxaemia, or in the efficacy of the aquatic exchanger in CO_2 removal. Taken together, however, all of these studies indicate that CO_2 is a minor contributor to the respiratory drive of air-breathing fish.

Amphibians

As in air-breathing fish, oxygen appears to play a dominant role in ventilatory control of amphibians. Even so, the addition of CO_2 to inspired air causes marked increases in ventilation in anuran, apodan and urodele amphibians including *Bufo marinus* (Macintyre and Toews 1976; Boutilier et al. 1979a; Jones 1982; West et al. 1987), *Rana catesbeiana* (Jackson and Braun 1979); *Xenopus laevis* (Boutilier 1988; Jones and Chu 1988); *Typhlonectes compressicauda* (Toews and Macintyre 1978) and *Cryptobranchus alleganiensis* (Boutilier and Toews 1981b; Boutilier et al. 1980). In most animals, it appears that the sensitivity of the ventilatory response is related to the severity of the associated respiratory acidosis, however, some species (e.g. *Amphiuma tridactylum*) are apparently insensitive to even quite large increases in aerial and/or aquatic CO_2 (Toews 1971). Thus, injections into the lungs of up to 15% CO_2 did not elicit any increased air-breathing response in *Amphiuma*, as was also observed in the lungfish *Neoceratodus* (Johansen et al. 1967), evidence of the substantial capacity for non-pulmonary CO_2 elimination in these animals. Moreover, aquatic hypercapnia was ineffective in eliciting any ventilatory response in *Amphiuma*, despite the fact that arterial PCO_2 levels rose from ~15 to ~40 Torr (Toews 1971). Certainly, in other amphibians, increases in arterial PCO_2 cause marked hyperventilation (Macintyre and Toews 1976; Boutilier and Shelton 1986b; Boutilier 1988; Fig. 6B).

7.1.5 Interactions Between O_2 and CO_2

The pattern generator for pulmonary ventilation in the lungfish evidently requires some degree of hypoxic stimulation, since ventilation with hyperoxic gases leads to a cessation of lung ventilation (Fishman et al. 1989). This is a response similar to that seen when the toad *Bufo marinus* was subjected to hyperoxic environments (Toews and Kirby 1985) or when unidirectionally ventilated with normoxic hypocapnic gas mixtures (West et al. 1987). This 'hypoxic drive' is common in many air-breathing fish (reviews by Johansen 1970; Smatresk 1988), as well as in several amphibians (Binet and Bochet 1963; Brett 1980; West and Burggren 1982; Kruhøffer et al. 1987; Boutilier 1988) and reptiles (Jackson 1978; Glass et al. 1983). Though aquatic hypercapnia is known to stimulate ventilation of the ABO in several air-breathing fish (Jesse et al. 1967; Johansen 1970; Burggren 1979; Glass et al. 1986), specific locations of CO_2-sensitive elements or of their interaction with O_2-sensitive structures are poorly known.

Recent studies have shown that hypoxic and hypercapnic stimuli may combine to drive lung ventilation in the anuran amphibians, *Bufo marinus* (Toews and Kirby 1985; West et al. 1987) and *Xenopus laevis* (Boutilier and Shelton 1986b, c; Boutilier 1988; Jones and Chu 1988). When the toad *Bufo* was exposed to hyperoxic environments (PO_2 ~ 340 Torr), lung ventilations ceased to occur, and the only remaining ventilatory activity was that of buccal oscillations (cf. Fig. 5), whose frequency were markedly reduced (Toews and Kirby 1985). The cessation of lung ventilations seen in dehydrated hyperoxic animals led to a progressive respiratory acidosis (Toews and Kirby 1985) that was not observed in their normoxic counterparts (Boutilier et al. 1979b). Moreover, the magnitude of the respiratory acidosis in the dehydrated hyperoxic animals would normally have stimulated increased lung ventilation in normoxic animals (Macintyre and Toews 1976). Similar ventilatory responses were observed when intact *Bufo marinus* were unidirectionally ventilated with various combinations of O_2, CO_2 and N_2 (West et al. 1987). As shown in Fig. 23, the hypoxaemic threshold (i.e. arterial PO_2) for lung ventilation increased from 54 Torr during hypocapnia to approximately 140 Torr during hypercapnia. This result, together with the hyperoxic-induced cessation of ventilation during respiratory acidosis (Toews and Kirby 1985), indicates that hypoxic

Fig. 23. Relationship between arterial O_2 tension (PaO_2) and *A* buccal pressure-time index, or *B* mean frequency of lung ventilation. Measurements were made in an individual, unidirectionally ventilated *Bufo marinus* at 21°–22°C. Arterial PCO_2 varied from 5 Torr when the animal was ventilated with 0% CO_2 to 18.4 Torr at 5% CO_2. Arterial pH correspondingly decreased from 8.27 to 7.65 over the same %CO_2 range (West et al. 1987)

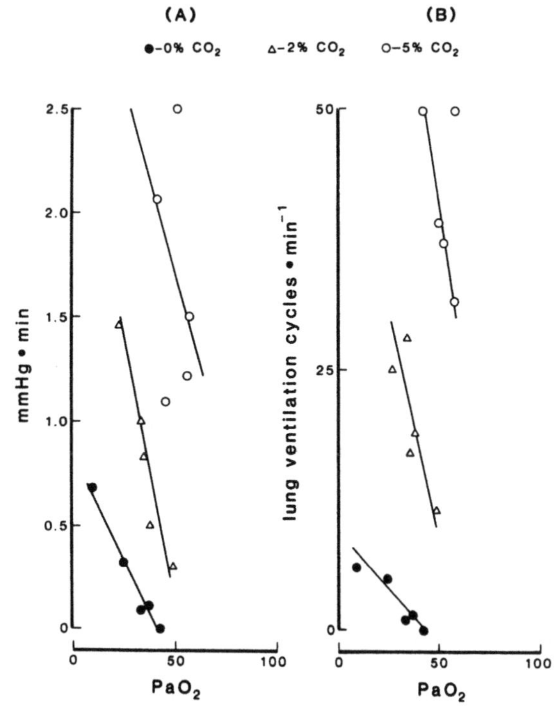

and hypercapnic stimuli summate to drive ventilation in the toad *Bufo marinus* (West et al. 1987). Similar interactions in *Xenopus* are known to bring dives to an end more quickly (Fig. 8), and it has been suggested that CO_2 alters the gain of a predominantly O_2-mediated response to initiate breathing (Boutilier and Shelton 1986c).

7.2 Bimodal Breathing Responses to Mechanical Stimuli

The discharge characteristics of certain mechanosensitive elements have been described for lungfish (DeLaney et al. 1983; Fishman et al. 1989), bowfin (Milsom and Jones 1985), gar (Smatresk and Azizi 1987) and frogs (Taglietti and Casella 1966, 1968; McKean 1969; Milsom and Jones 1977). While it is evident that these animals have the mechanisms to monitor ABO volume, as well as rates of inflation and deflation, very little is known about the role that mechanoreceptors play in the overall control of ventilation in these animals.

In many bimodal breathers, the ABO is used as an important store of O_2 during a dive, and so input from volume-sensitive receptors in the ABO may be important in determining the aerobic capacity of the diving animal. In addition, because the exchange ratio of the ABO declines during a dive (as more O_2 is taken up than replaced by CO_2; cf. Fig. 9), the volume of the ABO also decreases. Mechanisms for monitoring the volume of the ABO (i.e. the volume of the O_2 store) during the intervals between breaths could be important, therefore, in determining the length of a dive. In addition to monitoring changes in volume, many such receptors are also sensitive to CO_2 (DeLaney et al. 1983; Fishman et al. 1989; Milsom and Jones 1977; Kuhlmann and Fedde 1979), and could convey additional information about the respiratory status of the ABO. Even so, it is difficult to say whether it is the volume of the ABO itself, or the declining O_2 levels, that leads to the termination of the breath-hold. In unidirectionally ventilated lungfish, however, in which the lung and arterial blood PO_2 are maintained constant, decreases in intrapulmonary pressure cause marked decreases in the duration of the breath-hold (Fishman et al. 1989). This indicates that lung volume per se plays an important role in determining the overall pattern of air breathing in the lungfish. Inflation of the ABO of many air breathing fish and amphibians often causes an inhibition of air-breathing (Taglietti and Casella 1966, 1968; Lenfant et al. 1970; Toews 1971; Smatresk and Cameron 1982b; Pack et al. 1984), which could be related again either to the ABO volume or to the increased O_2. In *Amphiuma*, inflation of the lungs with pure nitrogen was effective in delaying the onset of breathing, suggesting that the ventilatory response can be abolished by lung distension alone (Toews 1971).

Mechanoreceptors also appear to operate as transducers for the reflex changes in ABO blood flow that occur during air breathing in several bimodal breathers; i.e. inflation of the ABO with air often leads to increases in cardiac output as well as ABO perfusion (Johansen 1966; Johansen et al. 1968a, b; Emilio and Shelton 1972; Lomholt and Johansen 1976; West and Burggren 1984; Fishman et al. 1985). In *Bufo marinus*, the mechanoreceptors play a major role in matching perfusion to ventilaton, through a reflexogenic vasodilation of the pulmonary vasculature (West and Burggren 1984). The efficacy with which this occurs will obviously affect the rates at which the gas stores of the body can be loaded and unloaded, and therefore the overall ventilatory patterns of the animal.

7.3 Mechanisms of Interaction Between Aquatic and Aerial Exchangers

In tadpoles, it appears that gill irrigation is reflexly mediated, since increased activity of pulmonary stretch receptors are implicated in the inhibitory action on the gills (West and Burggren 1983). This is evidently not the case for the lungfish *Protopterus* (Fishman et al. 1989); e.g. while afferent activity from pulmonary stretch receptors has a marked effect on lung ventilation, it has no effect on branchial pumping. Similarly, hypoxia evokes pronounced effects on lung ventilation but has essentially no influence on gill irrigation. In experiments in which the alveolar PO_2 and intrapulmonary pressure of lungfish could be maintained constant by unidirectional ventilation of the lungs, the animals still ventilated their lungs, indicating that there may be a central generator with an inherent rhythmicity (Fishman et al. 1989). Thus, while the duration of the non-ventilatory periods can be modulated by changes in arterial PO_2 (Johansen and Lenfant 1968) or through changes in lung volume (Fishman et al. 1989), air breathing is not completely abolished as long as there is sufficient tonic afferent stimulation. Under these same conditions, inhibition of gill irrigation still occurs after a spontaneous air breath. Fishman et al. (1989) thereby propose a model where there are separate pattern generators for pulmonary ventilation and gill irrigation. The system is hierarchical, in the sense that the action of the lung generator has an inhibitory effect on the pattern generator for branchial pumping. The inhibition is then gradually alleviated, restoring the normal activity of the branchial exchanger. As it presently stands, there in no evidence that the pattern generator for the gills is ever involved in inhibiting the action of the pulmonary system (Fishman et al. 1989).

8 Concluding Remarks

A generic model of the respiratory control system of a bimodal breather is shown in Fig. 24. It was assumed that the control system must be one that can regulate unevenly balanced exchanges of O_2 and CO_2 across respiratory organs that can

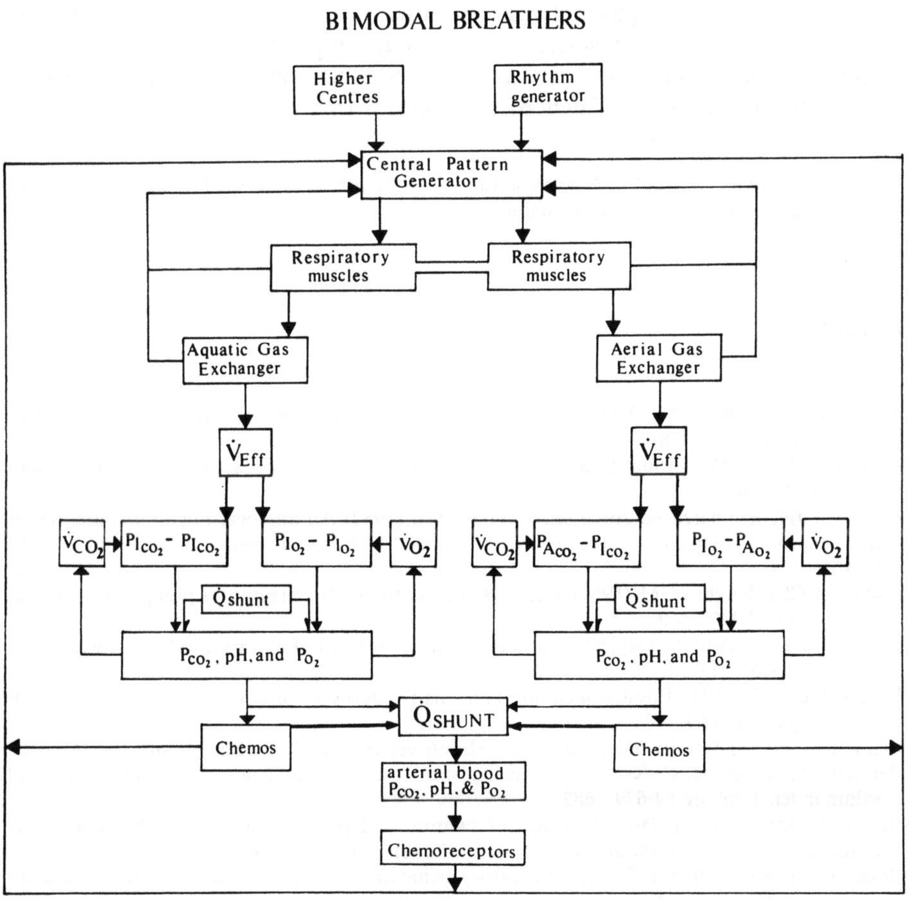

Fig. 24. Hypothetical model of the respiratory control system of a biomodal breather. See text for details

be intermittently ventilated and perfused. The central pattern generator is responsible for sorting afferent information from a number of peripheral receptors, integrating the information and then relaying commands to the aquatic and aerial exchangers via the respiratory muscles. The temporal output of each exchanger leads to an effective ventilation ($\dot{V}_{eff}$) which sets the levels of O_2 uptake and CO_2 removal. The PO_2, PCO_2 and pH of blood from each exchanger may or may not be detected by chemosensitive elements prior to mixing. Blood from the two exchangers may also be subject to central blood shunts, and the arterial blood that is formed can be detected by systemic chemoreceptors. The various chemoreceptor groups can thereby offer private information to the central pattern generator about the levels of gas exchange in the aquatic and aerial mode. Additional information may also be relayed to the pattern generator from the respiratory muscles associated with each exchanger and, depending on the inputs from other peripheral sources, could lead to shifts in emphasis between the two modes of gas exchange. Higher brain centres are considered capable of overriding suprathreshold input stimuli from receptors when certain behavioural activities take priority over the need for additional gas exchange. Although peripheral receptor systems and higher brain centres are considered to be of major importance in determining the overall patterns of dual gas exchange, the model also has provision for the patterns of breathing to be influenced by a central rhythm generator.

Acknowledgements. The author's work is supported by grants from the Natural Sciences and Engineering Research Council of Canada.

References

Ackerman RA, White FN (1979) Cyclic carbon dioxide exchange in the turtle *Pseudemys scripta*. Physiol Zool 52:378–389

Andersson M (1978) Optimal foraging area: size and allocation of search effort. Theor Popul Biol 13:397–409

Andresen JH, Ishimatzu A, Johansen K, Glass ML (1987) An angiocardiographic analysis of the central circulation in the air breathing teleost, *Channa argus*. Acta Zool (Stockh) 68:165–171

Ballintijn CM, Hughes GM (1965) The muscular basis of the respiratory pumps in the trout. J Exp Biol 43:349–362

Belkin DA (1968) Aquatic respiration and underwater survival of two freshwater turtles species. Respir Physiol 4:1–14

Berg T, Steen JB (1965) Physiological mechanisms for aerial respiration in the eel. Comp Biochem Physiol 15:469–484

Bennett AF (1978) Activity metabolism in the lower vertebrates. Annu Rev Physiol 400:447–469

Bennett AF, Houck LD (1983) The energetic cost of courtship and aggression in a plethodontid salamander. Ecology 64:679–683

Bevan DJ, Kramer DL (1986) The effect of swimming depth on respiratory behaviour of the honey gourami, *Colisa chuna* (Pisces, Belontiidae). Can J Zool 64:1893–1896

Bevan DJ, Kramer DL (1987) The respiratory behaviour of an air-breathing catfish, *Clarias macrocephalus* (Clariidae). Can J Zool 65:348–353

Bidani A, Crandall ED (1988) Velocity of CO_2 exchanges in the lungs. Annu Rev Physiol 50:639–652

Binet L, Bochet M (1963) Resistance a l'hyperoxie de la grenouille *Rana esculenta* maintenue dans l'oxygene a 100 pourcent durant 52 jours. J Physiol (Paris) 55:405–412
Boutilier RG (1984) Characterization of the intermittent breathing pattern in *Xenopus laevis*. J Exp Biol 110:291–309
Boutilier RG (1988) Control of arrhythmic breathing in bimodal breathers: Amphibia. Can J Zool 66:6–19
Boutilier RG (1989) Diving physiology: amphibians. In: Wood SC (ed) Comparative pulmonary physiology: current concepts, vol 39. Lung biology in health and disease. Marcel Dekker, New York, pp 677–695
Boutilier RG, Shelton G (1986a) Respiratory properties of blood from voluntarily and forcibly submerged *Xenopus laevis*. J Exp Biol 121:285–300
Boutilier RG, Shelton G (1986b) The effects of voluntary and forced diving on ventilation, blood gases and pH in *Xenopus laevis*. J Exp Biol 122:209–222
Boutilier RG, Shelton G (1986c) Gas exchange, storage and transport in voluntarily diving *Xenopus laevis*. J Exp Biol 126:133–155
Boutilier RG, Toews DP (1977) The effect of progressive hypoxia on respiration in the toad, *Bufo marinus*. J Exp Biol 68:99–107
Boutilier RG, Toews DP (1981a) Respiratory, circulatory and acid-base, adjustments to hypercapnia in a strictly aquatic and predominantly skin breathing urodele, *Cryptobranchus alleganiensis*. Respir Physiol 46:177–192
Boutilier RG, Toews DP (1981b) Respiratory properties of blood in a strictly aquatic and predominantly skin-breathing urodele *Cryptobranchus alleganiensis*. Respir Physiol 46:161–176
Boutilier RG, Randall DJ, Shelton G, Toews DP (1979a) Acid-base relationships in the blood of the toad *Bufo marinus*. I. The effects of environmental CO_2 J Exp Biol 82:331–344
Boutilier RG, Randall DJ, Shelton G, Toews DP (1979b) Acid-base relationships in the blood of the toad *Bufo marinus*. II. The effects of dehydration. J Exp Biol 82:345–355
Boutilier RG, McDonald DG, Toews DP (1980) The effects of enforced activity on ventilation, circulation and blood acid-base balance in the aquatic gill-less urodele, *Cryptobranchus alleganiensis*: a comparison with the semiterrestrial anuran, *Bufo marinus*. J Exp Biol 84:289–302
Boutilier RG, Glass ML, Heisler N (1986) The relative distribution of pulmocutaneous blood flow in *Rana catesbeiana*: effects of pulmonary or cutaneous hypoxia. J Exp Biol 126:33–39
Boutilier RG, Glass ML, Heisler N (1987) Blood gases, and extracellular/intracellular acid-base status as a function of temperature in the anuran amphibians *Xenopus laevis* and *Bufo marinus*. J Exp Biol 130:13–25
Brett SS (1980) Breathing and gas exchange in an aquatic amphibian, *Xenopus laevis*. PhD Dissertation, University of East Anglia, UK
Brett SS, Shelton G (1979) Ventilatory mechansisms of the amphibian, *Xenopus laevis*: the role of the buccal force pump. J Exp Biol 80:251–269
Bugge J (1960) The heart of the African lungfish, *Protopterus*. Vidensk Medd Dan Naturhist Foren 123:193–210
Burggren WW (1979) Bimodal gas exchange during variation in environmental oxygen and carbon dioxide in the air breathing fish *Trichogaster trichopterus*. J Exp Biol 82:197–213
Burggren WW (1988) Cardiovascular responses to diving and their relation to lung and blood oxygen stores in vertebrates. Can J Zool 66:20–28
Burggren WW (1989) Lung structure and function: Amphibians. In: Wood SC (ed) Comparative pulmonary physiology: current concepts, vol 39, Lung Biology in Health and Disease. Marcel Dekker, New York, pp 153–192
Burggren WW, Haswell MS (1979) Aerial CO_2 excretion in the obligate air breathing fish *Trichogaster trichopterus*. J Exp Biol 82:215–225
Burggren WW, Johansen K (1986) Circulation and respiration in lungfishes (Dipnoi). J Morphol (Suppl) 1:217–236
Burggren WW, Moalli R (1984) 'Active' regulation of cutaneous gas exchange by capillary recruitment in amphibians: experimental evidence and a revised model for skin respiration. Respir Physiol 55:379–392

Burggren WW, West NH (1982) Changing importance of gills, lungs and skin during metamorphosis in the bullfrog *Rana catesbeiana*. Respir Physiol 47:151–164
Burggren WW, Shelton G (1979) Gas exchange and transport during intermittent breathing in chelonian reptiles. J Exp Biol 82:75–92
Burggren WW, Feder ME, Pinder AW (1983) Temperature and the balance between aerial and aquatic respiration in larvae of *Rana berlandieri* and *Rana catesbeiana*. Physiol Zool 56:263–273
Carter GS, Beadle LC (1931) The fauna of the swamps of the Paraguayan Chaco in relation to its environment. II. Respiratory adaptations in the fishes. J Linn Soc Lond 37:327–366
Crandall ED, Bidani A (1981) Effect of red blood cell HCO_3^-/Cl^- exchange kinetics on lung CO_2 transfer: theory. J Appl Physiol 50:265–271
Crawford EC Jr, Schultetus RR (1970) Cutaneous gas exchange in the lizard, *Sauromalus obesus*. Copeia 1970:179–180
Czopek J (1959) Skin and lung capillaries in European common newts. Copeia 1959:91–96
Czopek J (1961) Vascularization of respiratory surfaces of some Plethodontidae. Zool Pol 11:131–148
Czopek J (1965) Quantitative studies on the morphology of respiratory surfaces in amphibians. Acta Anat 62:296–323
Davis JC, Cameron JN (1971) Water flow and gas exchange at the gills of rainbow trout, *Salmo gairdneri*. J Exp Biol 54:1–18
Daxboeck C, Barnard DK, Randall DJ (1981) Functional morphology of the gills of the bowfin, *Amia calva* L., with special reference to their significance during air exposure. Respir Physiol 9:349–364
DeGroodt M, Lagasse A, Sebruyns M (1960) Elektronenmikroskopische Morphologie der Lungenalveolen des *Protopterus* und *Amblystoma*. Proc 4th Internat Conf Electron Microscopy, Berlin 1958, vol 1. Springer, Berlin Göttingen Heidelberg, pp 418–421
DeJongh HJ (1972) Activity of the body wall musculature of the African clawed toad, *Xenopus laevis* (Daudin) during diving and respiration. Zool Med 47:135–144
DeJongh HJ, Gans C (1969) On the mechanism of respiration in the bullfrog, *Rana catesbeiana*: a reassessment. J Morphol 127:259–290
Dejours P (1973) Problems of control of breathing in fishes. In: Bolis L, Schmidt-Nielsen K, Madrell SHP (eds) Comparative physiology. Elsevier/North Holland, Amsterdam, pp 117–133
Dejours P (1981) Principles of comparative respiratory physiology, 2nd edn. Elsevier/North Holland, Amsterdam
DeLaney RG, Laurent P, Galante R, Pack AI, Fishman AP (1983) Pulmonary mechanoreceptors in the dipnoan lungfish *Protopterus* and *Lepidosiren*. Am J Physiol 244:R418–R428
de Saint-Aubain ML, Wingstrand K (1979) A sphincter in the pulmonary artery of the frog *Rana temporaria* and its influence on blood flow in skin and lungs. Acta Zool (Stockh) 60:163–172
Deyst KA, Liem KF (1985) The muscular basis of aerial ventilation of the primitive lung of *Amia calva*. Respir Physiol 59:213–223
Emilio MG, Shelton G (1972) Factors affecting blood flow to the lungs in the amphibian, *Xenopus laevis*. J Exp Biol 56:67–77
Evans BK, Shelton G (1984) Ventilation in *Xenopus laevis* after lung or carotid labyrinth denervation. In: Giles R (ed) First Congress of Comparative Physiology and Biochemistry, volume issued by the Eur Soc Comp Physiol Biochem, Liege, Belgium, pp A75
Farber J, Rahn H (1970) Gas exchange between air and water and the ventilation pattern in the electric eel. Respir Physiol 9:151–161
Farrell AP, Randall DJ (1978) Air-breathing mechanics in two Amazonian teleosts, *Arapaima gigas* and *Hoplerythrinus unitaeniatus*. Can J Zool 56:939–945
Feder ME, Burggren WW (1985) Cutaneous gas exchange in vertebrates: design, patterns, control and implications. Biol Rev 60:1–45
Feder ME, Pinder AW (1988) Ventilation and its effect on "infinite pool" exchangers. Am Zool 28:973–983
Fishman AP, DeLaney RG, Laurent P (1985) Circulatory adaptation to bimodal respiration in the dipnoan lungfish. J Appl Physiol 59:285–294

Fishman AP, Galante RJ, Pack AI (1989) Diving physiology: lungfish. In: Wood SC (ed) Comparative pulmonary physiology: current concepts, vol 39. Lung biology in health and disease. Marcel Dekker, New York, pp 645–676

Foxon GEH (1964) Blood and respiration. In: Moore JA (ed) Physiology of the Amphibia. Academic Press, New York, pp 151–209

Garey WF, Rahn H (1970) Normal arterial gas tensions and pH and the breathing frequency of the electric eel. Respir Physiol 9:141–150

Gee JH, Graham JB (1978) Respiratory and hydrostatic functions of the intestine of the catfishes *Hoplosternum thoracatum* and *Brochis splendens* (Callichthyidae). J Exp Biol 74:1–16

Girgis S (1961) Aquatic respiration in the common Nile turtle, *Trionyx triunguis* (Forskal). Comp Biochem Physiol 3:206–217

Glass ML, Johansen K (1976) Control of breathing in *Acrochordus javanicus*, an aquatic snake. Physiol Zool 49:328–340

Glass ML, Hicks JW, Riedesel ML (1979) Respiratory responses to long-term temperature exposure in the box turtle, *Terrapene ornata*. J Comp Physiol 131:353–359

Glass ML, Boutilier RG, Heisler N (1983) Ventilatory control of arterial PO_2 in the turtle, *Chrysemys picta bellii*: effects of temperature and hypoxia. J Comp Physiol 151:145–153

Glass ML, Ishimatsu A, Johansen K (1986) Responses of aerial ventilation to hypoxia and hypercapnia in *Channa argus*, an airbreathing fish. J Comp Physiol B 156:425–430

Graham JB (1973) Terrestrial life of the amphibious fish *Mnierpes macrocephalus*. Mar Biol (Berlin) 23:83–91

Graham JB (1974) Aquatic respiration in the sea snake *Pelamis platurus*. Respir Physiol 21:1–7

Graham JB (1976) Respiratory adaptations of marine air-breathing fishes. In: Hughes GM (ed) Respiration of amphibious vertebrates. Academic Press, New York, pp 165–187

Grahamm JB, Baird TA (1982) The transition to air breathing in fishes. I. Environmental effects on the facultative air breathing of *Ancistris chagresi* and *Hypostomus plecostomus* (Loricariidae). J Exp Biol 96:53–67

Graham JB, Kramer DL, Pineda E (1977) Respiration of the air-breathing fish *Piabucina festae*. J Comp Physiol 122:295–310

Guimond RW, Hutchison VH (1972) Pulmonary, branchial and cutaneous gas exchange in the mudpuppy, *Necturus maculosus maculosus* (Rafinesque). Comp Biochem Physiol 42A:367–392

Guimond RW, Hutchison VH (1973) Aquatic respiration: an unusual strategy in the hellbender *Cryptobranchus alleganiensis alleganiensis* (Daudin). Science 182:1263–1265

Guimond RW, Hutchison VH (1976) Gas exchange of the giant salamanders of North America. In: Hughes GM (ed) Respiration of amphibious vertebrates. Academic Press, New York, pp 313–338

Halliday TR, Sweatman HPA (1976) To breathe or not to breathe: the newt's problem. Anim Behav 24:551–561

Heath AG (1975) Respiratory responses to hypoxia by *Ambystoma tigrinum* larvae, paedomorphs, and metamorphosed adults. Comp Biochem Physiol 55A:45–49

Heatwole H, Seymour RS (1976) Respiration of marine snakes. In: Hughes GM (ed) Respiration of amphibious vertebrates. Academic Press, New York, pp 375–389

Heatwole H, Seymour RS (1978) Cutaneous oxygen uptake in three groups of aquatic snakes. Aust J Zool 26:481–486

Heisler N (1982) Intracellular and extracellular acid-base regulation in the tropical fresh-water teleosts *Symbranchus marmoratus* in response to the transition from water breathing to air breathing. J Exp Biol 99:9–28

Heming TA, Watson TA (1986) Activity and inhibition of carbonic anhydrase in *Amia calva*, a bimodal breathing holostean fish. J Fish Biol 28:385–392

Heming TA, Geers C, Gros G, Bidani A, Crandall ED (1986) Effects of dextran bound inhibitors on carbonic anhydrase activity in isolated rat lungs. J Appl Physiol 61:1849–1856

Hughes GM, Munshi JSD (1973) Nature of the air breathing organs of the Indian fishes *Channa*, *Amphipnous*, *Clarius* and *Saccobranchus* as shown by electron microscopy. J Zool Lond 170:245–270

Hughes GM, Singh BN (1971) Gas exchange with air and water in an air-breathing catfish *Saccobranchus fossilis*. J Exp Biol 53:281–298

Hutchison VH, Miller K (1979) Aerobic and anaerobic contributions to sustained activity in *Xenopus laevis*. Respir Physiol 38:93–103
Hutchison VH, Whitford WG, Kohl M (1968) Relation of body size and surface area to gas exchange in anurans. Physiol Zool 41:65–85
Hutchison VH, Haines HB, Engbretson G (1976) Aquatic life at high altitude: respiratory adaptations in the Lake Titicaca frog, *Telmatobius culeus*. Respir Physiol 27:115–129
Hyde DA, Moon TW, Perry SF (1987) Physiological consequences of prolonged aerial exposure in the American eel, *Anguilla rostrata*: blood respiratory and acid-base status. J Comp Physiol 157:635–642
Inger RF (1957) Ecological aspects of the origin of the tetrapods. Evolution 11:373–376
Ishii K, Ishii K (1976) The chemoreceptors of amphibians. In: Paintal AS (ed) Morphology and mechanisms of chemoreceptors. Vallabhbhai Patel Chest Inst, Delhi, India, pp 265–274
Ishii K, Kusakabe T (1982) The glomus cell of the carotid labyrinth of *Xenopus laevis*. Cell Tissue Res 224:459–463
Ishii K, Oosaki T (1969) Fine structure of the chemoreceptor cell in the amphibian carotid labyrinth. J Anat (Lond) 104:263–280
Ishii K, Honda K, Ishii K (1966) The function of the carotid labyrinth in the toad. Tohoku J Exp Med 88:103–116
Ishii K, Ishii K, Kusakabe T (1985) Chemo- and baroreceptor innervation of the aortic trunk of the toad *Bufo vulgaris*. Respir Physiol 60:365–375
Ishimatzu A, Itazawa Y (1981) Ventilation of the air-breathing organ in the snakehead *Channa argus*. Jpn J Ichthyol 28:276–282
Ishimatzu A, Itazawa Y (1983) Difference in blood O_2 levels in the outflow vessels of the heart of an air-breathing fish, *Channa argus*: do separate blood streams exist in a teleost heart. J Comp Physiol 149:435–440
Ishimatzu A, Itazawa Y, Takeda T (1979) On the circulatory systems of the snakeheads, *Channa maculata* and *C. argus* with reference to bimodal breathing. Jpn J Ichthyol 26:167–180
Ishimatzu A, Johansen K, Nilsson S (1986) Autonomic nervous control of the circulatory system in the air-breathing fish *Channa argus*. Comp Biochem Physiol 84C:55–60
Jackson DC (1973) Ventilatory response to hypoxia in turtles at various temperatures. Respir Physiol 18:178–187
Jackson DC (1976) Non-pulmonary CO_2 loss during diving in the turtle *Pseudemys scripta elegans*. Comp Biochem Physiol 55A:237–241
Jackson DC (1978) Respiratory control in air-breathing ectotherms. In: Davies DG, Barnes CD (eds) Regulation of ventilation and gas exchange. Academic Press, New York, pp 93–130
Jackson DC (1986) Acid-base regulation of reptiles. In: Heisler N (ed) Acid-base regulation in animals. Elsevier Science, Amsterdam, pp 235–263
Jackson DC, Braun BA (1979) Respiratory control in bullfrogs: cutaneous versus pulmonary response to selective CO_2 exposure. J Comp Physiol 129:339–342
Jackson DC, Allen J, Strupp PK (1976) The contribution of non-pulmonary surfaces to CO_2 loss in six species of turtles at 20 °C. Comp Biochem Physiol 55A:243–246
Jesse MJ, Shub C, Fishman AP (1967) Lung and gill ventilation of the African lung fish. Respir Physiol 3:267–287
Johansen K (1966) Airbreathing in the teleost *Synbranchus marmoratus*. Comp Biochem Physiol 18:383–395
Johansen K (1968) Air breathing fishes. Sci Am 219:102–111
Johansen K (1970) Air breathing in fishes. In: Hoar WS, Randall DJ (eds) Fish physiology, vol IV. Academic Press, New York, pp 361–411
Johansen K (1982) Blood circulation and the rise of air-breathing: passes and bipasses. In: Taylor CR, Johansen K, Bolis L (eds) A companion to animal physiology. Cambridge University Press, Cambridge, pp 91–105
Johansen K (1985) A phylogenetic overview of cardiovascular shunts. In: Johansen K, Burggren WW (eds) Cardiovascular shunts, Alfred Benzon Symposium 21. Munksgaard, Copenhagen pp 17–32
Johansen K, Burggren WW (1980) Cardiovascular function in the lower vertebrates. In: Bourne G (ed) Hearts and heart-like organs, vol 1. Academic Press, New York, pp 61–117

Johansen K, Lenfant C (1967) Respiratory function in the South American lungfish. J Exp Biol 46:205–218

Johansen K, Lenfant C (1968) Respiration in the African lungfish, *Protopterus aethiopicus*. II. Control of breathing. J Exp Biol 49:453–468

Johansen K, Lenfant C, Grigg GF (1967) Respiratory control in the lungfish *Neoceratodus forsteri* (Krefft). Comp Biochem Physiol 20:835–854

Johansen K, Lenfant C, Hanson D (1968a) Cardiovascular dynamics in the lungfishes. Z Vergl Physiol 59:157–186

Johansen K, Lenfant C, Schmidt-Nielsen K, Peterson JA (1968b) Gas exchange and control of breathing in the electric eel *Electrophorus electricus*. Z Vgl Physiol 61:137–163

Johansen K, Hanson D, Lenfant C (1970a) Respiration in a primitive air breather *Amia calva*. Respir Physiol 9:162–174

Johansen K, Lenfant C, Hanson D (1970b) Phylogenetic development of pulmonary circulation. Fed Proc Fed Am Soc Exp Biol 29:1135–1140

Jones DR, Chu C (1988) Effect of denervation of carotid labyrinths on breathing in unrestrained *Xenopus laevis*. Respir Physiol 73:243–256

Jones DR, Milsom WK (1982) Peripheral receptors affecting breathing and cardiovascular function in non-mammalian vertebrates. J Exp Biol 100:59–91

Jones RM (1982) How toads breathe: control of air flow to and from the lungs by the nares in *Bufo marinus*. Respir Physiol 49:251–265

Jordan J (1976) The influence of body weight on gas exchange in the air-breathing fish, *Clarius batrachus*. Comp Biochem Physiol 53A:305–310

Kern DM (1960) The hydration of carbon dioxide. J Chem Educ 37:14–23

Kramer DL (1978) Ventilation of the respiratory gas bladder in *Hoplerythrinus unitaeniatus* (Pisces, Characoidei, Erythrinidae). Can J Zool 56:931–938

Kramer DL (1983) The evolutionary ecology of respiratory mode in fishes: an analysis based on the costs of breathing. Environ Biol Fish 9:145–158

Kramer DL (1988) The behavioral ecology of air breathing by aquatic animals. Can J Zool 66:89–94

Kramer DL, McClure M (1981) The transit cost of aerial respiration in the catfish, *Corydoras aeneus* (Callichthyidae). Physiol Zool 54:189–194

Kruhøffer M, Glass ML, Abe AS, Johansen K (1987) Control of breathing in an amphibian *Bufo paracnemius*: effects of temperature and hypoxia. Respir Physiol 69:267–275

Kuhlmann WD, Fedde MR (1979) Intrapulmonary receptors in the bullfrog: sensitivity to CO_2. J Comp Physiol 132:69–75

Lahiri S, Szidon JP, Fishman AP (1970) Potential respiratory and circulatory adjustments to hypoxia in the African lungfish. Fed Proc 29:1141–1148

Laurent P, DeLaney RG, Fishman AP (1978) The vasculature of the gills in the aquatic and estivating lungfish, *Protopterus aethiopicus*. J Morphol 156:173–208

Legler JM, Cann J (1980) A new genus and species of chelid turtle from Queensland, Australia. Contrib Sci Nat Hist Mus Los Angeles County 324:1–18

Lenfant C, Johansen K (1967) Respiratory adaptations in selected amphibians. Respir Physiol 2:247–260

Lenfant C, Johansen K (1968) Respiration in an African lungfish, *Protopterus aethiopicus*: respiratory properties of blood and normal patterns of breathing and gas exchange. J Exp Biol 49:437–457

Lenfant C, Johansen K, Grigg GC (1966) Respiratory properties of blood and pattern of gas exchange in the lungfish *Neoceratodus forsteri* (Krefft). Respir Physiol 2:1–21

Lenfant C, Johansen K, Hanson D (1970) Bimodal gas exchange and ventilation-perfusison relationship in lower vertebrates. Fed Proc 29:1124–1129

Liem KF (1967) Functional morphology of the integumentary respiratory and digestive systems of the synbranchoid fish *Monopterus albus*. Copeia 1967:375–388

Liem KF (1980) Air ventilation in advanced teleosts: biomechanical and evolutionary aspects. In: Ali MA (ed) Environmental physiology of fishes. Plenum, New York, pp 57–91

Liem KF (1984) The muscular basis of aquatic and aerial ventilation in the air-breathing teleost fish *Channa*. J Exp Biol 113:1–18

Liem KF (1987) Functional design of the air ventilation apparatus and overland excursions by teleosts. Fieldiana Zool, New Ser 37:1–29

Lillo RS (1980) Localisation of chemoreceptors which may cause diving bradycardia in bullfrogs. Can J Zool 52:931–936

Lomholt JP, Johansen K (1974) Control of breathing in *Amphipnous cuchia*, an amphibious fish. Respir Physiol 21:325–340

Lomholt JP, Johansen K (1976) Gas exchange in the amphibious fish, *Amphipnous cuchia*. J Comp Physiol 107:141–157

Macintyre DH (1975) Respiratory mechanisms and acid-base regulation in *Bufo marinus marinus* (L.). M Sc dissertation, Acadia University, Canada

Macintyre DH, Toews DP (1976) The mechanics of lung ventilation and the effects of hypercapnia on respiration in the toad, *Bufo marinus*. Can J Zool 54:1364–1374

Mackenzie JA, Jackson DC (1978) The effect of temperature on cutaneous CO_2 loss and conductance in the bullfrog. Respir Physiol 32:313–323

Malvin GM (1985a) Vascular resistance and vasoactivity of gills and pulmonary artery of the salamander, *Ambystoma tigrinum*. J Comp Physiol 155:241–249

Malvin GM (1985b) Adrenoceptor types in the respiratory vasculature of the salamander gill. J Comp Physiol 155:591–596

Malvin GM (1985c) Cardiovascular shunting during amphibian metamorphosis. In: Johansen K, Burggren WW (eds) Cardiovascular shunts. Alfred Benzon Symposium 21. Munskgaard, Copenhagen, pp 163–178

Malvin GM (1988) Microvascular regulation of cutaneous gas exchange in amphibians. Am Zool 28:999–1007

Malvin GM (1989) Gill structure and function: Amphibian larvae. In: Wood SC (ed) Comparative pulmonary physiology: current concepts, vol 39. Lung biology in health and disease. Marcel Dekker, New York, pp 121–151

Malvin GM, Boutilier RG (1985) Ventilation-perfusion relationships in Amphibia. In: Giles R (ed) Circulation, respiration and metabolism. Springer, Berlin Heidelberg New York Tokyo, pp 114–124

Malvin GM, Hlastala MP (1986) Regulation of cutaneous gas exchange by environmental O_2 and CO_2 in the frog. Respir Physiol 65:99–111

Maren TH (1967) Carbonic anhydrase: chemistry, physiology, and inhibition. Physiol Rev 47:595–781

Martin KM, Hutchison VH (1979) Ventilatory activity in *Amphiuma tridactylum* and *Siren lacertina* (Amphibia, Caudata). J Herpetol 13:427–434

McDonald DG, Boutilier RG, Toews DP (1980) The effects of enforced activity on ventilation, circulation and blood acid-base balance in the semiterrestrial anuran, *Bufo marinus*. J Exp Biol 84:273–287

McKean T (1969) A linear approximation of the transfer function of pulmonary mechanoreceptors in the frog. J Appl Physiol 27:775–781

McMahon BR (1969) A functional analysis of the aquatic and aerial respiratory movements of an African lungfish, *Protopterus aethiopicus*, with reference to the evolution of the lung-ventilation mechanism in the vertebrates. J Exp Biol 51:407–430

McMahon BR, Burggren WW (1987) Respiratory physiology of intestinal air breathing in the teleost fish *Misgurnus anguillicaudatus*. J Exp Biol 133:371–393

Miller K, Hutchison VH (1979) Activity metabolism in the mudpuppy, *Necturus maculosus*. Physiol Zool 52:22–37

Milsom WK, Jones DR (1977) Carbon dioxide sensitivity of pulmonary receptors in the frog. Experientia 33:1167–1168

Milsom WK, Jones DR (1985) Characteristics of mechanoreceptors in the air-breathing organ of the holostean fish, *Amia calva*. J Exp Biol 117:389–399

Munshi JSD (1968) The accessory respiratory organs of *Anabas testudineus* (B.L.) Anabantidae Pisces. Proc Linn Soc Lond (Zool) 179:107–126

Munshi JSD, Singh BN (1968) On the respiratory organs of *Amphipnous cuchia* (Ham. Buch). J Morphol 124:423–444

Nikinmaa M, Railo E (1987) Anion movements across lamprey (*Lampetra fluviatilis*) red cell membrane. Biochim Biophys Acta 899:134–136

Nikinmaa M, Tufts BL (1989) Regulation of acid and ion transfer across the membrane of nucleated erythrocytes. Can J Zool 67 (in press)

Pack AI, Galante R, Fishman AP (1984) Breuer-Hering reflexes in the African lungfish (*Protopterus annectens*). Fed Proc Fed Am Soc Exp Biol 43:433 (Abstr)

Packard GC (1974) The evolution of air-breathing in Paleozoic gnathostome fishes. Evolution 28:320–325

Peters HM (1978) On the mechanism of air ventilation in anabantoids (Pisces: Teleostei). Zoomorphologie 89:93–123

Pettit MJ, Beitinger T (1985) Oxygen acquisition of the reedfish, *Erpetoichthys calabaracus*. J Exp Biol 114:289–306

Piiper J (1982) Respiratory gas exchange at lungs, gills and tissues: mechanisms and adjustments. J Exp Biol 100:5–22

Piiper J (1988) Models for cutaneous gas exchange and transport. Am Zool 28:963–972

Piiper J, Scheid P (1977) Comparative physiology of respiration: Functional analysis of gas exchange organs in vertebrates. In: Widdicombe JG (ed) International review of physiology, vol 14. Respiratory physiology, II. University Park Press, Baltimore, pp 219–253

Piiper J, Dejours P, Haab P, Rahn H (1971) Concepts and basic quantities in gas exchange physiology. Respir Physiol 13:292–304

Piiper J, Gatz RN, Crawford EC Jr (1976) Gas transport characteristics in an exclusively skin-breathing salamander, *Desmognathus fuscus* (Plethodontidae). In: Hughes GM (ed) Respiration in amphibious vertebrates. Academic Press, New York, pp 339–356

Pinder AW (1987) Cutaneous diffusing capacity increases during hypoxia in cold, submerged bullfrogs (*Rana catesbeiana*). Respir Physiol 70:85–95

Pinder AW, Burggren WW (1986) Ventilation and partitioning of oxygen uptake in the frog *Rana pipiens*: effects of hypoxia and activity. J Exp Biol 126:453–468

Poullet M (1977) A study of nerve endings in the carotid labyrinth of the toad (*Bufo bufo*) after degeneration. Biol Cell 28:75–80

Rahn H (1966) Aquatic gas exchange: theory. Respir Physiol 1:1–12

Rahn H (1967) Gas transport from the external environment of the cell. In: DeReuck AUS, Porter R (eds) Development of the lung. Ciba Found Symp, Churchill, London, pp 3–23

Rahn H, Howell BJ (1976) Bimodal gas exchange. In: Hughes GM (ed) Respiration of amphibious vertebrates. Academic Press, New York, pp 271–285

Rahn H, Rahn KB, Howell BJ, Gans C, Tenney SM (1971) Air-breathing of the garfish (*Lepisosteus osseus*). Respir Physiol 11:285–307

Randall DJ (1982a) The control of respiration and circulation in fish during exercise and hypoxia. J Exp Biol 100:275–288

Randall DJ (1982b) Blood flow through gills. In: Houlihan DF, Rankin JC, Shuttleworth TJ (eds) Gills. Soc Exp Biol Sem Ser 16, Cambridge University Press, Cambridge, pp 173–191

Randall DJ, Farrell AP, Haswell MS (1978a) Carbon dioxide excretion in the jeju *Hoplerythrinus unitaeniatus*, a facultative air-breathing teleost. Can J Zool 56:970–973

Randall DJ, Farrell AP, Haswell MS (1978b) Carbon dioxide excretion in the pirarucu (*Arapaima gigas*), an obligate air breathing fish. Can J Zool 56:977–982

Randall DJ, Burggren WW, Farrell AP, Haswell MS (1981) The evolution of air-breathing in vertebrates. Cambridge University Press, Cambridge

Rao HS, Hora LS (1938) On the ecology, bionomics and systematics of the blennid fishes of the genus *Andamia* Blyth. Rec Indian Mus 40:377–401

Rauch JC (1978) Integumentary blood vascular system in garter snakes (*Thamnophis sirtalis parietalis* and *Thamnophis radix*). Can J Zool 56:469–476

Rogers DC (1966) A histological and histochemical study of the carotid labyrinth in the anuran amphibians, *Bufo marinus*, *Hyla aurea* and *Neobatrachus pictus*. Acta Anat 63:249–280

Root RW (1949) Aquatic respiration in the musk turtle. Physiol Zool 22:172–178

Rosen S, Friedley NJ (1973) Carbonic anhydrase activity in *Rana pipiens* skin: biochemical and histochemical analysis. Histochemie 36:1–4

Rosenberg HI, Voris HK (1980) Cutaneous capillaries of sea snakes and their possible role in gas exchange. Am Zool 20:758

Sacca R, Burggren WW (1982) Oxygen uptake in air and water in the air-breathing reedfish *Calamoichthys calabaricus*: role of skin, gills and lungs. J Exp Biol 97:179–186

Scheid P (1982) A model for comparing gas-exchange systems in vertebrates. In: Taylor CR, Johansen K, Bolis L (eds) A companion to animal physiology. Cambridge University Press, Cambridge, pp 3–16
Schottle E (1932) Morphologie und Physiologie der Atmung bei wasser-, schlamm- und landlebenden Gobiiformes. Z Wiss Zool 140:1–114
Sewertzoff AN (1924) Die Entwicklung der Kiemen und Kiemenbogengefäße der Fische. Z Wiss Zool 121:494–556
Seymour RS (1982) Physiological adaptations to aquatic life. In: Gans C, Pough FH (eds) Biology of the Reptilia: physiological ecology, vol 13. Academic Press, New York, pp 1–51
Seymour RS (1989) Diving physiology: reptiles. In: Wood SC (ed) Comparative pulmonary physiology: current concepts, vol 39. Lung biology in health and disease. Marcel Dekker, New York, pp 677–695
Shannon P, Kramer DL (1988) Water depth alters respiratory behaviour of *Xenopus laevis*. J Exp Biol 137:597–602
Shelton G (1970) The effect of lung ventilation on blood flow to the lungs and body of the amphibian, *Xenopus laevis*. Respir Physiol 9:183–196
Shelton G (1985) Functional and evolutionary significance of cardiovascular shunts in the Amphibia. In: Johansen K, Burggren WW (eds) Cardiovascular shunts. Alfred Benzon Symposium 21. Munksgaard, Copenhagen, pp 100–120
Shelton G, Boutilier RG (1982) Apnoea in amphibians and reptiles. In: Butler PJ (ed) Control and co-ordination of respiration and circulation. J Exp Biol 100:245–273
Shelton G, Jones DR, Milsom WK (1986) Control of breathing in ectothermic vertebrates. In: Cherniack NS, Widdicombe JG (eds) Handbook of physiology, Sect 3. The respiratory system, vol 2. Control of breathing. American Physiological Society, Bethesda, MD, pp 857–909
Shield JW, Bentley PJ (1973) Respiration of some urodele and anuran amphibia. I. In water, role of the skin and gills. Comp Biochem Physiol 46A:17–28
Singh BN (1976) Balance between aquatic and aerial respiration. In: Hughes GM (ed) Respiration of amphibious vertebrates. Academic Press, New York, pp 125–164
Singh BN, Hughes GM (1971) Respiration of an air-breathing catfish *Clarias batrachus* (Linn). J Exp Biol 55:421–434
Singh BN, Hughes GM (1973) Cardiac and respiratory responses in the climbing perch *Anabas testudineus*. J Comp Physiol 84:205–226
Smatresk NJ (1986) Ventilatory and cardiovascular responses to hypoxia and NaCN in *Lepisosteus osseus*, an air-breathing fish. Physiol Zool 59:385–397
Smatresk NJ (1988) Control of the respiratory mode in air-breathing fishes. Can J Zool 66:144–151
Smatresk NJ, Azizi SQ (1987) Characteristics of lung mechanoreceptors in spotted gar, *Lepisosteus oculatus*. Am J Physiol 252:R1066–R1072
Smatresk NJ, Cameron JN (1982a) Respiration and acid-base physiology of the spotted gar, a bimodal breather. I. Normal values and the response to severe hypoxia. J Exp Biol 96:263–280
Smatresk NJ, Cameron JN (1982b) Respiration and acid-base physiology of the spotted gar, a bimodal breather. II. Responses to temperature change and hypercapnia. J Exp Biol 96:281–293
Smatresk NJ, Burleson ML, Azizi SQ (1986) Chemoreflexive responses to hypoxia and NaCN in longnose gar: evidence for two chemoreceptor loci. Am J Physiol 251:R116–R125
Smith DG (1976) The innervation of the cutaneous artery in the toad, *Bufo marinus*. Gen Pharmacol 7:404–409
Smith DG, Rapson L (1977) Differences in pulmonary microvascular anatomy between *Bufo marinus* and *Xenopus laevis*. Cell Tissue Res 178:1–15
Smyth DH (1939) The central and reflex control of respiration in the frog. J Physiol 95:305–327
Standaert T, Johansen K (1974) Cutaneous gas exchange in snakes. J Comp Physiol 89:313–320
Stevens ED, Holeton GF (1978) The partitioning of oxygen uptake from air and from water by erythrinids. Can J Zool 56:965–969
Taglietti V, Casella C (1966) Stretch receptors stimulation in frogs' lungs. Pflügers Arch 292:297–308
Taglietti V, Casella C (1968) Deflation receptors in frogs' lungs. Pflügers Arch 304:81–89

Tamura SO, Morii H, Yuzuriha M (1976) Respiration of the amphibious fishes, *Periophthalmus cantonensis* and *Boleophthalmus chinensis* in water and on land. J Exp Biol 65:97–107
Todd ES, Ebeling AW (1966) Aerial respiration in the long jaw mud-sucker *Gillichthys mirabilis* (teleostei: Gobiidae). Biol Bull 130:256–288
Toews DP (1971) Factors affecting the onset and termination of ventilation in the salamander, *Amphiuma tridactylum*. Can J Zool 49:1231–1237
Toews DP, Boutilier RG (1986) Acid-base regulation in the Amphibia. In: Heisler N (ed) Acid-base regulation in animals. Elsevier, Amsterdam, pp 265–308
Toews DP, Kirby S (1985) The ventilatory and acid-base physiology of the toad, *Bufo marinus*, during exposure to environmental hyperoxia. Respir Physiol 59:225–230
Toews DP, Macintyre D (1978) Respiration and circulation in an apodan amphibian. Can J Zool 56:998–1004
Toews D Boutilier R, Todd L, Fuller N (1978) Carbonic anhydrase in the Amphibia. Comp Biochem Physiol 59A:211–213
Toews DP, Shelton G, Randall DJ (1971) Gas tensions in the lungs and major blood vessels of the urodele amphibian, *Amphiuma tridactylum*. J Exp Biol 55:47–61
Tufts BL, Boutilier RG (1989) The absence of rapid chloride/bicarbonate exchange in lamprey erythrocytes: implications for CO_2 transport and ion distributions between plasma and erythrocytes in the blood of *Petromyzon marinus*. J Exp Biol 144:565–576
Van Vliet BN, West NH (1986) Cardiovascular responses to electrical stimulation of the recurrent laryngeal nerve in conscious toads (*Bufo marinus*). J Comp Physiol B156:363–375
Wasserzug RJ, Paul RD, Feder ME (1981) Cardiorespiratory syhchrony in anuran larvae (*Xenopus laevis*, *Pachymedusa dacnicolor*, and *Rana berlandieri*). Comp Biochem Physiol 70A:329–334
West NH, Burggren WW (1982) Gill and lung ventilatory responses to steady-state aquatic hypoxia and hyperoxia in the bullfrog tadpole. Respir Physiol 47:165–176
West NH, Burggren WW (1983) Reflex interactions between aerial and aquatic gas exchange organs in larval bullfrogs. Am J Physiol 244:R770–R777
West NH, Burggren WW (1984) Control of pulmonary and cutaneous blood flow in the toad, *Bufo marinus*. Am J Physiol 247:R884–R894
West NH, Jones DR (1975) Breathing movements in the frog *Rana pipiens*. I. The mechanical events associated with lung and buccal ventilation. Can J Zool 53:332–344
West NH, Topor ZL, Van Vliet BN (1987) Hypoxemic threshold for lung ventilation in the toad. Respir Physiol 70:377–390
White FN (1989) Carbon dioxide homeostasis. In: Wood SC (ed) Comparative pulmonary physiology: current concepts, vol 39. Lung biology in health and disease. Marcel Dekker, New York, pp 439–466
Winokur RM (1973) Adaptive modifications of buccal mucosae in turtles. Am Zool 13:1347–1348
Wood SC (1982) The effect of oxygen affinity on arterial PO_2 in animals with central vascular shunts. J Appl Physiol 53:1360–1364
Wood SC (1984) Cardiovascular shunts and oxygen transport in lower vertebrates. Am J Physiol 16:3–14
Wood SC, Hicks JW (1985) Oxygen homeostasis in vertebrates with cardiovascular shunts. In: Johansen K, Burggren WW (eds) Cardiovascular shunts. Alfred Benzon Symposium 21: Munksgaard, Copenhagen, pp 354–366
Wood SC, Lenfant CJM (1976) Respiration: mechanics, control, and gas exchange. In: Gans C, Dawson WR (eds) Biology of the Reptilia, vol 5. Physiology. Academic Press, New York, pp 225–274
Wright WG, Raymond JA (1978) Air-breathing in a California sculpin. J Exp Zool 203:171–176
Wright PA, Heming TA, Randall DJ (1986) Downstream pH changes in water flowing over the gills of rainbow trout. J Exp Biol 126:499–512
Yu K, Woo NY (1985) Effects of ambient oxygen tension and temperature on the bimodal respiration of an air-breathing teleost, *Channa maculata*. Physiol Zool 58:181–189
Zander CD (1972) Beziehungen zwischen Körperbau und Lebensweise bei Blenniidae (Pisces) aus dem Roten Meer. I. Äußere Morphologie. Mar Biol 13:238–246

Chapter 10

Control and Co-Ordination of Gas Exchange in Air Breathers

W. K. Milsom[1]

Contents

1 Introduction

A wealth of information exists on the control and co-ordination of gas exchange in air-breathing vertebrates. Fortunately, much of this literature has been reviewed, in depth, in the last few years[2]. These reviews give an excellent summary of existing knowledge and, consequently, the emphasis of the present chapter has been placed on a selective review of some of the more recent advances which have been made in the field, particularly emphasizing those areas about which little is yet known but which are now ripe for further study.

[1] Department of Zoology, University of British Columbia, Vancouver, British Columbia, Canada V6T 2A9

[2] Recent reviews include *(1) Mammals:* Fishman et al. (1986), Von Euler and Langercrantz (1986), Hornbein (1981), Feldman (1986), Long and Duffin (1986), *(2) Birds:* Sellers (1987), Piiper (1978), Scheid and Piiper (1986), *(3) Reptiles:* Wood (1989), Shelton et al. (1986), Glass and Wood (1983)

Advances in Comparative and Environmental Physiology
Vol. 6. Edited by R. G. Boutilier

The vast majority of studies on the control and co-ordination of gas exchange in air-breathing vertebrates have been conducted on a small handful of mammalian species. Furthermore, of those studies which have been conducted on other species, the vast majority have focused on stimulus/response relationships and little is yet known of the underlying mechanisms which connect stimulus to response. Such studies, however, have the potential to advance this field enormously. It has been recently argued that the various breathing patterns and ventilatory responses seen throughout all the air-breathing vertebrates are manifestations of a common control system (Ballintijn 1988; Milsom 1988). If so, any model which attempts to explain the control and co-ordination of gas exchange in mammals, including man, must be able to explain the various breathing patterns and ventilatory responses seen in these other groups as well. As a consequence, the periodic patterns of breathing seen throughout the reptiles and in mammals under conditions of reduced metabolic rate (such as hibernation), as well as the variations in ventilatory responses to hypoxia, hypercapnia, exercise etc. seen throughout the air-breathing vertebrates, may offer insights into the underlying control mechanisms regulating gas exchange, which might not be as apparent in studies on more classical mammals which exhibit much tighter 'apparent' homeostatic control. Furthermore, they most certainly offer much greater insight into the adaptability of the respiratory system for the control and co-ordination of gas exchange.

2 Central Rhythm Generators

The production, integration and optimization of respiratory movements in all air-breathing vertebrates results from the central processing of a wide variety of sensory inputs from different receptor groups. Certain inputs adjust the respiratory pattern so that appropriate ventilation is maintained for blood gas and acid-base homeostasis while other inputs integrate respiratory movements with other body movements such as feeding, vocalization, postural changes and locomotion. Yet other inputs may optimize the breathing pattern for minimal energy expenditure. These inputs are processed by a central pattern generator which determines the spatio-temporal output to the respiratory muscles (see Handbook of Physiology, Sect. 3, vol. 2 for reviews). It is still widely debated, however, whether this pattern generator also receives input from a central ryhythm generator. Is the rhythmic alternation of inspiration and expiration simply the result of the integration of peripheral inputs or is this the inherent property of a rhytm generator whose output characteristics (timing and burst generation) are modulated by peripheral inputs at a central pattern generator?

2.1 Continuous Breathers: Mammals, Birds and Fish

In vertebrate which express a continuous breathing pattern under resting conditions (mammals, birds and fishes), respiratory rhythm of a normal character per-

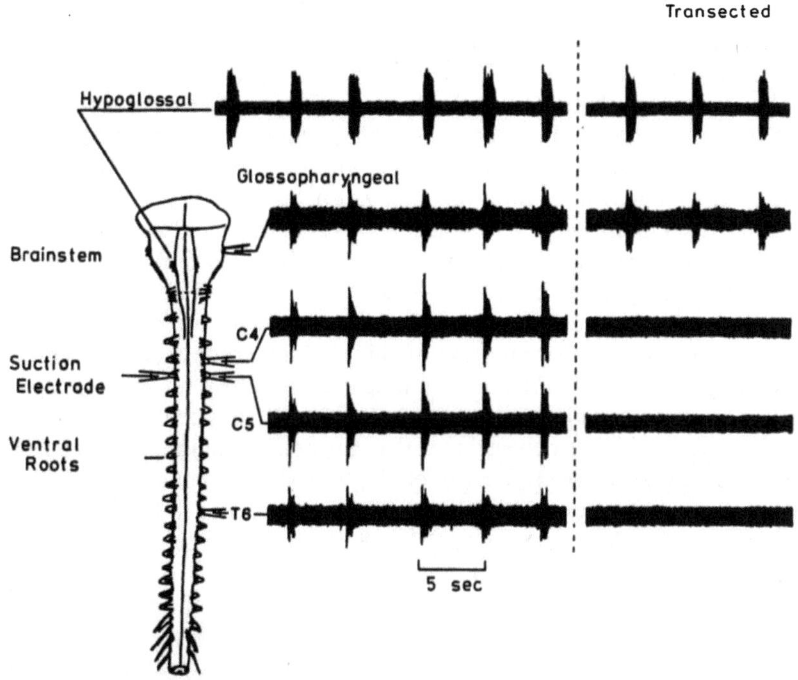

Fig. 1. Left: Neonatal rat spinal cord-brainstem preparation with suction electrode recordings from various nerve roots (*right*). Transection of the neuraxis at the spinomedullary junction abolishes spinal motoneuronal activity related to respiration, whereas cranial motoneuronal activity remains unaffected (Feldman et al. 1988)

sists following decerebration and decerebellation. Subsequent transection of the spino-medullary junction permanently abolishes respiratory rhythmic discharge of spinal motor nerves but does not markedly affect the respiratory activity of the cranial nerves (Hukuhara and Okada 1956; Rovainen 1974; Hukuhara 1976; Kawasaki 1979; St. John et al. 1981; Ballintijn 1982; Shelton et al. 1986). Similar results have also been produced in vitro (Adrian and Buytendijk 1931; Rovainen 1977; Suzue 1984; Russell 1986; Smith and Feldman 1987). Recordings from isolated brainstem-spinal cord preparations from lampreys and neonatal rats reveal rhythmic discharges in cranial and spinal respiratory motoneurons. Transections at the spino-medullary junctions in these preparations also abolish rhythmic respiratory discharge in spinal nerves without markedly affecting the rhythmic discharge of cranial nerves (Fig. 1). Such preparations are devoid of all afferent sensory feedback except potentially from the central CO_2 chemoreceptors. The sum of this evidence is taken to indicate that there is a separate site of respiratory rhythmogenesis in the brainstem, which functions independent of external input.

Phillipson and co-workers (Sullivan et al. 1978; Phillipson et al. 1981), however, found that if they maintained normal values of arterial P_{CO_2}, P_{O_2} and pH in sheep with an external membrane lung, removal of CO_2 from venous blood at a rate equal to its rate of metabolic production resulted in complete cessation of breathing. Similar results have been produced using the technique of artificial, unidirectional lung ventilation in birds (Peterson and Fedde 1968; Ray and Fedde 1969; Fedde and Peterson 1970). Such studies suggest that, in unaesthetized animals, respiratory rhythm is critically dependent on afferent respiratory stimuli. One interpretation of this data is that the respiratory neurons in the brainstem lack inherent automaticity (Phillipson et al. 1981; Kunz 1987). The concept of a central rhythm generator, however, does not necessarily imply that the respiratory neurons of the brainstem can by themselves generate an effective respiratory output. They may operate at a subthreshold level which requires some external stimulus to trigger respiratory events. In both cases mentioned above, a tonic respiratory input (such as elevated P_{CO_2}) is required to produce a phasic (rhythmic) respiratory output. It is the production of this rhythmic output by the central pattern generator, particularly in the in vitro preparations where no sensory afferent feedback other than a tonic input potentially arising from central CO_2 chemoreceptors is present, which is believed to reflect the additional input of a central rhythm generator (Kawasaki 1984; Shelton et al. 1986; Feldman et al. 1988). Similar conclusions have been drawn from parallel lines of evidence in birds and fish (Scheid and Piiper 1986; Davey and Seller 1987; Shelton 1970; Shelton et al. 1986; Ballintijn 1988).

Such neural rhythm may be explicable in terms of the membrane properties of special groups of cells (pacemaker cells) or may only emerge as a dynamic property of a complex neural network. Considerable interest has focused on the neural basis of respiratory rhythmogenesis and several mechanisms and models have been proposed (see Rovainen 1977, Kawasaki 1984, Feldman et al. 1988 for reviews). To date, however, no evidence exists to support the presence of pacemaker cells among the respiratory neurons in the brainstem, and arguments can be put forward against all mechanisms which have been suggested as contributing events of neural networks (Feldman et al. 1988). At present, therefore, these models are only useful in providing a framework in which to interpret data and for designing further experiments, as there is insufficient evidence to test such models adequately.

2.2 Periodic Breathers: Amphibia and Reptiles

Just as the presence of rhythmic ventilation has been used to argue the presence of a central rhythm generator in fish, birds and mammals, the periodic nature of ventilation in amphibians and reptiles has been used to argue its absence in these groups (Shelton et al. 1986). This is further supported by observations that artificial ventilation sufficient to maintain blood gases and pH at normal levels will completely suppress spontaneous ventilation (Kinney and White 1977). But, although pattern generation in amphibia and reptiles is more closely dependent on both peripheral receptor systems and higher centers than appears to be the case in fish, birds and mammals, two lines of evidence argue that the basic mechanisms underlying central pattern generation are common to all groups. The first is that, in these animals, changes in respiratory drive primarily result in changes in tidal volume and in the length of the interbreath interval or non-ventilatory period (T_{NVP}). The length of periods of inspiration and expiration change very little (Milsom 1988). This suggests that although peripheral inputs serve to regulate V_T and T_{NVP}, the timing of the respiratory events is under a separate (central) control. The second line of evidence comes from studies of hibernating mammals. During hibernation, mammals exhibit periodic breathing patterns which closely resemble those seen in reptiles, with the exception that the interbreath interval in mammals occurs at end-expiration whereas in reptiles it occurs at end-inspiration. All data collected to date, however, indicate that the control of the periodic breathing patterns seen in hibernating animals is identical to that seen in reptiles (Milsom 1988). It has been suggested that these similarities indicate that a common mechanism underlies the various arrhythmic breathing patterns seen under physiological conditions in all air-breathing vertebrates. Furthermore, the conversion of continuous breathing to periodic breathing in mammals entering hibernation suggests that both continuous and periodic breathing are manifestations of a common control system (Milsom 1988). These conclusions are further supported by observations that many species of fish which normally exhibit continuous ventilation will switch to periodic breathing under conditions of low metabolic rate in well-oxygenated water. Under these conditions, the control of this periodic breathing pattern is again similar to that seen in other vertebrate groups (Lomholt and Johansen 1979; Juch and Luiten 1981; Juch and Ballintijn 1983; Ballintijn 1988). These data further suggest that the basic mechanisms underlying the central nervous control of respiration are uniform throughout all vertebrate groups.

In summary, there is insufficient data at present to distinguish between three possible hypotheses regarding central rhythm generation. The first is that all vertebrates possess a central rhythm generator (either a pacemaker or a central neuronal network) which usually operates at subthreshold levels requiring different levels of peripheral or higher central input to bring to threshold, and which may be manifested as continuous or periodic motor outputs, depending on the overall level of respiratory drive. The second is that in all vertebrates, all aspects of respiratory pattern simply result from the integration of peripheral inputs. The third is that underlying control mechanisms are not common to all vertebrates and that central rhythm generators are present in animals which normally exhibit continuous breathing patterns (fish, birds and mammals) but not those which breathe in a periodic fashion (amphibia and reptiles). At present, based on recent evidence, the first of these hypotheses provides the most convincing and parsimonious explanation of the existing data, but much evidence is still required to elevate any of these possibilities above the level of speculation.

3 Feedforward and Feedback Control

3.1 Steady-State Relations Between Stimuli and Responses

3.1.1 The Feedback Loop

The primary function of the respiratory system in controlling and co-ordinating gas exchange is to maintain arterial blood gases and pH within 'homeostatic limits'. Figure 2 illustrates the scheme of the respiratory control system which I will use to discuss current understanding of the mechanisms involved in this regulation. On the right are the aspects of the control system associated with gas exchange, while on the left are those aspects associated with the production of the motor output to the gas exchanger. As mentioned earlier, the central pattern generator receives inputs from various receptor groups, which allow it to adjust the respiratory pattern to maintain blood gas and acid-base homeostasis (chemoreceptors), to integrate respiratory movements with other activities (higher centers and other, non-respiratory, inputs) and to optimize the respiratory pattern (respiratory muscles and gas exchanger). As discussed earlier, it may also receive input from a central rhythm generator. These inputs are processed to determine the spatio-temporal output to the respiratory muscles. These give rise to the effective ($\dot{V}_{Eff}$) and wasted, dead space ventilation ($\dot{V}_D$). Information related to overall lung movements is relayed back to the brainstem from receptors in the lungs, respiratory muscles and usually the chest wall, while information about the adequacy of the effective ventilation is provided by various chemoreceptor groups.

3.1.2 The Hypoxic Stimulus

Magnitude and Variability of the Hypoxic Ventilatory Response

There is abundant literature documenting the ventilatory responses of air-breathing vertebrates to hypoxia. Invariably, the relationship between ventilation and arterial P_{O2} is relatively flat over the physiological range, but substantial increases in ventilation do occur as hypoxia becomes more severe. The amount of variabil-

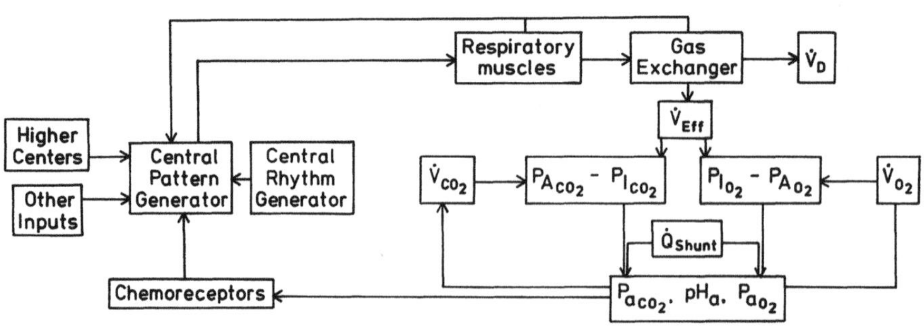

Fig. 2. Schematic diagram of respiratory control system. See text, for details

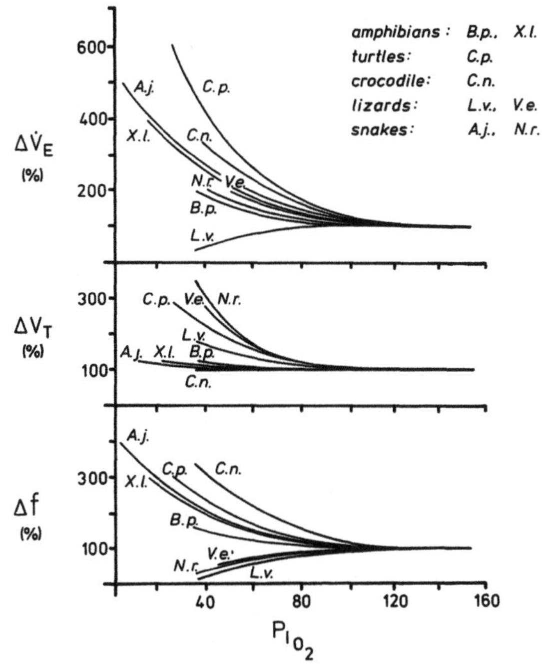

Fig. 3. Relationship between the percent change in minute ventilation ($\Delta\dot{V}_E$), tidal volume ($\Delta\dot{V}_T$) and breathing frequency (Δf) and the partial pressure of O_2 in inspired air ($P_{I_{O_2}}$) in various species of amphibian and reptile. Sources of data: *B.p.* = *Bufo paracnemius* (Kruhøffer et al. 1987); *X.l.* = *Xenopus laevis* (Jones and Chu 1988); *C.p.* = *Chrysemys picta* (Glass et al. 1983); *C.n.* = *Crocodilus niloticus*, (Davies 1978); *L.v.* = *Lacerta viridis* (Nielsen 1962); *V.e.* = *Varanus exanthematicus* (Mitchell et al. 1981); *A.j.* = *Acrochordus javanicus* (Glass and Johansen 1979); *N.r.* = *Natrix rhombifera* (Gratz 1979)

ity in this response from species to species, however, is enormous. Much of this variability undoubtedly reflects adaptation to differing environmental, behavioural or physiological demands. Thus the blunted hypoxic ventilatory responses (HVR) seen in fossorial (burrowing) birds and mammals (Boggs and Kilgore 1983; Boggs et al. 1984) are believed to be an adaptation which minimizes the use of increased air convection at the lung as a mechanism for dealing physiologically with the chronic hypoxia (and hypercarbia) of the burrow atmosphere. For much of the data, however, it is hard to draw such firm conclusions. Figure 3 illustrates the range of variability seen in the hypoxic ventilatory responses of various species of amphibia and reptiles. In general, large falls in inspired O_2 partial pressure (P_{IO2}) are required to produce relatively small increases (2–3×) in levels of minute ventilation ($\dot{V}_E$). The relative contributions of changes in the components of $\dot{V}_E$, respiratory frequency (f) and tidal volume (V_T) to these changes are also highly variable. Thus, a fall in $P_{I_{O_2}}$ to 40 mmHg results in an elevation of $\dot{V}_E$ to roughly four times resting levels, due exclusively to increases in breathing frequency in the Nile crocodile (*Crocodilus niloticus*) but to only twice resting levels, due exclusively to increases in tidal volume (breathing frequency actually decreases!), in the diamondback water snake (*Natrix rhombifera*). Much of this variability must represent adaptation to differing constraints placed upon the respiratory systems of these animals, but much may also reflect the manner in which these measurements were made and reported.

In many studies of hypoxic ventilatory responses, changes in $\dot{V}_E$ are measured as a function of changes in $P_{I_{O_2}}$ in animals breathing air diluted with nitrogen. Three aspects of such studies will lead to tremendous variability in the magnitude of the hypoxic ventilatory responses which are produced. The first stems from variability in the relationship between $P_{I_{O_2}}$ and arterial P_{O_2} (Pa_{O_2}). The latter is the true stimulus for the response but, paradoxically, the greater the ventilatory response, the smaller the change in pulmonary, and hence, arterial P_{O_2} for any given change in $P_{I_{O_2}}$. Thus, correlations of changes in ventilation with changes in $P_{I_{O_2}}$ will tend to underestimate differences in the HVR. A second problem with these studies is the use of $\dot{V}_E$ rather than $\dot{V}_{Eff}$ (effective ventilation) for determining the magnitude of the HVR. The effective ventilation is equal to $\dot{V}_E$ minus the dead space ($\dot{V}_D$) ventilation [$\dot{V}_{Eff} = f \cdot (V_T - V_D)$] and represents the amount of gas which comes into contact with the gas exchange surfaces. This is equivalent to the alveolar ventilation in mammals ($\dot{V}_A$) or the parabronchial ventilation in birds ($\dot{V}_{PB}$). Thus, in the comparison of the Nile crocodile and the diamondback water snake made above, although the HVR of the crocodile appears to be twice that of the snake when expressed as $\Delta\dot{V}_E/\Delta P_{I_{O_2}}$, because increasing V_T is far more effective at increasing $\dot{V}_{Eff}$ than increasing f (an increase in the latter will only partially increase $\dot{V}_{Eff}$ and partially increase dead space ventilation, $\dot{V}_D$), the HVR of the snake will most likely be larger when expressed as $\Delta\dot{V}_{Eff}/\Delta P_{I_{O_2}}$. Finally, one consequence of the HVR which is produced in these studies is that the blood of these animals will become hypocapnic. This will tend to offset the HVR by varying amounts which depend on the magnitude of the CO_2 sensitivity of each species; the greater the CO_2 sensitivity, the smaller the HVR.

Another factor which contributes to the variability in reports of hypoxic ventilatory responses is body size. As Tenney and Boggs (1986) point out, given the higher weight-specific metabolic and ventilatory rates of small relative to large animals, for any given degree of hypoxia a greater ventilatory response will be required to maintain arterial P_{O_2} in the smaller animals. To allow for this difference when comparing values measured from animals of differing sizes, therefore, they suggest that values either be normalized to body weight (but to the 0.75 power), normalized to metabolic rate [$\dot{V}_E/\dot{V}_{O_2}$, the air convection requirement (ACR)] or expressed as a percentage change (which normalizes the data to resting $\dot{V}_E$, a function of size and metabolic rate). As a consequence of the arguments outlined above, it is presently impossible to sort out how much of the variability observed in the magnitude of HVR in different species represents true physiological adaptation and how much is artifact. This will only be possible when studies are conducted which measure changes in weight-specific $\dot{V}_{Eff}$ as a function of changes in Pa_{O_2} under isocapnic conditions.

Interrelationships Between Pa_{O_2} and $\dot{V}_{O_2}$, $\dot{V}_{Eff}$ and $\dot{Q}_{Shunt}$

$\dot{V}_{O_2}$. In most air-breathing vertebrates, Pa_{O_2} is normally maintained within certain 'homeostatic' limits by altering $\dot{V}_{Eff}$ to match $\dot{V}_{O_2}$. It is becoming increasingly clear, however, that just as Pa_{O_2} is a function of the interaction between $\dot{V}_{Eff}$ and $\dot{V}_{O_2}$, $\dot{V}_{O_2}$ is also a function of Pa_{O_2}. Although this interrelationship has been documented in all air-breathing vertebrates (see Hochachka and Guppy 1987 for review), it is most evident from the data collected on reptiles. In this group, many studies have documented a 'critical P_{O_2}' below which further reductions

Table 1. Critical $P_{I_{O_2}}$ for Reptiles

Species	Temperature (°C)	Critical $P_{I_{O_2}}$ (Torr)[a]
Turtles		
Chelydra serpentina	25	<18
Chrysemys picta	10	12
	20	16
	30	32
Pseudemys scripta	22	8–12
	24	38
Sternothaerus minor	22	8
Lizards		
Dissosaurus dorsalis	25	74
Lacerta viridis	20	72
Sauromalus obesus	15–40	>74
Sceloporus occidentalis	15	<112
Tarentola mauritanica	20	72
Snakes		
Acrochordus javanicus	25	72
Farancia abacura	22	40
Heliops modestus	25	70
Liophis miliaris	25	70
Pituophis melanoleucus	25	74
Crocodilians		
Alligator mississippiensis	25	74

[a] $P_{I_{O_2}}$, partial pressure of inspired O_2.

in ambient P_{O_2} (and hence Pa_{O_2}) lead to a depression of metabolism (Table 1; Shelton et al. 1986). There appears to be an inverse correlation between the critical P_{O_2} and the hypoxia tolerance of each species.

The data further suggest that this critical P_{O_2} is closely correlated with the partial pressure of oxygen at which arterial blood is 50% saturated (P_{50}). Figure 4 shows the correlation between mean critical P_{O_2} and P_{50} for several species of reptile. A similar inverse relationship between oxyhemoglobin affinity and critical P_{O_2} has been reported for mammals (Hall 1966). This correlation suggests that the critical P_{O_2} is a function of a decrease in the oxygen transport capacity of the blood.

Finally, an inverse correlation also exists between the magnitude of the ventilatory response to hypoxia and the level of critical P_{O_2}. The greater the HVR, the more $F_{I_{O_2}}$ must fall to produce any given change in Pa_{O_2}. Thus, the greater the hypoxic ventilatory response and the higher the oxyhemoglobin affinity, the lower the critical P_{O_2} (i.e. the greater the fall in $P_{I_{O_2}}$ required to reduce Pa_{O_2} to a point where the O_2 transport capacity of the blood is reduced) and the greater the hypoxia tolerance of a species.

This is illustrated for three species with different degrees of hypoxia tolerance in Fig. 5 (turtle > snake > lizard). The turtle, *Chrysemys picta*, possesses a brisk hypoxic ventilatory response and a high affinity hemoglobin (Glass et al. 1983).

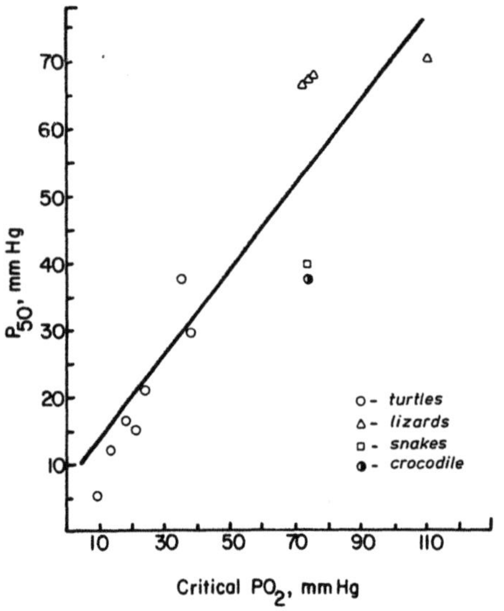

Fig. 4. Relationship between in vivo O_2 partial pressure (P_{O_2}) necessary for 50% saturation of hemoglobin and critical P_{O_2} in reptiles (Wood 1984)

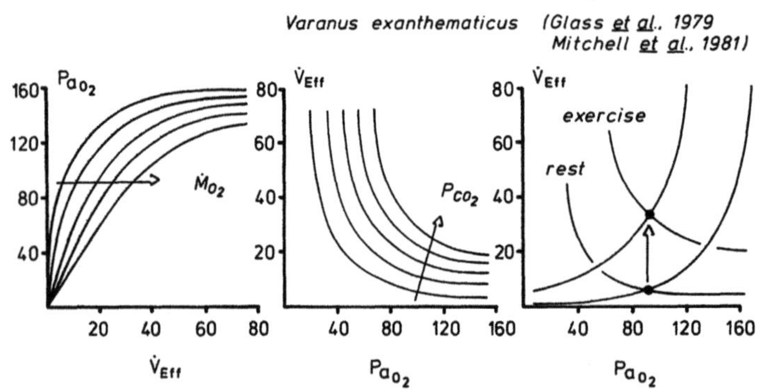

Fig. 5. Relationship between the percent change in minute ventilation ($\Delta\dot{V}_E$), air convection requirement (ΔACR) (*left axis*) and oxygen consumption ($\Delta\dot{V}_{O_2}$) (*right axis*) and the percent O_2 in inspired gas (data from Funk and Milsom 1987, Gratz 1979, Nielsen 1962)

The lizard, *Lacerta viridis*, on the other hand, possesses a very meager hypoxic ventilatory response and a lower affinity hemoglobin (Nielsen 1962). Values for the snake, *Natrix rhombifera*, lie in between (Gratz 1978, 1979). Because of the differences in hypoxic ventilatory responses, arterial P_{O_2} will fall less in the turtle for any given level of $F_{I_{O_2}}$ (which is reflected by the greater increase in the air convection requirement) and, because of the higher affinity hemoglobin, arterial P_{O_2} must fall lower in the turtle before the oxygen transport capacity of the blood is affected. As a consequence of both these factors, the critical P_{O_2} at which V_{O_2} begins to be depressed will be much lower, and the hypoxia tolerance higher, in the turtle.

$\dot{V}_{Eff}$. The interrelationship between $\dot{V}_{Eff}$ and Pa_{O_2} is the natural consequence of the negative feedback relationship between ventilation and blood gas concentra-

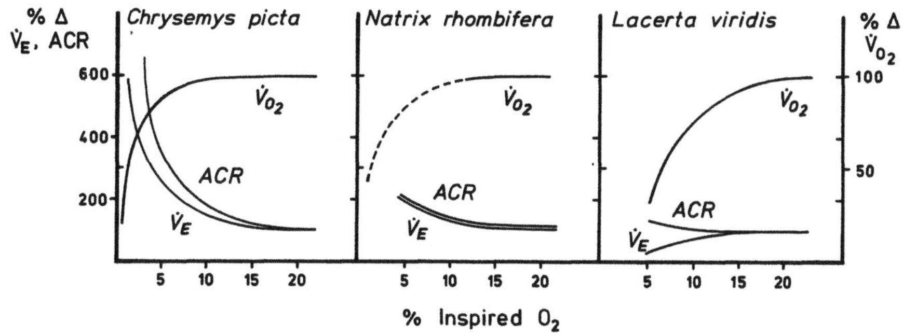

Fig. 6. Left Effect of changing effective minute ventilation ($\dot{V}_{Eff}$) on Pa_{O_2} at different levels of resting metabolic rate ($\dot{M}_{O_2}$). *Middle* Effect of changing Pa_{O_2} on $\dot{V}_{Eff}$ at different levels of Pa_{CO_2}. *Right* Effect of exercise on both of the relationships depicted in the other two panels in the lizard, *Varanus exanthematicus* (data from Glass et al. 1979, Mitchell et al. 1981)

tions. For any given metabolic rate, Pa_{O_2} will be proportional to $\dot{V}_{Eff}$. If $\dot{V}_{Eff}$ is reduced, Pa_{O_2} will fall, and vice versa. This gives rise to the metabolic hyperbolae illustrated in the left-hand panel of Fig. 6 (Pa_{O_2} is the dependent, and $\dot{V}_{Eff}$ the independent variable). At the same time, under any given set of conditions, the level of $\dot{V}_{Eff}$ will be determined by Pa_{O_2}. The lower the Pa_{O_2}, the greater the stimulus to breathe, and vice versa. This is illustrated by the series of oxygen response curves generated at different levels of Pa_{CO_2} in the middle panel of Fig. 6 ($\dot{V}_{Eff}$ is the dependent, and Pa_{O_2} the independent variable). The level of $\dot{V}_{Eff}$ and Pa_{O_2} found in any animal under resting conditions must be the result of these interactions and is represented graphically as the point at which the metabolic hyperbola and oxygen response curve cross. Such a solution must exist for all conditions, and the right hand panel of Fig. 6 illustrates two such solutions for a varanid lizard, during rest and exercise. The increase in metabolic rate during exercise produces both an upward shift of the metabolic hyperbola and a shift in the position of the oxygen response curve. As seen graphically in this figure, the common solution to these new series of relationships results in an elevation of $\dot{V}_{Eff}$ but maintenance of a constant level of Pa_{O_2}.

$\dot{Q}_{Shunt}$. Cardiovascular shunts of varying degrees (anatomical and/or physiological) exist in the pulmonary circulation of all vertebrates. One consequence of such shunts, however, is that the Pa_{O_2} of systemic arterial blood will be a function of the oxygen contents (not partial pressures) of the shunt and non-shunt fractions of the total pulmonary blood flow. As a consequence, the oxygen affinity of the blood, as well as the magnitude of the shunt, can also be key factors determining Pa_{O_2}.

Theoretical predictions of the effects of shunts on arterial P_{O_2} can be derived from the two-compartment model for O_2 transport shown in Fig. 7 (Wood 1984). In the 'perfect' compartment of the lung there is complete diffusion equilibrium for O_2, there is no regional heterogeneity and, consequently, the P_{O_2} of blood leaving the lung is equal to the P_{O_2} of gas within the lung (P_{LO_2}). P_{LO_2} is

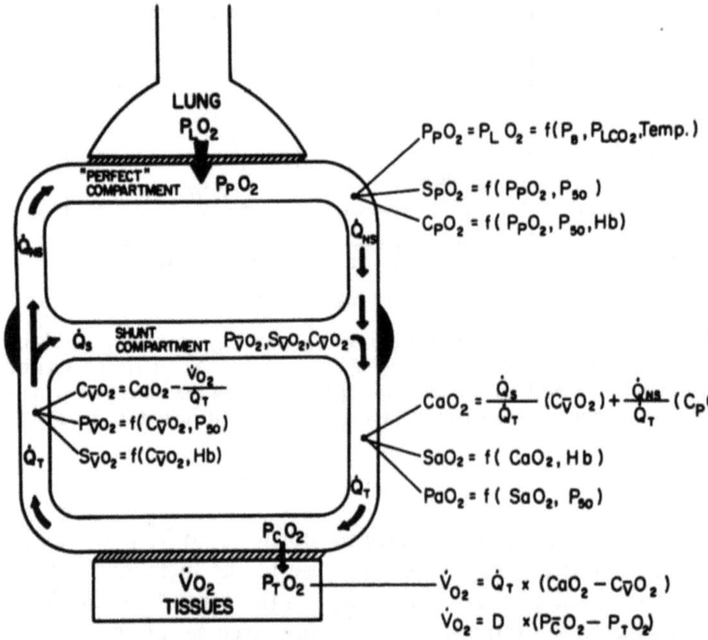

Fig. 7. Two-compartment model for O_2 transport in animals with cardiovascular shunts. See text, for details (Wood 1984)

determined by P_{LCO_2}, barometric pressure (P_B), the respiratory exchange ratio (R) and water vapour pressure (P_{H_2O}) according to the equation:

$$P_{LO_2} = (P_B - P_{H_2O})\ F_{IO_2} - (P_{LCO_2}/R).$$

Thus, the oxygen content of blood leaving the perfect compartment (Cp_{O_2}) is a dependent function of P_{LO2}, the oxygen affinity of the blood and the amount of hemoglobin in the blood.

In this model, systemic arterial blood flow ($\dot{Q}_T$) is the sum of blood flowing through the perfect, non-shunt compartment ($\dot{Q}_{NS}$) and blood flow from the shunt compartment ($\dot{Q}_s$). When these two flows mix, systemic Ca_{O_2} will be determined by the fraction of shunted blood and the O_2 contents of the shunted and non-shunted blood according to the equation:

$$Ca_{O_2} = \dot{Q}_S/\dot{Q}_T(Cv_{O_2}) + \dot{Q}_{NS}/\dot{Q}_T(Cp_{O_2}).$$

Systemic Pa_{O_2} is now a dependent function of Ca_{O_2} and the O_2 affinity of the blood (shape of the oxygen equilibrium curve (OEC)).

From the foregoing discussion, it can be seen that both the shunt fraction and the oxygen affinity of the blood can be key factors determining Pa_{O_2}. Changes in these factors under various physiological conditions have only been well studied in a few species and the extent to which they can be actively regulated is disputed (White 1976; Shelton and Burggren 1976; Heisler et al. 1983; Burggren 1985). Nonetheless, these factors appear to assume their greatest significance in the reptiles where the variations which naturally occur under different physiological con-

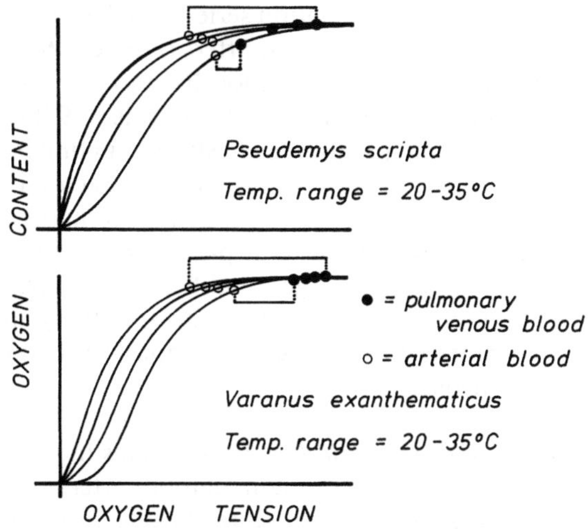

Fig. 8. Computer predictions of sea-level values for pulmonary venous O_2 content and O_2 tension (*closed circles*) and mixed, systemic arterial O_2 content and O_2 tension (*open circles*) in the turtle *Pseudemys scripta* and the lizard *Varanus exanthematicus*. O_2 equilibrium curves shift as a function of body temperature and concomitant changes in blood pH. See text, for discussion (after Wood 1984)

ditions are greatest. Changes in body temperature, for instance, alter both the shunt fraction and shape of the oxygen dissociation curve to different degrees in different species. The consequences of this have been derived for several species from this model by Hicks and Wood (Wood 1984; Fig. 8). In the turtle, *Pseudemys scripta*, increasing body temperature both decreases the oxygen affinity of the blood (right shift of the OEC) and the oxygen content of blood flowing through the perfect compartment (the increased lung oxygen extraction due to the increase in metabolic rate decreases P_{LO_2}). When combined with an increase in the right to left intracardiac shunt, these changes lead to a reduction in the oxygen content of the systemic arterial blood, accompanied, paradoxically, by an increase in Pa_{O_2} (Fig. 8). The displacement of the OEC with temperature over the same range is less in the lizard, *Varanus exanthematicus*, as is the fall in P_{LO_2}. Combined with a smaller shunt fraction, this leads to a smaller fall in systemic arterial oxygen content, but, nonetheless, still a substantial increase in Pa_{O_2} (Fig. 8).

The consequences of the three interrelationships just described will be more complex in animals which exhibit periodic breathing, due to the broad oscillations in Pa_{O_2} associated with such breathing patterns. They will be even further compounded in species utilizing multiple gas exchange sites. Again, however, they emphasize the need for careful measurement of $\dot{V}_{Eff}$ and Pa_{O_2} in studies of hypoxic ventilatory responses.

O_2 Chemoreception

In vertebrates, the tissue associated with O_2 chemoreception has a characteristic association of glomus cells (also known as chief cells or Type I cells), sustentacular cells (also known as Type II cells) and, usually, nerve terminals (see Jones and Milsom 1982 for review). The glomus cells are characterized by a profusion of cytoplasmic vesicles and other cytoplasmic organelles such as ribosomes, endoplasmic reticulum and mitochondria suggestive of endocrine tissue. It has been suggested that glomus cells are members of the APUD series (amine precursor uptake and decarboxylation series of cells) which produce peptides and/or amines as hormones or neurotransmitters (Pearse 1969). Certainly, a wide variety of peptides and amines have been localized to glomus cells and implicated in chemoreceptor function (Fidone and Gonzalez 1986). APUD cells are derived from neuroendocrine-programmed cells originating in the embryonic ectoblast and appear to disperse diffusely throughout the central cardiovascular area in vertebrates (Pearse and Polak 1978). This series is composed of many diverse cell groups, often collectively referred to as paraganglia, of which the carotid and aortic bodies of mammals are the largest and best studied groups (Fitzgerald and Lahiri 1986). The location, innervation, stimulus specificity and reflexogenic roles of the major O_2 chemoreceptive tissues in various vertebrate groups, however, is far from homogeneous.

Carotid and Aortic Bodies. In mammals the carotid bodies are situated at the bifurcations of the common carotid arteries into their internal and external branches. Afferent nerve fibres arising from these structures travel in the sinus nerve, a branch of the glossopharyngeal nerve. They only respond to decreases in Pa_{O_2} and are not affected by changes in Ca_{O_2} per se. They have been reported to contribute from 0% to 40% of the resting drive to breathe under eucapnic normoxia but may account entirely for the chemoreflex increase in ventilation during hypoxia in many species (Fitzgerald and Lahiri 1986). The aortic bodies, on the other hand, are located in the region of the aortic arch and the roots of the major arteries of the thorax. Their afferent fibres run in the aortic nerve, a branch of the vagus nerve. Their response to changes in Pa_{O_2} is less vigorous than that of the carotid bodies but, unlike the carotid bodies, they do respond to changes in Ca_{O_2} per se. They appear to make little contribution to the resting ventilatory drive in eucapnic normoxia and may not even contribute to the HVR in many species (goat, sheep, pony, rabbit, dog, cat, human) (Fitzgerald and Lahiri 1986). There have been recent reports, however, of species which still exhibit a brisk HVR following carotid body denervation (Webb and Milsom 1990), as well as of species which regain the HVR over time, following carotid body denervation, presumably due to other chemoreceptor groups (possibly the aortic bodies) assuming greater importance in ventilatory control (Fitzgerald and Lahiri 1986).

In the majority of lizards, the internal carotid artery arises from the common carotid arch by a variable number of openings forming communicating channels (Adams 1958; Rogers 1967). Between the main and collateral subdivisions of the common and internal carotid arteries, the adventitial connective tissue is highly vascular and is thickened into a wedge-shaped mass which is further characterized by the presence of scattered groups of glomus and sustentacular cells (Adams 1958; Rogers 1967; Kobayashi 1971). This region is profusely innervated by the superior laryngeal branch of the vagus nerve (Rogers 1967) and perhaps also from the glossopharyngeal nerve (Terni 1931; Adams 1958). Although these receptors have not been recorded from, lizards do increase their breathing when blood of low O_2 tension is injected into the junction of the carotid arch, and this response is abolished by denervation of this region (Courtice 1980).

In birds and turtles, chemoreceptive tissue is not present at the extant carotid bifurcation. During embryological development, the aortic arches retreat backwards and the carotid bifurcation remains close to the heart. The external carotids then atrophy and the internal carotids (now called common carotids) divide, secondarily, in the head region (West et al. 1981). In birds, the largest aggregations of chemoreceptive tissue are found in the central cardiovascular area close to the parathyroid and thyroid glands and are innervated by one or more vagal branches from the nodose ganglion (Abdel-Magied and King 1978). Given the weight of the evidence (innervation aside), these receptors are believed to be homologous to the carotid bodies of mammals (Muratori 1962). Glomus tissue which may be homologous to the aortic bodies of mammals has been described within the aortic walls of birds (Tcheng and Fu 1962; Tcheng et al. 1963), and a single recording of a putative aortic chemoreceptor has been reported by Nye and Powell (1984). All receptors respond to changes in Pa_{O_2}, but it is not known whether either group responds to changes in Ca_{O_2} per se. Birds do exhibit a ventilatory response to carbon monoxide inhalation, but it is very different (consisting solely of an increase in tidal volume) from the normal response to hypoxia (which is primarily a frequency response) (Tschorn and Fedde 1974). At the same time, however, there is a stronger correlation between the threshold for the HVR and Ca_{O_2} in some birds than with Pa_{O_2} (Boggs and Birchard 1983; Black and Tenney 1980). The carotid chemoreceptors do contribute to the resting ventilatory drive during eucapnic normoxia and may account exclusively for chemoreflex responses to hypoxia and hyperoxia (Bouverot 1978).

In turtles, on the other hand, the major aggregation of chemoreceptive tissue has been located on the aortic arch, innervated by both the superior and inferior truncal nerves (branches of the vagus) (Ishii et al. 1985). The former is believed to correspond to the aortic nerve of mammals, while the latter arises from the ganglion trunci of the vagus. These receptors have been shown to respond to changes in Pa_{O_2} but, again, it is not known whether they also respond to changes in Ca_{O_2} per se. As with birds, however, there is a better correlation between the threshold for the HVR and Ca_{O_2}, than with Pa_{O_2} (Glass et al. 1983), suggesting that such changes are perceived. The relative role of this specific receptor group in establishing resting ventilatory drive or hypoxic chemoreflex increases in ventilation is also still unknown.

Other Receptor Groups. The function of the other glomus tissue scattered throughout the central cardiovascular area is largely unknown. In turtles, O_2 chemoreceptors are also located on the pulmocutaneous artery, innervated by the inferior truncal branch of the vagus nerve (Adams 1962; Benchetrit and Dejours 1980; Ishii et al. 1985). Neuroepithelial bodies have also been reported in the lungs of mammals (Lauweryns and Cokelaere 1973), but neither their afferent innervation nor the chemoreflex effects of stimulation of either group have been determined. The paraganglia in the thorax and abdomen of some species of mammals will elicit reflex ventilatory responses, but only to severe stimuli (Hollinshead 1946; Howe et al. 1981; Fitzgerald and Lahiri 1986).

Stimulus Specificity and Mechanism of Transduction. The linkage between changes in arterial Pa_{O_2} and chemosensory discharge has remained an elusive one. Several hypotheses exist, which are not mutually exclusive, and which are thoroughly outlined in a recent review by Fidone and Gonzalez (1986). What has received less attention is the correlation between hypoxic ventilatory responses, chemosensory discharge and Ca_{O_2}. At present, only the aortic bodies of mammals have been shown to respond to changes in Ca_{O_2} although this does not seem to elicit any reflex ventilatory response (Hatcher et al. 1978; Lahiri et al. 1981). The responses of O_2 chemoreceptors in birds and reptiles to changes in Ca_{O_2} have not been studied, but these animals show much stronger correlations between the threshold for the hypoxic ventilatory response and Ca_{O_2} than with Pa_{O_2} (Boggs and Birchard 1983; Glass et al. 1983; Black and Tenney 1980). It has been argued that in animals where adaptive changes occur in the shape of the oxygen equilibrium curve (OEC), Ca_{O_2} would be a better indicator of the O_2 homeostasis of arterial blood than Pa_{O_2} (Wood 1984). Thus, in turtles and hibernating ground squirrels there are reductions in the Pa_{O_2} threshold of the hypoxic ventilatory response accompanying the temperature-induced left shift of the OEC as body temperature decreases (Glass et al. 1983; Milsom et al. 1986). This is also the case with animals which exhibit a left-shifted OEC as a consequence of chronic exposure to hypoxia, as occurs in animal burrows (Boggs and Birchard 1983) or at altitude (Brooks and Tenney 1968; Black and Tenney 1980). In all cases, if the HVR are plotted against the percent Hb saturation of arterial blood, ventilation begins to increase when saturation falls below 80%–85% (Wood 1984). It has been further shown that many species of ectotherm (four species of reptiles, two of amphibians and crayfish) behaviourly reduce their body temperatures under hypoxic conditions, increasing oxyhaemoglobin affinity such that levels of Ca_{O_2} are maintained despite the reduction in Pa_{O_2} (Wood, pers. comm.). Whether these responses reflect a more important role for aortic bodies and other 'secondary chemosensitive structures', or Ca_{O_2} sensitivity of carotid bodies in these species, remains an intriguing question. Furthermore, given the differences in location, innervation and possible stimulus modalities of different receptor groups, the validity of current homologies between non-mammalian O_2 chemoreceptors and mammalian carotid and aortic bodies is far from certain.

Central Integration

Central 'Respiratory Centers'. As mentioned earlier (central rhythm generators), O_2 chemosensory inputs must be processed, along with other sensory inputs, by a central pattern generator which determines the spatio-temporal output to the respiratory muscles and which may or may not receive input from a central rhythm generator. Available evidence suggests that the respiratory pattern generator (the major site(s) of central integration) is contained within the brainstem. Nuclei receiving afferent inputs, as well as those giving rise to motor outputs, have been isolated to the medulla, and neurons with respiratory-modulated patterns of discharge which may represent sites of integration and pattern generation have been found at several sites within this region. In the air-breathing vertebrates, these have been extensively documented only in mammals (Feldman 1986) although cursory studies suggest that the same neural organization and pattern of activation is found in turtles (Takeda et al. 1986) and birds (Davey and Seller 1987). The relative contributions of these populations of neurons to sensory integration and the genesis of the respiratory pattern remains far from clear, however, as do the mechanisms and pathways which underlie O_2-related changes in ventilation.

Integration with Other Inputs. Whatever the neural matrix for central integration of sensory inputs may be, the process of integration may be reviewed in terms of changes in the graphical solution of the metabolic hyperbola and oxygen response curve mentioned earlier. The integration of other inputs with O_2 chemosensory information will manifest itself as changes in the location and shape of either curve, which can be viewed as changes in 'setpoint' and 'gain'.

The setpoint of the oxygen response curve will be the level of $\dot{V}_{Eff}$ established during normoxia (or hyperoxia if there is a resting 'O_2-drive') (the asymptote of the oxygen response curve at elevated levels of Pa_{O_2}; Fig. 9, upper left), while that of the metabolic hyperbola will be the resting level of Pa_{O_2} produced by that level of $\dot{V}_{Eff}$ (the asymptote of the metabolic hyperbola at elevated levels of $\dot{V}_{Eff}$; Fig. 9, upper left). The gain of the oxygen response curve, on the other hand, is the slope of the relationship describing the changes in $\dot{V}_{Eff}$ produced by changing Pa_{O_2}, while that of the metabolic hyperbola is the slope of the relationship describing the changes in Pa_{O_2} produced by changing $\dot{V}_{Eff}$ (Fig. 9, upper right).

In the lower panel in Fig. 9, the effects of changes in body temperature on these parameters are shown for the turtle *Chrysemys picta*. Increasing body temperature from 10° to 30°C leads to an increase in resting metabolic rate which will cause a left shift of the metabolic hyperbola (change in setpoint; because of the elevation in $\dot{V}_{O_2}$, Pa_{O_2} will be reduced at any given levels of $\dot{V}_{Eff}$) and upward

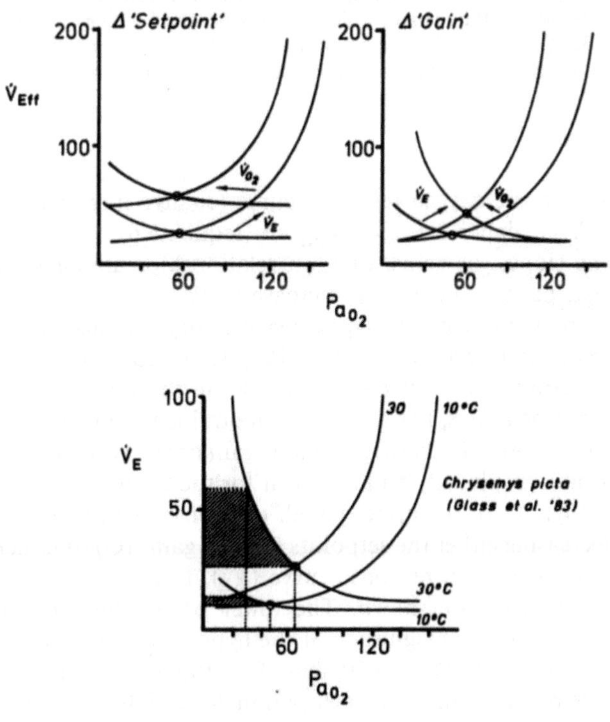

Fig. 9. Upper panels Effect of changing the setpoint and gain on the relationships between $\dot{V}_{Eff}$ and Pa_{O_2} ($\dot{V}_{O_2}$) and Pa_{O_2} and $\dot{V}_{Eff}$ ($\dot{V}_E$). *Lower panel* The changes in setpoint and gain which occur in these relationships with a change in body temperature from 10° to 30°C in the turtle, *Chrysemys picta* (data from Glass et al. 1983). The *stippled areas* indicate the effect of lowering Pa_{O_2} to 30 Torr on $\dot{V}_E$ at each temperature

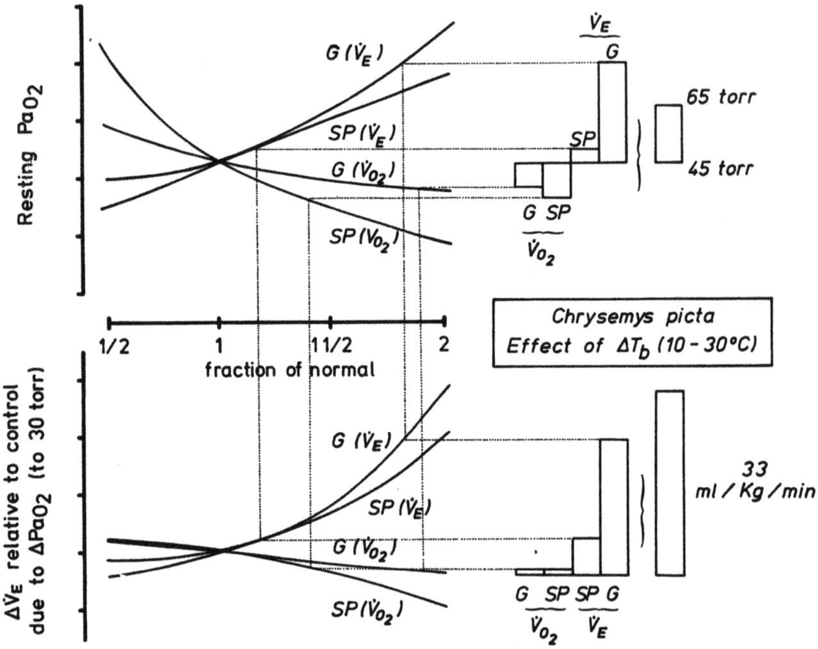

Fig. 10. The effect of changing the setpoint and gain of the relationship between $\dot{V}_{Eff}$ and Pa_{O_2} ($\dot{V}_{O_2}$) and Pa_{O_2} and $\dot{V}_{Eff}$ ($\dot{V}_E$) from 0.5 to 2× normal values on resting Pa_{O_2} (*upper panel*) and the change in minute ventilation which would result from lowering Pa_{O_2} to 30 Torr (*lower panel*) in the turtle *Chrysemys picta* (calculated from the data of Glass et al. 1983). The *dotted lines* indicate the changes which actually occurred with a change in body temperature from 10° to 30°C, and the histograms on the *right* illustrate the relative contributions of these to the resulting changes in Pa_{O_2} and the magnitude of the hypoxic ventilatory response (see text, for details)

shift in the oxygen response curve (change in setpoint; because of the elevation in $\dot{V}_{O_2}$, higher levels of $\dot{V}_{Eff}$ are required to maintain any given level of Pa_{O_2}). The slopes, or gains, of both relationships are also increased. The net effect is that, at rest, there is an increase in Pa_{O_2} of 20 Torr and a dramatic increase in the magnitude of the hypoxic ventilatory response (Fig. 9, stippled areas indicate predicted effects of lowering Pa_{O_2} to 30 Torr on $\dot{V}_{Eff}$). A similar graphical presentation of the effects of exercise on these relationships in the lizard, *Varanus exanthematicus*, has already been outlined in Fig. 6. The relative contributions of such changes in the setpoint and gain of the two relationships can also be demonstrated graphically and is shown for the change in body temperature in the turtle in Fig. 10. In the upper panel, the consequences of independently increasing or decreasing either the setpoints (SP) or gains (G) of either the metabolic hyperbola ($\dot{V}_{O_2}$) or oxygen response curve ($\dot{V}_E$), from one half to twice normal values, on resting Pa_{O_2} are shown. The lower panel illustrates the consequences of such changes on the magnitude of the hypoxic ventilatory response which would result from lowering Pa_{O_2} to 30 Torr. The dotted lines in this figure outline the magnitude of the changes in each parameter which did occur and show graphically, on the right, what the relative contributions of these were to the resulting changes in Pa_{O2} and the magnitude of the HVR. From this analysis it can be seen that it was the increase in the gain of the HVR which was primarily responsible for the observed changes.

Given our lack of knowledge of the neural mechanisms underlying the central integration of these different sensory inputs, such graphical representations can offer a useful way of interpreting the complex interactions which determine resting Pa_{O_2} and the HVR, and may provide some insight into the integrative processes which underlie these changes.

Temporal Output. The central pattern generator controls the motor output to the respiratory muscles and thus produces the pattern of respiratory movements. In mammals, this pattern has been viewed as consisting of an amplitude component (tidal volume, V_T) and a timing component (respiratory frequency, f) (Fig. 3) which itself has been viewed as consisting of an active, inspiratory phase (T_I) and a passive expiratory phase (T_E). As a consequence, breathing pattern was believed to be determined primarily by the magnitude of the inspiratory drive, as manifested in the mean inspiratory flow rate (V_T/T_I)', and the length of the active inspiratory phase (T_I) (Von Euler et al. 1976; Cunningham 1974). The expiratory duration was believed to be dependent on, and therefore determined by, the factors influencing the inspiratory duration (Clark and Von Euler 1972). The relative contributions of changes in the amplitude and timing components of motor output to ventilatory responses are not inconsequential, since the relative contributions of V_T and f to total $\dot{V}_E$ determine $\dot{V}_{Eff}$ and $\dot{V}_D$.

The concept of pattern regulation which is currently emerging is far more complex. Several lines of evidence suggest that there are more than two phases to the timing component of respiratory pattern:

1. Almost all reptiles exhibit periodic breathing. These patterns are of two basic types. In one, lung ventilations occur in episodes separated by relatively long periods of breath holding, whereas in the other, single breaths are separated by shorter intervals. In both cases, the respiratory pause (T_{NVP}) is a breathhold occurring at end-inspiration. The only respiratory muscles which remain active during this period are those associated with glottal closure. During episodes of breathing, expiration follows inspiration with little or no pause at all. Hypoxia-induced increases in minute ventilation are produced primarily by increases in f, resulting from reductions in T_{NVP} (not the duration of active inspiration) (Milsom 1988; Fig. 11). Marine mammals behave in much the same way (Kooyman and Cornell 1981). This data suggests that the end-inspiratory pause is a stable phase which is under separate control from the period of active inspiration but which can be completely suppressed, as during breathing episodes in periodic breathers or during eupneic breathing in most birds and mammals.
2. In birds and reptiles, expiration is an active process. In most mammals, even though expiratory muscles are not normally recruited during quiet eupneic breathing, expiration is still far from a passive event. Recent evidence suggests that there is increased activity in many airway muscles, which may be assisted by post-inspiratory activity in the inspiratory muscles (with associated activity in

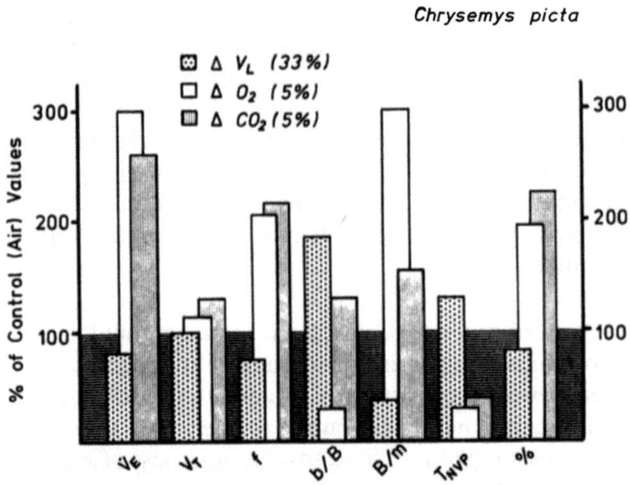

Fig. 11. The changes (as % of control values) which occurred in minute ventilation ($\dot{V}_E$), tidal volume ($\dot{V}_T$), breathing frequency (*f*), the number of breaths in each episode of breathing (*b/B*), the frequency of breathing episodes (*B/m*), the length of the interval between breathing episodes (T_{NVP}) and the percent of time spent actively breathing (%) in the turtle *Chrysemys picta*, due to either a 33% increase in resting lung volume (ΔV_L), or to inspiring 5% O_2 (ΔO_2) or 5% CO_2 (ΔCO_2) (data from Milsom and Chan 1986; Morgan and Milsom 1988)

medullary respiratory neurons) to produce braking or slowing of expiratory airflow (Richter et al. 1986). It has been suggested that this may reflect two independent and potentially stable phases of the ventilation cycle: Stage I, active slowing of expiration (post-inspiratory inspiratory activity, PIIA); Stage II, passive recoil or active expiratory recruitment (Richter et al. 1986). Alternatively, throughout the vertebrate classes, under different conditions, we can see a continuum from active slowing of expiration, through expiration by passive recoil only, to active expiration, suggesting that these variations are all manifestations of central integration of a large number of independent inputs within a single phase.

3. In mammals, during periods of reduced metabolism such as during sleep or hibernation, end-expiratory pauses (apneas) begin to appear. During hibernation, these periods can become prolonged, and episodic breathing patterns appear which are identical to those seen in reptiles, with the exception that the respiratory pause occurs following expiration. As in reptiles, it is this pause which becomes the primary regulated variable in eliciting ventilatory responses to hypoxic stimuli (Fig. 12; Phillipson and Bowes 1986; Milsom 1988). Again, as in reptiles, there is a continuous transition from expiration to inspiration during breathing episodes and breathing becomes continuous at elevated levels of respiratory drive. This data suggests that the end-expiratory pause is also an independent phase under separate control from the period of active (or passive) expiratory flow.

4. During periods of episodic breathing, breathing frequency may increase due to either an increase in the number of breaths in each breathing episode and/or a reduction in the nonventilatory period between episodes. These two options also appear to be under separate control. Thus, in the turtle, *Chrysemys picta* (Fig. 11),

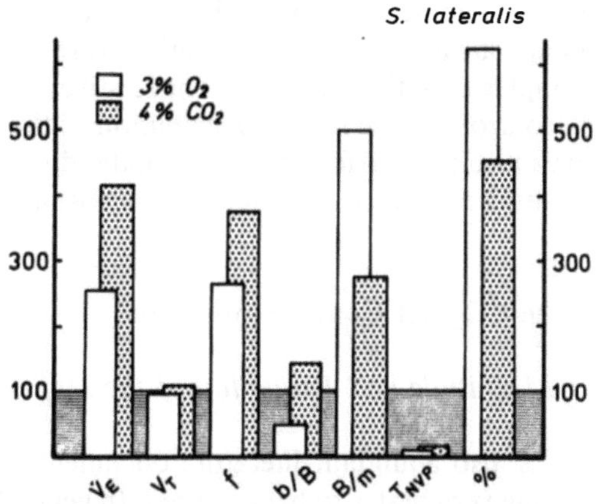

Fig. 12. The effects of inspiring 3% O_2 or 4% CO_2 on the same variables depicted in Fig. 11 during hibernation at 5 °C ambient temperature in the ground squirrel, *Spermophilus lateralis* (Webb and Milsom, data unpubl.)

hypoxia leads to a decrease in the number of breaths in each episode of breathing but to an increase in the number of breathing episodes, such that breathing frequency increases overall. Hypercapnia, on the other hand, leads to increases in both the number of breaths in each episode and the number of episodes per minute, whereas changes in resting lung volume which do not alter overall respiratory frequency lead to a change in breathing pattern, such that the number of breathing episodes decreases while the number of breaths in each episode increases tremendously (Milsom and Chan 1986; Morgan and Milsom 1988). Similar differences in the ventilatory responses to hypoxia and hypercapnia occur during periodic breathing in hibernating ground squirrels (Fig. 12; Webb and Milsom unpubl.).

Thus, there appear to be at least four phases to the ventilation cycle, each under separate control (five if the two stages of the expiratory phase are considered as separate phases). Two of these phases (the end-inspiratory and end-expiratory phases) are very labile and are usually not present in mammals and birds during eupnea. Interestingly, when they are present, as in reptiles or the homeotherms during periods of reduced metabolism, they are the primary controlled variable regulating breathing frequency. Consequently, the temporal output of the pattern generator appears to be far more complex than has been appreciated, and we are still far from understanding the integrative processes which underlie the timing components of ventilation.

Spatial Output. Ventilation is the consequence of the co-ordinated action of a series of muscles acting on the lungs and airways in all air-breathing vertebrates. In most species, both inspiration and expiration are active, further requiring the presence of antagonistic muscle pairs. Because of the different constraints placed on the respiratory systems of different species, the principle respiratory muscles which power ventilation vary tremendously, from the abdominal muscles associated with the inguinal pockets in turtles to the intercostal muscles in birds and the diaphragm in mammals. Furthermore, the relative contributions of various muscle groups varies tremendously, with such factors as arousal state, posture, respiratory drive and activity level. The use of different respiratory pumps for different exchange surfaces in bimodal breathers (previous chapter) adds yet another degree of complexity to this situation. The relative roles of various feedforward and feedback controls on the spatial distribution of the motor output to the various respiratory muscles in different species, under different conditions, remains one of the least studied aspects of vertebrate respiratory physiology.

3.2 The CO_2/pH Complex of Stimuli

3.2.1 Magnitude and Variability of the Hypercapnic Ventilatory Response

There is also abundant literature documenting the ventilatory responses of air-breathing vertebrates to hypercapnia. In general, the relationship between ventilation and arterial P_{CO_2} is roughly linear over the physiological range, but there are exceptions to this trend as well as tremendous variability in the slope of this relationship (Fig. 13). As with the hypoxic ventilatory response, much of this variability must reflect adaptation to different environmental, behavioural and physiological demands. Thus, as with their hypoxic ventilatory responses, the hypercapnic ventilatory responses of fossorial species such as the golden-mantled ground squirrel, pocket gopher and burrowing owl (Sl, Tb and Ac in Fig. 13) are believed to be an adaptation which minimizes the use of increased air convection at the lung as a mechanism for dealing with chronic hypercarbia (and hypoxia) of the burrow atmosphere. Such teleological explanations are less obvious for the blunted responses of the chicken and ducks (Gd, Cm and Ap; Fig. 13), however, not to mention the ventilatory depression seen in many species of reptiles (Fig. 13). The relative contributions of changes in V_T and f to the changes seen in $\dot{V}_E$ are equally varied (Fig. 13). In general, most species increase V_T proportionately more than f, and the species showing the briskest responses to hypercapnic stimuli exhibit proportionately larger increases in both variables.

As with the hypoxic ventilatory responses, much of this variability can probably also be attributed to the manner in which the measurements were made. Whereas most studies of hypercapnic ventilatory responses measure changes in $\dot{V}_E$ as a function of changes in $P_{I_{CO_2}}$ in animals breathing air containing various concentrations of CO_2, only measurements of weight-specific changes in $\dot{V}_{Eff}$ as a function of changes in Pa_{CO_2} under normoxic conditions will reflect the true hypercapnic ventilatory response. Furthermore, two additional factors may add to the variability observed in these responses; the types and stimulus specificities of the different chemoreceptors involved in eliciting these responses.

Firstly, intrapulmonary chemoreceptors (IPC) have been described in the lungs of birds and some reptiles (see Powell et al. 1988 for review) which exhibit discharge rates inversely proportional to the concentration of P_{CO_2} in the lungs (Fig. 14a). Because inhalation of CO_2-rich gas will reduce, eliminate or reverse the normal tidal fluctuations of CO_2 seen in the lungs during eupneic breathing, depending on the level of $P_{I_{CO_2}}$, it will also greatly alter the breath-by-breath discharge profile of these receptors. Although the physiological function of these receptors is not yet clear (see CO_2/pH chemoreception, below), it has been shown

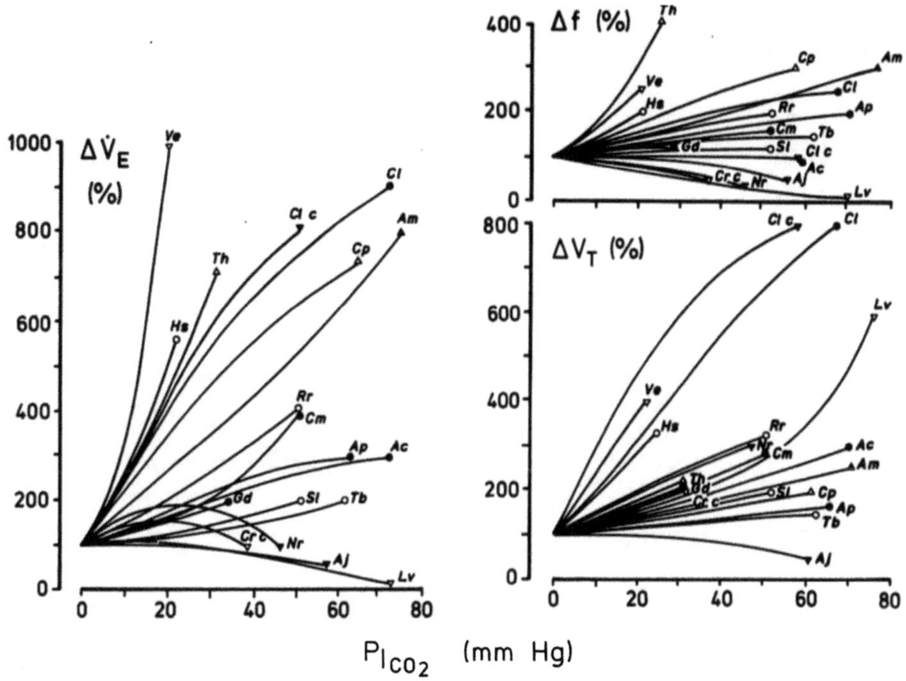

Fig. 13. Relationship between the percent change in minute ventilation ($\Delta\dot{V}_E$), tidal volume (ΔV_T) and breathing frequency (Δf) and the partial pressure of CO_2 in inspired air ($P_{I_{CO_2}}$) in various species. Sources of data: snakes: *Cl.c.* = *Coluber constrictor* (Nolan and Frankel 1982); *A.j.* = *Acrochordus javanicus* (Glass and Johansen 1976); *N.r.* = *Natrix rhombifera* (Gratz 1979). Lizards: *L.v.* = *Lacerta viridis* (Nielsen 1962); *V.e.* = *Varanus exanthematicus* (Mitchell et al. 1981); *Cr.c.* = *Crotaphytus collaris* (Templeton and Dawson 1963). Crocodilian: *A.m.* = *Alligator mississippiensis* (Davies 1978). Turtles: *C.p.* = *Chrysemys picta* (Funk and Milsom 1987); *T.h.* = *Testudo horsfieldi* (Benchtrit and Dejours 1980). Birds: *C.l.* = *Columbia livia*, *A.c.* = *Athene cunicularia* (Kilgore et al. 1985); *C.m.* = *Cairina moschata* (Powell et al. 1978); *G.d.* = *Gallus domesticus* (Mitchell and Osborne 1980); *A.p.* = *Anas platyrhynchos* (Milsom et al. 1981). Mammals: *H.s.* = *Homo sapiens* (Hey et al. 1966); *R.r.* = *Rattus rattus* (Cragg and Drysdale 1983); *S.l.* = *Spermophilus lateralis* Webb and Milsom unpubl.; *T.b.* = *Thomomys bottae* (Darden 1972)

that the inhalation of CO_2 in these species can produce very different results from those obtained in studies which elevate Pa_{CO_2} by other means (Fig. 14b; Milsom et al. 1981; Jones et al. 1985) and may account for the ventilatory depression seen at high levels of inspired CO_2 in many species of reptiles (Fig. 13; Shelton et al. 1986). As a consequence, although changing $F_{I_{CO_2}}$ may be suitable for studies of the effects of environmental hypercarbia on ventilation, it may be totally inappropriate for studies of the ventilatory responses to increased metabolic CO_2 production ($\dot{V}_{CO_2}$) or other forms of hypercapnia. Furthermore, recent studies have shown that some species which possess IPC may still show robust responses to inhaled CO_2 (Cl and Am in Fig. 13; Fig. 14c) suggesting that large species differences may exist in the relative contribution of IPC to the control of breathing pattern and total respiratory drive (Kilgore et al. 1985; Davies et al. 1982; Powell et al. 1988).

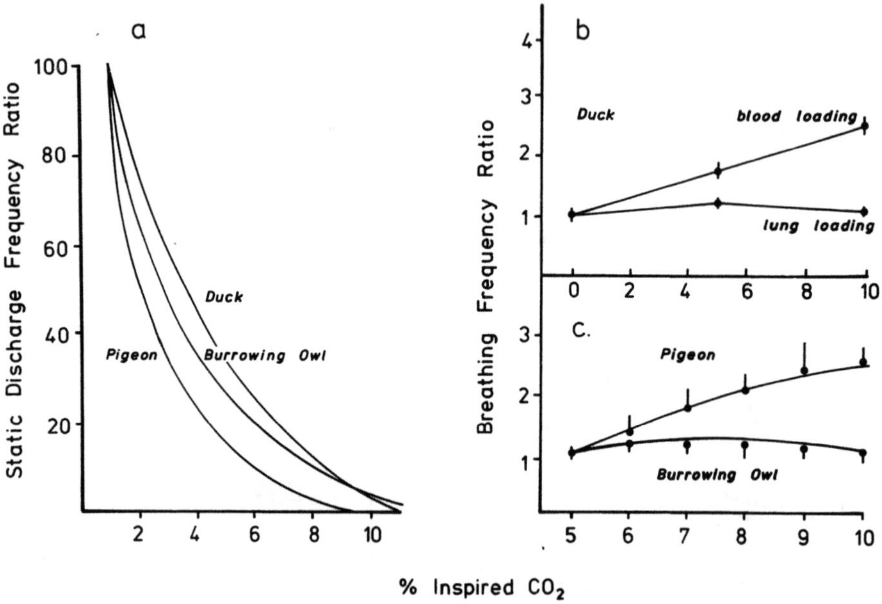

Fig. 14. a The change in static discharge frequency of intrapulmonary chemoreceptors in three species of birds as a function of the percentage of CO_2 in inspired gas. All values of discharge frequency are expressed as a fraction of the discharge occurring during inspiration of 1% CO_2. *b* The effects of changing arterial P_{CO_2} by either changing inspired CO_2 (lung loading) or venous P_{CO_2} (blood loading) on breathing frequency in the duck (all values of breathing frequency are expressed as the ratio of breathing frequency relative to that occurring with no CO_2 loading; Pa_{CO_2} expressed as the % CO_2 with which the air or blood was equilibrated). *c* The effects of changing inspired CO_2 on breathing frequency in the pigeon and burrowing owl (breathing frequency is expressed as the ratio of breathing frequency relative to that occurring when animals breathed air) (data from Fedde and Peterson 1970, Kilgore et al. 1985, Milsom et al. 1981)

Secondly, inhalation of CO_2-rich gases will produce both a hypercapnia and an acidosis. For any given change in Pa_{CO_2}, the extent to which pHa changes will be a function of the buffer properties of the blood. At present, it appears that peripheral chemoreceptors respond to changes in both pH and P_{CO_2} which, therefore, may act as independent stimuli (Fidone and Gonzalez 1986). Furthermore, although it is generally accepted that the central chemoreceptors respond to changes in pH, P_{CO_2} may have an independent effect and, due to the characteristics of the blood brain barrier, cerebral-spinal fluid pH and brain extracellular fluid pH reflect changes in Pa_{CO_2} more closely than in pHa (Bledsoe and Hornbein 1981) (see CO_2/pH chemoreception, below). As a consequence of these complex interactions, it is not clear at present to what extent Pa_{CO_2} reflects the true stimulus for the hypercapnic response.

Ventilatory responses to metabolic acidosis/alkalosis in air-breathing vertebrates have been less well studied. The relationship between ventilation and arterial pH is also roughly linear, however, and the variability seen in the slope of this relationship is subject to many of the same considerations just outlined for the hypercapnic ventilatory responses.

3.2.2 Interrelationship Between Pa_{CO_2}/pH and $\dot{V}_{O_2}$, $\dot{V}_{Eff}$ and $\dot{Q}_{Shunt}$

$\dot{V}_{O_2}$. As was the case for Pa_{O_2}, Pa_{CO_2} and pHa are also normally maintained within certain 'homeostatic' limits by altering $\dot{V}_{Eff}$ to match metabolic CO_2 production ($\dot{V}_{CO_2}$, which is related to $\dot{V}_{O_2}$ by the respiratory quotient). Once again, however, it has been shown that just as Pa_{CO_2} and pHa are functions of the interaction between $\dot{V}_{Eff}$ and $\dot{V}_{O_2}$, $\dot{V}_{O_2}$ is also a function of pHa and Pa_{CO_2}.

In mammals, acidosis has been shown to depress the metabolic rate of cells in tissue culture (Bussa and Nuccitelli 1984) and isolated limbs (Harken 1976) by 5% to 10% of resting $\dot{V}_{O_2}$ per 0.1 unit fall in pHa. In whole animals (Cain 1970), alkalosis greatly increases metabolic rate although acidosis has little effect. Changes in P_{CO_2} are without any direct effect. The physiological significance of the effect of changes in pHa on metabolism in mammals, thus, is unclear. At the cellular level it has been suggested that since intracellular pH, unlike other effectors which alter metabolic rate, does not require specialized receptors, it may function as a synergistic messenger providing a metabolic context within and through which the actions of other effectors are integrated (Bussa and Nuccitelli 1984).

At the whole animal level, it has been argued that the metabolic depression associated with the entrance into hibernation is the direct consequence of a respiratory acidosis. That is, the primary event associated with the entrance into hibernation is a reduction in ventilation leading to CO_2 retention, acidosis and then metabolic depression (Malan et al. 1973; Malan 1982; Bickler 1984; Geiser 1988). This latter claim is based on several trends which have been observed in animals during entrance into hibernation. (1) Ventilation and heart rate fall before recorded changes in metabolic rate or body temperature (Lyman and Chatfield 1955). (2) $\dot{V}_E/\dot{V}_{CO_2}$ is transiently reduced and thus CO_2 is retained in proportion to the increase in CO_2 solubility. As a consequence, plasma pH and Pa_{CO_2} are maintained relatively constant, but because of the change in neutral pH with temperature, the ratio of $[H^+]/[OH^-]$ in plasma increases (Malan et al. 1973; Malan 1982; Bickler 1984)[3]. This will alter the homeostasis of the dissociation ratio (alpha) of the predominant non-bicarbonate buffers, the imidazole groups of protein histidyl residues, and, consequently, will alter the electric charge-dependent properties of proteins, notably enzymes, altering their function. The degree of change in $[H^+]/[OH^-]$ seen in blood plasma is not transmitted to all tissues, however. The intracellular pH of such tissues as heart and liver increase markedly

[3] Respiratory acidosis has typically been defined in terms of the controlled variables involved in the regulation of ventilation under euthermic conditions in mammals, and refers to a decrese in pH, resulting from an elevation in Pa_{CO_2} due to hypoventilation (or an increase in inspired CO_2). In this instance, pHa and Pa_{CO_2} appear to be homeostatically maintained. One consequence of this is that there must be a transient decrease in $\dot{V}_E/\dot{V}_{O_2}$ to allow the CO_2 content of the blood to increase in proportion to the increase in CO_2 solubility (which produces the concomitant increase in $[H^+]/[OH^-]$. This does not represent a hypoventilation in terms of the standard controlled variables (pHa and Pa_{CO_2}) used in the 'classical' definition of respiratory acidosis. Appropriate terms to distinguish this respiratory-induced change in $[H^+]/[OH^-]$ due to pH/P_{CO_2} homeostasis with changing body temperature from respiratory-induced changes in $[H^+]/[OH^-]$ due to changes in pH/P_{CO_2} at constant body temperature remain to be defined.

during hibernation, such that the ratio of $[H^+]/[OH^-]$ increases only slightly (Malan et al. 1985). This may serve to maintain proper enzyme function in these tissues and restrict changes in the contractility of the heart as well as in the production and release of glucose, its primary source of energy, from the liver. The $[H^+]/[OH^-]$ ratio in skeletal muscle and brain, however, do increase significantly during entrance into hibernation, by an amount that could potentially cause metabolic rate to fall by 30% to 40% more than expected, due to changes in body temperature alone. This could contribute to the entrance into hibernation, either directly, by reducing general body metabolism (Malan et al. 1985) or indirectly, by its depressant effect on central integrative processes (Snapp and Heller 1981). (3) Consistent with the above, the reduction in metabolic rate measured in many studies exceeds that which would be expected, due to the fall in body temperature alone (i.e. $Q_{10}>3$; Malan 1982). (4) Changes in body temperature in torpid squirrels, accompanied by CO_2 retention, affect $\dot{V}_{O_2}$ more than changes which are not accompanied by CO_2 retention (Bickler 1984).

This hypothesis is enticing but, although the data which have just been presented are consistent with such a mechanism for the metabolic depression seen during the entrance into hibernation, they are not definitive. To begin with, due to body mass, even changes in $\dot{V}_E$ and heart rate, due to a reduction in metabolic rate, will appear before the changes in $\dot{V}_{O_2}$ and body temperature (T_b) which produce them, simply as a consequence of body stores and recording techniques. Furthermore, in a recent review of the literature, Geiser (1988) has shown that the effect of temperature on metabolic rate in hibernating animals is inversely proportional to body mass and thus Q_{10} values for the fall in metabolic rate during hibernation only approach and exceed 3 for small animals. Since the mechanisms which underlie the process of hibernation appear to be common to all hibernators (it is highly unlikely that the effects of pH on metabolism are size-dependent), the significance of Q_{10} values greater than 3 is far from clear. [This is particularly so if one considers that the Q_{10} for carotid body discharge in cats under basal conditions is 75.2! (Gallego et al. 1979).] Finally, almost all of the literature values are based on the differences computed between hibernating animals and active, euthermic animals. Snapp and Heller (1981) have argued that since hibernation is an extension of slow-wave sleep, comparisons should only be made between hibernating animals and sleeping, euthermic animals. When this is done, Q_{10} values never exceed 3. It has been argued that the reduction in metabolism which occurs during sleep is also due to a respiratory acidosis (Malan 1982), but even if this is so (which is far from clear), all further reductions which occur during the entrance into hibernation can be accounted for by the effects of changes in body temperature alone.

It should be noted that respiratory acidosis produced by increasing CO_2 in inspired gas from 0% to 6–8% has no significant effect on $\dot{V}_{O_2}$ in either euthermic or hibernating golden-mantled ground squirrels (Webb and Milsom 1989). This is also true of birds and reptiles (Jackson et al. 1974; Gratz 1978; Powell et al. 1978; Boggs and Kilgore 1983; Funk and Milsom 1987). It is conceivable that there is a slight metabolic depression which is masked by the increased metabolic demands associated with the higher levels of ventilation but, if so, it cannot be significant.

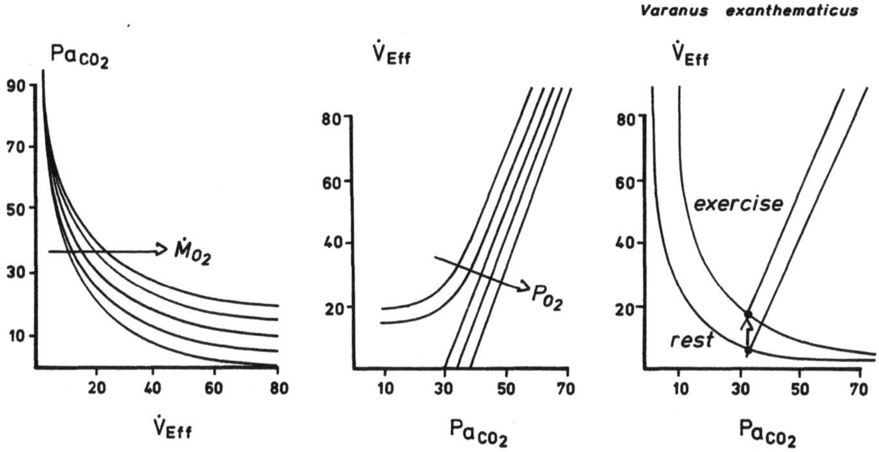

Fig. 15. Left Effect of changing effective minute ventilation ($\dot{V}_{Eff}$) on Pa_{CO_2} at different levels of resting metabolic rate ($\dot{M}_{O_2}$). *Middle*: Effect of changing Pa_{CO_2} on $\dot{V}_{Eff}$ at different levels of Pa_{O_2}. *Right* Effect of exercise on both of the relationships depicted in the other two panels in the lizard, *Varanus exanthematicus* (data from Glass et al. 1979, Mitchell et al. 1981)

At present, the equivocacy of this data makes firm conclusions about the physiological significance of the effects of P_{CO_2}/pH on $\dot{V}_{O_2}$ in whole animal studies impossible. Further investigation of the magnitude, distribution and physiological significance of the interaction between pHa and $\dot{V}_{O_2}$ is clearly needed.

$\dot{V}_{Eff}$. The interrelationship between $\dot{V}_{Eff}$ and Pa_{CO_2}/pH is the same as that described between $\dot{V}_{Eff}$ and Pa_{O_2}. Because of the negative feedback relationship between these two variables, if $\dot{V}_{Eff}$ is reduced, Pa_{CO_2} will rise (and pH fall), and vice versa. This gives rise to the metabolic hyperbolae illustrated in the left-hand panel of Fig. 15 (Pa_{CO_2} is the dependent and $\dot{V}_{Eff}$ the independent variable). At the same time, under any given set of conditions, the level of $\dot{V}_{eff}$ will be determined by Pa_{CO_2}. The higher the Pa_{CO_2}, the greater the stimulus to breathe, and vice versa. This is illustrated by the series of CO_2 response curves generated at different levels of Pa_{O_2} in the middle panel of Fig. 15 ($\dot{V}_{Eff}$ is the dependent and Pa_{CO_2} the independent variable). Note that the relationship between $\dot{V}_{Eff}$ and Pa_{CO_2} over the hypocapnic range is somewhat variable. In most air-breathing vertebrates, the relationship remains linear and breathing ceases if Pa_{CO_2} is reduced sufficiently (Kinney and White 1977; Ray and Fedde 1969; Nielsen and Smith 1952; Phillipson et al. 1981). In some mammals, however, at low levels of P_{O_2}, a threshold exists below which $\dot{V}_{Eff}$ is largely independent of Pa_{CO_2} (Fig. 15; see Cunningham et al. 1986 for review). The level of $\dot{V}_{Eff}$ and Pa_{CO_2} found under resting conditions must be the result of these interactions and is represented graphically as the point at which the metabolic hyperbolae and CO_2 response curves cross. Such a solution must exist for all conditions, and the right hand panel of Fig. 15 illustrates two such solutions for a varanid lizard, during rest and exercise. The increase in metabolic rate during exercise produces both an upward

shift of the metabolic hyperbola and a shift in the position of the CO_2 response curve. As seen graphically in this figure, the common solution to these new series of relationships is an elevation of $\dot{V}_{Eff}$ but maintenance of a constant level of Pa_{CO_2}.

$\dot{Q}_{Shunt}$. The consequence of central or pulmonary vascular shunts on CO_2 transport will be much less than on O_2 transport by the blood. Although the Pa_{CO_2} of systemic arterial blood will still be a function of the CO_2 contents (not partial pressures) of the shunt and non-shunt fractions of the total pulmonary blood flow and the CO_2 affinity of the blood, the blood-CO_2 equilibrium curve is linear over the physiological range, making the consequences of this much less than was the case for O_2. Nonetheless, factors which alter the size of the shunt fraction or the position of the CO_2 equilibrium curve (such as a change in body temperature or the Haldane effect) can dramatically alter the relationship between inspired or lung P_{CO_2} and Pa_{CO_2}.

As mentioned earlier, the consequences of such interrelationships will be more complex in animals which exhibit periodic breathing, due to the broad oscillations in Pa_{CO_2} and pHa associated with such breathing patterns, as well as in species utilizing multiple exchange sites. Again, however, they point out the need for careful measurements of $\dot{V}_{Eff}$ and Pa_{CO_2}/pHa in studies of ventilatory control.

3.2.3 CO_2/pH Chemoreception

The specific characteristics of tissue associated with CO_2/pH chemoreception in vertebrates have not been well defined. Certainly the glomus tissue associated with O_2 chemoreception in the carotid and aortic bodies is also responsive to changes in CO_2 and pH, but not all CO_2/pH chemosensors are derived from paraganglia. The structural details of other chemosensitive tissue remains elusive. As with O_2 chemoreception, however, the location, innervation, stimulus specificity and reflexogenic roles of major CO_2/pH chemoreceptive tissue in various vertebrate groups is far from homogeneous.

Central Chemoreceptors. Receptors responsive to changes in P_{CO_2}/pH have been described in mammals lying within 200 μm of the ventro-lateral surface of the medulla between the origins of the VIIth and XIIth cranial nerves. There appear to be two distinct populations (located in rostral and caudal areas), separated by an intermediate area which is the site of confluence of apparent inputs from both receptor populations, with synaptic relays to the deeper bulbar respiratory nuclei (see Loeschcke 1982; Dempsey and Forster 1982 for reviews). The afferent influences of the central chemoreceptors in mammals appear to be distributed to essentially all units of the dorsal and ventral respiratory groups (St. John and Wang 1977; St. John 1981). Although the presence of central chemoreceptors has not been determined directly in other vertebrate groups, it has been deduced from the results of studies of reflex ventilatory responses employing peripheral chemoreceptor denervation and/or cerebral- or cross-perfusion techniques (reptiles: Hitzig and Jackson 1978; Hitzig 1982; birds: Sebert 1979; Milsom et al. 1981).

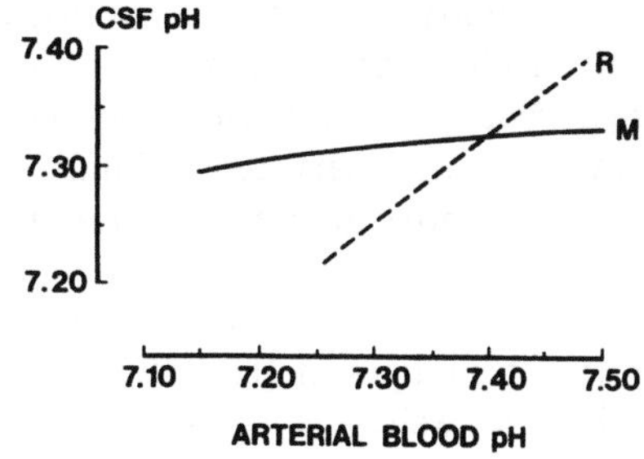

Fig. 16. Relationship between pH of arterial blood plasma and lumbar CSF in humans during stable acid-base disturbances of metabolic (*M*) and respiratory (*R*) origin (Fencl 1986)

It is generally accepted that the central chemoreceptors respond exclusively to changes in the pH of the cerebral interstitial fluid (ISF) which surrounds them. Although the evidence against an independent effect of changes in P_{CO_2} is equivocal, the effects of CO_2 are believed to be mediated by virtue of its acidity (see Loeschke 1982; Dempsey and Forster 1982 for reviews). Due to the characteristics of the blood brain barrier, however, cerebral interstitial fluid pH reflects changes in Pa_{CO_2} (CO_2 crosses the barrier easily) more closely than changes in $[H^+]_a$ (which does not). The correlation between changes in Pa_{CO_2} and ISF pH will be a consequence of local $\dot{V}_{CO_2}$, ISF $[HCO_3^-]$ (or strong ion difference: SID) and cerebral blood flow. The strength of the association between Pa_{CO_2} and ISF pH is reflected by the fact that in stable nonrespiratory (metabolic) acidosis and alkalosis, effective mechanisms minimize changes in cerebrospinal fluid (CSF) pH (and presumably ISF pH), while in stable respiratory acidosis and alkalosis, CSF pH changes are maintained (Fig. 16; Fencl 1986).

The reflex effects of central chemoreceptor stimulation by changes in pHa/Pa_{CO_2} appear to predominate over all other receptor inputs in mammals, birds and reptiles. They appear to be primarily responsible for both the resting respiratory drive as well as the respiratory responses to hypercapnia and acidosis. Inputs from peripheral chemoreceptors may account for 20% to 40% of the responses to changes in Pa_{CO_2} (range from 0% to 50%) (Heeringa et al. 1979; Berkenbosch et al. 1979; Kiwull et al. 1982; Milsom et al. 1981; Hitzig and Jackson 1978), but it is not clear to what extent this entirely represents the direct influence of these other inputs or to what extent it is due to synergistic (multiplicative) interactions between peripheral and central chemoreceptor inputs. Reports of additive and less than additive interactions between peripheral and central chemoreceptors also abound, and much of this difference in observed responses may be due to the species, preparation or anesthetic used in each study. In the case of metabolic acidosis, the data is also equivocal. It has been suggested that because of the diffusion barrier to H^+ between blood and brain tissue, central chemosensitive areas may play an increasing role in the ventilatory response to metabolic acidosis, which only becomes significant when pHa falls below approximately 7.35 (Mitchell 1965; Berger et al. 1977; Gonzalez et al. 1977; Bainton 1978).

There is preliminary evidence to suggest that there may be differences in the relative roles of the different central chemosensitive areas in these responses, which may vary between species. In rabbits, diethylpyrocarbonate (a chemical which binds to imidazole groups of histidine) appears to selectively block the rostral chemosensitive area of the medulla (Nattie 1986a). It also produces a hypoventilation and effectively blocks the ventilatory response to CO_2 in peripherally chemodenervated animals (Nattie 1986b). This suggests that the rostral chemosensitive area may play a predominant role in the control of ventilation in this species. As noted earlier, however, the pH and P_{CO_2} of both plasma and brain tissue are maintained relatively constant in hibernating animals, despite significant changes in $[H^+]/[OH^-]$, and hence alpha-imidazole. $\dot{V}_E/\dot{V}_{CO_2}$ is maintained as well as a brisk response to inspired CO_2. Since this is also true of carotid body denervated animals (Webb and Milsom 1990) this would suggest that, in these species, either the rostral chemosensitive areas do not depend on alpha-imidazole for normal function or that other central chemosensitive areas must be more important.

Carotid and Aortic Bodies. The distribution, location and innervation of carotid and aortic bodies in air breathing vertebrates have already been described in Sect. 3.1.2.3. The responses of these receptor groups to $CO_2/[H^+]$ as specific stimuli have only been well studied in a select group of mammals, but what evidence there is suggests that homologous structures in non-mammalian vertebrates behave in a very similar fashion (reptiles: Ishii et al. 1986; birds: Nye and Powell 1984).

The question of whether CO_2 has a direct effect on the carotid and aortic body chemoreceptors, independent of its effect through changes in pH, has been long-standing (see Fidone and Gonzalez 1986 for review). Work with artificially perfused or superfused preparations in vitro provide the most convincing evidence and suggest that it does (Eyzaguirre and Koyano 1965; Biscoe et al. 1970). These stimuli interact with changes in O_2 partial pressure at the receptor site in a synergistic (multiplicative) fashion (Fig. 17a; Lahiri and DeLaney 1976). The effects of changing CO_2 levels on the response of the chemoreceptors to O_2, however, is much greater than the effect of changing O_2 levels on the $CO_2/[H^+]$ response (Fig. 17a). Although the response of the aortic bodies to changes in $CO_2/[H^+]$ is much weaker than that of the carotid bodies, they do respond in a similar fashion (see Fitzgerald and Lahiri 1986 for review).

As mentioned earlier, the carotid bodies in all vertebrate groups appear to contribute from 20% to 40% of the ventilatory response to changes in inspired CO_2 (range from 0% to 50%) (Kiwull et al. 1982; Heeringa et al. 1979; Berkenbosch et al. 1979; Benchetrit and Dejours 1980; Milsom and Jones 1980; Milsom et al. 1981). It has been suggested that this contribution, however, may vary both over the time course of the response and as a function of the magnitude of the stimulus. Many studies suggest that the carotid bodies contribute more to the initial phase of the ventilatory response than to the steady-state changes. Certainly this corresponds with the much larger dynamic component of the receptor response (Black et al. 1971; Fig. 17b). Heymans and Neil (1958) and others (see Fidone and Gonzalez 1986 and Fitzgerald and Lahiri 1986 for reviews) further suggest that the carotid bodies only augment the drive of central respiratory neurons at moderate levels of hypercapnia. With higher levels of hypercapnia, the central actions of $CO_2/[H^+]$ are maximal and the carotid bodies can therefore contribute little more. This may also be the case with the ventilatory response to changes in

Fig. 17. A Three-dimensional graph of average carotid chemoreceptor response curves from 5 cats. Percent of maximum asphyxic response (% of MAR) is plotted as a function of arterial CO_2 tension (Pa_{CO_2}) at 4 levels of arterial O_2 tension (Pa_{O_2}) (Fitzgerald and Parks 1971). *B* Single chemoreceptor fiber. Response to sudden application of hypercapnic stimulus before (*upper graph*) and after (*lower graph*) inhibiting carbonic anhydrase of carotid body with acetazolamide (Diamox) (Black et al. 1971)

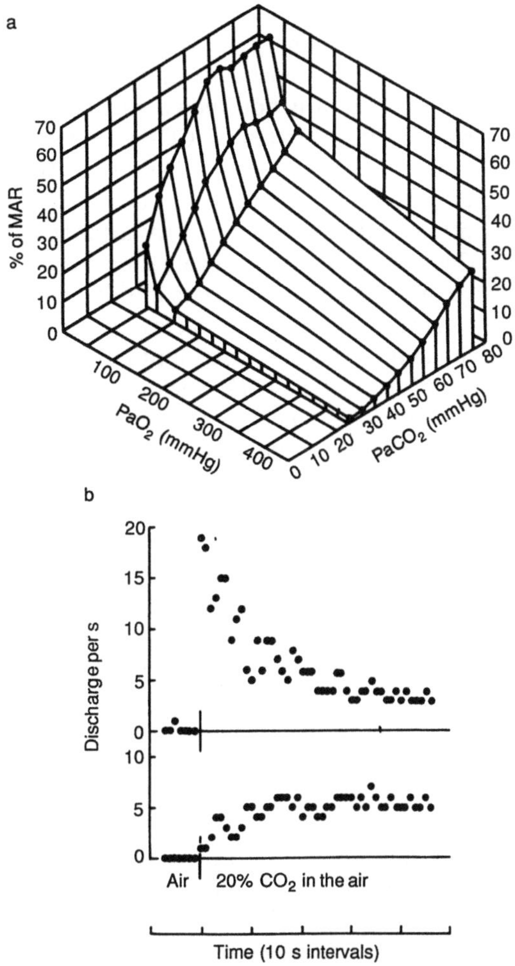

pHa due to metabolic acidosis. The carotid bodies may contribute to the response only until the changes in pH become severe enough to effect CSF pH (Hornbein and Roos 1963; Leusen 1954; Mitchell 1965; Fencl et al. 1966; Pappenheimer et al. 1965; Berger et al. 1977).

Although the aortic bodies exhibit a weak response to changes in arterial P_{CO_2}/pH, they do not appear to contribute to reflex ventilatory responses (Fitzgerald and Lahiri 1986).

Lung Receptors. There are three groups of receptors found within the lungs of different species of air-breathing vertebrates, which have been implicated in ventilatory responses to hypercapnic stimuli:

1. Pulmonary stretch receptors: The distribution, location and innervation of pulmonary stretch receptors (PSR) in these species are described in Sect. 3.3.1. Although the primary stimulus modalities of these receptors are changes in lung

volume, pressure or wall tension, increasing pulmonary P_{CO_2} has been shown to inhibit receptor discharge by varying degrees in reptiles and mammals. This inhibition is largely independent of the effects of the CO_2 on pulmonary smooth muscle tone, but it is not known whether these effects are due to the changes in pH or the direct effects of CO_2 (see Jones and Milsom 1982; Coleridge and Coleridge 1986 for reviews). The reflex effects of CO_2 modulation of PSR discharge have been best studied in dogs on cardiopulmonary bypass. Under these conditions, increasing pulmonary arterial P_{CO_2} (and hence alveolar P_{CO_2}) from low (hypocapnic) to normal values decreased both inspiratory and expiratory discharge and resulted in increases in tidal volume and breathing frequency. Raising P_{CO_2} above normal values decreased expiratory activity more than inspiratory activity and further increased ventilation, primarily by increasing respiratory frequency (Bartoli et al. 1973; Sheldon and Green 1982; Green et al. 1986; Schertel et al. 1988). These increases are not particularly significant but do suggest that CO_2-modulation of PSR discharge may play some physiological role, particularly in the ventilatory response to hypocapnia (Coleridge and Coleridge 1986; Schertel et al. 1988).

2. Pulmonary (mixed venous) chemoreceptors: The presence of chemoreceptors within the lungs responsive to changes in P_{CO_2} or CO_2 content of pulmonary arterial (mixed venous) blood has been debated for many years. The strongest arguments for their presence come from observations of isocapnic or hypocapnic hyperpnea during exercise (Wasserman et al. 1986; Dempsey and Forster 1982) and isocapnic hyperpnea or hypopnea (including apnea) during venous CO_2 loading or unloading (Yamamoto and Edwards 1960; Yamamoto 1981; Phillipson et al. 1981). In all of these cases, ventilatory responses are elicited which are proportional to the changes in $\dot{V}_{CO_2}$ or CO_2 delivery to the lungs, despite the fact there is no apparent increase in Pa_{CO_2} or decrease in pHa. Despite intense investigation, however, CO_2-sensitive receptors have only been found in the lungs of birds and reptiles. These receptors have discharge which is inhibited by CO_2 and appear to represent a distinct receptor group. Their distribution, functional characteristics and postulated reflexogenic roles are considered in the next section (Intrapulmonary Chemoreceptors). In mammals, most investigators currently believe that all changes in ventilation specifically due to changes in pulmonary arterial P_{CO_2} can be accounted for by stimulation of systemic arterial or central chemoreceptors (Coleridge et al. 1967; Cropp and Comroe 1961; Gonzalez et al. 1977; Ponte and Purves 1978; Phillipson et al. 1981; Sylvester et al. 1973) as well as inhibition of PSR discharge (see previous section). The control of breathing during exercise is still hotly debated and is further discussed in Sect. 3.3.3.1.

There is some evidence that pulmonary C fibers in mammals may evoke a tachypnea when stimulated by elevated CO_2, but the magnitude of the changes in discharge and rate of adaptation of these fibers during CO_2 exposure suggest their possible contribution to reflex changes in $\dot{V}_E$ is probably very small (see Coleridge and Coleridge 1986 for review).

In Chelonia (turtles and tortoises), 'Heymans-type' chemoreceptors have been reported to occur in the truncal region, including the pulmonary artery. These receptors are innervated by the inferior truncal branch of the vagus nerve and are stimulated by changes in both P_{O_2} and P_{CO_2} (Ishii et al. 1985, 1986). They ap-

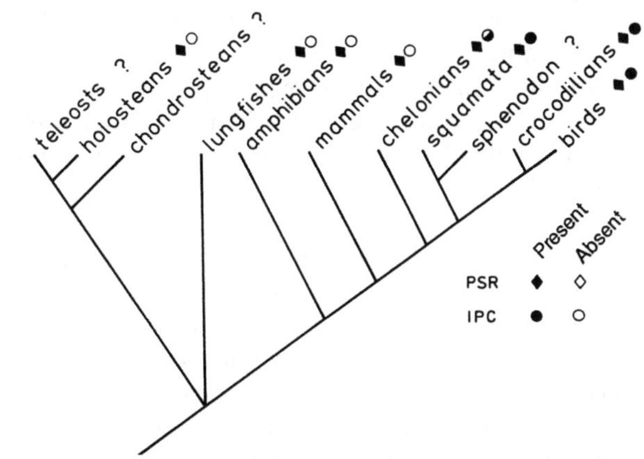

Fig. 18. Cladogram illustrating the phylogenetic relationships between various vertebrate groups. ***Symbols*** indicate the presence or absence of slowly adapting pulmonary stretch receptors (***PSR***) or intrapulmonary chemoreceptors (***IPC***) in each group. The ***half-filled symbol*** for IPC for the chelonians stems from the presence of IPC in *Testudo hermani* (Ishii et al. 1986) but not *Chrysemys picta* (Jones and Milsom 1982)

pear to produce significant reflex ventilatory responses to stimulation by NaCN (Benchetrit et al. 1977), but their contribution to hypercapnic ventilatory responses in undetermined. Bilateral vagotomy, which will eliminate the influences of CO_2-sensitive PSR and aortic chemoreceptors (and also intrapulmonary chemoreceptors in tortoises; see next section) as well as pulmonary artery chemoreceptors, can reduce the hypercapnic ventilatory response by up to 50% (Benchetrit and Dejours 1980; Milsom and Jones 1980).

3. Intrapulmonary chemoreceptors: Chemoreceptors have been described in the lungs of birds and some reptiles which respond primarily to changes in CO_2 (Fig. 18). They are located within the lung, innervated by the vagus nerve (and, to some extent, the cardiac sympathetic nerve in birds; Burger and Estavillo 1978) and have discharge which is inversely proportional to P_{CO_2} (Fig. 14a). They are insensitive to changing P_{O_2} and their responses to changes in lung volume or pressure are not significant within physiological limits (Fedde et al. 1977; Scheid and Piiper 1986; Furilla and Bartlett 1987; Douse and Mitchell 1988; Powell et al. 1988). It is currently believed that both CO_2 and H^+ exert independent effects on IPC activity with the sensitivity to CO_2 being somewhat greater (Burger et al. 1974; Powell et al. 1978; Scheid et al. 1978). The location, innervation and functional characteristics of avian IPC have been most extensively studied and an excellent review of this literature by Scheid and Piiper (1986) was published recently.

The reflex roles of this receptor group remain somewhat controversial. As noted earlier, because inhalation of CO_2-rich gas will reduce, eliminate or reverse the normal tidal fluctuations of tidal CO_2 in the lungs during eupneic breathing, altering $F_{I_{CO_2}}$ is only appropriate for studying ventilatory responses to environmental hypercapnia. It does not mimic the effects of increasing $\dot{V}_{CO_2}$ or hypoventilation (which will increase the tidal CO_2 fluctuations in the lungs), and, consequently, these phenomena are best studied using techniques for venous CO_2 loading.

In birds, environmental hypercapnia increases ventilation primarily by an increase in V_T, while breathing frequency increases slightly (Powell et al. 1978; Kilgore et al. 1985), decreases (Bouverot and Leitner 1972) or remains unaltered (Ray and Fedde 1969; Jones and Purves 1970; Milsom et al. 1981; Kilgore et al. 1985) (Figs. 13, 14). IPC appear to be predominant in eliciting the reflex responses produced over the hypocapnic range (Osborne et al. 1977) and can evoke maximal responses, alone, in the hypercapnic range (Fedde et al. 1982). In the latter case a near maximal response can also be produced in their absence (Milsom et al. 1981), suggesting that their unique role in ventilatory responses to inspired CO_2 may occur primarily over the hypocapnic range.

With venous loading of CO_2, however, much larger increases in $\dot{V}_E$ are produced for similar changes in Pa_{CO_2}, which are primarily the consequence of larger increases in breathing frequency. Under these circumstances, IPC appear to be important in adjusting the pattern of breathing but play no apparent role in the ventilatory response to changes in Pa_{CO_2} per se (Milsom et al. 1981; Tallman and Grodins 1982a, b; Jones et al. 1985). As such, they may act in a homologous fashion to PSR using the rate and extent of CO_2 washout/accumulation during inspiration/expiration to monitor the rate of lung inflation/deflation (Burger et al. 1974; Fedde et al. 1974a, b; Molony 1974; Banzett and Burger 1977; Burger et al. 1978; Milsom et al. 1981).

The combined effects of CO_2 breathing and venous loading experiments suggest that the tonic and phasic components of IPC discharge may play very different roles in the control of breathing, but this has yet to be adequately examined.

Similar results to those described in birds have also been obtained in reptiles possessing IPC. Environmental hypercapnia produces variable effects on breathing frequency, and hence $\dot{V}_E$ (see Shelton et al. 1986 for review), while hypercapnia produced by other means consistently produces a brisk ventilatory response with significant increases in breathing frequency (Gatz et al. 1975; Ballam and Donaldson 1989). The presence of IPC in these animals, which probably rarely encounter environmental hypercapnia, and also PSR which can adequately monitor changes in lung volume, suggests that they must have an alternate function. Recently, Powell et al. (1988) have outlined the differences in the discharge profiles of IPC and PSR which would occur during an episode of breathing and the ensuing breath-hold period in the alligator (Fig. 19). They suggest that in animals with periodic breathing patterns, IPC discharge does not accurately reflect lung volume but may convey information about changes in lung CO_2 stores, which might play a significant role in the control of the length of both the breathing episode and the breath-hold interval.

Mechanism of Transduction. At present, little is known of the mechanism by which changes in P_{CO_2}/pH alter chemoreceptor discharge. Certainly the mechanism cannot be a common one, since the response of IPC to changes in P_{CO_2}/pH is opposite to that exhibited by other chemoreceptor groups. Furthermore, as noted earlier, Nattie (1986b) has recently shown that there are even differences in the chemosensory mechanisms of the caudal and rostral populations of central

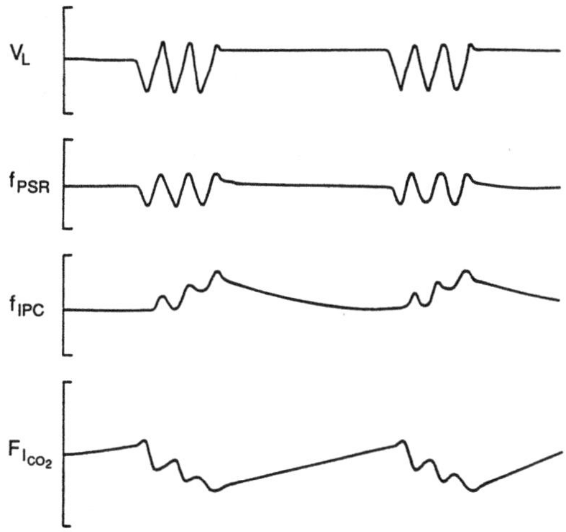

Fig. 19. Theoretical changes in lung volume (V_L), PSR discharge rate (f_{PSR}), IPC discharge rate (f_{IPC}) and lung CO_2 ($F_{I_{CO_2}}$) in an alligator during a breath-hold and two bouts of breathing. Note that PSR activity shows a constant relation to V_L during a breath-hold or breathing bout, but IPC discharge changes as CO_2 is washed out during breathing or accumulates during the breath-hold (Powell et al. 1989)

chemoreceptors[4]. Finally, current theories link the mechanisms of transduction of hypoxic and hypercapnic stimuli in the carotid and aortic bodies (other CO_2/H^+ chemosensitive tissue is not affected by hypoxia; Fidone and Gonzalez 1986), although there is recent evidence which suggests that the mechanisms involved are independent (Mulligan and Lahiri 1982; Milsom and Sadig 1983). Obviously, much further research is required to delineate the mechanisms which link changes in P_{CO_2}/pH to changes in receptor discharge.

3.2.4 *Central Integration*

Just as was the case for O_2-related changes in ventilation, the central respiratory centers are implicated as the site of integration of CO_2/H^+-related afferent information. The mechanisms and pathways which underlie this integration and the genesis of CO_2/H^+-related changes in ventilation, however, are unclear.

Integration with Other Inputs. Once again, whatever the neural matrix for central integration of sensory inputs may be, the process of integration can be viewed in terms of changes in the graphical solution of the metabolic hyperbola and, this time, the hypercapnic response curve. The integration of other inputs with CO_2/H^+-chemosensory information manifests itself as changes in the setpoint or gain of the relationships, as described earlier.

The setpoint of the CO_2 response curve, for instance, is usually its intercept, i.e. the Pa_{CO_2} at which ventilation falls to zero. In those species which exhibit a

[4] The results of Nattie's study suggest that a protein governed by imidazole-histidine dissociation is involved in the receptor mechanism of the rostral chemoreceptive area in the medulla. Whether this protein is an enzyme or is part of a cell surface receptor or an ionic channel is unknown. The response to changes in function of this protein must be to bring about a change in Pa_{CO_2}, however, since only this change will alter the initial chemically-induced perturbation in the protein.

threshold below which $\dot{V}_{Eff}$ becomes independent of Pa_{CO_2} (usually only present when Pa_{O_2} is reduced), the setpoint will be this threshold value (Fig. 15). The setpoint of the metabolic hyperbola is the asymptotic value of Pa_{CO_2} produced at elevated levels of $\dot{V}_{Eff}$ (Fig. 15). The gain of the CO_2 response is the slope of the relationship describing the changes in $\dot{V}_{Eff}$ produced by changing Pa_{CO_2}, while that of the metabolic hyperbola is the slope of the relationship describing the changes in Pa_{CO_2} produced by changing $\dot{V}_{Eff}$ (Fig. 15).

In Fig. 20, the effects of changes in body temperature on these parameters are shown for the turtle *Chrysemys picta*. Increasing body temperature from 10° to 30 °C leads to an increase in resting metabolic rate, which causes a right, upward shift of the metabolic hyperbola (change in setpoint; because of the elevation in metabolic rate, Pa_{CO_2} will be increased at any given level of $\dot{V}_{Eff}$) and a right shift in the CO_2 response curve (increase in setpoint). The slopes or gains of both relationships are also increased. The net effect is that, at rest, there is an increase in Pa_{CO_2} of 9.1 Torr and a dramatic increase in the magnitude of the hypercapnic ventilatory response (Fig. 20, stippled areas indicate predicted effects of raising Pa_{CO_2} to 35 Torr on $\dot{V}_{Eff}$). A similar graphical presentation of the effects of exercise on these relationships in the lizard, *Varanus exanthematicus*, has already been documented in Fig. 15. Relative contributions of the changes in the setpoint and gain of each relationship to the net changes in Pa_{CO_2} and $\dot{V}_{Eff}$ can be calculated and shown graphically, as they were for Pa_{O_2} and $\dot{V}_{Eff}$ in Fig. 10. Again, although such presentations reveal nothing of the mechanisms underlying the central integration of these sensory inputs, they do offer a way of interpreting the complex interactions that determine resting Pa_{CO_2}/pHa and the hypercapnic ventilatory response, and may provide some insight into the integrative processes which underlie these changes.

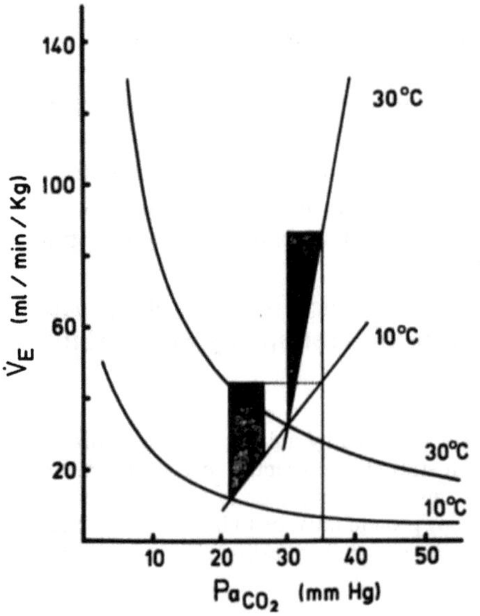

Fig. 20. The changes in setpoint and gain which occur in the relationships between $\dot{V}_{Eff}$ and Pa_{CO_2}, and Pa_{CO_2} and $\dot{V}_{Eff}$ with a change in body temperature from 10° to 30 °C in the turtle *Chrysemys picta* (data from Funk and Milsom 1987). The *stippled areas* indicate the effect of raising Pa_{CO_2} to 35 Torr on $\dot{V}_{Eff}$ at each temperature

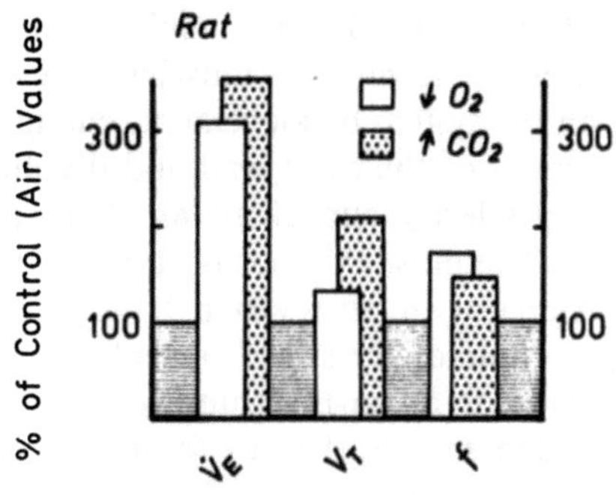

Fig. 21. The effects of reducing arterial P_{O_2} to 40 torr or increasing P_{CO_2} to 60 torr on minute ventilation ($\dot{V}_E$), tidal volume (V_T) and breathing frequency (*f*) in the rat (Cragg and Drysdale 1983)

Temporal and Spatial Outputs. Earlier, evidence was presented which suggested that the ventilation cycle consisted of four phases, each under separate control, and that the tidal volume of each breath was a function of the inspiratory drive, as manifested in the mean inspiratory flow rate (V_T/T_I) and the length of the active inspiratory phase (T_I). Thus, the temporal output of the pattern generator in determining the length of each phase determined f, V_T, $\dot{V}_{Eff}$ and $\dot{V}_D$. Although we are far from understanding the integrative processes which underlie the timing components of ventilation, a quick glance at Figs. 11, 12 and 21 reveals that the central integration of sensory information pertaining to changes in Pa_{O_2} and Pa_{CO_2}/pHa produces distinctly different responses. Differences in spatial output between species have already been described (see Sect. 3.1.2.4) and the consequences of changes in Pa_{CO_2}/pHa on spatial output to different muscle groups, within any given species, have not really been investigated.

3.3 Other Stimuli

Referring back to Fig. 2, we can see that although other inputs can influence the control system regulating ventilation at any of a number of sites, they can be considered to influence the system either by their effects on blood gases and acid-base balance, or by independent effects on sites of central integration. In the latter case, their effects will be manifested as changes in (1) the setpoint or gain of the ventilatory responses to changes in blood gases or pH and/or (2) the temporal output and/or the spatial output of the controller, both under eupneic conditions and in response to hypoxic and/or hypercapnic stimuli.

3.3.1 Mechanical Stimuli

Receptors Involved. There are a number of receptor groups associated with the chest wall, upper respiratory tract, tracheobronchial tree and lungs of air-breathing vertebrates, which respond to mechanical stimuli. The locations, functional characteristics and reflex roles of all have been reviewed recently (Coleridge and Coleridge 1986; Scheid and Piiper 1986; Shelton et al. 1986). Many are involved in defense reflexes that protect the respiratory tract (Rapidly Adapting Receptors

(RAR) and C Fibers), while others regulate the relative forces developed by the different respiratory muscles in response to changes in the various loads placed on the system by movement, changes in posture, etc. (muscle spindle and tendon organs of the chest wall and diaphragm). Most of these have only been studied in a select group of mammalian species.

The most extensively studied group, however, are those receptors involved in regulatory reflexes which determine breathing pattern. These are the slowly adapting pulmonary stretch receptors (SAR), whose afferent fibers run in the vagus nerves in all air-breathing vertebrates which have been studied to date. In reptiles and mammals, they are found throughout the tracheobronchial tree and lungs, although it appears that the extra- and intrapulmonary receptors may represent two distinct populations in terms of their response characteristics and reflex roles (Jones and Milsom 1982; Coleridge and Coleridge 1986). In birds, their location is not well known. They seem to be absent from the relatively non-expansible lungs but appear to be located in the lower respiratory tract, in the thoracic septa or in other associated membranes of the thoracoabdominal cavity (see Bouverot 1978 for review). There is some variation in the exact nature of the stimulus modality for these receptors from species to species, ranging from the changes in volume, pressure or wall tension of the lungs and/or respiratory passages. In all cases, however, they convey information which monitors the rate and extent of inflation and deflation of the lungs. As mentioned earlier, the intrapulmonary chemoreceptors (IPC) described in birds (but not those in reptiles) which respond to the rate and extent of CO_2 accumulation/washout during expiration/inspiration may also act in an analogous fashion.

Effects on Central Integration

1. Set point/gain. There is a growing consensus that pulmonary vagal afferent influences carried in the cervical vagus of mammals modulate the ventilatory responsiveness to CO_2 and possibly hypoxia (Kiwull-Schöne and Kiwull 1979). The underlying mechanisms for this, however, are far from clear. These inputs appear to depress the response to low to moderate levels of hypercapnia, and this effect has been ascribed to the inspiratory-inhibiting action of SAR (which will restrict the increase in V_T) (Richardson and Widdicombe 1969; Wiemer and Kiwull 1972). At higher levels of hypercapnia, however, vagal pulmonary input facilitates the ventilatory response, and this has been attributed, at least in part, to the tonic, late-expiratory activity of SAR (Kiwull-Schöne et al. 1981) as well as to the activation of lung-irritant receptors (Phillipson et al. 1970; Kiwull-Schöne et al. 1981). These effects have not been widely studied, even in mammals, and their physiological significance is far from clear.

2. Temporal and spatial output. It is generally accepted that SAR provide the primary sensory input involved in the control of breathing pattern. The mechanisms by which this is achieved have received considerable attention in recent years, revealing a complex series of interactions for what was once considered a simple process. It was once believed that normal respiratory patterns were simply the consequence of the phasic inhibitory activity of SAR generated during inspiration. It is now known that SAR activity may play no role in determining eupneic breathing patterns in some species, while in others, as well as in all species under conditions of elevated respiratory drive, both tonic and phasic SAR activity may have independent roles in determining breathing pattern (Coleridge and Coleridge 1986). Furthermore, the influence of phasic inspiratory activity varies dramatically over the course of the respiratory cycle.

Tonic Input. In mammals (Woldring 1965; Grunstein et al. 1975; D'Angelo and Agostoni 1975; Finkler and Iscoe 1984), an increase in the functional residual capacity of the lungs (FRC), which increases the tonic discharge in SAR occurring at end-expiration, leads to a decrease in breathing frequency and usually to a decrease in $\dot{V}_E$. This is primarily due to an expiratory-excitatory reflex which leads to prolongation of expiration and hence the reduction in f (Finkler and Iscoe 1984). Both hypercapnia (Mitchell et al. 1982) and hypoxia (Mitchell 1987) attenuate the effects of lung inflation on expiratory duration by carotid chemoreflex effects (Mitchell and Selby 1987). Such an increase in FRC will also elicit the Breuer-Hering inspiratory-inhibitory reflex (due to an increase in the phasic activity associated with lung inflation), which will act to shorten inspiration. This would act to offset the prolongation of expiration and thus maintain the total breath length and breathing frequency constant. It has been shown, however, that both adaptation of SAR and central habituation of phasic afferent input occur and act to offset completely the inhibition of inspiration (D'Angelo and Agostoni 1975; Muza and Frazier 1983; Finkler and Iscoe 1984). Thus, under steady-state conditions, only a prolongation of expiration and a depression in breathing frequency are seen. There are no direct effects of the increase in tonic end-expiratory discharge on T_I or V_T (Bartoli et al. 1973; Phillipson 1974). The fall in $\dot{V}_E$ which follows may cause an increased chemoreflex drive which can act to increase V_T, but this compensatory effect is only sufficient to restore $\dot{V}_E$ to resting levels if the increase in FRC is small (Finkler and Iscoe 1984).

In turtles, similar increases in resting lung volume also lead to changes in breathing pattern. These animals exhibit periodic breathing patterns, as described earlier and, whereas increasing V_{Lr} has no effect on V_T or $\dot{V}_E$, it increases the number of breaths in each period of breathing but decreases the incidence of episodes of breathing proportionately (Fig. 11; Milsom and Chan 1986). Thus, the normal respiratory pause is prolonged in both turtles and mammals, even though this pause coincides with end-inspiration in turtles but end-expiration in mammals. Again, as in mammals, increasing respiratory drive by hypoxia or hypercapnia will attenuate or even eliminate this effect (Milsom and Chan 1986).

Phasic Input. It is unfortunate that, despite the many advances which have been made in understanding the role of phasic afferent information from SAR in the control of the breathing pattern in the last decade, the Hering-Breuer inflation reflex (HBIR) continues to be the only response acknowledged by many physiologists. It now appears that phasic information from SAR during the first 30% of inspiration does not influence the inspiratory interval at all. Phasic information within the cycle, however, is excitatory to inspiratory activity (positive feedback) (Fig. 22; Bartoli et al. 1975; Cross et al. 1980), while only that arriving during a substantial period preceding the end of inspiration is inhibitory to inspiratory motor output (this latter effect is the HBIR; Younes et al. 1978; Cross et al. 1980).

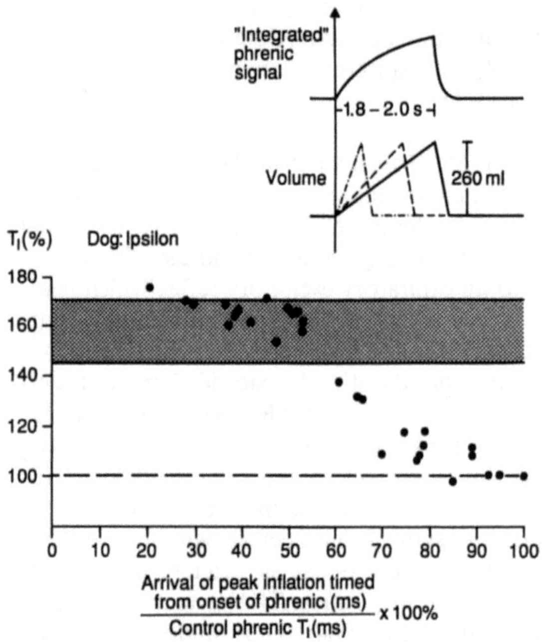

Fig. 22. Relationship between the change in the inspiratory interval (T_I) and the arrival of peak inflation timed from the onset of the phrenic signal and expressed as a % of the control T_I. In the control situation $T_I\% = 100$ and the arrival time for peak inflation = 100%. A diagram of the experimental design is given in the *top right-hand corner*, the control V_T and T_I for this animal (dog ipsilon) are also given. The *stippled area* of the graph refers to the range of values of $T_I\%$ obtained when inflation was withheld for one breath (Cross et al. 1980)

The rate of inflation does not appear to contribute to any of these responses (Cross et al. 1980). Increasing respiratory drive by chemoreceptor stimulation leads to an increase in V_T, in part by a direct inhibitory effect on SAR discharge, and in part by reducing the inhibitory effect of the phasic vagal input on inspiratory motor activity (see Von Euler 1986 for review).

The physiological significance of this input in regulating respiratory pattern has been hypothesized to be the minimization of either the work (Otis et al. 1950) or force of the respiratory muscles (Mead 1960) required to effect ventilation. Although the mechanisms underlying such ergometric minimization remain unclear, as does the extent of the receptor groups involved, vagal afferent information clearly plays a role (Agostoni et al. 1970). What studies have been performed on nonmammalian vertebrates are consistent with these views (Shelton et al. 1986; Milsom 1984, 1988).

3.3.2 Temperature

Temperature effects ventilation in air-breathing vertebrates, both by its effects on blood gases and acid-base balance as well as by its effects on the central integration of various afferent inputs. Its primary effects on blood gases and acid-base balance result from its effects on the affinity of hemoglobin for O_2 and CO_2, as manifested in the shapes and positions of the O_2 and CO_2 equilibrium curves. In lower vertebrates, it will also effect blood gases by its effects on central cardiac shunting. These have been described, in part, in earlier sections of this review (hypoxic stimulus, Fig. 8; CO_2/H^+ complex of stimuli). Its primary effects on central integration are described more fully below.

Effects on Central Integration

1. Setpoint/gain. The effects of changing temperature on the setpoint and gain of the hypoxic and hypercapnic ventilatory responses of the turtle *Chrysemys picta* have also been described already (Figs. 9, 20). Considerable variability has been noted in the magnitude of the changes in these parameters, which occur with changing temperature from species to species, and particular attention has focused on the changes which occur in the Pa_{CO_2} setpoint in lower vertebrates. Considerable controversy has arisen over possible explanations of these changes which warrant further attention here.

In most lower vertebrates, resting Pa_{CO_2} rises and pHa falls with increasing temperature. In early studies, the slope of the relationship between pH and temperature was found to be practically identical to that observed in in vitro closed systems (Robin 1962) as well as to that observed in the neutral pH of water (the change in pH required to maintain the ratio of $OH^-:H^+$ constant) (Rahn 1967). One consequence of such changes is that the degree of dissociation of protein moieties will remain constant, and this led Reeves (1972, 1977) to suggest that perhaps ectotherms regulated their acid-base balance with changing temperature to maintain constant the fractional dissociation (alpha) of the most important protein buffer group, the imidazole of histidine residues. This suggestion has become known as the '(imidazole) alphastat hypothesis'. The physiological significance of such a regulatory mechanism would be the maintenance of net protein charge, ensuring constant Donnan ratios and unchanged passive distribution of ions and water across membranes. It would also maintain protein configuration and hence the functioning of such regulatory proteins as enzymes, receptor molecules and ion channels.

Initially, alphastat was believed to be maintained exclusively by ventilatory adjustments. Thus, the $OH^- : H^+$ ratio was maintained constant with changing temperature by the appropriate respiratory control of Pa_{CO_2}. As temperature increased, $\dot{V}_E$ must increase at a slower rate than $\dot{V}_{CO_2}$ leading to a fall in the air convection requirement, CO_2 retention, a constant $OH^-:H^+$ ratio, alphastat and an apparent resetting of the Pa_{CO_2} setpoint. This was certainly consistent with the initial data selected to test this hypothesis (Reeves 1972; Malan et al. 1976). Although the mechanisms by which such regulation was achieved were far from clear, it was not difficult to imagine various scenarios by which it might occur. Changes in the degree of dissociation of imidazole histidine residues (αIm) could act directly by altering the activity of specific receptors (an αIm receptor) or, more likely, indirectly by altering the functional characteristics of other receptor groups such as the CO_2/H^+ receptors themselves. In this way, their setpoint might be changed in such a way that alphastat occurred. Such changes could also act at the level of the central integrator, altering the activity of key regulatory proteins involved somehow in sensory processing to achieve the same effect (Nattie 1986a, b).

One of the major weaknesses of this hypothesis, however, was that it was an hypothesis for acid-base regulation which depended exclusively on the control of ventilation for its proper function. In terms of ventilatory control, however, pH (and hence αIm) is only one of a host of afferent inputs which influence ventilation. Thus, not surprisingly, not all species exhibited alphastat regulation, and even in those species which did, this regulation broke down under hypoxic or hypercapnic conditions when the need to maintain O_2 supply or CO_2 excretion took precedence over pH/alphastat regulation (Jackson 1973, 1978; Glass et al. 1979, 1985). It now appears that even during quiet, eupneic breathing, only roughly one-half of all species studied show pH changes with changing temperature within the relatively narrow limits compatible with alphastat regulation (Shelton et al. 1986; Truchot 1987). While a few of the other species regulate pH

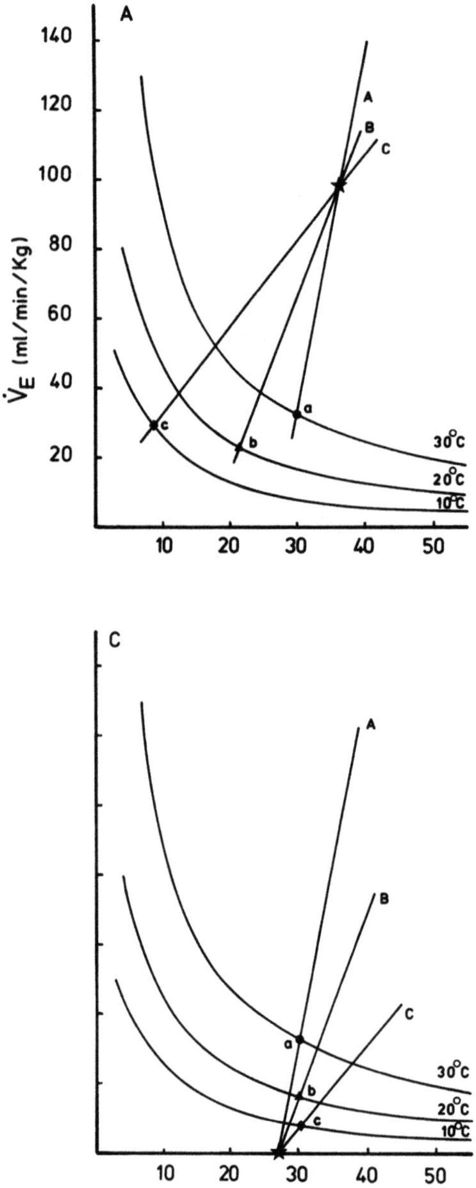

Fig. 23. Graphical illustration of the effect of a Q_{10} of 2 of changing temperature on metabolic rate and respiratory sensitivity to CO_2 on the control of breathing. *Lines* labelled *30 °*, *20 °* and *10 °C* represent the metabolic hyperbolae (iso$\dot{V}_{CO_2}$ lines) for these respective temperatures. *Points* labelled *a*, *b* and *c* represent resting (air-breathing) values of $\dot{V}_E$ and Pa_{CO_2} at 30 °, 20 ° and 10 °C respectively, while *lines* labelled *A*, *B* and *C* represent the CO_2 response curves of animals at these respective temperatures. The *three panels* show the effect of alphastat, 'half-a-stat' and pHstat regulation on the temperature-independent combination of $\dot{V}_E$ and Pa_{CO_2} common to all response curves. (Data in *top left panel* is for *Pseudemys scripta* (Jackson et al. 1974); *top right panel* is for *Chrysemys picta* (Funk and Milsom 1987); *bottom panel* is hypothetical

constant with changing temperature (pHstat), the majority show changes in pH which are roughly half-way between alphastat and pHstat regulation. This data has led to the suggestion that, at best, the ability of lower vertebrates to use the respiratory system primarily to regulate acid-base balance with changing body temperature is restricted to sluggish species with low aerobic scope under resting conditions at preferred body temperatures (Shelton et al. 1986).

Acid-base balance, and hence αIm, can, of course, be regulated by other means as temperature changes. The extent to which this occurs, however, remains quite controversial and has been discussed recently in several excellent reviews (Heisler 1986; Truchot 1987).

The effects of temperature on both metabolic rate and ventilatory sensitivity to changing Pa_{CO_2} (and hence pH) appear to be roughly equal (Q_{10} of 2 to 3) in those species in which this has been examined, regardless of the nature of the changes in their acid-base balance. What is different is the manner in which Pa_{CO_2} and $\dot{V}_E$ change with body temperature. These similarities and differences can be clearly seen in the interactions between the metabolic hyperbolae and the hypercapnic ventilatory responses, at each temperature, for species exhibiting different acid-base strategies (Fig. 23). It is interesting that on such plots, the ventilatory response curves all intersect at one temperature-independent point and that it is this point which varies from species to species. The significance of this temperature-independent point is unclear, other than as a reflection of the interaction between acid-base balance and blood gas homeostasis in the different species. Thus, despite similar temperature effects on $\dot{V}_{CO_2}$ and $\Delta\dot{V}_E/\Delta Pa_{CO_2}$, in alphastat regulators, $\dot{V}_E$ remains constant while Pa_{CO_2} falls with decreasing temperature (Fig. 23, top left panel), while in pHstat regulators, $\dot{V}_E$ falls, while Pa_{CO_2} remains constant (Fig. 23, lower panel).

2. *Temporal and spatial output.* Although decreases in body temperature may have differential effects on the temporal and spatial outputs of the respiratory controller, such changes are most evident, and best studied during increases in body temperature. This is particularly true in birds and mammals, where changes in breathing pattern appear as the respiratory system is recruited by the thermoregulatory control system to provide evaporative cooling. Various aspects of this topic have also been reviewed recently (Richards and Avery 1978; Cooper and Veale 1986; Tenney and Boggs 1986).

3.3.3 Other Inputs

Exercise. Exercise effects ventilation in air-breathing vertebrates by its effects on the central integration of afferent inputs. Blood gases and acid-base balance are either maintained homeostatically, or more typically, overcompensated (such that Pa_{O_2} and pHa increase while Pa_{CO_2} falls), at all but the highest levels of exercise (Dempsey et al. 1985; Wasserman et al. 1986). In terms of central integration of afferent inputs, the effects of exercise on the gain of the hypoxic and hypercapnic ventilatory responses of the lizard, *Varanus exanthematicus*, have already been described (Figs. 6, 15) and are typical of all air-breathing vertebrates. The effects of exercise on temporal and spatial outputs have not been as well studied, although such studies have provided some rewarding insights. The locomotor and respiratory systems are sometimes very tightly coupled during locomotion in a variety of animals, especially those which employ synchronous (in-phase) movements of the limbs (i.e. birds in flight, Hart and Roy 1966; rabbits hopping, Viala et al. 1987; wallabies bounding, Baudinette et al. 1987; horses galloping, Bramble and Carrier 1983). In lizards, on the other hand, where locomotion employs a sprawling posture and lateral vertebral bending (and hence synchronous contraction of both inspiratory and expiratory muscles on one side of the body and relaxation of those on the other side), ventilation is constrained. Minute ventilation decreases at higher running speeds and may be temporarily disrupted coincident with locomotor movements (Carrier 1987). Increases in body temperature coincident with exercise may further influence the temporal and spatial outputs of the respiratory controller, illustrating both the flexibility and complexity of ventilatory control (Kiley et al. 1979; Brackenbury and Gleeson 1983).

Ion Regulation. Although physiologists have always been aware of the interaction between ion regulation and acid-base balance in vertebrates, and the possible influence of such interactions on ventilatory control, the terminology employed in describing such interactions, as well as the level of precision employed in analyzing them, has varied. The current use of strong ion difference (SID) analysis is appealing, since all dependent system variables are derived from the true independent variables: the P_{CO2}, total weak acid concentration, and the strong ion difference (Stewart 1981). There are certainly strengths and limitations to this approach to analyzing acid-base changes (Cameron 1984), but in terms of respiratory control, regardless of the method used, these will generally appear as changes in the CO_2/H^+ complex of stimuli.

Higher Centers. Although all physiologists acknowledge that changes in arousal state due to changes in the sleep/wakefulness cycle, or in levels of internal or external stimuli, affect breathing, this area has received little attention until very recently. Even so, the recent attention focuses primarily on the changes which occur during sleep in man (Remmers 1981; Orem 1986; Phillipson and Bowes 1986), with little attempt to put these studies into the broader context of arousal state or higher-center influences on ventilatory control in all air-breathing vertebrates (Orem 1987). The ability of the higher centers to alter the setpoint, gain, temporal and spatial output of the respiratory controller, as well as to over-ride homeostatic control to varying degrees, however, indicates the extent to which such studies are essential to understanding normal ventilatory control.

4 Homeostatic Regulation

This review has considered the steady-state relations between stimuli and responses in terms of the feedback control system outlined in Fig. 2. With this scheme, the primary function of the respiratory system in controlling and co-ordinating gas exchange is to maintain blood gases and pH within homeostatic limits. As a final note, it should be pointed out that these limits, and hence the set-

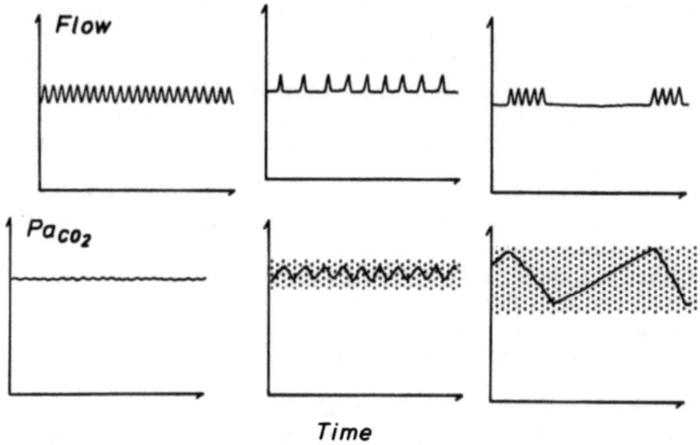

Fig. 24. Theoretical changes in arterial P_{CO_2} (*lower panels*), which would be associated with the various breathing patterns exhibited in the air-flow traces in the *upper panels*. *Stippled* areas indicate the homeostatic range within which Pa_{CO2} must be regulated in each case

point of the control system, have been difficult to define precisely in the lower vertebrates and in diving birds and mammals, as well as in hibernating mammals, primarily because of the presence of periodic breathing patterns. In these groups, there appears to be a homeostatic range determined by separate 'on' and 'off thresholds' (Fig. 24). Given the variety of different afferent inputs which affect ventilation, these thresholds appear to be quite labile and give rise to much variability in measurements of resting ventilation and ventilatory responses. With increases in respiratory drive, however, in all species, this homeostatic range becomes progressively narrower until breathing becomes continuous, and this range is replaced by a setpoint which can be clearly defined (Fig. 24). This simply reflects differences in the interaction of various afferent inputs in determining ventilatory pattern when the net respiratory drive is above or below the level required to produce continuous ventilation.

Acknowledgements. It would like to thank my graduate students for their patience, advice and assistance in the writing of this review. This work was funded in part by the NSERC of Canada.

References

Abdel-Magied EM, King AS (1978) The topographical anatomy and blood supply of the carotid body region of domestic fowl. J Anat 126:535–546

Adams WE (1958) The comparative morphology of the carotid body and carotid sinus. Thomas, Springfield, IL

Adams WE (1962) The carotid sinus-carotid body problem in the chelonia (with a note on a Foramen of Panizza in *Dermochelys*). Arch Int Pharmacodyn 139:28–37

Adrian ED, Buytendijk FJJ (1931) Potential changes in the isolated brain stem of goldfish. J Physiol 71:121–135

Agostoni E, Campbell GJM, Freedman S (1970) Energetics. In: Campbell GJM, Agostoni E, Newsom Davis J (eds) The respiratory muscles. Lloyd-Luke, London

Bainton CR (1978) Canine ventilation after acid-base infusions, exercise, and carotid body denervation. J Appl Physiol 44:28–35
Ballam GD, Donaldson LA (1988) Effect of venous (gut) CO_2 loading on intrapulmonary gas fractions and ventilation in the Tegu lizard. J Comp Physiol 158:591–600
Ballintijn CM (1982) Neural control of respiration in fishes and mammals. In: Addink ASF, Spronk N (eds) Exogenous and endogenous influences on metabolic and neural control, vol 1. Pergamon, New York
Ballintijn CM (1988) Evolution of central nervous control of ventilation in vertebrates. In: Taylor EW (ed) The neurobiology of the cardio-respiratory system. Manchester University Press, Manchester
Banzett RB, Burger RE (1977) Responses of avian intrapulmonary chemoreceptors to venous CO_2 and ventilatory gas flow. Respir Physiol 29:63–72
Bartoli A, Bystrzycka E, Guz A, Jain SK, Noble MIM, Trenchard DW (1973) The effect of carbon dioxide in the airways and alveoli on ventilation; a vagal refex studied in the dog. J Physiol 230:449–465
Bartoli A, Cross BA, Guz A, Huszczuk A, Jeffries R (1975) The effect of varying tidal volume on the associated phrenic motoneurone output: studies of vagal and chemical feedback. Respir Physiol 25:136–155
Baudinette RV, Gannon BJ, Runciman WB, Wells S, Love JB (1987) Do cardiorespiratory frequencies show entrainment with hopping in the tammar wallaby? J Exp Biol 129:251–263
Benchetrit G, Dejours P (1980) Ventilatory CO_2 drive in the tortoise *Testudo horsfieldi.* J Exp Biol 87:229–236
Benchetrit G, Armand J, Dejours P (1977) Ventilatory chemoreflex drive in the tortoise *Testudo horsfieldi.* Respir Physiol 31:183–191
Berger AJ, Mitchell RAM, Severinghaus JW (1977) Regulation of respiration. N Engl J Med 297:194–201
Berkenbosch A, Heringe J, Olievier CN, Kruyt EW (1979) Artificial perfusion of the pontomedullary region of cats, a method for separation of central and peripheral effects of chemical stimulation of ventilation. Respir Physiol 37:381–390
Bickler PE (1984) CO_2 balance of a heterothermic rodent: comparison of sleep, torpor, and awake states. Am J Physiol 246:R49–R55
Biscoe TJ, Purves MJ, Sampson SR (1970) The frequency of nerve impulses in single carotid body chemoreceptor afferent fibers recorded in vivo with intact circulation. J Physiol 208:121–131
Black AMS, McCloskey DI, Torrance RW (1971) The responses of carotid body chemoreceptors in the cat to sudden changes of hypercapnic and hypoxic stimuli. Respir Physiol 13:36–49
Black CP, Tenney SM (1980) Pulmonary hemodynamic responses to acute and chronic hypoxia in two waterfowl species. Comp Biochem Physiol 67A:291–293
Bledsoe SW, Hornbein TF (1981) Central chemoreceptors and the regulation of their chemical environment. In: Lenfant C (ed) Lung biology in health and disease, vol 17. Regulation of breathing, part 1. Marcel Dekker, New York
Boggs DF, Birchard GF (1983) Relationship between haemoglobin O_2 affinity and the ventilatory response to hypoxia in the rhea and pheasant. J Exp Biol 102:347–352
Boggs DF, Kilgore DL (1983) Ventilatory responses of the burrowing owl and bobwhite to hypercarbia and hypoxia. J Comp Physiol 149:527–533
Boggs DF, Kilgore DL, Birchard GF (1984) Minireview: respiratory physiology of burrowing animals and birds. Comp Biochem Physiol 77A:1–7
Bouverot P (1978) Control of breathing in birds compared with mammals. Physiol Rev 58:604–655
Bouverot P, Leitner LM (1972) Arterial chemoreceptors in the domestic fowl. Respir Physiol 15:310–320
Brackenbury JH, Gleeson M (1983) Effects of P_{CO_2} on respiratory pattern during thermal and exercise hyperventilation in domestic fowl. Respir Physiol 54:109–119
Bramble DM, Carrier DR (1983) Running and breathing in mammals. Science 219:251–256
Brooks JG III, Tenney SM (1968) Ventilatory responses of llama to hypoxia at sea level and high altitude. Respir Physiol 5:269–278

Burger RE, Barker MR, Nye PCG, Powell FL (1978) Effects of intrapulmonary chemoreceptors in perfused and non-perfused lungs. In: Piiper J (ed) Respiratory function in birds. Adult and embryonic. Springer, Berlin Heidelberg New York

Burger RE, Estavillo JA (1978) The alteration of CO_2 respiratory sensitivity in chickens by thoracic visceral denervation. Respir Physiol 32:251–263

Burger RE, Osborne JL, Banzett RB (1974) Intrapulmonary chemoreceptors in gallus domesticus: adequate stimulus and functional localization. Respir Physiol 22:87–97

Burggren W (1985) Hemodynamics and regulation of central cardiovascular shunts in reptiles. In: Johansen K, Burggren WW (eds) Cardiovascular shunts. Alfred Benzon Symposium 21. Munksgaard, Copenhagen

Busa WB, Nuccitelli R (1984) Metabolic regulation via intracellular pH. Am J Physiol 246:R409–R438

Cain SM (1970) Increased oxygen uptake with passive hyperventilation of dogs. J Appl Physiol 28:4–7

Cameron JN (1984) Acid-base status of fish at different temperatures. Am J Physiol 246:R452–R459

Carrier DR (1987) Lung ventilation during walking and running in four species of lizards. Exp Biol 47:33–42

Clark FJ, Von Euler C (1972) On the regulation of depth and rate of breathing. J Physiol 222:267–295

Coleridge HM, Coleridge JCG (1986) Reflexes evoked from tracheobronchial tree and lungs. In: Fishman AP, Cherniak NS, Widdicombe JG, Geiger SR (eds) Handbook of physiology, section 3. The respiratory system, vol II. Control of breathing, part 2. American Physiological Society, Bethesda

Coleridge HM, Coleridge JCG, Howe A (1967) A search for pulmonary arterial chemoreceptors in the cat, with a comparison of the blood supply of the aortic bodies in the newborn and adult mammal. J Physiol 191:353–374

Cooper KG, Veale WL (1986) Effects of temperature on breathing. In: Fishman AP, Cherniak NS, Widdicombe JG, Geiger SR (eds) Handbook of physiology, section 3. The respiratory system, vol II. Control of breathing, part 2. American Physiological Society, Bethesda

Courtice GP (1980) Stimulation of carotid arterial chemoreceptors by hypoxia and hypercapnia in a lizard. In: Proc 27th Int Congr Physiol Sci, Budapest 14:368

Cragg PA, Drysdale DB (1983) Interaction of hypoxia and hypercapnia on ventilation, tidal volume and respiratory frequency in the anesthetized rat. J Physiol 341:477–493

Cropp GJA, Comroe JH Jr (1961) Role of mixed venous blood P_{CO_2} in respiratory control. J Appl Physiol 16:1029–1033

Cross BA, Jones PW, Guz A (1980) The role of vagal afferent information during inspiration in determining phrenic motoneurone output. Respir Physiol 39:149–167

Cunningham DJC (1974) Integrative aspects of the regulation of breathing in man. Q Rev Biophys 6:433–483

Cunningham DJC, Robbins PA, Wolff CB (1986) Integration of respiratory responses to changes in alveolar partial pressures of CO_2 and O_2 and in arterial pH. In: Fishman AP, Cherniak NS, Widdicombe JG, Geiger SR (eds) Handbook of physiology, section 3. The respiratory system, vol II. Control of breathing, part 2. American Physiological Society, Bethesda

D'Angelo E, Agostoni E (1975) Tonic vagal influences on inspiratory duration. Respir Physiol 24:287–302

Darden TR (1972) Respiratory adaptations of a fossorial mammal, the pocket gopher (*Thomomys bottae*). J Comp Physiol 78:121–137

Davey NJ, Seller TJ (1987) Brain mechanisms for respiratory control. In: Seller T (ed) Bird respiration. CRC Press, Boca Raton, FL

Davies DG (1978) Temperature-induced changes in blood acid base status in the alligator, *Alligator mississipiensis*. J Appl Physiol 45:922–926

Davies DG, Thomas JL, Smith EN (1982) Effect of body temperature on the ventilatory response to hypercapnia in the awake alligator. J Appl Physiol 52:114–118

Dempsey JA, Forster HV (1982) Mediation of ventilatory adaptations. Physiol Rev 62:262–346

Dempsey JA, Vidruk EH, Mitchell GS (1985) Pulmonary control systems in exercise: update. Fed Proc 44:2260–2270
Douse MA, Mitchell GS (1988) Temperature effects on CO_2-sensitive intrapulmonary chemoreceptors in the lizard *Tupinambis nigropunctatus*. Respir Physiol 72:327–342
Eyzaguirre C, Koyano H (1965) Effects of hypoxia, hypercapnia and pH on the chemoreceptor activity of the carotid body in vitro. J Physiol 178:385–409
Fedde MR, Peterson DF (1970) Intrapulmonary receptor response to changes in airway-gas composition in *Gallus domesticus*. J Physiol 209:609–625
Fedde MR, Gatz RN, Slama H, Scheid P (1974a) Intrapulmonary CO_2 receptors in the duck. I. Stimulus specificity. Respir Physiol 22:99–114
Fedde MR, Gatz RN, Slama H, Scheid P (1974b) Intrapulmonary CO_2 receptors in the duck. II. Comparison with mechanoreceptors. Respir Physiol 22:115–121
Fedde MR, Kuhlmann WD, Scheid P (1977) Intrapulmonary receptors in the tegu lizard. I. Sensitivity to CO_2. Respir Physiol 29:35–48
Fedde MR, Kiley JP, Powell FL, Scheid P (1982) Intrapulmonary CO_2 receptors and control of breathing in ducks: effects of prolonged circulation time to carotid bodies and brain. Respir Physiol 47:121–140
Feldman JL (1986) Neurophysiology of respiration in mammals. In: Bloom FE (ed) Handbook of physiology, section 1. The nervous system, vol IV. American Physiological Society, Bethesda
Feldman JL, Smith JC, McCrimmon DR, Ellenberger HH, Speck DF (1988) Generation of respiratory pattern in mammals. In: Cohen A (ed) Neural control of rhythmic movements in vertebrates. John Wiley, New York
Fencl V (1986) Acid-base balance in cerebral fluid. In: Fishman AP, Cherniak NS, Widdicombe JG, Geiger SR (eds) Handbook of physiology, section 3. The respiratory system, vol II. Control of breathing, part 1. American Physiological Society, Bethesda
Fencl V, Miller TB, Pappenheimer JR (1966) Studies on the respiratory response to disturbances of acid-base balance with deductions concerning ionic composition of cerebral interstitial fluid. Am J Physiol 210:459–472
Fidone SJ, Gonzalez C (1986) Initiation and control of chemoreceptor activity in the carotid body. In: Fishman AP, Cherniak NS, Widdicombe JG, Geiger SR (eds) Handbook of physiology, section 3. The respiratory system, vol II. Control of breathing, part 1. American Physiological Society, Bethesda
Finkler J, Iscoe S (1984) Control of breathing at elevated lung volumes in anesthetized cats. J Appl Physiol 56:839–844
Fishman AP, Cherniak NS, Widdicombe JG, Geiger SR (eds) (1986) Handbook of physiology, section 3. The respiratory system, vol II. Control of breathing, parts I and II. American Physiological Society, Bethesda
Fitzgerald RS, Lahiri S (1986) Reflex responses to chemoreceptor stimulation. In: Fishman AP, Cherniak NS, Widdicombe JG, Geiger SR (eds) Handbook of physiology, section 3. The respiratory system, vol II. Control of breathing, part 1. American Physiological Society, Bethesda
Fitzgerald RS, Parks DC (1971) Effect of hypoxia on carotid chemoreceptor response to carbon dioxide in cats. Respir Physiol 12:218–229
Funk GD, Milsom WK (1987) Changes in ventilation and breathing pattern produced by changing body temperature and inspired CO_2 concentration in turtles. Respir Physiol 67:37–51
Furilla RA, Bartlett D Jr (1987) Intrapulmonary receptors in the garter snake, *Thamnophis sirtalis*. Fed Proc 46:793
Gallego R, Eyzaguirre C, Monti-Bloch L (1979) Thermal and osmotic responses of arterial chemoreceptors. J Neurophysiol 42:665–680
Gatz RN, Fedde MR, Crawford EC Jr (1975) Lizard lungs: CO_2-sensitive receptors in *Tupinambis nigropunctatus*. Experientia 31:455–456
Geiser F (1988) Reduction of metabolism during hibernation and daily torpor in mammals and birds: temperature effect or physiological inhibition. J Comp Physiol B 158:25–37
Glass ML, Johansen K (1976) Control of breathing in *Acrochordus javanicus*, an aquatic snake. Physiol Zool 49:328–340

Glass ML, Johansen K (1979) Periodic breathing in the crocodile, *Crocodylus niloticus*: consequences for the gas exchange ratio and control of breathing. J Exp Zool 208:319–326
Glass ML, Wood SC, Hoyt RW, Johansen K (1979) Chemical control of breathing in the lizard, *Varanus exanthematicus*. Comp Biochem Physiol 62A:999–1003
Glass ML, Wood SC (1983) Gas exchange and control of breathing in reptiles. Physiol Rev 63:232-260
Glass ML, Boutilier RG, Heisler N (1983) Ventilatory control of arterial P_{O2} in the turtle *Chrysemys picta bellii*: effects of temperature and hypoxia. J Comp Physiol B 151:145–153
Glass ML, Boutilier RG, Heisler N (1985) Effects of body temperature on respiration, blood gases and acid-base status in the turtle *Chrysemys picta bellii*. J Exp Biol 114:37–51
Gonzalez F Jr, Fordyce WE, Grodins FS (1977) Mechanism of respiratory responses to intravenous $NaHCO_2$, HCl and KCN. J Appl Physiol 34:1075–1079
Gratz RK (1978) Ventilation and gas exchange in the diamondback water snake, *Natrix rhombifera*. J Comp Physiol 127:299–305
Gratz RK (1979) Ventilatory response of the diamondback water snake *Natrix rhombifera* to hypoxia, hypercapnia and increased oxygen demand. J Comp Physiol 129:105–110
Green JF, Schertel ER, Coleridge HM, Coleridge JCG (1986) Effects of pulmonary arterial P_{CO2} on slowly adapting pulmonary stretch receptors. J Appl Physiol 60:2048–2055
Grunstein MM, Wyszogrodski I, Milic-Emili J (1975) Regulation of frequency and depth of breathing during expiratory threshold loading in cats. J Appl Physiol 39:395–404
Hall FG (1966) Minimal utilizable oxygen and the oxygen dissociation curve of blood of rodents. J Appl Physiol 21:375–378
Harken AH (1976) Hydrogen ion concentration and oxygen uptake in an isolated canine hindlimb. J Appl Physiol 40:1–5
Hart JS, Roy OZ (1966) Respiratory and cardiac responses to flight in pigeons. Physiol Zool 39:291–306
Hatcher JD, Chiu LK, Jennings DB (1978) Anemia is a stimulus to aortic and carotid chemoreceptors in the cat. J Appl Physiol 44:696–702
Heeringa MS, Berkenbosch A, De Goede J, Olievier CN (1979) Relative contribution of central and peripheral chemoreceptors to the ventilatory response to CO_2 during hyperoxia. Respir Physiol 37:365–379
Heisler N (1986) Acid-base regulation in fishes: In: Heisler N (ed) Acid-base regulation in animals. Elsevier Biomedical, Amsterdam
Heisler N, Neumann P, Maloiy GMO (1983) The mechanism of intracardiac shunting in the lizard *Varanus exanthematicus*. J Exp Biol 105:15–31
Hey EN, Lloyd BB, Cunningham DJC, Jukes MGM, Bolton DPG (1966) Effects of various respiratory stimuli on the depth and frequency of breathing in man. Respir Physiol 1: 193–205
Heymans C, Neil E (1958) Reflexogenic areas of the cardiovascular system. Churchill, London
Hitzig BM (1982) Temperature-induced changes in turtle CSF pH and central control of ventilation. Respir Physiol 49:205–222
Hitzig BM, Jackson DC (1978) Central chemical control of ventilation in the unanaesthetized turtle. Am J Physiol 235:R257–R264
Hochachka PW, Guppy M (1987) Metabolic arrest and the control of biological time. Harvard University Press, Cambridge MA
Hollinshead WH (1946) The function of the abdominal chemoreceptors of the rat and mouse. Am J Physiol 147:654–660
Hornbein TF (ed) (1981) Regulation of breathing, parts I and II, vol 17. Lung Biology in health and disease. Marcel Dekker, New York
Hornbein TF, Roos A (1963) Specificity of H^+ ion concentration as a carotid chemoreceptors stimulus. J Appl Physiol 18:580–584
Howe A, Pack RJ, Wise JCM (1981) Arterial chemoreceptors in the abdominal vagus of the rat. J Physiol 320:309–318
Hukuhara T (1976) Functional organization of brain stem respiratory neurons and its afferences. In: Duran B (ed) Respiratory centers and afferent systems. INSERM, Paris
Hukuhara T, Okada H (1956) On automaticity of the respiratory centers of the catfish and crucian carp. Jpn J Physiol 6:313–320

Ishii K, Ishii K, Kusakabe T (1985) Electrophysiological aspects of reflexogenic area in the chelonian, *Geoclemmys reevesii.* Respir Physiol 59:45–54

Ishii K, Ishii K, Dejours P (1986) Activity of vagal afferent fibers innervating CO_2-sensitive receptors in the tortoise, *Testudo hermanni.* Jpn J Physiol 36:1015–1026

Jackson DC (1973) Ventilatory response to hypoxia in turtles at various temperatures. Respir Physiol 18:178–187

Jackson DC (1978) Respiratory control in air-breathing ectotherms. In: Davies DG, Barnes CD (eds) Regulation of ventilation and gas exchange . Academic Press, New York

Jackson DC, Palmer SE, Meadow WL (1974) The effects of temperature and carbon dioxide breathing on ventilation and acid-base status of turtles. Respir Physiol 20:131–146

Jones DR, Chu C (1988) Effect of denervation of carotid labyrinths on breathing in unrestrained *Xenopus laevis.* Respir Physiol 73:243–256

Jones DR, Milsom WK (1982) Peripheral receptors affecting breathing and cardiovascular function in nonmammalian vertebrates. J Exp Biol 100:59–91

Jones DR, Purves MJ (1970) The effect of carotid body denervation upon the respiratory response to hypoxia and hypercapnia in the duck. J Physiol 211:279–294

Jones DR, Milsom WK, Butler PJ (1985) Ventilatory response to venous CO_2 loading by gut ventilation in ducks. Can J Zool 63:1232–1236

Juch PJW, Ballintijn CM (1983) Tegmental neurons controlling medullary respiratory centre activity in the carp. Respir Physiol 51:95–107

Juch PJW, Luiten PGM (1981) Anatomy of respiratory rhythmic systems in brainstem and cerebellum. Brain Res 230:51–64

Kawasaki R (1979) Breathing rhythm-generation in the adult lamprey, *Entosphenus japonicus.* Jpn J Physiol 29:327–338

Kawasaki R (1984) Breathing rhythm-generation mechanism in the adult lamprey, *Lampetra japonica.* Jpn J Physiol 34:319–335

Kiley JP, Kuhlmann WD, Fedde MR (1979) Respiratory and cardiovascular responses to exercise in the duck. J Appl Physiol 47:927–933

Kilgore DL Jr, Faraci FM, Fedde MR (1985) Ventilatory and intrapulmonary chemoreceptor sensitivity to CO_2 in the burrowing owl. Respir Physiol 62:325–340

Kinney JL, White FN (1977) Oxidative cost of ventilation in a turtle, *Pseudemys floridana.* Respir Physiol 31:327–332

Kiwull P, Wiemer W, Schöne H (1982) The role of the carotid chemoreceptors in the CO_2-hyperpnea under hyperoxia. Pflügers Arch 336:171–186

Kiwull-Schöne H, Kiwull P (1979) The role of the vagus nerves in the ventilatory response to lowered PaO_2 with intact and eliminated carotid chemoreflexes. Pflügers Arch 381:1–9

Kiwull-Schöne H, Ward SA, Kiwull P (1981) The involvement of expiratory termination in the vagally mediated facilitation of ventilatory CO_2 responsiveness during hyperoxia. Pflügers Arch 390:63–69

Kobayashi S (1971) Comparative cytological studies of the carotid body. 1. Demonstration of monoamine-storing cells by correlated chromaffin reaction and fluorescence histochemistry. Arch Histol Jpn 31:9–19

Kooyman GL, Cornell LH (1981) Flow properties of expiration and inspiration in a trained bottle-nosed porpoise. Physiol Zool 54:55–61

Kruhøffer M, Glass ML, Abe AS, Johansen K (1987) Control of breathing in an amphibian *Bufo paracnemius*: effects of temperature and hypoxia. Respir Physiol 69:267–275

Kunz AL (1987) Peripheral mechanisms in the control of breathing. In: Seller T (ed) Bird respiration. CRC Press, Boca Raton, FL

Lahiri S, DeLaney RG (1976) The nature of response of single chemoreceptor fibers of carotid body to changes in arterial P_{O_2} and P_{CO_2}-H^+. In: Paintal AS (ed) Morphology and mechanisms of chemoreceptors. Navchetan, New Delhi

Lahiri S, Mulligan E, Nishino T, Mokashi A, Davies RO (1981) Relative responses of aortic body and carotid body chemoreceptors to carboxyhemoglobinemia. J Appl Physiol 50:580–586

Lauweryns JM, Cokelaere M (1973) Intrapulmonary neuro-epithelial bodies: hypoxia-sensitive neuro(chemo)receptors. Experientia 29:1384–1386

Leusen IR (1954) Chemosensitivity of the respiratory center. Influences of changes in the H^+ and total buffer concentrations in the cerebral ventricles on respiration. Am J Physiol 176:45–51

Loeschke HH (1982) Central chemosensitivity and the reaction theory. J Physiol 322:1–24

Lomholt JP, Johansen K (1979) Hypoxia acclimation in carp – how it effects O_2 uptake, ventilation, and O_2 extraction from water. Physiol Zool 52:38–49

Long S, Duffin J (1986) The neuronal determinants of respiratory rhythm. Prog Neurobiol 27:101–182

Lyman CP, Chatfield PO (1955) Physiology of hibernation in animals. Physiol Rev 35:403–425

Malan A (1982) Respiration and acid-base state in hibernation. In: Lyman CP, Willis JS, Malan A, Wang LCH (eds) Hibernation and torpor in mammals and birds. Academic Press, New York

Malan A, Arens H, Waechter A (1973) Pulmonary respiration and acid-base state in hibernating marmots and hamsters. Respir Physiol 17:450–461

Malan A, Wilson TL, Reeves RB (1976) Intracellular pH in cold-blooded vertebrates as a function of body temperature. Respir Physiol 28:29–47

Malan A, Rodeau JL, Daull F (1985) Intracellular pH in hibernation and respiratory acidosis in the European hamster. J Comp Physiol 156B:251–258

Mead J (1960) Control of respiration frequency. J Appl Physiol 15:325–327

Milsom WK (1984) The interrelationship between pulmonary mechanics and the spontaneous breathing pattern in the Tokay lizard, *Gecko gecko*. J Exp Biol 113:203–214

Milsom WK (1988) Control of arrhythmic breathing in aerial brèathers. Can J Zool 66:99–108

Milsom WK, Chan P (1986) The relationship between lung volume, respiratory drive and breathing pattern in the turtle, *Chrysemys picta*. J Exp Biol 120:233–247

Milsom WK, Jones DR (1980) The role of vagal afferent information and hypercapnia in control of the breathing pattern in chelonia. J Exp Biol 87:53–63

Milsom WK, Sadig T (1983) Interaction between norepinephrine and hypoxia on carotid body chemoreceptors in rabbits. J Appl Physiol 55:1893–1898

Milsom WK, Jones DR, Gabbot GRJ (1981) On chemoreceptor control of ventilatory responses to CO_2 in unanesthetized ducks. J Appl Physiol 50:R1121–R1128

Milsom WK, McArthur MD, Webb CL (1986) Control of breathing in hibernating ground squirrels. In: Heller HC, Musacchia XJ, Wang LCH (eds) Living in the cold: physiological and biochemical adaptations. Elsevier, New York

Mitchell GS (1987) Effects of hypoxemia on phrenic nerve responses to static lung inflation in anesthetized dogs. Respir Physiol 67:183–195

Mitchell GS, Osborne JL (1980) A comparison between carbon dioxide inhalation and increased dead space ventilation in chickens. Respir Physiol 40:227–239

Mitchell GS, Selby BD (1987) Effects of carotid denervation on interactions between lung inflation and Pa_{CO_2} in modulating phrenic activity. Respir Physiol 67:367–378

Mitchell GS, Gleeson TT, Bennett AF (1981) Ventilation and acid-base balance during graded activity in lizards. Am J Physiol 240:R29–R37

Mitchell GS, Cross BA, Hiramoto T, Scheid P (1982) Interactions between lung stretch and Pa_{CO2} in modulating ventilation in dogs. J Appl Physiol 53:185–191

Mitchell RA (1965) The regulation of respiration in metabolic acidosis and alkalosis. In: Brooks C, Kao F, Lloyd BB (eds) Cerebrospinal fluid and the regulation of respiration. Blackwell, Oxford

Molony V (1974) Classification of vagal afferents firing in phase with breathing in *Gallus domesticus*. Respir Physiol 22:57–76

Morgan MN, Milsom WK (1990) Mechanical and oxidative costs and efficiency of breathing in semi-aquatic turtles (*Chrysemys* sp.) J Exp Biol (in press)

Mulligan E, Lahiri S (1982) Separation of carotid body chemoreceptor responses to O_2 and CO_2 by oligomycin and by antimycin A. Am J Physiol 242:C200–C206

Muratori G (1962) Histological observations on the cervico-thoracic paraganglia of amniotes. Arch Int Pharmadyn Ther 140:217–226

Muza SR, Frazier DT (1983) Response of pulmonary stretch receptors to shifts of functional residual capacity. Respir Physiol 52:371–386

Nattie E (1986a) Intracisternal diethylpyrocarbonate inhibits central chemosensitivity in conscious rabbits. Respir Physiol 64:161–176
Nattie E (1986b) Diethyl pyrocarbonate (an imidazole binding substance) inhibits rostral VLM CO_2 sensitivity. J Appl Physiol 61:843–850
Nielsen B (1962) On the regulation of respiration in reptiles. II. The effect of hypoxia with and without moderate hypercapnia on the respiration and metabolism of lizards. J Exp Biol 39:107–117
Nielsen M, Smith H (1952) Studies on the regulation of respiration in acute hypoxia. Acta Physiol Scand 24:293–313
Nolan WF, Frankel HM (1982) Ventilatory responses to CO_2 at different body temperatures in the snake, *Coluber constrictor.* Experientia 38:943–944
Nye PCG, Powell FL (1984) Steady-state discharge and bursting of arterial chemoreceptors in the duck. Respir Physiol 50:335–350
Orem JM (1986) Respiratory neuronal activity in sleep. In: Edelman NH, Santiago TV (eds) Breathing disorders of sleep. Churchill Livingstone, New York
Orem JM (1987) Inspiratory neurons that are activated when inspiration is inhibited behaviorally. Neurosci Lett 83:282–286
Osborne JL, Mitchell GS, Powell F (1977) Ventilatory responses to CO_2 in the chicken: intrapulmonary and systemic chemoreceptors. Respir Physiol 30:368–382
Otis AB, Fenn WO, Rahn H (1950) Mechanics of breathing in man. J Appl Physiol 2:592–607
Pappenheimer JR, Fencl V, Heisey SR, Held D (1965) Role of cerebral fluids in control of respiration as studied in unanesthetized goats. Am J Physiol 208:242–250
Pearse AGE (1969) The cytochemistry and ultrastructure of polypeptide hormone-producing cells of the APUD series and the embryologic, physiologic and pathologic implications of the concept. J Histochem Cytochem 17:303–313
Pearse AGE, Polak JM (1978) The diffuse neuroendocrine system and the APUD concept. In: Bloom SR (ed) Gut hormones. Churchill, London
Peterson DF, Fedde MR (1968) Receptors sensitive to carbon dioxide in lungs of chicken. Science 162:1449–1501
Phillipson EA (1974) Vagal control of breathing pattern independent of lung inflation in conscious dogs. J Appl Physiol 37:183–189
Phillipson EA, Bowes G (1986) Control of breathing during sleep. In: Fishman AP, Cherniak NS, Widdicombe JG, Geiger SR (eds) Handbook of physiology, section 3. The Respiratory System, vol II. Control of breathing, part 1. American Physiological Society, Bethesda
Phillipson EA, Hickey RF, Bainton CR, Nadel JA (1970) Effect of vagal blockade on regulation of breathing in conscious dogs. J Appl Physiol 29:475–479
Phillipson EA, Duffin J, Cooper JD (1981) Critical dependence of respiratory rhythmicity on metabolic CO_2 load. J Appl Physiol 50:45–54
Piiper J (ed) (1978) Respiratory function in birds, adult and embryonic. Springer, Berlin Heidelberg New York
Ponte J, Purves MJ (1978) Carbon dioxide and venous return and their interaction as stimuli to ventilation in the cat. J Physiol 274:44
Powell FL, Fedde MR, Gratz RK, Scheid P (1978) Ventilatory response to CO_2 in birds. I. Measurements in the unanesthetized duck. Respir Physiol 35:349–359
Powell FL, Milsom WK, Mitchell GS (1988) Effects of intrapulmonary CO_2 and airway pressure on pulmonary vagal afferent activity in the alligator. Respir Physiol 74:285–298
Rahn H (1967) Gas transport from the external environment to the cell. In: De Reuck AUS, Porter R (eds) Development of the lung. Ciba Found Symp Churchill, London
Ray PJ, Fedde MR (1969) Response to alterations in respiratory P_{O_2} and P_{CO_2} in the chicken. Respir Physiol 6:135–143
Reeves RB (1972) An imidazole alphastat hypothesis for vertebrate acid-base regulation: tissue carbon dioxide content and body temperature in bullfrogs. Respir Physiol 14:219–236
Reeves RB (1977) The interaction of body temperature and acid-base balance in ectothermic vertebrates. Annu Rev Physiol 39:559–586
Remmers JE (1981) Control of breathing during sleep. In: Hornbein T (ed) Regulation of breathing, part II, vol 17. Lung biology in health and disease. Marcel Dekker, New York

Richards SA, Avery P (1978) Central nervous mechanisms regulating thermal panting. In: Piiper J (ed) Respiratory function in birds: adult and embryonic. Springer, Berlin Heidelberg New York

Richardson PS, Widdicombe JG (1969) The role of the vagus nerves in the ventilatory responses to hypercapnia and hypoxia in anaesthetized and unanesthetized rabbits. Respir Physiol 7:122–135

Richter DW, Ballantyne D, Remmers JE (1986) How is the respiratory rhythm generated? A Model. News. Physiol Sci 1:109–112

Robin ED (1962) Relationship between temperature and plasma pH and carbon dioxide tension in the turtle. Nature 195:249–251

Rogers DC (1967) The structure of the carotid bifurcation in the lizards *Tiliqua occipitalis* and *Trachysaurus rugosus*. J Morphol 122:115–130

Rovainen CM (1974) Respiratory motoneurons in lampreys. J Comp Physiol 94:57–68

Rovainen CM (1977) Neural control of ventilation in the lamprey. Fed Proc 36:2386–2389

Russell DF (1986) Respiratory pattern generation in adult lamprey (*Lampetra fluviatilis*) interneurons and burst resetting. J Comp Physiol 158:91–102

Scheid P, Piiper J (1986) Control of breathing in birds. In: Fishman AP, Cherniak NS, Widdicombe JG, Geiger SR (eds) Handbook of physiology, section 3. The respiratory system, vol II. Control of breathing, part 1. American Physiological Society, Bethesda

Scheid P, Gratz RK, Powell FL, Fedde MR (1978) Ventilatory response to CO_2 in birds. II. Contribution by intrapulmonary CO_2 receptors. Respir Physiol 35:361–372

Schertel ER, Schneider DA, Adams L, Green JF (1988) Effect of pulmonary arterial PCO_2 on breathing pattern. J Appl Physiol 64:1844–1850

Sebert P (1979) Mise en evidence de l'action centrale du stimulus $CO_2(H^+)$ de la ventilation chez le canard pekin. J Physiol 75:902–909

Sellers TS (ed) (1987) Bird respiration. CRC Press Inc, Boca Raton, FL

Sheldon MI, Green JF (1982) Evidence for pulmonary CO_2 chemosensitivity: effects on ventilation. J Appl Physiol 52:1192–1197

Shelton G (1970) The regulation of breathing. In: Hoar WS, Randall DJ (eds) Fish physiology, vol 4. The nervous system, circulation and respiration. Academic Press, London

Shelton G, Burggren W (1976) Cardiovascular dynamics of the chelonia during apnoea and lung ventilation. J Exp Biol 100:245–273

Shelton G, Jones DR, Milsom WK (1986) Control of breathing in ectothermic vertebrates. In: Fishman AP, Cherniak NS, Widdicombe JG, Geiger SR (eds) Handbook of physiology, section 3. The respiratory system, vol II. Control of breathing, part 1. American Physiological Society, Bethesda

Smith JC, Feldman JL (1987) Central respiratory pattern generation studied in an in vitro mammalian brainstem-spinal cord preparation. In: Sieck GC, Gondevia S, Cameron WC (eds) Respiratory muscles and their neuromotor control. AR Liss, New York

Snapp BD, Heller HC (1981) Suppression of metabolism during hibernation in ground squirrels (*Citellus lateralis*). Physiol Zool 54:297–307

Stewart PA (1981) How to understand acid-base: A quantitative acid-base primer to biology and medicine. Elsevier Biomedical, Amsterdam

St John WM (1981) Respiratory neuron responses to hypercapnia and carotid chemoreceptor stimulation. J Appl Physiol 51:816–822

St John WM, Wang SC (1977) Response of medullary respiratory neurons to hypercapnia and isocapnic hypoxia. J Appl Physiol 43:812–821

St John WM, Bartlett D Jr, Knuth V, Hwang JC (1981) Brain stem genesis of automatic ventilatory patterns independent of spinal mechanism. J Appl Physiol 51:204–210

Sullivan CE, Kozar F, Murphy E, Phillipson EA (1978) Primary role of respiratory afferents in sustaining breathing rhythm. J Appl Physiol 45:11–17

Suzue T (1984) Respiratory rhythm generation in the in vitro brain stem-spinal cord preparation of the neonatal rat. J Physiol 354:135–152

Sylvester JT, Whipp BJ, Wasserman K (1973) Ventilatory control during brief infusions of CO_2-laden blood in the awake dog. J Appl Physiol 35:178–186

Takeda R, Remmers JE, Baker JP, Madden KP, Farber JP (1986) Postsynaptic potentials of bulbar respiratory neurons of the turtle. Respir Physiol 64:149–160

Tallman RD Jr, Grodins FS (1982a) Intrapulmonary CO_2 receptors and ventilatory response to lung CO_2 loading. J Appl Physiol 52:1272–1277
Tallman RD Jr, Grodins FS (1982b) Intrapulmonary CO_2 receptor discharge at different levels of venous P_{CO_2}. J Appl Physiol 53:1386–1391
Tcheng KT, Fu SK (1962) The structure and innervation of the aortic body of the yellow-breasted bunting. Sci Sin 11:221–232
Tcheng KT, Fu SK, Chen TY (1963) Supracardial encapsulated receptors of the aorta and the pulmonary artery in birds. Sci Sin 12:73–81
Templeton JR, Dawson WR (1963) Respiration in the lizard *Crotaphytus collaris*. Physiol Zool 36:104–121
Tenney SM, Boggs DF (1986) Comparative mammalian respiratory control. In: Fishman AP, Cherniak NS, Widdicombe JG, Geiger SR (eds) Handbook of physiology, section 3. The respiratory system, vol II. Control of breathing, part 1. American Physiological Society, Bethesda
Terni T (1931) Il simpatico cervicale degli amnioti (Ricerche di morfologia comparata). Z Anat Entwicklungsgesch 96:289–426
Truchot JP (1987) Comparative aspects of extracellular acid-base balance. Springer, Berlin Heidelberg New York Tokyo
Tschorn RR, Fedde MR (1974) Effects of carbon monoxide on avian intrapulmonary carbon dioxide-sensitive receptors. Respir Physiol 20:303–311
Viala D, Persegol L, Palisses R (1987) Relationship between phrenic and hindlimb extensor activities during fictive locomotion. Neurosci Lett 74:49–52
Von Euler C (1986) Brain stem mechanisms for generation and control of breathing pattern. Section 3, The respiratory system, vol II. Control of Breathing, part 1. American Physiological Society, Bethesda
Von Euler C, Lagercrantz H (eds) (1986) Neurobiology of the control of Breathing. Raven Press, New York
Von Euler C, Martila I, Remmers JE, Trippenbach T (1976) Effects of lesions in the parabranchial nucleus on the mechanisms for central and reflex termination of inspiration in the cat. Acta Physiol Scand 96:324–337
Wasserman K, Whipp BJ, Casaburi R (1986) Respiratory control during exercise. In: Fishman AP, Cherniak NS, Widdicombe JG, Geiger SR (eds) Handbook of physiology, section 3. The respiratory system, vol II. Control of breathing, part 1. American Physiological Society, Bethesda
Webb CL, Milsom WK (1990) Carotid body contribution to hypoxic ventilatory responses in euthermic and hibernating ground squirrels. In: Eyzaguirre C, Fidone S (eds) Arterial Chemoreception. Springer, New York
West NH, Langille BL, Jones DR (1981) Cardiovascular system. In: King AS, McLelland J (eds) Form and function in birds, vol 2. Academic Press, London
White FN (1976) Circulation. In: Gans C, Davidson WR (eds) Biology of the reptilia, vol 5. Academic Press, New York
Wiemer W, Kiwull P (1972) The role of the vagus nerves in the respiratory response to CO_2 under hyperoxic conditions. Pflügers Arch 336:147–170
Woldring S (1965) Interrelationship between lung volume, arterial CO_2 tension, and respiratory activity. J Appl Physiol 20:647–652
Wood SC (1984) Cardiovascular shunts and oxygen transport in lower vertebrates. Am J Physiol 247:R3–R14
Wood SC (ed) (1989) Comparative pulmonary physiology: current concepts, vol 39. Lung biology in health and disease. Marcel Dekker, New York
Yamamoto WS (1981) Computer simulation of experiments in response to intravenous and inhaled CO_2. J Appl Physiol 50:835–843
Yamamoto WS, Edwards MW Jr (1960) Homeostasis of carbon dioxide during intra-venous infusion of carbon dioxide. J Appl Physiol 15:807–818
Younes MK, Remmers JE, Baker J (1978) Characteristics of inspiratory inhibition by phasic volume feedback in cats. J Appl Physiol 45:80–86

Subject Index

MIX
Papier aus verantwortungsvollen Quellen
Paper from responsible sources
FSC® C105338

If you have any concerns about our products,
you can contact us on
ProductSafety@springernature.com

In case Publisher is established outside the EU,
the EU authorized representative is:
Springer Nature Customer Service Center GmbH
Europaplatz 3, 69115 Heidelberg, Germany

Printed by Libri Plureos GmbH
in Hamburg, Germany